STATA BASE REFERENCE MANUAL
VOLUME 1
A–F
RELEASE 12

A Stata Press Publication
StataCorp LP
College Station, Texas

Copyright © 1985–2011 StataCorp LP
All rights reserved
Version 12

Published by Stata Press, 4905 Lakeway Drive, College Station, Texas 77845
Typeset in T_EX
Printed in the United States of America

10 9 8 7 6 5 4 3 2 1

ISBN-10: 1-59718-098-X (volumes 1–4)
ISBN-10: 1-59718-099-8 (volume 1)
ISBN-10: 1-59718-100-5 (volume 2)
ISBN-10: 1-59718-101-3 (volume 3)
ISBN-10: 1-59718-102-1 (volume 4)
ISBN-13: 978-1-59718-098-6 (volumes 1–4)
ISBN-13: 978-1-59718-099-3 (volume 1)
ISBN-13: 978-1-59718-100-6 (volume 2)
ISBN-13: 978-1-59718-101-3 (volume 3)
ISBN-13: 978-1-59718-102-0 (volume 4)

The suggested citation for this software is

StataCorp. 2011. *Stata: Release 12*. Statistical Software. College Station, TX: StataCorp LP.

Table of contents

Contents of Reference Volume 1

Contents of Reference Volume 2

Contents of Reference Volume 3

Contents of Reference Volume 4

Cross-referencing the documentation

When reading this manual, you will find references to other Stata manuals. For example,

[U] **26 Overview of Stata estimation commands**
[XT] **xtabond**
[D] **reshape**

The first example is a reference to chapter 26, *Overview of Stata estimation commands*, in the *User's Guide*; the second is a reference to the xtabond entry in the *Longitudinal-Data/Panel-Data Reference Manual*; and the third is a reference to the reshape entry in the *Data-Management Reference Manual*.

All the manuals in the Stata Documentation have a shorthand notation:

[GSM]	*Getting Started with Stata for Mac*
[GSU]	*Getting Started with Stata for Unix*
[GSW]	*Getting Started with Stata for Windows*
[U]	*Stata User's Guide*
[R]	*Stata Base Reference Manual*
[D]	*Stata Data-Management Reference Manual*
[G]	*Stata Graphics Reference Manual*
[XT]	*Stata Longitudinal-Data/Panel-Data Reference Manual*
[MI]	*Stata Multiple-Imputation Reference Manual*
[MV]	*Stata Multivariate Statistics Reference Manual*
[P]	*Stata Programming Reference Manual*
[SEM]	*Stata Structural Equation Modeling Reference Manual*
[SVY]	*Stata Survey Data Reference Manual*
[ST]	*Stata Survival Analysis and Epidemiological Tables Reference Manual*
[TS]	*Stata Time-Series Reference Manual*
[I]	*Stata Quick Reference and Index*
[M]	*Mata Reference Manual*

Detailed information about each of these manuals may be found online at

http://www.stata-press.com/manuals/

Title

> **intro** — Introduction to base reference manual

Description

This entry describes the organization of the reference manuals.

Remarks

The complete list of reference manuals is as follows:

[R]	*Stata Base Reference Manual*
	Volume 1, A–F
	Volume 2, G–M
	Volume 3, N–R
	Volume 4, S–Z
[D]	*Stata Data-Management Reference Manual*
[G]	*Stata Graphics Reference Manual*
[XT]	*Stata Longitudinal-Data/Panel-Data Reference Manual*
[MI]	*Stata Multiple-Imputation Reference Manual*
[MV]	*Stata Multivariate Statistics Reference Manual*
[P]	*Stata Programming Reference Manual*
[SEM]	*Stata Structural Equation Modeling Reference Manual*
[SVY]	*Stata Survey Data Reference Manual*
[ST]	*Stata Survival Analysis and Epidemiological Tables Reference Manual*
[TS]	*Stata Time-Series Reference Manual*
[I]	*Stata Quick Reference and Index*
[M]	*Mata Reference Manual*

When we refer to "reference manuals", we mean all manuals listed above.

When we refer to the *Base Reference Manual*, we mean just the four-volume *Base Reference Manual*, known as [R].

When we refer to the specialty manuals, we mean all the manuals listed above except [R] and [I], the *Stata Quick Reference and Index*.

Detailed information about each of these manuals can be found online at

http://www.stata-press.com/manuals/

Arrangement of the reference manuals

Each manual contains the following sections:

- Table of contents.
  At the beginning of volume 1 of [R], the *Base Reference Manual*, is a table of contents for the four volumes.

- Cross-referencing the documentation.
  This section lists all the manuals and explains how they are cross-referenced.

- Introduction.
  This entry—usually called intro—provides an overview of the manual. In the specialty manuals, this introduction suggests entries that you might want to read first and provides information about new features.

  Each specialty manual contains an overview of the commands described in it.

- Entries.
  Entries are arranged in alphabetical order. Most entries describe Stata commands, but some entries discuss concepts, and others provide overviews.

  Entries that describe estimation commands are followed by an entry discussing postestimation commands that are available for use after the estimation command. For example, the **xtlogit** entry in the [XT] manual is followed by the **xtlogit postestimation** entry.

- Index.
  At the end of each manual is an index. The index for the entire four-volume *Base Reference Manual* is found at the end of the fourth volume.

  The *Quick Reference and Index*, [I], contains a combined index for all the manuals and a subject table of contents for all the manuals and the *User's Guide*. It also contains quick-reference information on many subjects, such as the estimation commands.

  To find information and commands quickly, use Stata's search command; see [R] **search** (see the entry search in the [R] manual). You can broaden your search to the Internet by using search, all to find commands and extensions written by Stata users.

Arrangement of each entry

Entries in the Stata reference manuals, except the [M] and [SEM] manuals, generally contain the following sections, which are explained below:

> *Syntax*
> *Menu*
> *Description*
> *Options*
> *Remarks*
> *Saved results*
> *Methods and formulas*
> *References*
> *Also see*

Syntax

A command's syntax diagram shows how to type the command, indicates all possible options, and gives the minimal allowed abbreviations for all the items in the command. For instance, the syntax diagram for the summarize command is

summarize $\left[\textit{varlist}\right]$ $\left[\textit{if}\right]$ $\left[\textit{in}\right]$ $\left[\textit{weight}\right]$ $\left[,\ \textit{options}\right]$

options	Description
Main	
detail	display additional statistics
meanonly	suppress the display; calculate only the mean; programmer's option
format	use variable's display format
separator(#)	draw separator line after every # variables; default is separator(5)
display_options	control spacing and base and empty cells

varlist may contain factor variables; see [U] **11.4.3 Factor variables**.

varlist may contain time-series operators; see [U] **11.4.4 Time-series varlists**.

by is allowed; see [D] **by**.

aweights, fweights, and iweights are allowed. However, iweights may not be used with the detail option; see [U] **11.1.6 weight**.

Items in the typewriter-style font should be typed exactly as they appear in the diagram, although they may be abbreviated. Underlining indicates the shortest abbreviations where abbreviations are allowed. For instance, summarize may be abbreviated su, sum, summ, etc., or it may be spelled out completely. Items in the typewriter font that are not underlined may not be abbreviated.

Square brackets denote optional items. In the syntax diagram above, *varlist*, *if*, *in*, *weight*, and the *options* are optional.

The *options* are listed in a table immediately following the diagram, along with a brief description of each.

Items typed in *italics* represent arguments for which you are to substitute variable names, observation numbers, and the like.

The diagrams use the following symbols:

#	Indicates a literal number, for example, 5; see [U] **12.2 Numbers**.
[ ]	Anything enclosed in brackets is optional.
{ }	At least one of the items enclosed in braces must appear.
|	The vertical bar separates alternatives.
%fmt	Any Stata format, for example, %8.2f; see [U] **12.5 Formats: Controlling how data are displayed**.
depvar	The dependent variable in an estimation command; see [U] **20 Estimation and postestimation commands**.
exp	Any algebraic expression, for example, (5+myvar)/2; see [U] **13 Functions and expressions**.
filename	Any filename; see [U] **11.6 Filenaming conventions**.

indepvars   The independent variables in an estimation command; see [U] **20 Estimation and postestimation commands**.

newvar   A variable that will be created by the current command; see [U] **11.4.2 Lists of new variables**.

numlist   A list of numbers; see [U] **11.1.8 numlist**.

oldvar   A previously created variable; see [U] **11.4.1 Lists of existing variables**.

options   A list of options; see [U] **11.1.7 options**.

range   An observation range, for example, 5/20; see [U] **11.1.4 in range**.

"string"   Any string of characters enclosed in double quotes; see [U] **12.4 Strings**.

varlist   A list of variable names; see [U] **11.4 varlists**. If *varlist* allows factor variables, a note to that effect will be shown below the syntax diagram; see [U] **11.4.3 Factor variables**. If *varlist* allows time-series operators, a note to that effect will be shown below the syntax diagram; see [U] **11.4.4 Time-series varlists**.

varname   A variable name; see [U] **11.3 Naming conventions**.

weight   A [*wgttype=exp*] modifier; see [U] **11.1.6 weight** and [U] **20.22 Weighted estimation**.

xvar   The variable to be displayed on the horizontal axis.

yvar   The variable to be displayed on the vertical axis.

The *Syntax* section will indicate whether factor variables or time-series operators may be used with a command. summarize allows factor variables and time-series operators.

If a command allows prefix commands, this will be indicated immediately following the table of options. summarize allows by.

If a command allows weights, the types of weights allowed will be specified, with the default weight listed first. summarize allows aweights, fweights, and iweights, and if the type of weight is not specified, the default is aweights.

Menu

A menu indicates how the dialog box for the command may be accessed using the menu system.

Description

Following the syntax diagram is a brief description of the purpose of the command.

Options

If the command allows any options, they are explained here, and for dialog users the location of the options in the dialog is indicated. For instance, in the **logistic** entry in this manual, the *Options* section looks like this:

⌐ Model ⌐

. . .

⌐ SE/Robust ⌐

. . .

Remarks

The explanations under *Description* and *Options* are exceedingly brief and technical; they are designed to provide a quick summary. The remarks explain in English what the preceding technical jargon means. Examples are used to illustrate the command.

Saved results

Commands are classified as e-class, r-class, s-class, or n-class, according to whether they save calculated results in e(), r(), s(), or not at all. These results can then be used in subroutines by other programs (ado-files). Such saved results are documented here; see [U] **18.8 Accessing results calculated by other programs** and [U] **18.9 Accessing results calculated by estimation commands**.

Methods and formulas

The techniques and formulas used in obtaining the results are described here as tersely and technically as possible. If a command is implemented as an ado-file, that is indicated here.

References

Published sources are listed that either were directly referenced in the preceding text or might be of interest.

Also see

Other manual entries relating to this entry are listed that might also interest you.

Also see

[U] **1.1 Getting Started with Stata**

Title

> **about** — Display information about your Stata

Syntax

```
about
```

Menu

Help > About

Description

about displays information about your version of Stata.

Remarks

about displays information about the release number, flavor, serial number, and license for your Stata. If you are running Stata for Windows, information about memory is also displayed:

```
. about

Stata/MP 12.0 for Windows (64-bit x86-64)
Revision 24 Aug 2011
Copyright 1985-2011 StataCorp LP

Total physical memory:     8388608 KB
Available physical memory:  937932 KB

10-user 32-core Stata network perpetual license:
       Serial number:  5012041234
         Licensed to:  Alan R. Riley
                       StataCorp
```

Also see

[R] **which** — Display location and version for an ado-file

[U] **5 Flavors of Stata**

Title

> **adoupdate** — Update user-written ado-files

Syntax

> adoupdate [*pkglist*] [, *options*]

options	Description
update	perform update; default is to list packages that have updates, but not to update them
all	include packages that might have updates; default is to list or update only packages that are known to have updates
ssconly	check only packages obtained from SSC; default is to check all installed packages
dir(*dir*)	check packages installed in *dir*; default is to check those installed in PLUS
verbose	provide output to assist in debugging network problems

Description

User-written additions to Stata are called packages. These packages can add remarkable abilities to Stata. Packages are found and installed by using ssc, findit, and net; see [R] ssc, [R] search, and [R] net.

User-written packages are updated by their developers, just as official Stata software is updated by StataCorp.

To determine whether your official Stata software is up to date, and to update it if it is not, you use update; see [R] update.

To determine whether your user-written additions are up to date, and to update them if they are not, you use adoupdate.

Options

update specifies that packages with updates be updated. The default is simply to list the packages that could be updated without actually performing the update.

The first time you adoupdate, do not specify this option. Once you see adoupdate work, you will be more comfortable with it. Then type

```
. adoupdate, update
```

The packages that can be updated will be listed and updated.

all is rarely specified. Sometimes, adoupdate cannot determine whether a package you previously installed has been updated. adoupdate can determine that the package is still available over the web but is unsure whether the package has changed. Usually, the package has not changed, but if you want to be certain that you are using the latest version, reinstall from the source.

Specifying all does this. Typing

```
. adoupdate, all
```

7

adds such packages to the displayed list as needing updating but does not update them. Typing

```
. adoupdate, update all
```

lists such packages and updates them.

ssconly is a popular option. Many packages are available from the Statistical Software Components (SSC) archive—often called the Boston College Archive—which is provided at http://repec.org. Many users find most of what they want there. See [R] **ssc** for more information on the SSC.

ssconly specifies that adoupdate check only packages obtained from that source. Specifying this option is popular because SSC always provides distribution dates, and so adoupdate can be certain whether an update exists.

dir(*dir*) specifies which installed packages be checked. The default is dir(PLUS), and that is probably correct. If you are responsible for maintaining a large system, however, you may have previously installed packages in dir(SITE), where they are shared across users. See [P] **sysdir** for an explanation of these directory codewords. You may also specify an actual directory name, such as C:\mydir.

verbose is specified when you suspect network problems. It provides more detailed output that may help you diagnose the problem.

Remarks

Do not confuse adoupdate with update. Use adoupdate to update user-written files. Use update to update the components (including ado-files) of the official Stata software. To use either command, you must be connected to the Internet.

Remarks are presented under the following headings:

> *Using adoupdate*
> *Possible problem the first time you run adoupdate and the solution*
> *Notes for developers*

Using adoupdate

The first time you try adoupdate, type

```
. adoupdate
```

That is, do not specify the update option. adoupdate without update produces a report but does not update any files. The first time you run adoupdate, you may see messages such as

```
. adoupdate
(note:  package utx was installed more than once; older copy removed)
(remaining output omitted)
```

Having the same packages installed multiple times is common; adoupdate cleans that up.

The second time you run adoupdate, pick one package to update. Suppose that the report indicates that package st0008 has an update available. Type

```
. adoupdate st0008, update
```

You can specify one or many packages after the adoupdate command. You can even use wildcards such as st* to mean all packages that start with st or st*8 to mean all packages that start with st and end with 8. You can do that with or without the update option.

Finally, you can let `adoupdate` update all your user-written additions:

```
. adoupdate, update
```

Possible problem the first time you run adoupdate and the solution

The first time you run `adoupdate`, you might get many duplicate messages:

```
. adoupdate
(note: package ___ installed more than once; older copy removed)
(note: package ___ installed more than once; older copy removed)
(note: package ___ installed more than once; older copy removed)
...
(note: package ___ installed more than once; older copy removed)
(remaining output omitted)
```

Some users have hundreds of duplicates. You might even see the same package name repeated more than once:

```
(note: package stylus installed more than once; older copy removed)
(note: package stylus installed more than once; older copy removed)
```

That means that the package was duplicated twice.

Stata tolerates duplicates, and you did nothing wrong when you previously installed and updated packages. `adoupdate`, however, needs the duplicates removed, mainly so that it does not keep checking the same files.

The solution is to just let `adoupdate` run. `adoupdate` will run faster next time, when there are no (or just a few) duplicates.

Notes for developers

`adoupdate` reports whether an installed package is up to date by comparing its distribution date with that of the package available over the web.

If you are distributing software, include the line

```
d Distribution-Date: date
```

somewhere in your `.pkg` file. The capitalization of `Distribution-Date` does not matter, but include the hyphen and the colon as shown. Code the date in either of two formats:

all numeric:	*yyyymmdd*, for example, 20110701
Stata standard:	*ddMONyyyy*, for example, `01jul2011`

Saved results

`adoupdate` saves the following in `r()`:

Macros
   `r(pkglist)`   a space-separated list of package names that need updating (`update` not specified) or that were updated (`update` specified)

Methods and formulas

adoupdate is implemented as an ado-file.

Also see

[R] **ssc** — Install and uninstall packages from SSC

[R] **search** — Search Stata documentation

[R] **net** — Install and manage user-written additions from the Internet

[R] **update** — Update Stata

Title

alpha — Compute interitem correlations (covariances) and Cronbach's alpha

Syntax

alpha *varlist* [*if*] [*in*] [*, options*]

options	Description
Options	
<u>a</u>sis	take sign of each item as is
<u>c</u>asewise	delete cases with missing values
<u>d</u>etail	list individual interitem correlations and covariances
<u>g</u>enerate(*newvar*)	save the generated scale in *newvar*
<u>i</u>tem	display item-test and item-rest correlations
<u>l</u>abel	include variable labels in output table
<u>m</u>in(*#*)	must have at least *#* observations for inclusion
<u>r</u>everse(*varlist*)	reverse signs of these variables
<u>s</u>td	standardize items in the scale to mean 0, variance 1

by is allowed; see [D] **by**.

Menu

Statistics > Multivariate analysis > Cronbach's alpha

Description

alpha computes the interitem correlations or covariances for all pairs of variables in *varlist* and Cronbach's α statistic for the scale formed from them. At least two variables must be specified with alpha.

Options

Options

asis specifies that the sense (sign) of each item be taken as presented in the data. The default is to determine the sense empirically and reverse the scorings for any that enter negatively.

casewise specifies that cases with missing values be deleted listwise. The default is pairwise computation of covariances and correlations.

detail lists the individual interitem correlations and covariances.

generate(*newvar*) specifies that the scale constructed from *varlist* be stored in *newvar*. Unless asis is specified, the sense of items entering negatively is automatically reversed. If std is also specified, the scale is constructed by using standardized (mean 0, variance 1) values of the individual items. Unlike most Stata commands, generate() does not use casewise deletion. A score is created for every observation for which there is a response to at least one item (one variable in *varlist* is not missing). The summative score is divided by the number of items over which the sum is calculated.

item specifies that item-test and item-rest correlations and the effects of removing an item from the scale be displayed. item is valid only when more than two variables are specified in *varlist*.

label requests that the detailed output table be displayed in a compact format that enables the inclusion of variable labels.

min(#) specifies that only cases with at least # observations be included in the computations. casewise is a shorthand for min(k), where k is the number of variables in *varlist*.

reverse(*varlist*) specifies that the signs (directions) of the variables (items) in *varlist* be reversed. Any variables specified in reverse() that are not also included in alpha's *varlist* are ignored.

std specifies that the items in the scale be standardized (mean 0, variance 1) before summing.

Remarks

Cronbach's alpha (Cronbach 1951) assesses the reliability of a summative rating (Likert 1932) scale composed of the variables (called *items*) specified. The set of items is often called a *test* or *battery*. A scale is simply the sum of the individual item scores, reversing the scoring for statements that have negative correlations with the factor (for example, attitude) being measured. Scales can be formed by using the raw item scores or standardized item scores.

The reliability α is defined as the square of the correlation between the measured scale and the underlying factor. If you think of a test as being composed of a random sample of items from a hypothetical domain of items designed to measure the same thing, α represents the expected correlation of one test with an alternative form containing the same number of items. The square root of α is the estimated correlation of a test with errorless true scores (Nunnally and Bernstein 1994, 235).

In addition to reporting α, alpha generates the summative scale from the items (variables) specified and automatically reverses the sense of any when necessary. Stata's decision can be overridden by specifying the reverse(*varlist*) option.

Because it concerns reliability in measuring an unobserved factor, α is related to factor analysis. The test should be designed to measure one factor, and, because the scale will be composed of an unweighted sum, the factor loadings should all contribute roughly equal information to the score. Both of these assumptions can be verified with factor; see [MV] **factor**. Equality of factor loadings can also be assessed by using the item option.

▷ Example 1

To illustrate alpha, we apply it, first without and then with the item option, to the automobile dataset after randomly introducing missing values:

```
. use http://www.stata-press.com/data/r12/automiss
(1978 Automobile Data)

. alpha price headroom rep78 trunk weight length turn displ, std
Test scale = mean(standardized items)
Reversed item: rep78
Average interitem correlation:      0.5251
Number of items in the scale:            8
Scale reliability coefficient:      0.8984
```

The scale derived from our somewhat arbitrarily chosen automobile items (variables) appears to be reasonable because the estimated correlation between it and the underlying factor it measures is $\sqrt{0.8984} \approx 0.9478$ and the estimated correlation between this battery of eight items and all other eight-item batteries from the same domain is 0.8984. Because the "items" are not on the same scale, it is important that std was specified so that the scale and its reliability were based on the sum of standardized variables. We could obtain the scale in a new variable called sc with the gen(sc) option.

Though the scale appears reasonable, we include the item option to determine if all the items fit the scale:

```
. alpha price headroom rep78 trunk weight length turn displ, std item
Test scale = mean(standardized items)
```

Item	Obs	Sign	item-test correlation	item-rest correlation	average interitem correlation	alpha
price	70	+	0.5260	0.3719	0.5993	0.9128
headroom	66	+	0.6716	0.5497	0.5542	0.8969
rep78	61	−	0.4874	0.3398	0.6040	0.9143
trunk	69	+	0.7979	0.7144	0.5159	0.8818
weight	64	+	0.9404	0.9096	0.4747	0.8635
length	69	+	0.9382	0.9076	0.4725	0.8625
turn	66	+	0.8678	0.8071	0.4948	0.8727
displacement	63	+	0.8992	0.8496	0.4852	0.8684
Test scale					0.5251	0.8984

"Test" denotes the additive scale; here 0.5251 is the average interitem correlation, and 0.8984 is the alpha coefficient for a test scale based on all items.

"Obs" shows the number of nonmissing values of the items; "Sign" indicates the direction in which an item variable entered the scale; "−" denotes that the item was reversed. The remaining four columns in the table provide information on the effect of one item on the scale.

Column four gives the item-test correlations. Apart from the sign of the correlation for items that entered the scale in reversed order, these correlations are the same numbers as those computed by the commands

```
. alpha price headroom rep78 trunk weight length turn displ, std gen(sc)
. pwcorr sc price headroom rep78 trunk weight length turn displ
```

Typically, the item-test correlations should be roughly the same for all items. Item-test correlations may not be adequate to detect items that fit poorly because the poorly fitting items may distort the scale. Accordingly, it may be more useful to consider item-rest correlations (Nunnally and Bernstein 1994), that is, the correlation between an item and the scale that is formed by all other items. The average interitem correlations (covariances if std is omitted) of all items, excluding one, are shown in column six. Finally, column seven gives Cronbach's α for the test scale, which consists of all but the one item.

Here neither the price item nor the rep78 item seems to fit well in the scale in all respects. The item-test and item-rest correlations of price and rep78 are much lower than those of the other items. The average interitem correlation increases substantially by removing either price or rep78; apparently, they do not correlate strongly with the other items. Finally, we see that Cronbach's α coefficient will increase from 0.8984 to 0.9128 if the price item is dropped, and it will increase from 0.8984 to 0.9143 if rep78 is dropped. For well-fitting items, we would of course expect that α decreases by shortening the test.

◁

▷ Example 2

The variable names for the automobile data are reasonably informative. This may not always be true; items in batteries commonly used to measure personality traits, attitudes, values, etc., are usually named with indexed names such as `item12a`, `item12b`, etc. The `label` option forces `alpha` to produce the same statistical information in a more compact format that leaves room to include variable (item) labels. In this compact format, `alpha` excludes the number of nonmissing values of the items, displays the statistics using fewer digits, and uses somewhat cryptic headers:

```
. alpha price headroom rep78 trunk weight length turn displ, std item label detail
Test scale = mean(standardized items)
```

Items	S	it-cor	ir-cor	ii-cor	alpha	label
price	+	0.526	0.372	0.599	0.913	Price
headroom	+	0.672	0.550	0.554	0.897	Headroom (in.)
rep78	−	0.487	0.340	0.604	0.914	Repair Record 1978
trunk	+	0.798	0.714	0.516	0.882	Trunk space (cu. ft.)
weight	+	0.940	0.910	0.475	0.863	Weight (lbs.)
length	+	0.938	0.908	0.473	0.862	Length (in.)
turn	+	0.868	0.807	0.495	0.873	Turn Circle (ft.)
displacement	+	0.899	0.850	0.485	0.868	Displacement (cu. in.)
Test scale				0.525	0.898	mean(standardized items)

Interitem correlations (reverse applied) (obs=pairwise, see below)

	price	headroom	rep78	trunk
price	1.0000			
headroom	0.1174	1.0000		
rep78	−0.0479	0.1955	1.0000	
trunk	0.2748	0.6841	0.2777	1.0000
weight	0.5093	0.5464	0.3624	0.6486
length	0.4511	0.5823	0.3162	0.7404
turn	0.3528	0.4067	0.4715	0.5900
displacement	0.5537	0.5166	0.3391	0.6471

	weight	length	turn	displacement
weight	1.0000			
length	0.9425	1.0000		
turn	0.8712	0.8589	1.0000	
displacement	0.8753	0.8422	0.7723	1.0000

Pairwise number of observations

	price	headroom	rep78	trunk
price	70			
headroom	62	66		
rep78	59	54	61	
trunk	65	61	59	69
weight	60	56	52	60
length	66	61	58	64
turn	62	58	56	62
displacement	59	58	51	58

	weight	length	turn	displacement
weight	64			
length	60	69		
turn	57	61	66	
displacement	54	58	56	63

Because the `detail` option was also specified, the interitem correlation matrix was printed, together with the number of observations used for each entry (because these varied across the matrix). Note the negative sign attached to `rep78` in the output, indicating the sense in which it entered the scale.

Better-looking output with less-cryptic headers is produced if the linesize is set to a value of at least 100:

```
. set linesize 100
. alpha price headroom rep78 trunk weight length turn displ, std item label
Test scale = mean(standardized items)
```

Item	Obs	Sign	item-test corr.	item-rest corr.	interitem corr.	alpha	Label
price	70	+	0.5260	0.3719	0.5993	0.9128	Price
headroom	62	+	0.6716	0.5497	0.5542	0.8969	Headroom (in.)
rep78	59	−	0.4874	0.3398	0.6040	0.9143	Repair Record 1978
trunk	65	+	0.7979	0.7144	0.5159	0.8818	Trunk space (cu. ft.)
weight	60	+	0.9404	0.9096	0.4747	0.8635	Weight (lbs.)
length	66	+	0.9382	0.9076	0.4725	0.8625	Length (in.)
turn	62	+	0.8678	0.8071	0.4948	0.8727	Turn Circle (ft.)
displacement	59	+	0.8992	0.8496	0.4852	0.8684	Displacement (cu. in.)
Test scale					0.5251	0.8984	mean(standardized items)

◁

Users of alpha require some standard for judging values of α. We paraphrase Nunnally and Bernstein (1994, 265): In the early stages of research, modest reliability of 0.70 or higher will suffice; values in excess of 0.80 often waste time and funds. In contrast, where measurements on individuals are of interest, a reliability of 0.80 may not be nearly high enough. Even with a reliability of 0.90, the standard error of measurement is almost one-third as large as the standard deviation of test scores; a reliability of 0.90 is the minimum that should be tolerated, and a reliability of 0.95 should be considered the desirable standard.

Saved results

alpha saves the following in r():

Scalars
r(alpha)	scale reliability coefficient
r(k)	number of items in the scale
r(cov)	average interitem covariance
r(rho)	average interitem correlation if std is specified

Matrices
r(Alpha)	scale reliability coefficient
r(ItemTestCorr)	item-test correlation
r(ItemRestCorr)	item-rest correlation
r(MeanInterItemCov)	average interitem covariance
r(MeanInterItemCorr)	average interitem correlation if std is specified

If the item option is specified, results are saved as row matrices for the k subscales when one variable is removed.

Methods and formulas

`alpha` is implemented as an ado-file.

Let x_i, $i = 1, \ldots, k$, be the variables over which α is to be calculated. Let s_i be the sign with which x_i enters the scale. If `asis` is specified, $s_i = 1$ for all i. Otherwise, principal-factor analysis is performed on x_i, and the first factor's score is predicted; see [MV] **factor**. s_i is -1 if correlation of the x_i and the predicted score is negative and $+1$ otherwise.

Let r_{ij} be the correlation between x_i and x_j, c_{ij} be the covariance, and n_{ij} be the number of observations used in calculating the correlation or covariance. The average correlation is

$$\overline{r} = \frac{\displaystyle\sum_{i=2}^{k} \sum_{j=1}^{i-1} s_i s_j n_{ij} r_{ij}}{\displaystyle\sum_{i=2}^{k} \sum_{j=1}^{i-1} n_{ij}}$$

and the average covariance similarly is

$$\overline{c} = \frac{\displaystyle\sum_{i=2}^{k} \sum_{j=1}^{i-1} s_i s_j n_{ij} c_{ij}}{\displaystyle\sum_{i=2}^{k} \sum_{j=1}^{i-1} n_{ij}}$$

Let c_{ii} denote the variance of x_i, and define the average variance as

$$\overline{v} = \frac{\displaystyle\sum_{i=1}^{k} n_{ii} c_{ii}}{\displaystyle\sum_{i=1}^{k} n_{ii}}$$

If `std` is specified, the scale reliability α is calculated as defined by the general form of the Spearman–Brown Prophecy Formula (Nunnally and Bernstein 1994, 232; Allen and Yen 1979, 85–88):

$$\alpha = \frac{k\overline{r}}{1 + (k-1)\overline{r}}$$

This expression corresponds to α under the assumption that the summative rating is the sum of the standardized variables (Nunnally and Bernstein 1994, 234). If `std` is not specified, α is defined (Nunnally and Bernstein 1994, 232 and 234) as

$$\alpha = \frac{k\overline{c}}{\overline{v} + (k-1)\overline{c}}$$

Let x_{ij} reflect the value of item i in the jth observation. If `std` is specified, the jth value of the scale computed from the k x_{ij} items is

$$S_j = \frac{1}{k_j} \sum_{i=1}^{k} s_i S(x_{ij})$$

where $S()$ is the function that returns the standardized (mean 0, variance 1) value if x_{ij} is not missing and returns zero if x_{ij} is missing. k_j is the number of nonmissing values in x_{ij}, $i = 1, \ldots, k$. If std is not specified, $S()$ is the function that returns x_{ij} or returns missing if x_{ij} is missing.

Lee Joseph Cronbach (1916–2001) was an American psychometrician and educational psychologist who worked principally on measurement theory, program evaluation, and instruction. He taught and researched at the State College of Washington, the University of Chicago, the University of Illinois, and Stanford University. Cronbach's initial paper on alpha led to a theory of test reliability.

Acknowledgment

This improved version of alpha was written by Jeroen Weesie, Department of Sociology, Utrecht University, The Netherlands.

References

Acock, A. C. 2010. *A Gentle Introduction to Stata*. 3rd ed. College Station, TX: Stata Press.

Allen, M. J., and W. M. Yen. 1979. *Introduction to Measurement Theory*. Monterey, CA: Brooks/Cole.

Bleda, M.-J., and A. Tobías. 2000. sg143: Cronbach's alpha one-sided confidence interval. *Stata Technical Bulletin* 56: 26–27. Reprinted in *Stata Technical Bulletin Reprints*, vol. 10, pp. 187–189. College Station, TX: Stata Press.

Cronbach, L. J. 1951. Coefficient alpha and the internal structure of tests. *Psychometrika* 16: 297–334.

Likert, R. A. 1932. A technique for the measurement of attitudes. *Archives of Psychology* 140: 5–55.

Nunnally, J. C., and I. H. Bernstein. 1994. *Psychometric Theory*. 3rd ed. New York: McGraw–Hill.

Shavelson, R. J., and G. Gleser. 2002. Lee J. Cronbach (1916–2001). *American Psychologist* 57: 360–361.

Tarlov, A. R., J. E. Ware, Jr., S. Greenfield, E. C. Nelson, E. Perrin, and M. Zubkoff. 1989. The medical outcomes study. An application of methods for monitoring the results of medical care. *Journal of the American Medical Association* 262: 925–930.

Weesie, J. 1997. sg66: Enhancements to the alpha command. *Stata Technical Bulletin* 35: 32–34. Reprinted in *Stata Technical Bulletin Reprints*, vol. 6, pp. 176–179. College Station, TX: Stata Press.

Also see

[MV] **factor** — Factor analysis

Title

> **ameans** — Arithmetic, geometric, and harmonic means

Syntax

ameans $[varlist]$ $[if]$ $[in]$ $[weight]$ $[,\ options]$

options	Description
Main	
<u>a</u>dd(#)	add # to each variable in *varlist*
<u>on</u>ly	add # only to variables with nonpositive values
<u>l</u>evel(#)	set confidence level; default is level(95)

by is allowed; see [D] **by**.
aweights and fweights are allowed; see [U] **11.1.6 weight**.

Menu

Statistics > Summaries, tables, and tests > Summary and descriptive statistics > Arith./geometric/harmonic means

Description

ameans computes the arithmetic, geometric, and harmonic means, with their corresponding confidence intervals, for each variable in *varlist* or for all the variables in the data if *varlist* is not specified. gmeans and hmeans are synonyms for ameans.

If you simply want arithmetic means and corresponding confidence intervals, see [R] **ci**.

Options

> **Main**

add(#) adds the value # to each variable in *varlist* before computing the means and confidence intervals. This option is useful when analyzing variables with nonpositive values.

only modifies the action of the add(#) option so that it adds # only to variables with at least one nonpositive value.

level(#) specifies the confidence level, as a percentage, for confidence intervals. The default is level(95) or as set by set level; see [U] **20.7 Specifying the width of confidence intervals**.

Remarks

▷ Example 1

We have a dataset containing 8 observations on a variable named x. The eight values are 5, 4, −4, −5, 0, 0, *missing*, and 7.

```
. ameans x
```

Variable	Type	Obs	Mean	[95% Conf. Interval]	
x	Arithmetic	7	1	-3.204405	5.204405
	Geometric	3	5.192494	2.57899	10.45448
	Harmonic	3	5.060241	3.023008	15.5179

```
. ameans x, add(5)
```

Variable	Type	Obs	Mean	[95% Conf. Interval]	
x	Arithmetic	7	6	1.795595	10.2044 *
	Geometric	6	5.477226	2.1096	14.22071 *
	Harmonic	6	3.540984	.	. *

(*) 5 was added to the variables prior to calculating the results.
Missing values in confidence intervals for harmonic mean indicate
that confidence interval is undefined for corresponding variables.
Consult Reference Manual for details.

The number of observations displayed for the arithmetic mean is the number of nonmissing observations. The number of observations displayed for the geometric and harmonic means is the number of nonmissing, positive observations. Specifying the add(5) option produces 3 more positive observations. The confidence interval for the harmonic mean is not reported; see *Methods and formulas* below.

◁

Saved results

ameans saves the following in r():

Scalars

r(N)	number of nonmissing observations; used for arithmetic mean
r(N_pos)	number of nonmissing positive observations; used for geometric and harmonic means
r(mean)	arithmetic mean
r(lb)	lower bound of confidence interval for arithmetic mean
r(ub)	upper bound of confidence interval for arithmetic mean
r(Var)	variance of untransformed data
r(mean_g)	geometric mean
r(lb_g)	lower bound of confidence interval for geometric mean
r(ub_g)	upper bound of confidence interval for geometric mean
r(Var_g)	variance of $\ln x_i$
r(mean_h)	harmonic mean
r(lb_h)	lower bound of confidence interval for harmonic mean
r(ub_h)	upper bound of confidence interval for harmonic mean
r(Var_h)	variance of $1/x_i$

Methods and formulas

ameans is implemented as an ado-file.

See Armitage, Berry, and Matthews (2002) or Snedecor and Cochran (1989). For a history of the concept of the mean, see Plackett (1958).

When restricted to the same set of values (that is, to positive values), the arithmetic mean ($\bar{x}$) is greater than or equal to the geometric mean, which in turn is greater than or equal to the harmonic mean. Equality holds only if all values within a sample are equal to a positive constant.

The arithmetic mean and its confidence interval are identical to those provided by ci; see [R] **ci**.

To compute the geometric mean, ameans first creates $u_j = \ln x_j$ for all positive x_j. The arithmetic mean of the u_j and its confidence interval are then computed as in ci. Let $\bar{u}$ be the resulting mean, and let $[L, U]$ be the corresponding confidence interval. The geometric mean is then $\exp(\bar{u})$, and its confidence interval is $[\exp(L), \exp(U)]$.

The same procedure is followed for the harmonic mean, except that then $u_j = 1/x_j$. The harmonic mean is then $1/\bar{u}$, and its confidence interval is $[1/U, 1/L]$ if L is greater than zero. If L is not greater than zero, this confidence interval is not defined, and missing values are reported.

When weights are specified, ameans applies the weights to the transformed values, $u_j = \ln x_j$ and $u_j = 1/x_j$, respectively, when computing the geometric and harmonic means. For details on how the weights are used to compute the mean and variance of the u_j, see [R] **summarize**. Without weights, the formula for the geometric mean reduces to

$$\exp\left\{\frac{1}{n} \sum_j \ln(x_j)\right\}$$

Without weights, the formula for the harmonic mean is

$$\frac{n}{\sum_j \dfrac{1}{x_j}}$$

Acknowledgments

This improved version of ameans is based on the gmci command (Carlin, Vidmar, and Ramalheira 1998) and was written by John Carlin, University of Melbourne, Australia; Suzanna Vidmar, University of Melbourne, Australia; and Carlos Ramalheira, Coimbra University Hospital, Portugal.

References

Armitage, P., G. Berry, and J. N. S. Matthews. 2002. *Statistical Methods in Medical Research*. 4th ed. Oxford: Blackwell.

Carlin, J., S. Vidmar, and C. Ramalheira. 1998. sg75: Geometric means and confidence intervals. *Stata Technical Bulletin* 41: 23–25. Reprinted in *Stata Technical Bulletin Reprints*, vol. 7, pp. 197–199. College Station, TX: Stata Press.

Keynes, J. M. 1911. The principal averages and the laws of error which lead to them. *Journal of the Royal Statistical Society* 74: 322–331.

Plackett, R. L. 1958. Studies in the history of probability and statistics: VII. The principle of the arithmetic mean. *Biometrika* 45: 130–135.

Snedecor, G. W., and W. G. Cochran. 1989. *Statistical Methods*. 8th ed. Ames, IA: Iowa State University Press.

Stigler, S. M. 1985. Arithmetic means. In Vol. 1 of *Encyclopedia of Statistical Sciences*, ed. S. Kotz and N. L. Johnson, 126–129. New York: Wiley.

Also see

[R] **ci** — Confidence intervals for means, proportions, and counts

[R] **mean** — Estimate means

[R] **summarize** — Summary statistics

[SVY] **svy estimation** — Estimation commands for survey data

Title

> **anova** — Analysis of variance and covariance

Syntax

$\underline{\text{an}}\text{ova}$ *varname* [termlist] [*if*] [*in*] [*weight*] [, *options*]

where *termlist* is a factor-variable list (see [U] **11.4.3 Factor variables**) with the following additional features:

- Variables are assumed to be categorical; use the c. factor-variable operator to override this.

- The | symbol (indicating nesting) may be used in place of the # symbol (indicating interaction).

- The / symbol is allowed after a term and indicates that the following term is the error term for the preceding terms.

options	Description
Model	
repeated(*varlist*)	variables in *term*s that are repeated-measures variables
partial	use partial (or marginal) sums of squares
sequential	use sequential sums of squares
noconstant	suppress constant term
dropemptycells	drop empty cells from the design matrix
Adv. model	
bse(*term*)	between-subjects error term in repeated-measures ANOVA
bseunit(*varname*)	variable representing lowest unit in the between-subjects error term
grouping(*varname*)	grouping variable for computing pooled covariance matrix

bootstrap, by, jackknife, and statsby are allowed; see [U] **11.1.10 Prefix commands**.
aweights and fweights are allowed; see [U] **11.1.6 weight**.
See [U] **20 Estimation and postestimation commands** for more capabilities of estimation commands.

Menu

Statistics > Linear models and related > ANOVA/MANOVA > Analysis of variance and covariance

Description

The anova command fits analysis-of-variance (ANOVA) and analysis-of-covariance (ANCOVA) models for balanced and unbalanced designs, including designs with missing cells; for repeated-measures ANOVA; and for factorial, nested, or mixed designs.

The regress command (see [R] **regress**) will display the coefficients, standard errors, etc., of the regression model underlying the last run of anova.

If you want to fit one-way ANOVA models, you may find the oneway or loneway command more convenient; see [R] **oneway** and [R] **loneway**. If you are interested in MANOVA or MANCOVA, see [MV] **manova**.

Options

⌐‾‾‾| Model |‾‾

repeated(*varlist*) indicates the names of the categorical variables in the *term*s that are to be treated as repeated-measures variables in a repeated-measures ANOVA or ANCOVA.

partial presents the ANOVA table using partial (or marginal) sums of squares. This setting is the default. Also see the sequential option.

sequential presents the ANOVA table using sequential sums of squares.

noconstant suppresses the constant term (intercept) from the ANOVA or regression model.

dropemptycells drops empty cells from the design matrix. If c(emptycells) is set to keep (see [R] **set emptycells**), this option temporarily resets it to drop before running the ANOVA model. If c(emptycells) is already set to drop, this option does nothing.

⌐‾‾‾| Adv. model |‾‾‾

bse(*term*) indicates the between-subjects error term in a repeated-measures ANOVA. This option is needed only in the rare case when the anova command cannot automatically determine the between-subjects error term.

bseunit(*varname*) indicates the variable representing the lowest unit in the between-subjects error term in a repeated-measures ANOVA. This option is rarely needed because the anova command automatically selects the first variable listed in the between-subjects error term as the default for this option.

grouping(*varname*) indicates a variable that determines which observations are grouped together in computing the covariance matrices that will be pooled and used in a repeated-measures ANOVA. This option is rarely needed because the anova command automatically selects the combination of all variables except the first (or as specified in the bseunit() option) in the between-subjects error term as the default for grouping observations.

Remarks

Remarks are presented under the following headings:

> *Introduction*
> *One-way ANOVA*
> *Two-way ANOVA*
> *N-way ANOVA*
> *Weighted data*
> *ANCOVA*
> *Nested designs*
> *Mixed designs*
> *Latin-square designs*
> *Repeated-measures ANOVA*

Introduction

anova uses least squares to fit the linear models known as ANOVA or ANCOVA (henceforth referred to simply as ANOVA models).

If your interest is in one-way ANOVA, you may find the oneway command to be more convenient; see [R] **oneway**.

Structural equation modeling provides a more general framework for fitting ANOVA models; see the *Stata Structural Equation Modeling Reference Manual.*

ANOVA was pioneered by Fisher. It features prominently in his texts on statistical methods and his design of experiments (1925, 1935). Many books discuss ANOVA; see, for instance, Altman (1991); van Belle et al. (2004); Cobb (1998); Snedecor and Cochran (1989); or Winer, Brown, and Michels (1991). For a classic source, see Scheffé (1959). Kennedy and Gentle (1980) discuss ANOVA's computing problems. Edwards (1985) is concerned primarily with the relationship between multiple regression and ANOVA. Acock (2010, chap. 9) illustrates his discussion with Stata output. Repeated-measures ANOVA is discussed in Winer, Brown, and Michels (1991); Kuehl (2000); and Milliken and Johnson (2009). Pioneering work in repeated-measures ANOVA can be found in Box (1954); Geisser and Greenhouse (1958); Huynh and Feldt (1976); and Huynh (1978).

One-way ANOVA

anova, entered without options, performs and reports standard ANOVA. For instance, to perform a one-way layout of a variable called endog on exog, you would type anova endog exog.

▷ Example 1

We run an experiment varying the amount of fertilizer used in growing apple trees. We test four concentrations, using each concentration in three groves of 12 trees each. Later in the year, we measure the average weight of the fruit.

If all had gone well, we would have had 3 observations on the average weight for each of the four concentrations. Instead, two of the groves were mistakenly leveled by a confused man on a large bulldozer. We are left with the following data:

```
. use http://www.stata-press.com/data/r12/apple
(Apple trees)
. list, abbrev(10) sepby(treatment)
```

	treatment	weight
1.	1	117.5
2.	1	113.8
3.	1	104.4
4.	2	48.9
5.	2	50.4
6.	2	58.9
7.	3	70.4
8.	3	86.9
9.	4	87.7
10.	4	67.3

To obtain one-way ANOVA results, we type

. anova weight treatment

| | | | | Number of obs = | 10 | R-squared = | 0.9147 |
| | | | | Root MSE = 9.07002 | | Adj R-squared = | 0.8721 |
Source	Partial SS	df	MS			F	Prob > F
Model	5295.54433	3	1765.18144			21.46	0.0013
treatment	5295.54433	3	1765.18144			21.46	0.0013
Residual	493.591667	6	82.2652778				
Total	5789.136	9	643.237333				

We find significant (at better than the 1% level) differences among the four concentrations.

Although the output is a usual ANOVA table, let's run through it anyway. Above the table is a summary of the underlying regression. The model was fit on 10 observations, and the root mean squared error (Root MSE) is 9.07. The R^2 for the model is 0.9147, and the adjusted R^2 is 0.8721.

The first line of the table summarizes the model. The sum of squares (Partial SS) for the model is 5295.5 with 3 degrees of freedom (df). This line results in a mean square (MS) of $5295.5/3 \approx 1765.2$. The corresponding F statistic is 21.46 and has a significance level of 0.0013. Thus the model appears to be significant at the 0.13% level.

The next line summarizes the first (and only) term in the model, treatment. Because there is only one term, the line is identical to that for the overall model.

The third line summarizes the residual. The residual sum of squares is 493.59 with 6 degrees of freedom, resulting in a mean squared error of 82.27. The square root of this latter number is reported as the Root MSE.

The model plus the residual sum of squares equals the total sum of squares, which is reported as 5789.1 in the last line of the table. This is the total sum of squares of weight after removal of the mean. Similarly, the model plus the residual degrees of freedom sum to the total degrees of freedom, 9. Remember that there are 10 observations. Subtracting 1 for the mean, we are left with 9 total degrees of freedom.

◁

❑ Technical note

Rather than using the anova command, we could have performed this analysis by using the oneway command. Example 1 in [R] **oneway** repeats this same analysis. You may wish to compare the output.

❑

Type regress to see the underlying regression model corresponding to an ANOVA model fit using the anova command.

▷ Example 2

Returning to the apple tree experiment, we found that the fertilizer concentration appears to significantly affect the average weight of the fruit. Although that finding is interesting, we next want to know which concentration appears to grow the heaviest fruit. One way to find out is by examining the underlying regression coefficients.

```
. regress, baselevels
```

Source	SS	df	MS
Model	5295.54433	3	1765.18144
Residual	493.591667	6	82.2652778
Total	5789.136	9	643.237333

```
Number of obs =      10
F(  3,    6) =   21.46
Prob > F      =  0.0013
R-squared     =  0.9147
Adj R-squared =  0.8721
Root MSE      =    9.07
```

weight	Coef.	Std. Err.	t	P>|t|	[95% Conf. Interval]	
treatment						
1	0	(base)				
2	-59.16667	7.405641	-7.99	0.000	-77.28762	-41.04572
3	-33.25	8.279758	-4.02	0.007	-53.50984	-12.99016
4	-34.4	8.279758	-4.15	0.006	-54.65984	-14.14016
_cons	111.9	5.236579	21.37	0.000	99.08655	124.7134

See [R] **regress** for an explanation of how to read this table. The baselevels option of regress displays a row indicating the base category for our categorical variable, treatment. In summary, we find that concentration 1, the base (omitted) group, produces significantly heavier fruits than concentration 2, 3, and 4; concentration 2 produces the lightest fruits; and concentrations 3 and 4 appear to be roughly equivalent.

◁

▷ Example 3

We previously typed anova weight treatment to produce and display the ANOVA table for our apple tree experiment. Typing regress displays the regression coefficients. We can redisplay the ANOVA table by typing anova without arguments:

```
. anova
```

```
Number of obs =      10     R-squared     =  0.9147
Root MSE      = 9.07002     Adj R-squared =  0.8721
```

Source	Partial SS	df	MS	F	Prob > F
Model	5295.54433	3	1765.18144	21.46	0.0013
treatment	5295.54433	3	1765.18144	21.46	0.0013
Residual	493.591667	6	82.2652778		
Total	5789.136	9	643.237333		

◁

Two-way ANOVA

You can include multiple explanatory variables with the anova command, and you can specify interactions by placing '#' between the variable names. For instance, typing anova y a b performs a two-way layout of y on a and b. Typing anova y a b a#b performs a full two-way factorial layout. The shorthand anova y a##b does the same.

With the default partial sums of squares, when you specify interacted terms, the order of the terms does not matter. Typing anova y a b a#b is the same as typing anova y b a b#a.

▷ Example 4

The classic two-way factorial ANOVA problem, at least as far as computer manuals are concerned, is a two-way ANOVA design from Afifi and Azen (1979).

Fifty-eight patients, each suffering from one of three different diseases, were randomly assigned to one of four different drug treatments, and the change in their systolic blood pressure was recorded. Here are the data:

	Disease 1	Disease 2	Disease 3
Drug 1	42, 44, 36 13, 19, 22	33, 26, 33 21	31, −3, 25 25, 24
Drug 2	28, 23, 34 42, 13	34, 33, 31 36	3, 26, 28 32, 4, 16
Drug 3	1, 29, 19	11, 9, 7 1, −6	21, 1, 9 3
Drug 4	24, 9, 22 −2, 15	27, 12, 12 −5, 16, 15	22, 7, 25 5, 12

Let's assume that we have entered these data into Stata and stored the data as `systolic.dta`. Below we `use` the data, `list` the first 10 observations, `summarize` the variables, and `tabulate` the control variables:

```
. use http://www.stata-press.com/data/r12/systolic
(Systolic Blood Pressure Data)

. list in 1/10
```

	drug	disease	systolic
1.	1	1	42
2.	1	1	44
3.	1	1	36
4.	1	1	13
5.	1	1	19
6.	1	1	22
7.	1	2	33
8.	1	2	26
9.	1	2	33
10.	1	2	21

```
. summarize
```

Variable	Obs	Mean	Std. Dev.	Min	Max
drug	58	2.5	1.158493	1	4
disease	58	2.017241	.8269873	1	3
systolic	58	18.87931	12.80087	−6	44

```
. tabulate drug disease
```

Drug Used	Patient's Disease			Total
	1	2	3	
1	6	4	5	15
2	5	4	6	15
3	3	5	4	12
4	5	6	5	16
Total	19	19	20	58

Each observation in our data corresponds to one patient, and for each patient we record drug, disease, and the increase in the systolic blood pressure, systolic. The tabulation reveals that the data are not balanced—there are not equal numbers of patients in each drug–disease cell. Stata does not require that the data be balanced. We can perform a two-way factorial ANOVA by typing

```
. anova systolic drug disease drug#disease
                            Number of obs =        58    R-squared      =  0.4560
                            Root MSE      =  10.5096    Adj R-squared  =  0.3259

        Source |   Partial SS     df        MS             F      Prob > F

         Model |   4259.33851     11    387.212591          3.51    0.0013

          drug |   2997.47186      3    999.157287          9.05    0.0001
       disease |   415.873046      2    207.936523          1.88    0.1637
  drug#disease |   707.266259      6    117.87771           1.07    0.3958

      Residual |   5080.81667     46    110.452536

         Total |   9340.15517     57    163.862371
```

Although Stata's table command does not perform ANOVA, it can produce useful summary tables of your data (see [R] **table**):

```
. table drug disease, c(mean systolic) row col f(%8.2f)

            |    Patient's Disease
  Drug Used |     1       2       3     Total

          1 |  29.33   28.25   20.40    26.07
          2 |  28.00   33.50   18.17    25.53
          3 |  16.33    4.40    8.50     8.75
          4 |  13.60   12.83   14.20    13.50

      Total |  22.79   18.21   15.80    18.88
```

These are simple means and are not influenced by our anova model. More useful is the margins command (see [R] **margins**) that provides marginal means and adjusted predictions. Because drug is the only significant factor in our ANOVA, we now examine the adjusted marginal means for drug.

```
. margins drug, asbalanced
Adjusted predictions                             Number of obs     =        58
Expression   : Linear prediction, predict()
at           : drug                  (asbalanced)
               disease               (asbalanced)

                       Delta-method
            |   Margin   Std. Err.      z    P>|z|     [95% Conf. Interval]

       drug |
          1 | 25.99444   2.751008     9.45   0.000    20.60257    31.38632
          2 | 26.55556   2.751008     9.65   0.000    21.16368    31.94743
          3 | 9.744444   3.100558     3.14   0.002    3.667462    15.82143
          4 | 13.54444   2.637123     5.14   0.000    8.375778    18.71311
```

These adjusted marginal predictions are not equal to the simple drug means (see the total column from the table command); they are based upon predictions from our ANOVA model. The asbalanced option of margins corresponds with the interpretation of the F statistic produced by ANOVA—each cell is given equal weight regardless of its sample size (see the following three technical notes). You can omit the `asbalanced` option and obtain predictive margins that take into account the unequal sample sizes of the cells.

```
. margins drug

Predictive margins                              Number of obs    =         58
Expression     : Linear prediction, predict()
```

	Margin	Delta-method Std. Err.	z	P>|z|	[95% Conf. Interval]
drug						
1	25.89799	2.750533	9.42	0.000	20.50704	31.28893
2	26.41092	2.742762	9.63	0.000	21.0352	31.78664
3	9.722989	3.099185	3.14	0.002	3.648697	15.79728
4	13.55575	2.640602	5.13	0.000	8.380261	18.73123

◁

❏ Technical note

How do you interpret the significance of terms like `drug` and `disease` in unbalanced data? If you are familiar with SAS, the sums of squares and the F statistic reported by Stata correspond to SAS type III sums of squares. (Stata can also calculate sequential sums of squares, but we will postpone that topic for now.)

Let's think in terms of the following table:

	Disease 1	Disease 2	Disease 3	
Drug 1	μ_{11}	μ_{12}	μ_{13}	$\mu_{1\cdot}$
Drug 2	μ_{21}	μ_{22}	μ_{23}	$\mu_{2\cdot}$
Drug 3	μ_{31}	μ_{32}	μ_{33}	$\mu_{3\cdot}$
Drug 4	μ_{41}	μ_{42}	μ_{43}	$\mu_{4\cdot}$
	$\mu_{\cdot 1}$	$\mu_{\cdot 2}$	$\mu_{\cdot 3}$	$\mu_{\cdot\cdot}$

In this table, μ_{ij} is the mean increase in systolic blood pressure associated with drug i and disease j, while $\mu_{i\cdot}$ is the mean for drug i, $\mu_{\cdot j}$ is the mean for disease j, and $\mu_{\cdot\cdot}$ is the overall mean.

If the data are balanced, meaning that there are equal numbers of observations going into the calculation of each mean μ_{ij}, the row means, $\mu_{i\cdot}$, are given by

$$\mu_{i\cdot} = \frac{\mu_{i1} + \mu_{i2} + \mu_{i3}}{3}$$

In our case, the data are not balanced, but we define the $\mu_{i\cdot}$ according to that formula anyway. The test for the main effect of drug is the test that

$$\mu_{1\cdot} = \mu_{2\cdot} = \mu_{3\cdot} = \mu_{4\cdot}$$

To be absolutely clear, the F test of the term `drug`, called the *main effect* of drug, is formally equivalent to the test of the three constraints:

$$\frac{\mu_{11} + \mu_{12} + \mu_{13}}{3} = \frac{\mu_{21} + \mu_{22} + \mu_{23}}{3}$$

$$\frac{\mu_{11} + \mu_{12} + \mu_{13}}{3} = \frac{\mu_{31} + \mu_{32} + \mu_{33}}{3}$$

$$\frac{\mu_{11} + \mu_{12} + \mu_{13}}{3} = \frac{\mu_{41} + \mu_{42} + \mu_{43}}{3}$$

In our data, we obtain a significant F statistic of 9.05 and thus reject those constraints.

❑

❑ Technical note

Stata can display the symbolic form underlying the test statistics it presents, as well as display other test statistics and their symbolic forms; see *Obtaining symbolic forms* in [R] **anova postestimation**. Here is the result of requesting the symbolic form for the main effect of drug in our data:

```
. test drug, symbolic
drug
               1    -(r2+r3+r4)
               2    r2
               3    r3
               4    r4
disease
               1    0
               2    0
               3    0
drug#disease
             1  1   -1/3 (r2+r3+r4)
             1  2   -1/3 (r2+r3+r4)
             1  3   -1/3 (r2+r3+r4)
             2  1   1/3 r2
             2  2   1/3 r2
             2  3   1/3 r2
             3  1   1/3 r3
             3  2   1/3 r3
             3  3   1/3 r3
             4  1   1/3 r4
             4  2   1/3 r4
             4  3   1/3 r4
   _cons           0
```

This says exactly what we said in the previous technical note.

❑

❑ Technical note

Saying that there is no main effect of a variable is not the same as saying that it has no effect at all. Stata's ability to perform ANOVA on unbalanced data can easily be put to ill use.

For example, consider the following table of the probability of surviving a bout with one of two diseases according to the drug administered to you:

	Disease 1	Disease 2
Drug 1	1	0
Drug 2	0	1

If you have disease 1 and are administered drug 1, you live. If you have disease 2 and are administered drug 2, you live. In all other cases, you die.

This table has no main effects of either drug or disease, although there is a large interaction effect. You might now be tempted to reason that because there is only an interaction effect, you would be indifferent between the two drugs in the absence of knowledge about which disease infects you. Given an equal chance of having either disease, you reason that it does not matter which drug is administered to you—either way, your chances of surviving are 0.5.

You may not, however, have an equal chance of having either disease. If you knew that disease 1 was 100 times more likely to occur in the population, and if you knew that you had one of the two diseases, you would express a strong preference for receiving drug 1.

When you calculate the significance of main effects on unbalanced data, you must ask yourself why the data are unbalanced. If the data are unbalanced for random reasons and you are making predictions for a balanced population, the test of the main effect makes perfect sense. If, however, the data are unbalanced because the underlying populations are unbalanced and you are making predictions for such unbalanced populations, the test of the main effect may be practically—if not statistically—meaningless.

❑

▷ Example 5

Stata can perform ANOVA not only on unbalanced populations, but also on populations that are so unbalanced that entire cells are missing. For instance, using our systolic blood pressure data, let's refit the model eliminating the drug 1–disease 1 cell. Because anova follows the same syntax as all other Stata commands, we can explicitly specify the data to be used by typing the if qualifier at the end of the anova command. Here we want to use the data that are not for drug 1 and disease 1:

```
. anova systolic drug##disease if !(drug==1 & disease==1)
```

	Number of obs =	52	R-squared	= 0.4545
	Root MSE	= 10.1615	Adj R-squared =	0.3215

Source	Partial SS	df	MS	F	Prob > F
Model	3527.95897	10	352.795897	3.42	0.0025
drug	2686.57832	3	895.526107	8.67	0.0001
disease	327.792598	2	163.896299	1.59	0.2168
drug#disease	703.007602	5	140.60152	1.36	0.2586
Residual	4233.48333	41	103.255691		
Total	7761.44231	51	152.185143		

Here we used drug##disease as a shorthand for drug disease drug#disease.

◁

❑ Technical note

The test of the main effect of drug in the presence of missing cells is more complicated than that for unbalanced data. Our underlying tableau now has the following form:

	Disease 1	Disease 2	Disease 3	
Drug 1		μ_{12}	μ_{13}	
Drug 2	μ_{21}	μ_{22}	μ_{23}	$\mu_{2\cdot}$
Drug 3	μ_{31}	μ_{32}	μ_{33}	$\mu_{3\cdot}$
Drug 4	μ_{41}	μ_{42}	μ_{43}	$\mu_{4\cdot}$
		$\mu_{\cdot 2}$	$\mu_{\cdot 3}$	

The hole in the drug 1–disease 1 cell indicates that the mean is unobserved. Considering the main effect of drug, the test is unchanged for the rows in which all the cells are defined:

$$\mu_{2\cdot} = \mu_{3\cdot} = \mu_{4\cdot}$$

The first row, however, requires special attention. Here we want the average outcome for drug 1, which is averaged only over diseases 2 and 3, to be equal to the average values of all other drugs averaged over those same two diseases:

$$\frac{\mu_{12} + \mu_{13}}{2} = \frac{(\mu_{22} + \mu_{23})/2 + (\mu_{32} + \mu_{33})/2 + (\mu_{42} + \mu_{43})/2}{3}$$

Thus the test contains three constraints:

$$\frac{\mu_{21} + \mu_{22} + \mu_{23}}{3} = \frac{\mu_{31} + \mu_{32} + \mu_{33}}{3}$$

$$\frac{\mu_{21} + \mu_{22} + \mu_{23}}{3} = \frac{\mu_{41} + \mu_{42} + \mu_{43}}{3}$$

$$\frac{\mu_{12} + \mu_{13}}{2} = \frac{\mu_{22} + \mu_{23} + \mu_{32} + \mu_{33} + \mu_{42} + \mu_{43}}{6}$$

❑

Stata can calculate two types of sums of squares, *partial* and *sequential*. If you do not specify which sums of squares to calculate, Stata calculates partial sums of squares. The technical notes above have gone into great detail about the definition and use of partial sums of squares. Use the sequential option to obtain sequential sums of squares.

❑ Technical note

Before we illustrate sequential sums of squares, consider one more feature of the partial sums. If you know how such things are calculated, you may worry that the terms must be specified in some particular order, that Stata would balk or, even worse, produce different results if you typed, say, anova drug#disease drug disease rather than anova drug disease drug#disease. We assure you that is not the case.

When you type a model, Stata internally reorganizes the terms, forms the cross-product matrix, inverts it, converts the result to an upper-Hermite form, and then performs the hypothesis tests. As a final touch, Stata reports the results in the same order that you typed the terms.

❑

▷ Example 6

We wish to estimate the effects on systolic blood pressure of drug and disease by using sequential sums of squares. We want to introduce disease first, then drug, and finally, the interaction of drug and disease:

```
. anova systolic disease drug disease#drug, sequential
```

| | Number of obs = | | 58 | R-squared | = | 0.4560 |
| | Root MSE | = 10.5096 | | Adj R-squared = | | 0.3259 |

Source	Seq. SS	df	MS	F	Prob > F
Model	4259.33851	11	387.212591	3.51	0.0013
disease	488.639383	2	244.319691	2.21	0.1210
drug	3063.43286	3	1021.14429	9.25	0.0001
disease#drug	707.266259	6	117.87771	1.07	0.3958
Residual	5080.81667	46	110.452536		
Total	9340.15517	57	163.862371		

The F statistic on disease is now 2.21. When we fit this same model by using partial sums of squares, the statistic was 1.88.

◁

N-way ANOVA

You may include high-order interaction terms, such as a third-order interaction between the variables A, B, and C, by typing A#B#C.

▷ Example 7

We wish to determine the operating conditions that maximize yield for a manufacturing process. There are three temperature settings, two chemical supply companies, and two mixing methods under investigation. Three observations are obtained for each combination of these three factors.

```
. use http://www.stata-press.com/data/r12/manuf
(manufacturing process data)
. describe
Contains data from http://www.stata-press.com/data/r12/manuf.dta
  obs:          36                          manufacturing process data
 vars:           4                          2 Jan 2011 13:28
 size:         144
```

variable name	storage type	display format	value label	variable label
temperature	byte	%9.0g	temp	machine temperature setting
chemical	byte	%9.0g	supplier	chemical supplier
method	byte	%9.0g	meth	mixing method
yield	byte	%9.0g		product yield

```
Sorted by:
```

We wish to perform a three-way factorial ANOVA. We could type

```
. anova yield temp chem temp#chem meth temp#meth chem#meth temp#chem#meth
```

but prefer to use the **##** factor-variable operator for brevity.

```
. anova yield temp##chem##meth
```

| | Number of obs = 36 | | R-squared = 0.5474 | | |
| | Root MSE = 2.62996 | | Adj R-squared = 0.3399 | | |

Source	Partial SS	df	MS	F	Prob > F
Model	200.75	11	18.25	2.64	0.0227
temperature	30.5	2	15.25	2.20	0.1321
chemical	12.25	1	12.25	1.77	0.1958
temperature#chemical	24.5	2	12.25	1.77	0.1917
method	42.25	1	42.25	6.11	0.0209
temperature#method	87.5	2	43.75	6.33	0.0062
chemical#method	.25	1	.25	0.04	0.8508
temperature#chemical# method	3.5	2	1.75	0.25	0.7785
Residual	166	24	6.91666667		
Total	366.75	35	10.4785714		

The interaction between temperature and method appears to be the important story in these data. A table of means for this interaction is given below.

```
. table method temp, c(mean yield) row col f(%8.2f)
```

mixing method	machine temperature setting			
	low	medium	high	Total
stir	7.50	6.00	6.00	6.50
fold	5.50	9.00	11.50	8.67
Total	6.50	7.50	8.75	7.58

Here our ANOVA is balanced (each cell has the same number of observations), and we obtain the same values as in the table above (but with additional information such as confidence intervals) by using the margins command. Because our ANOVA is balanced, using the asbalanced option with margins would not produce different results. We request the predictive margins for the two terms that appear significant in our ANOVA: temperature#method and method.

```
. margins temperature#method method
Predictive margins                              Number of obs    =         36
Expression    : Linear prediction, predict()
```

	Margin	Delta-method Std. Err.	z	P>|z|	[95% Conf. Interval]	
temperature# method						
1 1	7.5	1.073675	6.99	0.000	5.395636	9.604364
1 2	5.5	1.073675	5.12	0.000	3.395636	7.604364
2 1	6	1.073675	5.59	0.000	3.895636	8.104364
2 2	9	1.073675	8.38	0.000	6.895636	11.10436
3 1	6	1.073675	5.59	0.000	3.895636	8.104364
3 2	11.5	1.073675	10.71	0.000	9.395636	13.60436
method						
1	6.5	.6198865	10.49	0.000	5.285045	7.714955
2	8.666667	.6198865	13.98	0.000	7.451711	9.881622

We decide to use the folding method of mixing and a high temperature in our manufacturing process.

◁

Weighted data

Like all estimation commands, anova can produce estimates on weighted data. See [U] **11.1.6 weight** for details on specifying the weight.

▷ Example 8

We wish to investigate the prevalence of byssinosis, a form of pneumoconiosis that can afflict workers exposed to cotton dust. We have data on 5,419 workers in a large cotton mill. We know whether each worker smokes, his or her race, and the dustiness of the work area. The variables are

smokes	smoker or nonsmoker in the last five years
race	white or other
workplace	1 (most dusty), 2 (less dusty), 3 (least dusty)

We wish to fit an ANOVA model explaining the prevalence of byssinosis according to a full factorial model of smokes, race, and workplace.

The data are unbalanced. Moreover, although we have data on 5,419 workers, the data are grouped according to the explanatory variables, along with some other variables, resulting in 72 observations. For each observation, we know the number of workers in the group (pop), the prevalence of byssinosis (prob), and the values of the three explanatory variables. Thus we wish to fit a three-way factorial model on grouped data.

We begin by showing a bit of the data, which are from Higgins and Koch (1977).

```
. use http://www.stata-press.com/data/r12/byssin
(Byssinosis incidence)

. describe

Contains data from http://www.stata-press.com/data/r12/byssin.dta
  obs:            72                          Byssinosis incidence
  vars:            5                          19 Dec 2010 07:04
  size:          864
```

variable name	storage type	display format	value label	variable label
smokes	int	%8.0g	smokes	Smokes
race	int	%8.0g	race	Race
workplace	int	%8.0g	workplace	Dustiness of workplace
pop	int	%8.0g		Population size
prob	float	%9.0g		Prevalence of byssinosis

```
Sorted by:

. list in 1/5, abbrev(10) divider
```

	smokes	race	workplace	pop	prob
1.	yes	white	most	40	.075
2.	yes	white	less	74	0
3.	yes	white	least	260	.0076923
4.	yes	other	most	164	.152439
5.	yes	other	less	88	0

The first observation in the data represents a group of 40 white workers who smoke and work in a "most" dusty work area. Of those 40 workers, 7.5% have byssinosis. The second observation represents a group of 74 white workers who also smoke but who work in a "less" dusty environment. None of those workers has byssinosis.

Almost every Stata command allows weights. Here we want to weight the data by pop. We can, for instance, make a table of the number of workers by their smoking status and race:

```
. tabulate smokes race [fw=pop]
```

	Race		
Smokes	other	white	Total
no	799	1,431	2,230
yes	1,104	2,085	3,189
Total	1,903	3,516	5,419

The [fw=pop] at the end of the `tabulate` command tells Stata to count each observation as representing pop persons. When making the tally, `tabulate` treats the first observation as representing 40 workers, the second as representing 74 workers, and so on.

Similarly, we can make a table of the dustiness of the workplace:

```
. tabulate workplace [fw=pop]
```

Dustiness of workplace	Freq.	Percent	Cum.
least	3,450	63.66	63.66
less	1,300	23.99	87.65
most	669	12.35	100.00
Total	5,419	100.00	

We can discover the average incidence of byssinosis among these workers by typing

```
. summarize prob [fw=pop]
```

Variable	Obs	Mean	Std. Dev.	Min	Max
prob	5419	.0304484	.0567373	0	.287037

We discover that 3.04% of these workers have byssinosis. Across all cells, the byssinosis rates vary from 0 to 28.7%. Just to prove that there might be something here, let's obtain the average incidence rates according to the dustiness of the workplace:

```
. table workplace smokes race [fw=pop], c(mean prob)
```

Dustiness of workplace	Race and Smokes			
	——— other ———		——— white ———	
	no	yes	no	yes
least	.0107527	.0101523	.0081549	.0162774
less	.02	.0081633	.0136612	.0143149
most	.0820896	.1679105	.0833333	.2295082

Let's now fit the ANOVA model.

```
. anova prob workplace smokes race workplace#smokes workplace#race
> smokes#race workplace#smokes#race [aweight=pop]
(sum of wgt is   5.4190e+03)
```

```
                          Number of obs =      65    R-squared     =  0.8300
                          Root MSE      = .025902    Adj R-squared =  0.7948
```

Source	Partial SS	df	MS	F	Prob > F
Model	.173646538	11	.015786049	23.53	0.0000
workplace	.097625175	2	.048812588	72.76	0.0000
smokes	.013030812	1	.013030812	19.42	0.0001
race	.001094723	1	.001094723	1.63	0.2070
workplace#smokes	.019690342	2	.009845171	14.67	0.0000
workplace#race	.001352516	2	.000676258	1.01	0.3718
smokes#race	.001662874	1	.001662874	2.48	0.1214
workplace#smokes#race	.000950841	2	.00047542	0.71	0.4969
Residual	.035557766	53	.000670901		
Total	.209204304	64	.003268817		

Of course, if we want to see the underlying regression, we could type regress.

Above we examined simple means of the cells of workplace#smokes#race. Our ANOVA shows workplace, smokes, and their interaction as being the only significant factors in our model. We now examine the predictive marginal mean byssinosis rates for these terms.

```
. margins workplace#smokes workplace smokes
Predictive margins                              Number of obs   =      65
Expression    : Linear prediction, predict()
```

	Margin	Delta-method Std. Err.	z	P>|z|	[95% Conf. Interval]	
workplace# smokes						
1 1	.0090672	.0062319	1.45	0.146	−.003147	.0212814
1 2	.0141264	.0053231	2.65	0.008	.0036934	.0245595
2 1	.0158872	.009941	1.60	0.110	−.0035967	.0353711
2 2	.0121546	.0087353	1.39	0.164	−.0049663	.0292756
3 1	.0828966	.0182151	4.55	0.000	.0471957	.1185975
3 2	.2078768	.012426	16.73	0.000	.1835222	.2322314
workplace						
1	.0120701	.0040471	2.98	0.003	.0041379	.0200022
2	.0137273	.0065685	2.09	0.037	.0008533	.0266012
3	.1566225	.0104602	14.97	0.000	.1361208	.1771241
smokes						
1	.0196915	.0050298	3.91	0.000	.0098332	.0295498
2	.0358626	.0041949	8.55	0.000	.0276408	.0440844

Smoking combined with the most dusty workplace produces the highest byssinosis rates.

◁

Ronald Aylmer Fisher (1890–1962) (Sir Ronald from 1952) studied mathematics at Cambridge. Even before he finished his studies, he had published on statistics. He worked as a statistician at Rothamsted Experimental Station (1919–1933), as professor of eugenics at University College London (1933–1943), as professor of genetics at Cambridge (1943–1957), and in retirement at the CSIRO Division of Mathematical Statistics in Adelaide. His many fundamental and applied contributions to statistics and genetics mark him as one of the greatest statisticians of all time, including original work on tests of significance, distribution theory, theory of estimation, fiducial inference, and design of experiments.

ANCOVA

You can include multiple explanatory variables with the anova command, but unless you explicitly state otherwise by using the c. factor-variable operator, all the variables are interpreted as *categorical variables*. Using the c. operator, you can designate variables as *continuous* and thus perform ANCOVA.

▷ Example 9

We have census data recording the death rate (drate) and median age (age) for each state. The dataset also includes the region of the country in which each state is located (region):

```
. use http://www.stata-press.com/data/r12/census2
(1980 Census data by state)
. summarize drate age region
```

Variable	Obs	Mean	Std. Dev.	Min	Max
drate	50	84.3	13.07318	40	107
age	50	29.5	1.752549	24	35
region	50	2.66	1.061574	1	4

age is coded in integral years from 24 to 35, and `region` is coded from 1 to 4, with 1 standing for the Northeast, 2 for the North Central, 3 for the South, and 4 for the West.

When we examine the data more closely, we discover large differences in the death rate across regions of the country:

```
. tabulate region, summarize(drate)
```

Census region	Summary of Death Rate Mean	Std. Dev.	Freq.
NE	93.444444	7.0553368	9
N Cntrl	88.916667	5.5833899	12
South	88.3125	8.5457104	16
West	68.769231	13.342625	13
Total	84.3	13.073185	50

Naturally, we wonder if these differences might not be explained by differences in the median ages of the populations. To find out, we fit a regression model (via `anova`) of `drate` on `region` and `age`. In the `anova` example below, we treat `age` as a categorical variable.

```
. anova drate region age
```

```
                          Number of obs =      50    R-squared     = 0.7927
                          Root MSE      = 6.7583    Adj R-squared = 0.7328
```

Source	Partial SS	df	MS	F	Prob > F
Model	6638.86529	11	603.533208	13.21	0.0000
region	1320.00973	3	440.003244	9.63	0.0001
age	2237.24937	8	279.656171	6.12	0.0000
Residual	1735.63471	38	45.6745977		
Total	8374.5	49	170.908163		

We have the answer to our question: differences in median ages do not eliminate the differences in death rates across the four regions. The ANOVA table summarizes the two terms in the model, `region` and `age`. The `region` term contains 3 degrees of freedom, and the `age` term contains 8 degrees of freedom. Both are significant at better than the 1% level.

The `age` term contains 8 degrees of freedom. Because we did not explicitly indicate that `age` was to be treated as a continuous variable, it was treated as *categorical*, meaning that unique coefficients were estimated for each level of age. The only clue of this labeling is that the number of degrees of freedom associated with the `age` term exceeds 1. The labeling becomes more obvious if we review the regression coefficients:

```
. regress, baselevels
```

Source	SS	df	MS		
Model	6638.86529	11	603.533208		
Residual	1735.63471	38	45.6745977		
Total	8374.5	49	170.908163		

	Number of obs =	50
	F(11, 38) =	13.21
	Prob > F =	0.0000
	R-squared =	0.7927
	Adj R-squared =	0.7328
	Root MSE =	6.7583

drate	Coef.	Std. Err.	t	P>|t|	[95% Conf. Interval]	
region						
1	0	(base)				
2	.4428387	3.983664	0.11	0.912	-7.621668	8.507345
3	-.2964637	3.934766	-0.08	0.940	-8.261981	7.669054
4	-13.37147	4.195344	-3.19	0.003	-21.8645	-4.878439
age						
24	0	(base)				
26	-15	9.557677	-1.57	0.125	-34.34851	4.348506
27	14.30833	7.857378	1.82	0.076	-1.598099	30.21476
28	12.66011	7.495513	1.69	0.099	-2.51376	27.83399
29	18.861	7.28918	2.59	0.014	4.104825	33.61717
30	20.87003	7.210148	2.89	0.006	6.273847	35.46621
31	29.91307	8.242741	3.63	0.001	13.22652	46.59963
32	27.02853	8.509432	3.18	0.003	9.802089	44.25498
35	38.925	9.944825	3.91	0.000	18.79275	59.05724
_cons	68.37147	7.95459	8.60	0.000	52.26824	84.47469

The `regress` command displayed the `anova` model as a regression table. We used the `baselevels` option to display the dropped level (or base) for each term.

If we want to treat `age` as a continuous variable, we must prepend `c.` to `age` in our `anova`.

```
. anova drate region c.age
```

	Number of obs =	50	R-squared =	0.7203
	Root MSE = 7.21483		Adj R-squared =	0.6954

Source	Partial SS	df	MS	F	Prob > F
Model	6032.08254	4	1508.02064	28.97	0.0000
region	1645.66228	3	548.554092	10.54	0.0000
age	1630.46662	1	1630.46662	31.32	0.0000
Residual	2342.41746	45	52.0537213		
Total	8374.5	49	170.908163		

The `age` term now has 1 degree of freedom. The regression coefficients are

```
. regress, baselevels
```

Source	SS	df	MS
Model	6032.08254	4	1508.02064
Residual	2342.41746	45	52.0537213
Total	8374.5	49	170.908163

Number of obs =	50
F(4, 45) =	28.97
Prob > F =	0.0000
R-squared =	0.7203
Adj R-squared =	0.6954
Root MSE =	7.2148

drate	Coef.	Std. Err.	t	P>|t|	[95% Conf. Interval]	
region						
1	0	(base)				
2	1.792526	3.375925	0.53	0.598	-5.006935	8.591988
3	.6979912	3.18154	0.22	0.827	-5.70996	7.105942
4	-13.37578	3.723447	-3.59	0.001	-20.87519	-5.876377
age	3.922947	.7009425	5.60	0.000	2.511177	5.334718
_cons	-28.60281	21.93931	-1.30	0.199	-72.79085	15.58524

Although we started analyzing these data to explain the regional differences in death rate, let's focus on the effect of age for a moment. In our first model, each level of age had a unique death rate associated with it. For instance, the predicted death rate in a north central state with a median age of 28 was

$$0.44 + 12.66 + 68.37 \approx 81.47$$

whereas the predicted death rate from our current model is

$$1.79 + 3.92 \times 28 - 28.60 \approx 82.95$$

Our previous model had an R^2 of 0.7927, whereas our current model has an R^2 of 0.7203. This "small" loss of predictive power accompanies a gain of 7 degrees of freedom, so we suspect that the continuous-age model is as good as the discrete-age model.

◁

❏ Technical note

There is enough information in the two ANOVA tables to attach a statistical significance to our suspicion that the loss of predictive power is offset by the savings in degrees of freedom. Because the continuous-age model is nested within the discrete-age model, we can perform a standard Chow test. For those of us who know such formulas off the top of our heads, the F statistic is

$$\frac{(2342.41746 - 1735.63471)/7}{45.6745977} = 1.90$$

There is, however, a better way.

We can find out whether our continuous model is as good as our discrete model by putting age in the model twice: once as a continuous variable and once as a categorical variable. The categorical variable will then measure deviations around the straight line implied by the continuous variable, and the F test for the significance of the categorical variable will test whether those deviations are jointly zero.

```
. anova drate region c.age age
```

		Number of obs =	50	R-squared	= 0.7927
		Root MSE = 6.7583		Adj R-squared	= 0.7328

Source	Partial SS	df	MS	F	Prob > F
Model	6638.86529	11	603.533208	13.21	0.0000
region	1320.00973	3	440.003244	9.63	0.0001
age	699.74137	1	699.74137	15.32	0.0004
age	606.782747	7	86.6832496	1.90	0.0970
Residual	1735.63471	38	45.6745977		
Total	8374.5	49	170.908163		

We find that the F test for the significance of the (categorical) age variable is 1.90, just as we calculated above. It is significant at the 9.7% level. If we hold to a 5% significance level, we cannot reject the null hypothesis that the effect of age is linear.

❑

▷ Example 10

In our census data, we still find significant differences across the regions after controlling for the median age of the population. We might now wonder whether the regional differences are differences in level—independent of age—or are instead differences in the regional effects of age. Just as we can interact categorical variables with other categorical variables, we can interact categorical variables with continuous variables.

```
. anova drate region c.age region#c.age
```

		Number of obs =	50	R-squared	= 0.7365
		Root MSE = 7.24852		Adj R-squared	= 0.6926

Source	Partial SS	df	MS	F	Prob > F
Model	6167.7737	7	881.110529	16.77	0.0000
region	188.713602	3	62.9045339	1.20	0.3225
age	873.425599	1	873.425599	16.62	0.0002
region#age	135.691162	3	45.2303874	0.86	0.4689
Residual	2206.7263	42	52.5411023		
Total	8374.5	49	170.908163		

The region#c.age term in our model measures the differences in slopes across the regions. We cannot reject the null hypothesis that there are no such differences. The region effect is now "insignificant". This status does not mean that there are no regional differences in death rates because each test is a *marginal* or *partial* test. Here, with region#c.age included in the model, region is being tested at the point where age is zero. Apart from this value not existing in the dataset, it is also a long way from the mean value of age, so the test of region at this point is meaningless (although it is valid if you acknowledge what is being tested).

To obtain a more sensible test of region, we can subtract the mean from the age variable and use this in the model.

```
. quietly summarize age
. generate mage = age - r(mean)
```

```
. anova drate region c.mage region#c.mage
```

	Number of obs =		50	R-squared	= 0.7365
	Root MSE	= 7.24852		Adj R-squared =	0.6926

Source	Partial SS	df	MS	F	Prob > F
Model	6167.7737	7	881.110529	16.77	0.0000
region	1166.14735	3	388.715783	7.40	0.0004
mage	873.425599	1	873.425599	16.62	0.0002
region#mage	135.691162	3	45.2303874	0.86	0.4689
Residual	2206.7263	42	52.5411023		
Total	8374.5	49	170.908163		

region is significant when tested at the mean of the age variable.

◁

Remember that we can specify interactions by typing *varname*#*varname*. We have seen examples of interacting categorical variables with categorical variables and, in the examples above, a categorical variable (region) with a continuous variable (age or mage).

We can also interact continuous variables with continuous variables. To include an age^2 term in our model, we could type c.age#c.age. If we also wanted to interact the categorical variable region with the age^2 term, we could type region#c.age#c.age (or even c.age#region#c.age).

Nested designs

In addition to specifying interaction terms, nested terms can also be specified in an ANOVA. A vertical bar is used to indicate nesting: A|B is read as A nested within B. A|B|C is read as A nested within B, which is nested within C. A|B#C is read as A is nested within the interaction of B and C. A#B|C is read as the interaction of A and B, which is nested within C.

Different error terms can be specified for different parts of the model. The forward slash is used to indicate that the next term in the model is the error term for what precedes it. For instance, anova y A / B|A indicates that the F test for A is to be tested by using the mean square from B|A in the denominator. Error terms (terms following the slash) are generally not tested unless they are themselves followed by a slash. Residual error is the default error term.

For example, consider A / B / C, where A, B, and C may be arbitrarily complex terms. Then anova will report A tested by B and B tested by C. If we add one more slash on the end to form A / B / C /, then anova will also report C tested by the residual error.

▷ Example 11

We have collected data from a manufacturer that is evaluating which of five different brands of machinery to buy to perform a particular function in an assembly line. Twenty assembly-line employees were selected at random for training on these machines, with four employees assigned to learn a particular machine. The output from each employee (operator) on the brand of machine for which he trained was measured during four trial periods. In this example, the operator is nested within machine. Because of sickness and employee resignations, the final data are not balanced. The following table gives the mean output and sample size for each machine and operator combination.

```
. use http://www.stata-press.com/data/r12/machine, clear
(machine data)
```

```
. table machine operator, c(mean output n output) col f(%8.2f)
```

| five brands of machine | operator nested in machine | | | | |
	1	2	3	4	Total
1	9.15	9.48	8.27	8.20	8.75
	2	4	3	4	13
2	15.03	11.55	11.45	11.52	12.47
	3	2	2	4	11
3	11.27	10.13	11.13		10.84
	3	3	3		9
4	16.10	18.97	15.35	16.60	16.65
	3	3	4	3	13
5	15.30	14.35	10.43		13.63
	4	4	3		11

Assuming that operator is random (that is, we wish to infer to the larger population of possible operators) and machine is fixed (that is, only these five machines are of interest), the typical test for machine uses operator nested within machine as the error term. operator nested within machine can be tested by residual error. Our earlier warning concerning designs with either unplanned missing cells or unbalanced cell sizes, or both, also applies to interpreting the ANOVA results from this unbalanced nested example.

```
. anova output machine / operator|machine /
```

| | | Number of obs = | 57 | R-squared | = 0.8661 |
| | | Root MSE | = 1.47089 | Adj R-squared = | 0.8077 |
Source	Partial SS	df	MS	F	Prob > F
Model	545.822288	17	32.1071934	14.84	0.0000
machine	430.980792	4	107.745198	13.82	0.0001
operator|machine	101.353804	13	7.79644648		
operator|machine	101.353804	13	7.79644648	3.60	0.0009
Residual	84.3766582	39	2.16350406		
Total	630.198947	56	11.2535526		

operator|machine is preceded by a slash, indicating that it is the error term for the terms before it (here machine). operator|machine is also followed by a slash that indicates it should be tested with residual error. The output lists the operator|machine term twice, once as the error term for machine, and again as a term tested by residual error. A line is placed in the ANOVA table to separate the two. In general, a dividing line is placed in the output to separate the terms into groups that are tested with the same error term. The overall model is tested by residual error and is separated from the rest of the table by a blank line at the top of the table.

The results indicate that the machines are not all equal and that there are significant differences between operators.

◁

▷ Example 12

Your company builds and operates sewage treatment facilities. You want to compare two particulate solutions during the particulate reduction step of the sewage treatment process. For each solution, two area managers are randomly selected to implement and oversee the change to the new treatment process in two of their randomly chosen facilities. Two workers at each of these facilities are trained to operate the new process. A measure of particulate reduction is recorded at various times during the month at each facility for each worker. The data are described below.

```
. use http://www.stata-press.com/data/r12/sewage
(Sewage treatment)

. describe

Contains data from http://www.stata-press.com/data/r12/sewage.dta
  obs:            64                          Sewage treatment
 vars:             5                          9 May 2011 12:43
 size:           320
```

variable name	storage type	display format	value label	variable label
particulate	byte	%9.0g		particulate reduction
solution	byte	%9.0g		2 particulate solutions
manager	byte	%9.0g		2 managers per solution
facility	byte	%9.0g		2 facilities per manager
worker	byte	%9.0g		2 workers per facility

```
Sorted by:  solution  manager  facility  worker
```

You want to determine if the two particulate solutions provide significantly different particulate reduction. You would also like to know if `manager`, `facility`, and `worker` are significant effects. `solution` is a fixed factor, whereas `manager`, `facility`, and `worker` are random factors.

In the following `anova` command, we use abbreviations for the variable names, which can sometimes make long ANOVA model statements easier to read.

```
. anova particulate s / m|s / f|m|s / w|f|m|s /, dropemptycells
```

		Number of obs = 64		R-squared = 0.6338	
		Root MSE = 12.7445		Adj R-squared = 0.5194	
Source	Partial SS	df	MS	F	Prob > F
Model	13493.6094	15	899.573958	5.54	0.0000
solution	7203.76563	1	7203.76563	17.19	0.0536
manager|solution	838.28125	2	419.140625		
manager|solution	838.28125	2	419.140625	0.55	0.6166
facility|manager|solution	3064.9375	4	766.234375		
facility|manager|solution	3064.9375	4	766.234375	2.57	0.1193
worker|facility|manager|solution	2386.625	8	298.328125		
worker|facility|manager|solution	2386.625	8	298.328125	1.84	0.0931
Residual	7796.25	48	162.421875		
Total	21289.8594	63	337.934276		

While `solution` is not declared significant at the 5% significance level, it is near enough to that threshold to warrant further investigation (see example 3 in [R] **anova postestimation** for a continuation of the analysis of these data).

◁

❏ Technical note

Why did we use the `dropemptycells` option with the previous `anova`? By default, Stata retains empty cells when building the design matrix and currently treats | and # the same in how it determines the possible number of cells. Retaining empty cells in an ANOVA with nested terms can cause your design matrix to become too large. In example 12, there are $1024 = 2 \times 4 \times 8 \times 16$ cells that are considered possible for the `worker|facility|manager|solution` term because the `worker`, `facility`, and `manager` variables are uniquely numbered. With the `dropemptycells` option, the `worker|facility|manager|solution` term requires just 16 columns in the design matrix (corresponding to the 16 unique workers).

Why did we not use the `dropemptycells` option in example 11, where `operator` is nested in `machine`? If you look at the table presented at the beginning of that example, you will see that `operator` is compactly instead of uniquely numbered (you need both `operator` number and `machine` number to determine the `operator`). Here the `dropemptycells` option would have only reduced our design matrix from 26 columns down to 24 columns (because there were only 3 operators instead of 4 for `machines` 3 and 5).

We suggest that you specify `dropemptycells` when there are nested terms in your ANOVA. You could also use the `set emptycells drop` command to accomplish the same thing; see [R] **set**.

❏

Mixed designs

An ANOVA can consist of both nested and crossed terms. A split-plot ANOVA design provides an example.

▷ Example 13

Two reading programs and three skill-enhancement techniques are under investigation. Ten classes of first-grade students were randomly assigned so that five classes were taught with one reading program and another five classes were taught with the other. The 30 students in each class were divided into six groups with 5 students each. Within each class, the six groups were divided randomly so that each of the three skill-enhancement techniques was taught to two of the groups within each class. At the end of the school year, a reading assessment test was administered to all the students. In this split-plot ANOVA, the whole-plot treatment is the two reading programs, and the split-plot treatment is the three skill-enhancement techniques.

```
. use http://www.stata-press.com/data/r12/reading
(Reading experiment data)
```

```
. describe

Contains data from http://www.stata-press.com/data/r12/reading.dta
  obs:           300                          Reading experiment data
 vars:             5                          9 Mar 2011 18:57
 size:         1,500                          (_dta has notes)
```

variable name	storage type	display format	value label	variable label
score	byte	%9.0g		reading score
program	byte	%9.0g		reading program
class	byte	%9.0g		class nested in program
skill	byte	%9.0g		skill enhancement technique
group	byte	%9.0g		group nested in class and skill

```
Sorted by:
```

In this split-plot ANOVA, the error term for `program` is `class` nested within `program`. The error term for `skill` and the `program` by `skill` interaction is the `class` by `skill` interaction nested within `program`. Other terms are also involved in the model and can be seen below.

Our `anova` command is too long to fit on one line of this manual. Where we have chosen to break the command into multiple lines is arbitrary. If we were typing this command into Stata, we would just type along and let Stata automatically wrap across lines, as necessary.

```
. anova score prog / class|prog skill prog#skill / class#skill|prog
> / group|class#skill|prog /, dropemptycells
```

	Number of obs = 300		R-squared = 0.3738		
	Root MSE = 14.6268		Adj R-squared = 0.2199		
Source	Partial SS	df	MS	F	Prob > F
Model	30656.5167	59	519.601977	2.43	0.0000
program	4493.07	1	4493.07	8.73	0.0183
class|program	4116.61333	8	514.576667		
skill	1122.64667	2	561.323333	1.54	0.2450
program#skill	5694.62	2	2847.31	7.80	0.0043
class#skill|program	5841.46667	16	365.091667		
class#skill|program	5841.46667	16	365.091667	1.17	0.3463
group|class#skill|program	9388.1	30	312.936667		
group|class#skill|program	9388.1	30	312.936667	1.46	0.0636
Residual	51346.4	240	213.943333		
Total	82002.9167	299	274.257246		

The `program#skill` term is significant, as is the `program` term. Let's look at the predictive margins for these two terms and at a `marginsplot` for the first term.

```
. margins, within(program skill)
Predictive margins                                Number of obs    =        300
Expression    : Linear prediction, predict()
within        : program skill
Empty cells   : reweight
```

	Margin	Delta-method Std. Err.	z	P>|z|	[95% Conf. Interval]	
program# skill						
1 1	68.16	2.068542	32.95	0.000	64.10573	72.21427
1 2	52.86	2.068542	25.55	0.000	48.80573	56.91427
1 3	61.54	2.068542	29.75	0.000	57.48573	65.59427
2 1	50.7	2.068542	24.51	0.000	46.64573	54.75427
2 2	56.54	2.068542	27.33	0.000	52.48573	60.59427
2 3	52.1	2.068542	25.19	0.000	48.04573	56.15427

```
. marginsplot, plot2opts(lp(dash) m(D)) plot3opts(lp(dot) m(T))
Variables that uniquely identify margins: program skill
```

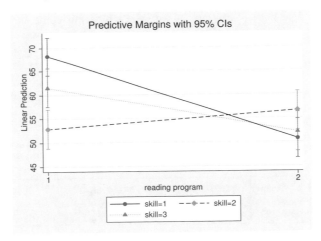

```
. margins, within(program)
Predictive margins                                Number of obs    =        300
Expression    : Linear prediction, predict()
within        : program
Empty cells   : reweight
```

	Margin	Delta-method Std. Err.	z	P>|z|	[95% Conf. Interval]	
program						
1	60.85333	1.194273	50.95	0.000	58.5126	63.19407
2	53.11333	1.194273	44.47	0.000	50.7726	55.45407

Because our ANOVA involves nested terms, we used the within() option of margins; see [R] **margins**.

skill 2 produces a low score when combined with program 1 and a high score when combined with program 2, demonstrating the interaction between the reading program and the skill-enhancement technique. You might conclude that the first reading program and the first skill-enhancement technique perform best when combined. However, notice the overlapping confidence interval for the first reading program and the third skill-enhancement technique.

◁

❑ Technical note

There are several valid ways to write complicated anova terms. In the reading experiment example (example 13), we had a term group|class#skill|program. This term can be read as group nested within both class and skill and further nested within program. You can also write this term as group|class#skill#program or group|program#class#skill or group|skill#class|program, etc. All variations will produce the same result. Some people prefer having only one '|' in a term and would use group|class#skill#program, which is read as group nested within class, skill, and program.

❑

Gertrude Mary Cox (1900–1978) was born on a farm near Dayton, Iowa. Initially intending to become superintendent of an orphanage, she enrolled at Iowa State College. There she majored in mathematics and attained the college's first Master's degree in statistics. After working on her PhD in psychological statistics for two years at the University of California–Berkeley, she decided to go back to Iowa State to work with George W. Snedecor. There she pursued her interest in and taught a course in design of experiments. That work led to her collaboration with W. G. Cochran, which produced a classic text. In 1940, when Snedecor shared with her his list of men he was nominating to head the statistics department at North Carolina State College, she wanted to know why she had not been included. He added her name, she won the position, and she built an outstanding department at North Carolina State. Cox retired early so she could work at the Research Triangle Institute in North Carolina. She consulted widely, served as editor of *Biometrics*, and was elected to the National Academy of Sciences.

Latin-square designs

You can use anova to analyze a Latin-square design. Consider the following example, published in Snedecor and Cochran (1989).

▷ Example 14

Data from a Latin-square design are as follows:

Row	Column 1	Column 2	Column 3	Column 4	Column 5
1	257(B)	230(E)	279(A)	287(C)	202(D)
2	245(D)	283(A)	245(E)	280(B)	260(C)
3	182(E)	252(B)	280(C)	246(D)	250(A)
4	203(A)	204(C)	227(D)	193(E)	259(B)
5	231(C)	271(D)	266(B)	334(A)	338(E)

In Stata, the data might appear as follows:

```
. use http://www.stata-press.com/data/r12/latinsq
. list
```

	row	c1	c2	c3	c4	c5
1.	1	257	230	279	287	202
2.	2	245	283	245	280	260
3.	3	182	252	280	246	250
4.	4	203	204	227	193	259
5.	5	231	271	266	334	338

Before anova can be used on these data, the data must be organized so that the outcome measurement is in one column. reshape is inadequate for this task because there is information about the treatments in the sequence of these observations. pkshape is designed to reshape this type of data; see [R] **pkshape**.

```
. pkshape row row c1-c5, order(beacd daebc ebcda acdeb cdbae)
. list
```

	sequence	outcome	treat	carry	period
1.	1	257	1	0	1
2.	2	245	5	0	1
3.	3	182	2	0	1
4.	4	203	3	0	1
5.	5	231	4	0	1
6.	1	230	2	1	2
7.	2	283	3	5	2
8.	3	252	1	2	2
9.	4	204	4	3	2
10.	5	271	5	4	2
11.	1	279	3	2	3
12.	2	245	2	3	3
13.	3	280	4	1	3
14.	4	227	5	4	3
15.	5	266	1	5	3
16.	1	287	4	3	4
17.	2	280	1	2	4
18.	3	246	5	4	4
19.	4	193	2	5	4
20.	5	334	3	1	4
21.	1	202	5	4	5
22.	2	260	4	1	5
23.	3	250	3	5	5
24.	4	259	1	2	5
25.	5	338	2	3	5

```
. anova outcome sequence period treat
```

| | Number of obs = | 25 | R-squared | = 0.6536 | |
| | Root MSE | = 32.4901 | Adj R-squared | = 0.3073 | |
Source	Partial SS	df	MS	F	Prob > F
Model	23904.08	12	1992.00667	1.89	0.1426
sequence	13601.36	4	3400.34	3.22	0.0516
period	6146.16	4	1536.54	1.46	0.2758
treat	4156.56	4	1039.14	0.98	0.4523
Residual	12667.28	12	1055.60667		
Total	36571.36	24	1523.80667		

◁

These methods will work with any type of Latin-square design, including those with replicated measurements. For more information, see [R] **pk**, [R] **pkcross**, and [R] **pkshape**.

Repeated-measures ANOVA

One approach for analyzing repeated-measures data is to use multivariate ANOVA (MANOVA); see [MV] **manova**. In this approach, the data are placed in wide form (see [D] **reshape**), and the repeated measures enter the MANOVA as dependent variables.

A second approach for analyzing repeated measures is to use anova. However, one of the underlying assumptions for the F tests in ANOVA is independence of observations. In a repeated-measures design, this assumption is almost certainly violated or is at least suspect. In a repeated-measures ANOVA, the subjects (or whatever the experimental units are called) are observed for each level of one or more of the other categorical variables in the model. These variables are called the repeated-measure variables. Observations from the same subject are likely to be correlated.

The approach used in repeated-measures ANOVA to correct for this lack of independence is to apply a correction to the degrees of freedom of the F test for terms in the model that involve repeated measures. This correction factor, ϵ, lies between the reciprocal of the degrees of freedom for the repeated term and 1. Box (1954) provided the pioneering work in this area. Milliken and Johnson (2009) refer to the lower bound of this correction factor as Box's conservative correction factor. Winer, Brown, and Michels (1991) call it simply the conservative correction factor.

Geisser and Greenhouse (1958) provide an estimate for the correction factor called the Greenhouse–Geisser ϵ. This value is estimated from the data. Huynh and Feldt (1976) show that the Greenhouse–Geisser ϵ tends to be conservatively biased. They provide a revised correction factor called the Huynh–Feldt ϵ. When the Huynh–Feldt ϵ exceeds 1, it is set to 1. Thus there is a natural ordering for these correction factors:

$$\text{Box's conservative } \epsilon \leq \text{Greenhouse–Geisser } \epsilon \leq \text{Huynh–Feldt } \epsilon \leq 1$$

A correction factor of 1 is the same as no correction.

anova with the repeated() option computes these correction factors and displays the revised test results in a table that follows the standard ANOVA table. In the resulting table, H-F stands for Huynh–Feldt, G-G stands for Greenhouse–Geisser, and Box stands for Box's conservative ϵ.

▷ Example 15

This example is taken from table 4.3 of Winer, Brown, and Michels (1991). The reaction time for five subjects each tested with four drugs was recorded in the variable `score`. Here is a table of the data (see [P] **tabdisp** if you are unfamiliar with `tabdisp`):

```
. use http://www.stata-press.com/data/r12/t43, clear
(T4.3 -- Winer, Brown, Michels)
. tabdisp person drug, cellvar(score)
```

		drug		
person	1	2	3	4
1	30	28	16	34
2	14	18	10	22
3	24	20	18	30
4	38	34	20	44
5	26	28	14	30

`drug` is the repeated variable in this simple repeated-measures ANOVA example. The ANOVA is specified as follows:

```
. anova score person drug, repeated(drug)
```

```
                        Number of obs =      20     R-squared     = 0.9244
                        Root MSE       = 3.06594     Adj R-squared = 0.8803
```

Source	Partial SS	df	MS	F	Prob > F
Model	1379	7	197	20.96	0.0000
person	680.8	4	170.2	18.11	0.0001
drug	698.2	3	232.733333	24.76	0.0000
Residual	112.8	12	9.4		
Total	1491.8	19	78.5157895		

```
Between-subjects error term:  person
                     Levels:  5           (4 df)
        Lowest b.s.e. variable:  person
Repeated variable: drug
```

```
                            Huynh-Feldt epsilon       = 1.0789
                            *Huynh-Feldt epsilon reset to 1.0000
                            Greenhouse-Geisser epsilon = 0.6049
                            Box's conservative epsilon = 0.3333
```

				Prob > F		
Source	df	F	Regular	H-F	G-G	Box
drug	3	24.76	0.0000	0.0000	0.0006	0.0076
Residual	12					

Here the Huynh–Feldt ϵ is 1.0789, which is larger than 1. It is reset to 1, which is the same as making no adjustment to the standard test computed in the main ANOVA table. The Greenhouse–Geisser ϵ is 0.6049, and its associated p-value is computed from an F ratio of 24.76 using 1.8147 ($= 3\epsilon$) and 7.2588 ($= 12\epsilon$) degrees of freedom. Box's conservative ϵ is set equal to the reciprocal of the degrees of freedom for the repeated term. Here it is 1/3, so Box's conservative test is computed using 1 and 4 degrees of freedom for the observed F ratio of 24.76.

Even for Box's conservative ϵ, drug is significant with a p-value of 0.0076. The following table gives the predictive marginal mean score (that is, response time) for each of the four drugs:

```
. margins drug
Predictive margins                              Number of obs    =         20
Expression    : Linear prediction, predict()
```

	Margin	Delta-method Std. Err.	z	P>|z|	[95% Conf. Interval]	
drug						
1	26.4	1.371131	19.25	0.000	23.71263	29.08737
2	25.6	1.371131	18.67	0.000	22.91263	28.28737
3	15.6	1.371131	11.38	0.000	12.91263	18.28737
4	32	1.371131	23.34	0.000	29.31263	34.68737

The ANOVA table for this example provides an F test for person, but you should ignore it. An appropriate test for person would require replication (that is, multiple measurements for person and drug combinations). Also, without replication there is no test available for investigating the interaction between person and drug.

◁

▷ Example 16

Table 7.7 of Winer, Brown, and Michels (1991) provides another repeated-measures ANOVA example. There are four dial shapes and two methods for calibrating dials. Subjects are nested within calibration method, and an accuracy score is obtained. The data are shown below.

```
. use http://www.stata-press.com/data/r12/t77
(T7.7 -- Winer, Brown, Michels)
. tabdisp shape subject calib, cell(score)
```

4 dial shapes	2 methods for calibrating dials and subject nested in calib					
	——— 1 ———			——— 2 ———		
	1	2	3	1	2	3
1	0	3	4	4	5	7
2	0	1	3	2	4	5
3	5	5	6	7	6	8
4	3	4	2	8	6	9

The calibration method and dial shapes are fixed factors, whereas subjects are random. The appropriate test for calibration method uses the nested subject term as the error term. Both the dial shape and the interaction between dial shape and calibration method are tested with the dial shape by subject interaction nested within calibration method. Here we drop this term from the anova command, and it becomes residual error. The dial shape is the repeated variable because each subject is tested with all four dial shapes. Here is the anova command that produces the desired results:

```
. anova score calib / subject|calib shape calib#shape, repeated(shape)
```

```
                           Number of obs =        24    R-squared     =  0.8925
                           Root MSE      = 1.11181    Adj R-squared =  0.7939

            Source |   Partial SS    df       MS             F      Prob > F
```

Source	Partial SS	df	MS	F	Prob > F
Model	123.125	11	11.1931818	9.06	0.0003
calib	51.0416667	1	51.0416667	11.89	0.0261
subject|calib	17.1666667	4	4.29166667		
shape	47.4583333	3	15.8194444	12.80	0.0005
calib#shape	7.45833333	3	2.48611111	2.01	0.1662
Residual	14.8333333	12	1.23611111		
Total	137.958333	23	5.99818841		

```
Between-subjects error term:  subject|calib
                   Levels:  6          (4 df)
      Lowest b.s.e. variable:  subject
      Covariance pooled over:  calib      (for repeated variable)
Repeated variable: shape
                                    Huynh-Feldt epsilon          =  0.8483
                                    Greenhouse-Geisser epsilon =  0.4751
                                    Box's conservative epsilon =  0.3333
```

				———— Prob > F ————		
Source	df	F	Regular	H-F	G-G	Box
shape	3	12.80	0.0005	0.0011	0.0099	0.0232
calib#shape	3	2.01	0.1662	0.1791	0.2152	0.2291
Residual	12					

The repeated-measure ϵ corrections are applied to any terms that are tested in the main ANOVA table and have the repeated variable in the term. These ϵ corrections are given in a table below the main ANOVA table. Here the repeated-measures tests for `shape` and `calib#shape` are presented.

Calibration method is significant, as is dial shape. The interaction between calibration method and dial shape is not significant. The repeated-measure ϵ corrections do not change these conclusions, but they do change the significance level for the tests on `shape` and `calib#shape`. Here, though, unlike in the previous example, the Huynh–Feldt ϵ is less than 1.

Here are the predictive marginal mean scores for calibration method and dial shapes. Because the interaction was not significant, we request only the `calib` and `shape` predictive margins.

```
. margins, within(calib)
Predictive margins                              Number of obs  =         24
Expression   : Linear prediction, predict()
within       : calib
Empty cells  : reweight
```

		Delta-method				
	Margin	Std. Err.	z	P>|z|	[95% Conf. Interval]	
calib						
1	3	.3209506	9.35	0.000	2.370948	3.629052
2	5.916667	.3209506	18.43	0.000	5.287615	6.545718

```
. margins, within(shape)
```

Predictive margins   Number of obs   =   24

Expression   : Linear prediction, predict()
within   : shape
Empty cells   : reweight

		Delta-method				
	Margin	Std. Err.	z	P>|z|	[95% Conf.	Interval]
shape						
1	3.833333	.4538926	8.45	0.000	2.94372	4.722947
2	2.5	.4538926	5.51	0.000	1.610387	3.389613
3	6.166667	.4538926	13.59	0.000	5.277053	7.05628
4	5.333333	.4538926	11.75	0.000	4.44372	6.222947

◁

❏ Technical note

The computation of the Greenhouse–Geisser and Huynh–Feldt epsilons in a repeated-measures ANOVA requires the number of levels and degrees of freedom for the between-subjects error term, as well as a value computed from a pooled covariance matrix. The observations are grouped based on all but the lowest-level variable in the between-subjects error term. The covariance over the repeated variables is computed for each resulting group, and then these covariance matrices are pooled. The dimension of the pooled covariance matrix is the number of levels of the repeated variable (or combination of levels for multiple repeated variables). In example 16, there are four levels of the repeated variable (shape), so the resulting covariance matrix is 4×4.

The anova command automatically attempts to determine the between-subjects error term and the lowest-level variable in the between-subjects error term to group the observations for computation of the pooled covariance matrix. anova issues an error message indicating that the bse() or bseunit() option is required when anova cannot determine them. You may override the default selections of anova by specifying the bse(), bseunit(), or grouping() option. The term specified in the bse() option must be a term in the ANOVA model.

The default selection for the between-subjects error term (the bse() option) is the interaction of the nonrepeated categorical variables in the ANOVA model. The first variable listed in the between-subjects error term is automatically selected as the lowest-level variable in the between-subjects error term but can be overridden with the bseunit(*varname*) option. *varname* is often a term, such as subject or subsample within subject, and is most often listed first in the term because of the nesting notation of ANOVA. This term makes sense in most repeated-measures ANOVA designs when the terms of the model are written in standard form. For instance, in example 16, there were three categorical variables (subject, calib, and shape), with shape being the repeated variable. Here anova looked for a term involving only subject and calib to determine the between-subjects error term. It found subject|calib as the term with six levels and 4 degrees of freedom. anova then picked subject as the default for the bseunit() option (the lowest variable in the between-subjects error term) because it was listed first in the term.

The grouping of observations proceeds, based on the different combinations of values of the variables in the between-subjects error term, excluding the lowest level variable (as found by default or as specified with the bseunit() option). You may specify the grouping() option to change the default grouping used in computing the pooled covariance matrix.

The between-subjects error term, number of levels, degrees of freedom, lowest variable in the term, and grouping information are presented after the main ANOVA table and before the rest of the repeated-measures output.

❏

▷ Example 17

Data with two repeated variables are given in table 7.13 of Winer, Brown, and Michels (1991). The accuracy scores of subjects making adjustments to three dials during three different periods are recorded. Three subjects are exposed to a certain noise background level, whereas a different set of three subjects is exposed to a different noise background level. Here is a table of accuracy scores for the noise, subject, period, and dial variables:

```
. use http://www.stata-press.com/data/r12/t713
(T7.13 -- Winer, Brown, Michels)
. tabdisp subject dial period, by(noise) cell(score) stubwidth(11)
```

noise background and subject nested in noise	10 minute time periods and dial								
	——— 1 ———			——— 2 ———			——— 3 ———		
	1	2	3	1	2	3	1	2	3
1									
1	45	53	60	40	52	57	28	37	46
2	35	41	50	30	37	47	25	32	41
3	60	65	75	58	54	70	40	47	50
2									
1	50	48	61	25	34	51	16	23	35
2	42	45	55	30	37	43	22	27	37
3	56	60	77	40	39	57	31	29	46

noise, period, and dial are fixed, whereas subject is random. Both period and dial are repeated variables. The ANOVA for this example is specified next.

```
. anova score noise / subject|noise period noise#period
> / period#subject|noise  dial noise#dial / dial#subject|noise
> period#dial noise#period#dial, repeated(period dial)
```

	Number of obs =	54	R-squared	= 0.9872
	Root MSE	= 2.81859	Adj R-squared =	0.9576

Source	Partial SS	df	MS	F	Prob > F
Model	9797.72222	37	264.803303	33.33	0.0000
noise	468.166667	1	468.166667	0.75	0.4348
subject|noise	2491.11111	4	622.777778		
period	3722.33333	2	1861.16667	63.39	0.0000
noise#period	333	2	166.5	5.67	0.0293
period#subject|noise	234.888889	8	29.3611111		
dial	2370.33333	2	1185.16667	89.82	0.0000
noise#dial	50.3333333	2	25.1666667	1.91	0.2102
dial#subject|noise	105.555556	8	13.1944444		
period#dial	10.6666667	4	2.66666667	0.34	0.8499
noise#period#dial	11.3333333	4	2.83333333	0.36	0.8357
Residual	127.111111	16	7.94444444		
Total	9924.83333	53	187.261006		

```
Between-subjects error term:  subject|noise
                     Levels:  6          (4 df)
      Lowest b.s.e. variable:  subject
       Covariance pooled over:  noise      (for repeated variables)
Repeated variable: period
```

				Huynh-Feldt epsilon	= 1.0668
				*Huynh-Feldt epsilon reset to	1.0000
				Greenhouse-Geisser epsilon =	0.6476
				Box's conservative epsilon =	0.5000

Source	df	F	Regular	H-F	G-G	Box
period	2	63.39	0.0000	0.0000	0.0003	0.0013
noise#period	2	5.67	0.0293	0.0293	0.0569	0.0759
period#subject|noise	8					

```
Repeated variable: dial
```

				Huynh-Feldt epsilon	= 2.0788
				*Huynh-Feldt epsilon reset to	1.0000
				Greenhouse-Geisser epsilon =	0.9171
				Box's conservative epsilon =	0.5000

Source	df	F	Regular	H-F	G-G	Box
dial	2	89.82	0.0000	0.0000	0.0000	0.0007
noise#dial	2	1.91	0.2102	0.2102	0.2152	0.2394
dial#subject|noise	8					

```
Repeated variables: period#dial
                                  Huynh-Feldt epsilon      =  1.3258
                                 *Huynh-Feldt epsilon reset to  1.0000
                                  Greenhouse-Geisser epsilon =  0.5134
                                  Box's conservative epsilon =  0.2500
                                 ───────────── Prob > F ─────────────
            Source │  df     F    Regular    H-F      G-G      Box
        ───────────┼──────────────────────────────────────────────
        period#dial │   4    0.34   0.8499   0.8499   0.7295   0.5934
  noise#period#dial │   4    0.36   0.8357   0.8357   0.7156   0.5825
           Residual │  16
```

For each repeated variable and for each combination of interactions of repeated variables, there are different ϵ correction values. The anova command produces tables for each applicable combination.

The two most significant factors in this model appear to be dial and period. The noise by period interaction may also be significant, depending on the correction factor you use. Below is a table of predictive margins for the accuracy score for dial, period, and noise by period.

```
. margins, within(dial)

Predictive margins                        Number of obs   =        54

Expression    : Linear prediction, predict()
within        : dial
Empty cells   : reweight
```

		Delta-method				
	Margin	Std. Err.	z	P>|z|	[95% Conf.	Interval]
dial						
1	37.38889	.6643478	56.28	0.000	36.08679	38.69099
2	42.22222	.6643478	63.55	0.000	40.92012	43.52432
3	53.22222	.6643478	80.11	0.000	51.92012	54.52432

```
. margins, within(period)

Predictive margins                        Number of obs   =        54

Expression    : Linear prediction, predict()
within        : period
Empty cells   : reweight
```

		Delta-method				
	Margin	Std. Err.	z	P>|z|	[95% Conf.	Interval]
period						
1	54.33333	.6643478	81.78	0.000	53.03124	55.63543
2	44.5	.6643478	66.98	0.000	43.1979	45.8021
3	34	.6643478	51.18	0.000	32.6979	35.3021

```
. margins, within(noise period)
Predictive margins                              Number of obs   =        54
Expression    : Linear prediction, predict()
within        : noise period
Empty cells   : reweight
```

	Margin	Delta-method Std. Err.	z	P>|z|	[95% Conf. Interval]	
noise#period						
1 1	53.77778	.9395297	57.24	0.000	51.93633	55.61922
1 2	49.44444	.9395297	52.63	0.000	47.603	51.28589
1 3	38.44444	.9395297	40.92	0.000	36.603	40.28589
2 1	54.88889	.9395297	58.42	0.000	53.04744	56.73033
2 2	39.55556	.9395297	42.10	0.000	37.71411	41.397
2 3	29.55556	.9395297	31.46	0.000	27.71411	31.397

Dial shape 3 produces the highest score, and scores decrease over the periods.

◁

Example 17 had two repeated-measurement variables. Up to four repeated-measurement variables may be specified in the anova command.

Saved results

anova saves the following in e():

Scalars
e(N)	number of observations
e(mss)	model sum of squares
e(df_m)	model degrees of freedom
e(rss)	residual sum of squares
e(df_r)	residual degrees of freedom
e(r2)	R-squared
e(r2_a)	adjusted R-squared
e(F)	F statistic
e(rmse)	root mean squared error
e(ll)	log likelihood
e(ll_0)	log likelihood, constant-only model
e(ss_#)	sum of squares for term #
e(df_#)	numerator degrees of freedom for term #
e(ssdenom_#)	denominator sum of squares for term # (when using nonresidual error)
e(dfdenom_#)	denominator degrees of freedom for term # (when using nonresidual error)
e(F_#)	F statistic for term # (if computed)
e(N_bse)	number of levels of the between-subjects error term
e(df_bse)	degrees of freedom for the between-subjects error term
e(box#)	Box's conservative epsilon for a particular combination of repeated variables (repeated() only)
e(gg#)	Greenhouse–Geisser epsilon for a particular combination of repeated variables (repeated() only)
e(hf#)	Huynh–Feldt epsilon for a particular combination of repeated variables (repeated() only)
e(rank)	rank of e(V)

Macros
e(cmd)	anova
e(cmdline)	command as typed
e(depvar)	name of dependent variable
e(varnames)	names of the right-hand-side variables
e(term_#)	term #
e(errorterm_#)	error term for term # (when using nonresidual error)
e(sstype)	type of sum of squares; sequential or partial
e(repvars)	names of repeated variables (repeated() only)
e(repvar#)	names of repeated variables for a particular combination (repeated() only)
e(model)	ols
e(wtype)	weight type
e(wexp)	weight expression
e(properties)	b V
e(estat_cmd)	program used to implement estat
e(predict)	program used to implement predict
e(asbalanced)	factor variables fvset as asbalanced
e(asobserved)	factor variables fvset as asobserved

Matrices
e(b)	coefficient vector
e(V)	variance–covariance matrix of the estimators
e(Srep)	covariance matrix based on repeated measures (repeated() only)

Functions
e(sample)	marks estimation sample

Methods and formulas

anova is implemented as an ado-file.

References

Acock, A. C. 2010. *A Gentle Introduction to Stata*. 3rd ed. College Station, TX: Stata Press.

Afifi, A. A., and S. P. Azen. 1979. *Statistical Analysis: A Computer Oriented Approach*. 2nd ed. New York: Academic Press.

Altman, D. G. 1991. *Practical Statistics for Medical Research*. London: Chapman & Hall/CRC.

Anderson, R. L. 1990. Gertrude Mary Cox 1900–1978. *Biographical Memoirs, National Academy of Sciences* 59: 116–132.

Box, G. E. P. 1954. Some theorems on quadratic forms applied in the study of analysis of variance problems, I. Effect of inequality of variance in the one-way classification. *Annals of Mathematical Statistics* 25: 290–302.

Box, J. F. 1978. *R. A. Fisher: The Life of a Scientist*. New York: Wiley.

Chatfield, M., and A. Mander. 2009. The Skillings–Mack test (Friedman test when there are missing data). *Stata Journal* 9: 299–305.

Cobb, G. W. 1998. *Introduction to Design and Analysis of Experiments*. New York: Springer.

Edwards, A. L. 1985. *Multiple Regression and the Analysis of Variance and Covariance*. 2nd ed. New York: Freeman.

Fisher, R. A. 1925. *Statistical Methods for Research Workers*. Edinburgh: Oliver & Boyd.

——. 1935. *The Design of Experiments*. Edinburgh: Oliver & Boyd.

——. 1990. *Statistical Methods, Experimental Design, and Scientific Inference*. Oxford: Oxford University Press.

Geisser, S., and S. W. Greenhouse. 1958. An extension of Box's results on the use of the F distribution in multivariate analysis. *Annals of Mathematical Statistics* 29: 885–891.

Gleason, J. R. 1999. sg103: Within subjects (repeated measures) ANOVA, including between subjects factors. *Stata Technical Bulletin* 47: 40–45. Reprinted in *Stata Technical Bulletin Reprints*, vol. 8, pp. 236–243. College Station, TX: Stata Press.

——. 2000. sg132: Analysis of variance from summary statistics. *Stata Technical Bulletin* 54: 42–46. Reprinted in *Stata Technical Bulletin Reprints*, vol. 9, pp. 328–332. College Station, TX: Stata Press.

Hall, N. S. 2010. Ronald Fisher and Gertrude Cox: Two statistical pioneers sometimes cooperate and sometimes collide. *American Statistician* 64: 212–220.

Higgins, J. E., and G. G. Koch. 1977. Variable selection and generalized chi-square analysis of categorical data applied to a large cross-sectional occupational health survey. *International Statistical Review* 45: 51–62.

Huynh, H. 1978. Some approximate tests for repeated measurement designs. *Psychometrika* 43: 161–175.

Huynh, H., and L. S. Feldt. 1976. Estimation of the Box correction for degrees of freedom from sample data in randomized block and split-plot designs. *Journal of Educational Statistics* 1: 69–82.

Kennedy, W. J., Jr., and J. E. Gentle. 1980. *Statistical Computing.* New York: Dekker.

Kuehl, R. O. 2000. *Design of Experiments: Statistical Principles of Research Design and Analysis.* 2nd ed. Belmont, CA: Duxbury.

Marchenko, Y. V. 2006. Estimating variance components in Stata. *Stata Journal* 6: 1–21.

Milliken, G. A., and D. E. Johnson. 2009. *Analysis of Messy Data, Volume 1: Designed Experiments.* 2nd ed. Boca Raton, FL: CRC Press.

Scheffé, H. 1959. *The Analysis of Variance.* New York: Wiley.

Snedecor, G. W., and W. G. Cochran. 1989. *Statistical Methods.* 8th ed. Ames, IA: Iowa State University Press.

van Belle, G., L. D. Fisher, P. J. Heagerty, and T. S. Lumley. 2004. *Biostatistics: A Methodology for the Health Sciences.* 2nd ed. New York: Wiley.

Winer, B. J., D. R. Brown, and K. M. Michels. 1991. *Statistical Principles in Experimental Design.* 3rd ed. New York: McGraw–Hill.

Also see

[R] **anova postestimation** — Postestimation tools for anova

[R] **contrast** — Contrasts and linear hypothesis tests after estimation

[R] **loneway** — Large one-way ANOVA, random effects, and reliability

[R] **oneway** — One-way analysis of variance

[R] **regress** — Linear regression

[MV] **manova** — Multivariate analysis of variance and covariance

Stata Structural Equation Modeling Reference Manual

Title

---
anova postestimation — Postestimation tools for anova

Description

The following postestimation commands are of special interest after `anova`:

Command	Description
dfbeta	DFBETA influence statistics
estat hettest	tests for heteroskedasticity
estat imtest	information matrix test
estat ovtest	Ramsey regression specification-error test for omitted variables
estat szroeter	Szroeter's rank test for heteroskedasticity
estat vif	variance inflation factors for the independent variables
acprplot	augmented component-plus-residual plot
avplot	added-variable plot
avplots	all added-variable plots in one image
cprplot	component-plus-residual plot
lvr2plot	leverage-versus-squared-residual plot
rvfplot	residual-versus-fitted plot
rvpplot	residual-versus-predictor plot

For information about these commands, see [R] **regress postestimation**.

The following standard postestimation commands are also available:

Command	Description
contrast	contrasts and ANOVA-style joint tests of estimates
estat	AIC, BIC, VCE, and estimation sample summary
estimates	cataloging estimation results
hausman	Hausman's specification test
lincom	point estimates, standard errors, testing, and inference for linear combinations of coefficients
linktest	link test for model specification
lrtest	likelihood-ratio test
margins	marginal means, predictive margins, marginal effects, and average marginal effects
marginsplot	graph the results from margins (profile plots, interaction plots, etc.)
nlcom	point estimates, standard errors, testing, and inference for nonlinear combinations of coefficients
predict	predictions, residuals, influence statistics, and other diagnostic measures
predictnl	point estimates, standard errors, testing, and inference for generalized predictions
pwcompare	pairwise comparisons of estimates
suest	seemingly unrelated estimation
test	Wald tests of simple and composite linear hypotheses
testnl	Wald tests of nonlinear hypotheses

See the corresponding entries in the *Base Reference Manual* for details.

Special-interest postestimation commands

In addition to the common estat commands (see [R] **estat**), estat hettest, estat imtest, estat ovtest, estat szroeter, and estat vif are also available. dfbeta is also available. The syntax for dfbeta and these estat commands is the same as after regress; see [R] **regress postestimation**.

In addition to the standard syntax of test (see [R] **test**), test after anova has three additionally allowed syntaxes; see below. test performs Wald tests of expressions involving the coefficients of the underlying regression model. Simple and composite linear hypotheses are possible.

Syntax for predict

predict after anova follows the same syntax as predict after regress and can provide predictions, residuals, standardized residuals, Studentized residuals, the standard error of the residuals, the standard error of the prediction, the diagonal elements of the projection (hat) matrix, and Cook's D. See [R] **regress postestimation** for details.

Syntax for test after anova

In addition to the standard syntax of test (see [R] **test**), test after anova also allows the following:

<u>te</u>st, test(*matname*) $\left[\underline{\text{m}}\text{test}\left[(opt)\right]\text{ matvlc}(matname)\right]$	syntax a
<u>te</u>st, showorder	syntax b
<u>te</u>st $\left[term\left[term\ \ldots\right]\right]\left[/\ term\left[term\ \ldots\right]\right]\left[,\ \underline{\text{s}}\text{ymbolic}\right]$	syntax c

syntax a   test expression involving the coefficients of the underlying regression model; you provide information as a matrix

syntax b   show underlying order of design matrix, which is useful when constructing *matname* argument of the test() option

syntax c   test effects and show symbolic forms

Menu

Statistics > Linear models and related > ANOVA/MANOVA > Test linear hypotheses after anova

Options for test after anova

test(*matname*) is required with syntax a of test. The rows of *matname* specify linear combinations of the underlying design matrix of the ANOVA that are to be jointly tested. The columns correspond to the underlying design matrix (including the constant if it has not been suppressed). The column and row names of *matname* are ignored.

A listing of the constraints imposed by the test() option is presented before the table containing the tests. You should examine this table to verify that you have applied the linear combinations you desired. Typing test, showorder allows you to examine the ordering of the columns for the design matrix from the ANOVA.

mtest [(*opt*)] specifies that tests are performed for each condition separately. *opt* specifies the method for adjusting *p*-values for multiple testing. Valid values for *opt* are

bonferroni	Bonferroni's method
holm	Holm's method
sidak	Šidák's method
noadjust	no adjustment is to be made

Specifying mtest with no argument is equivalent to mtest(noadjust).

matvlc(*matname*), a programmer's option, saves the variance–covariance matrix of the linear combinations involved in the suite of tests. For the test **Lb** = **c**, what is returned in *matname* is **LVL′**, where V is the estimated variance–covariance matrix of **b**.

showorder causes test to list the definition of each column in the design matrix. showorder is not allowed with any other option.

symbolic requests the symbolic form of the test rather than the test statistic. When this option is specified with no terms (test, symbolic), the symbolic form of the estimable functions is displayed.

Remarks

Remarks are presented under the following headings:

> *Testing effects*
> *Obtaining symbolic forms*
> *Testing coefficients and contrasts of margins*

See examples 4, 7, 8, 13, 15, 16, and 17 in [R] **anova** for examples that use the margins command.

Testing effects

After fitting a model using anova, you can test for the significance of effects in the ANOVA table, as well as for effects that are not reported in the ANOVA table, by using the test or contrast command. You follow test or contrast by the list of effects that you wish to test. By default, these commands use the residual mean squared error in the denominator of the F ratio. You can specify other error terms by using the slash notation, just as you would with anova. See [R] **contrast** for details on this command.

▷ Example 1

Recall our byssinosis example (example 8) in [R] **anova**:

```
. anova prob workplace smokes race workplace#smokes workplace#race
> smokes#race workplace#smokes#race [aweight=pop]
(sum of wgt is    5.4190e+03)
```

		Number of obs =		65	R-squared	=	0.8300
		Root MSE	= .025902		Adj R-squared =		0.7948
Source	Partial SS	df	MS		F		Prob > F
Model	.173646538	11	.015786049		23.53		0.0000
workplace	.097625175	2	.048812588		72.76		0.0000
smokes	.013030812	1	.013030812		19.42		0.0001
race	.001094723	1	.001094723		1.63		0.2070
workplace#smokes	.019690342	2	.009845171		14.67		0.0000
workplace#race	.001352516	2	.000676258		1.01		0.3718
smokes#race	.001662874	1	.001662874		2.48		0.1214
workplace#smokes#race	.000950841	2	.00047542		0.71		0.4969
Residual	.035557766	53	.000670901				
Total	.209204304	64	.003268817				

We can easily obtain a test on a particular term from the ANOVA table. Here are two examples:

```
. test smokes
```

Source	Partial SS	df	MS	F	Prob > F
smokes	.013030812	1	.013030812	19.42	0.0001
Residual	.035557766	53	.000670901		

```
. test smokes#race
```

Source	Partial SS	df	MS	F	Prob > F
smokes#race	.001662874	1	.001662874	2.48	0.1214
Residual	.035557766	53	.000670901		

Both of these tests use residual error by default and agree with the ANOVA table produced earlier.

We could have performed these same tests with `contrast`:

```
. contrast smokes
Contrasts of marginal linear predictions
Margins       : asbalanced
```

	df	F	P>F
smokes	1	19.42	0.0001
Residual	53		

```
. contrast smokes#race
Contrasts of marginal linear predictions
Margins       : asbalanced
```

	df	F	P>F
smokes#race	1	2.48	0.1214
Residual	53		

❑ Technical note

    After `anova`, you can use the '/' syntax in `test` or `contrast` to perform tests with a variety of non-$\sigma^2\mathbf{I}$ error structures. However, in most unbalanced models, the mean squares are not independent and do not have equal expectations under the null hypothesis. Also, be warned that you assume responsibility for the validity of the test statistic.

❑

▷ Example 2

    We return to the nested ANOVA example (example 11) in [R] **anova**, where five brands of machinery were compared in an assembly line. We can obtain appropriate tests for the nested terms using `test`, even if we had run the `anova` command without initially indicating the proper error terms.

```
. use http://www.stata-press.com/data/r12/machine
(machine data)
. anova output machine operator|machine
```

		Number of obs =	57	R-squared	=	0.8661
		Root MSE = 1.47089		Adj R-squared =		0.8077
Source	Partial SS	df	MS	F		Prob > F
Model	545.822288	17	32.1071934	14.84		0.0000
machine	430.980792	4	107.745198	49.80		0.0000
operator|machine	101.353804	13	7.79644648	3.60		0.0009
Residual	84.3766582	39	2.16350406			
Total	630.198947	56	11.2535526			

In this ANOVA table, `machine` is tested with residual error. With this particular nested design, the appropriate error term for testing `machine` is `operator` nested within `machine`, which is easily obtained from `test`.

```
. test machine / operator|machine
```

Source	Partial SS	df	MS	F	Prob > F
machine	430.980792	4	107.745198	13.82	0.0001
operator|machine	101.353804	13	7.79644648		

This result from `test` matches what we obtained from our `anova` command.

◁

▷ Example 3

    The other nested ANOVA example (example 12) in [R] **anova** was based on the sewage data. The ANOVA table is presented here again. As before, we will use abbreviations of variable names in typing the commands.

```
. use http://www.stata-press.com/data/r12/sewage
(Sewage treatment)
```

```
. anova particulate s / m|s / f|m|s / w|f|m|s /, dropemptycells
```

| | Number of obs = | | 64 | R-squared | = 0.6338 |
| | Root MSE | = 12.7445 | | Adj R-squared | = 0.5194 |

Source	Partial SS	df	MS	F	Prob > F
Model	13493.6094	15	899.573958	5.54	0.0000
solution	7203.76563	1	7203.76563	17.19	0.0536
manager|solution	838.28125	2	419.140625		
manager|solution	838.28125	2	419.140625	0.55	0.6166
facility|manager| solution	3064.9375	4	766.234375		
facility|manager| solution	3064.9375	4	766.234375	2.57	0.1193
worker|facility| manager|solution	2386.625	8	298.328125		
worker|facility| manager|solution	2386.625	8	298.328125	1.84	0.0931
Residual	7796.25	48	162.421875		
Total	21289.8594	63	337.934276		

In practice, it is often beneficial to pool nonsignificant nested terms to increase the power of tests on remaining terms. One rule of thumb is to allow the pooling of a term whose p-value is larger than 0.25. In this sewage example, the p-value for the test of manager is 0.6166. This value indicates that the manager effect is negligible and might be ignored. Currently, solution is tested by manager\|solution, which has only 2 degrees of freedom. If we pool the manager and facility terms and use this pooled estimate as the error term for solution, we would have a term with 6 degrees of freedom.

Below are two tests: a test of solution with the pooled manager and facility terms and a test of this pooled term by worker.

```
. test s / m|s f|m|s
```

Source	Partial SS	df	MS	F	Prob > F
solution	7203.76563	1	7203.76563	11.07	0.0159
manager|solution facility|manager| solution	3903.21875	6	650.536458		

```
. test m|s f|m|s / w|f|m|s
```

Source	Partial SS	df	MS	F	Prob > F
manager|solution facility|manager| solution	3903.21875	6	650.536458	2.18	0.1520
worker|facility|manager| solution	2386.625	8	298.328125		

In the first test, we included two terms after the forward slash (m\|s and f\|m\|s). test after anova allows multiple terms both before and after the slash. The terms before the slash are combined and are then tested by the combined terms that follow the slash (or residual error if no slash is present).

The p-value for solution using the pooled term is 0.0159. Originally, it was 0.0536. The increase in the power of the test is due to the increase in degrees of freedom for the pooled error term.

We can get identical results if we drop `manager` from the `anova` model. (This dataset has unique numbers for each facility so that there is no confusion of facilities when `manager` is dropped.)

```
. anova particulate s / f|s / w|f|s /, dropemptycells
```

	Number of obs =		64	R-squared	=	0.6338
	Root MSE	= 12.7445		Adj R-squared =		0.5194

Source	Partial SS	df	MS	F	Prob > F
Model	13493.6094	15	899.573958	5.54	0.0000
solution	7203.76563	1	7203.76563	11.07	0.0159
facility|solution	3903.21875	6	650.536458		
facility|solution	3903.21875	6	650.536458	2.18	0.1520
worker|facility| solution	2386.625	8	298.328125		
worker|facility| solution	2386.625	8	298.328125	1.84	0.0931
Residual	7796.25	48	162.421875		
Total	21289.8594	63	337.934276		

This output agrees with our earlier `test` results.

◁

In the following example, two terms from the `anova` are jointly tested (pooled).

▷ Example 4

In example 10 of [R] **anova**, we fit the model `anova drate region c.mage region#c.mage`. Now we use the `contrast` command to test for the overall significance of `region`.

```
. contrast region region#c.mage, overall
Contrasts of marginal linear predictions
Margins        : asbalanced
```

	df	F	P>F
region	3	7.40	0.0004
region#c.mage	3	0.86	0.4689
Overall	6	5.65	0.0002
Residual	42		

The overall F statistic associated with the `region` and `region#c.mage` terms is 5.65, and it is significant at the 0.02% level.

In the ANOVA output, the `region` term, by itself, had a sum of squares of 1166.15, which, based on 3 degrees of freedom, yielded an F statistic of 7.40 and a significance level of 0.0004. This is the same test that is reported by `contrast` in the row labeled `region`. Likewise, the test from the ANOVA output for the `region#c.mage` term is reproduced in the second row of the `contrast` output.

◁

Obtaining symbolic forms

test can produce the symbolic form of the estimable functions and symbolic forms for particular tests.

▷ Example 5

After fitting an ANOVA model, we type test, symbolic to obtain the symbolic form of the estimable functions. For instance, returning to our blood pressure data introduced in example 4 of [R] **anova**, let's begin by reestimating systolic on drug, disease, and drug#disease:

```
. use http://www.stata-press.com/data/r12/systolic, clear
(Systolic Blood Pressure Data)
. anova systolic drug##disease
```

| | | Number of obs = 58 | | | R-squared = 0.4560 | |
| | | Root MSE = 10.5096 | | | Adj R-squared = 0.3259 | |

Source	Partial SS	df	MS	F	Prob > F
Model	4259.33851	11	387.212591	3.51	0.0013
drug	2997.47186	3	999.157287	9.05	0.0001
disease	415.873046	2	207.936523	1.88	0.1637
drug#disease	707.266259	6	117.87771	1.07	0.3958
Residual	5080.81667	46	110.452536		
Total	9340.15517	57	163.862371		

To obtain the symbolic form of the estimable functions, type

```
. test, symbolic
drug
            1   -(r2+r3+r4-r0)
            2    r2
            3    r3
            4    r4
disease
            1   -(r6+r7-r0)
            2    r6
            3    r7
drug#disease
       1    1   -(r2+r3+r4+r6+r7-r12-r13-r15-r16-r18-r19-r0)
       1    2    r6 - (r12+r15+r18)
       1    3    r7 - (r13+r16+r19)
       2    1    r2 - (r12+r13)
       2    2    r12
       2    3    r13
       3    1    r3 - (r15+r16)
       3    2    r15
       3    3    r16
       4    1    r4 - (r18+r19)
       4    2    r18
       4    3    r19
  _cons          r0
```

◁

▷ Example 6

To obtain the symbolic form for a particular test, we type test *term* [*term* ...], symbolic. For instance, the symbolic form for the test of the main effect of drug is

```
. test drug, symbolic
drug
            1  -(r2+r3+r4)
            2   r2
            3   r3
            4   r4
disease
            1  0
            2  0
            3  0
drug#disease
         1  1  -1/3 (r2+r3+r4)
         1  2  -1/3 (r2+r3+r4)
         1  3  -1/3 (r2+r3+r4)
         2  1   1/3 r2
         2  2   1/3 r2
         2  3   1/3 r2
         3  1   1/3 r3
         3  2   1/3 r3
         3  3   1/3 r3
         4  1   1/3 r4
         4  2   1/3 r4
         4  3   1/3 r4
   _cons        0
```

If we omit the symbolic option, we instead see the result of the test:

```
. test drug
```

Source	Partial SS	df	MS	F	Prob > F
drug	2997.47186	3	999.157287	9.05	0.0001
Residual	5080.81667	46	110.452536		

◁

Testing coefficients and contrasts of margins

The test command allows you to perform tests directly on the coefficients of the underlying regression model. For instance, the coefficient on the third drug and the second disease is referred to as 3.drug#2.disease. This could also be written as i3.drug#i2.disease, or _b[3.drug#2.disease], or even _coef[i3.drug#i2.disease]; see [U] **13.5 Accessing coefficients and standard errors**.

▷ Example 7

Let's begin by testing whether the coefficient on the third drug is equal to the coefficient on the fourth in our blood pressure data. We have already fit the model anova systolic drug##disease (equivalent to anova systolic drug disease drug#disease), and you can see the results of that estimation in example 5. Even though we have performed many tasks since we fit the model, Stata still remembers, and we can perform tests at any time.

```
. test 3.drug = 4.drug
 ( 1)   3.drug - 4.drug = 0
       F(   1,    46) =     0.13
             Prob > F =     0.7234
```

We find that the two coefficients are not significantly different, at least at any significance level smaller than 73%.

For more complex tests, the `contrast` command often provides a more concise way to specify the test we are interested in and prevents us from having to write the tests in terms of the regression coefficients. With `contrast`, we instead specify our tests in terms of differences in the marginal means for the levels of a particular factor. For example, if we want to compare the third and fourth drugs, we can test the difference in the mean impact on systolic blood pressure separately for each disease using the @ operator. We also use the reverse adjacent operator, `ar.`, to compare the fourth level of drug with the previous level.

```
. contrast ar4.drug@disease
Contrasts of marginal linear predictions
Margins       : asbalanced
```

	df	F	P>F
drug@disease			
(4 vs 3) 1	1	0.13	0.7234
(4 vs 3) 2	1	1.76	0.1917
(4 vs 3) 3	1	0.65	0.4230
Joint	3	0.85	0.4761
Residual	46		

	Contrast	Std. Err.	[95% Conf.	Interval]
drug@disease				
(4 vs 3) 1	-2.733333	7.675156	-18.18262	12.71595
(4 vs 3) 2	8.433333	6.363903	-4.376539	21.24321
(4 vs 3) 3	5.7	7.050081	-8.491077	19.89108

None of the individual contrasts shows significant differences between the third drug and the fourth drug. Likewise, the overall F statistic is 0.85, which is hardly significant. We cannot reject the hypothesis that the third drug has the same effect as the fourth drug.

◁

❏ Technical note

Alternatively, we could have specified these tests based on the coefficients of the underlying regression model using the `test` command. We would have needed to perform tests on the coefficients for `drug` and for the coefficients on `drug` interacted with `disease` in order to test for differences in the means mentioned above. To do this, we start with our previous `test` command:

```
. test 3.drug = 4.drug
```

Notice that the F statistic for this test is equivalent to the test labeled (4 vs 3) 1 in the `contrast` output. Let's now add the constraint that the coefficient on the third drug interacted with the third disease is equal to the coefficient on the fourth drug, again interacted with the third disease. We do that by typing the new constraint and adding the `accumulate` option:

```
. test 3.drug#3.disease = 4.drug#3.disease, accumulate

 ( 1)   3.drug - 4.drug = 0
 ( 2)   3.drug#3.disease - 4.drug#3.disease = 0

       F(  2,    46) =      0.39
            Prob > F =      0.6791
```

So far, our test includes the equality of the two drug coefficients, along with the equality of the two drug coefficients when interacted with the third disease. Now we add two more equations, one for each of the remaining two diseases:

```
. test 3.drug#2.disease = 4.drug#2.disease, accumulate

 ( 1)   3.drug - 4.drug = 0
 ( 2)   3.drug#3.disease - 4.drug#3.disease = 0
 ( 3)   3.drug#2.disease - 4.drug#2.disease = 0

       F(  3,    46) =      0.85
            Prob > F =      0.4761

. test 3.drug#1.disease = 4.drug#1.disease, accumulate

 ( 1)   3.drug - 4.drug = 0
 ( 2)   3.drug#3.disease - 4.drug#3.disease = 0
 ( 3)   3.drug#2.disease - 4.drug#2.disease = 0
 ( 4)   3o.drug#1b.disease - 4o.drug#1b.disease = 0
        Constraint 4 dropped

       F(  3,    46) =      0.85
            Prob > F =      0.4761
```

The overall F statistic reproduces the one from the joint test in the contrast output.

You may notice that we also got the message "Constraint 4 dropped". For the technically inclined, this constraint was unnecessary, given the normalization of the model. If we specify all the constraints involved in our test or use contrast, we need not worry about the normalization because Stata handles this automatically.

❑

The test() option of test provides another alternative for testing coefficients. Instead of spelling out each coefficient involved in the test, a matrix representing the test provides the needed information. test, showorder shows the order of the terms in the ANOVA corresponding to the order of the columns for the matrix argument of test().

▷ Example 8

We repeat the last test of example 7 above with the test() option. First, we view the definition and order of the columns underlying the ANOVA performed on the systolic data.

```
. test, showorder
Order of columns in the design matrix
      1: (drug==1)
      2: (drug==2)
      3: (drug==3)
      4: (drug==4)
      5: (disease==1)
      6: (disease==2)
      7: (disease==3)
      8: (drug==1)*(disease==1)
      9: (drug==1)*(disease==2)
     10: (drug==1)*(disease==3)
     11: (drug==2)*(disease==1)
     12: (drug==2)*(disease==2)
     13: (drug==2)*(disease==3)
     14: (drug==3)*(disease==1)
     15: (drug==3)*(disease==2)
     16: (drug==3)*(disease==3)
     17: (drug==4)*(disease==1)
     18: (drug==4)*(disease==2)
     19: (drug==4)*(disease==3)
     20: _cons
```

Columns 1–4 correspond to the four levels of drug. Columns 5–7 correspond to the three levels of disease. Columns 8–19 correspond to the interaction of drug and disease. The last column corresponds to _cons, the constant in the model.

We construct the matrix dr3vs4 with the same four constraints as the last test shown in example 7 and then use the test(dr3vs4) option to perform the test.

```
. mat dr3vs4 = (0,0,1,-1,  0,0,0,  0,0,0,0,0,0,0,0,0,  0, 0, 0,  0 \
>               0,0,0, 0,  0,0,0,  0,0,0,0,0,0,0,0,1,  0, 0,-1,  0 \
>               0,0,0, 0,  0,0,0,  0,0,0,0,0,0,1,0,  0,-1, 0,  0 \
>               0,0,0, 0,  0,0,0,  0,0,0,0,0,1,0,0,-1, 0, 0,  0)
. test, test(dr3vs4)
 ( 1)  3.drug - 4.drug = 0
 ( 2)  3.drug#3.disease - 4.drug#3.disease = 0
 ( 3)  3.drug#2.disease - 4.drug#2.disease = 0
 ( 4)  3o.drug#1b.disease - 4o.drug#1b.disease = 0
       Constraint 4 dropped
       F( 3,   46) =    0.85
            Prob > F =    0.4761
```

Here the effort involved with spelling out the coefficients is similar to that of constructing a matrix and using it in the test() option. When the test involving coefficients is more complicated, the test() option may be more convenient than specifying the coefficients directly in test. However, as previously demonstrated, contrast may provide an even simpler method for testing the same hypothesis.

◁

After fitting an ANOVA model, various contrasts (1-degree-of-freedom tests comparing different levels of a categorical variable) are often of interest. contrast can perform each 1-degree-of-freedom test in addition to the combined test, even in cases in which the contrasts do not correspond to one of the contrast operators.

▷ Example 9

Rencher and Schaalje (2008) illustrate 1-degree-of-freedom contrasts for an ANOVA comparing the net weight of cans filled by five machines (labeled A–E). The data were originally obtained from Ostle and Mensing (1975). Rencher and Schaalje use a cell-means ANOVA model approach for this problem. We could do the same by using the noconstant option of anova; see [R] anova. Instead, we obtain the same results by using the standard overparameterized ANOVA approach (that is, we keep the constant in the model).

```
. use http://www.stata-press.com/data/r12/canfill
(Can Fill Data)
. list, sepby(machine)
```

	machine	weight
1.	A	11.95
2.	A	12.00
3.	A	12.25
4.	A	12.10
5.	B	12.18
6.	B	12.11
7.	C	12.16
8.	C	12.15
9.	C	12.08
10.	D	12.25
11.	D	12.30
12.	D	12.10
13.	E	12.10
14.	E	12.04
15.	E	12.02
16.	E	12.02

```
. anova weight machine
```

			Number of obs =	16	R-squared	= 0.4123
			Root MSE	= .087758	Adj R-squared	= 0.1986
Source	Partial SS	df	MS		F	Prob > F
Model	.059426993	4	.014856748		1.93	0.1757
machine	.059426993	4	.014856748		1.93	0.1757
Residual	.084716701	11	.007701518			
Total	.144143694	15	.00960958			

The four 1-degree-of-freedom tests of interest among the five machines are A and D versus B, C, and E; B and E versus C; A versus D; and B versus E. We can specify these tests as user-defined contrasts by placing the corresponding contrast coefficients into positions related to the five levels of machine as described in *User-defined contrasts* of [R] **contrast**.

```
. contrast {machine  3 -2 -2   3 -2}
>          {machine  0  1 -2   0  1}
>          {machine  1  0  0  -1  0}
>          {machine  0  1  0   0 -1}, noeffects
```
Contrasts of marginal linear predictions

Margins   : asbalanced

	df	F	P>F
machine			
(1)	1	0.75	0.4055
(2)	1	0.31	0.5916
(3)	1	4.47	0.0582
(4)	1	1.73	0.2150
Joint	4	1.93	0.1757
Residual	11		

contrast produces a 1-degree-of-freedom test for each of the specified contrasts as well as a joint test. We included the noeffects option so that the table displaying the values of the individual contrasts with their confidence intervals was suppressed.

The significance values above are not adjusted for multiple comparisons. We could have produced the Bonferroni-adjusted significance values by using the mcompare(bonferroni) option.

```
. contrast {machine  3 -2 -2   3 -2}
>          {machine  0  1 -2   0  1}
>          {machine  1  0  0  -1  0}
>          {machine  0  1  0   0 -1}, mcompare(bonferroni) noeffects
```
Contrasts of marginal linear predictions

Margins   : asbalanced

	df	F	P>F	Bonferroni P>F
machine				
(1)	1	0.75	0.4055	1.0000
(2)	1	0.31	0.5916	1.0000
(3)	1	4.47	0.0582	0.2329
(4)	1	1.73	0.2150	0.8601
Joint	4	1.93	0.1757	
Residual	11			

Note: Bonferroni-adjusted p-values are reported for tests
      on individual contrasts only.

◁

▷ Example 10

Here there are two factors, A and B, each with three levels. The levels are quantitative so that linear and quadratic contrasts are of interest.

```
. use http://www.stata-press.com/data/r12/twowaytrend
. anova Y A B A#B
```

| | | Number of obs = | 36 | R-squared | = | 0.9304 |
| | | Root MSE   = | 2.6736 | Adj R-squared = | 0.9097 |
Source	Partial SS	df	MS	F	Prob > F
Model	2578.55556	8	322.319444	45.09	0.0000
A	2026.72222	2	1013.36111	141.77	0.0000
B	383.722222	2	191.861111	26.84	0.0000
A#B	168.111111	4	42.0277778	5.88	0.0015
Residual	193	27	7.14814815		
Total	2771.55556	35	79.1873016		

We can use the p. contrast operator to obtain the 1-degree-of-freedom tests for the linear and quadratic effects of A and B.

```
. contrast p.A p.B, noeffects
Contrasts of marginal linear predictions
Margins        : asbalanced
```

	df	F	P>F
A			
(linear)	1	212.65	0.0000
(quadratic)	1	70.88	0.0000
Joint	2	141.77	0.0000
B			
(linear)	1	26.17	0.0000
(quadratic)	1	27.51	0.0000
Joint	2	26.84	0.0000
Residual	27		

All the above tests appear to be significant. In addition to presenting the 1-degree-of-freedom tests, the combined tests for A and B are produced and agree with the original ANOVA results.

Now we explore the interaction between A and B.

```
. contrast p.A#p1.B, noeffects
Contrasts of marginal linear predictions
Margins        : asbalanced
```

	df	F	P>F
A#B			
(linear) (linear)	1	17.71	0.0003
(quadratic) (linear)	1	0.07	0.7893
Joint	2	8.89	0.0011
Residual	27		

The 2-degrees-of-freedom test of the interaction of A with the linear components of B is significant at the 0.0011 level. But, when we examine the two 1-degree-of-freedom tests that compose this result, the significance is due to the linear A by linear B contrast (significance level of 0.0003). A significance value of 0.7893 for the quadratic A by linear B indicates that this factor is not significant for these data.

```
. contrast p.A#p2.B, noeffects
Contrasts of marginal linear predictions
Margins      : asbalanced
```

	df	F	P>F
A#B			
(linear) (quadratic)	1	2.80	0.1058
(quadratic) (quadratic)	1	2.94	0.0979
Joint	2	2.87	0.0741
Residual	27		

The test of A with the quadratic components of B does not fall below the 0.05 significance level.

◁

Methods and formulas

All postestimation commands listed above are implemented as ado-files.

References

Ostle, B., and R. W. Mensing. 1975. *Statistics in Research*. 3rd ed. Ames, IA: Iowa State University Press.

Rencher, A. C., and G. B. Schaalje. 2008. *Linear Models in Statistics*. 2nd ed. New York: Wiley.

Also see

[R] **anova** — Analysis of variance and covariance

[R] **regress postestimation** — Postestimation tools for regress

[U] **20 Estimation and postestimation commands**

Title

areg — Linear regression with a large dummy-variable set

Syntax

areg *depvar* [*indepvars*] [*if*] [*in*] [*weight*] , <u>a</u>bsorb(*varname*) [*options*]

options	Description
Model	
* <u>a</u>bsorb(*varname*)	categorical variable to be absorbed
SE/Robust	
vce(*vcetype*)	*vcetype* may be ols, <u>r</u>obust, <u>c</u>luster *clustvar*, <u>boot</u>strap, or jackknife
Reporting	
<u>l</u>evel(#)	set confidence level; default is level(95)
display_options	control column formats, row spacing, line width, and display of omitted variables and base and empty cells
<u>coefl</u>egend	display legend instead of statistics

* absorb(*varname*) is required.

indepvars may contain factor variables; see [U] **11.4.3 Factor variables**.

depvar and *indepvars* may contain time-series operators; see [U] **11.4.4 Time-series varlists**.

bootstrap, by, jackknife, mi estimate, rolling, and statsby are allowed; see [U] **11.1.10 Prefix commands**.

vce(bootstrap) and vce(jackknife) are not allowed with the mi estimate prefix; see [MI] **mi estimate**.

Weights are not allowed with the bootstrap prefix; see [R] **bootstrap**.

aweights are not allowed with the jackknife prefix; see [R] **jackknife**.

aweights, fweights, and pweights are allowed; see [U] **11.1.6 weight**.

coeflegend does not appear in the dialog box.

See [U] **20 Estimation and postestimation commands** for more capabilities of estimation commands.

Menu

Statistics > Linear models and related > Other > Linear regression absorbing one cat. variable

Description

areg fits a linear regression absorbing one categorical factor. **areg** is designed for datasets with many groups, but not a number of groups that increases with the sample size. See the xtreg, fe command in [XT] **xtreg** for an estimator that handles the case in which the number of groups increases with the sample size.

Options

absorb(*varname*) specifies the categorical variable, which is to be included in the regression as if it were specified by dummy variables. absorb() is required.

vce(*vcetype*) specifies the type of standard error reported, which includes types that are derived from asymptotic theory, that are robust to some kinds of misspecification, that allow for intragroup correlation, and that use bootstrap or jackknife methods; see [R] *vce_option*.

vce(ols), the default, uses the standard variance estimator for ordinary least-squares regression.

Exercise caution when using the vce(cluster *clustvar*) option with areg. The effective number of degrees of freedom for the robust variance estimator is $n_g - 1$, where n_g is the number of clusters. Thus the number of levels of the absorb() variable should not exceed the number of clusters.

level(*#*); see [R] **estimation options**.

display_options: <u>noomitted</u>, vsquish, <u>noemptycells</u>, <u>baselevels</u>, <u>allbaselevels</u>, cformat(*%fmt*), pformat(*%fmt*), sformat(*%fmt*), and nolstretch; see [R] **estimation options**.

The following option is available with areg but is not shown in the dialog box:

coeflegend; see [R] **estimation options**.

Remarks

Suppose that you have a regression model that includes among the explanatory variables a large number, k, of mutually exclusive and exhaustive dummies:

$$\mathbf{y} = \mathbf{X}\beta + \mathbf{d}_1\gamma_1 + \mathbf{d}_2\gamma_2 + \cdots + \mathbf{d}_k\gamma_k + \epsilon$$

For instance, the dummy variables, $\mathbf{d}_i$, might indicate countries in the world or states of the United States. One solution would be to fit the model with regress, but this solution is possible only if k is small enough so that the total number of variables (the number of columns of $\mathbf{X}$ plus the number of $\mathbf{d}_i$'s plus one for $\mathbf{y}$) is sufficiently small—meaning less than matsize (see [R] **matsize**). For problems with more variables than the largest possible value of matsize (100 for Small Stata, 800 for Stata/IC, and 11,000 for Stata/SE and Stata/MP), regress will not work. areg provides a way of obtaining estimates of β—but not the γ_i's—in these cases. The effects of the dummy variables are said to be absorbed.

▷ Example 1

So that we can compare the results produced by areg with Stata's other regression commands, we will fit a model in which k is small. areg's real use, however, is when k is large.

In our automobile data, we have a variable called rep78 that is coded 1, 2, 3, 4, and 5, where 1 means poor and 5 means excellent. Let's assume that we wish to fit a regression of mpg on weight, gear_ratio, and rep78 (parameterized as a set of dummies).

```
. use http://www.stata-press.com/data/r12/auto
(1978 Automobile Data)
. regress mpg weight gear_ratio b5.rep78
```

Source	SS	df	MS
Model	1575.97621	6	262.662702
Residual	764.226686	62	12.3262369
Total	2340.2029	68	34.4147485

Number of obs =	69
F(6, 62) =	21.31
Prob > F =	0.0000
R-squared =	0.6734
Adj R-squared =	0.6418
Root MSE =	3.5109

mpg	Coef.	Std. Err.	t	P>|t|	[95% Conf. Interval]	
weight	-.0051031	.0009206	-5.54	0.000	-.0069433	-.003263
gear_ratio	.901478	1.565552	0.58	0.567	-2.228015	4.030971
rep78						
1	-2.036937	2.740728	-0.74	0.460	-7.515574	3.4417
2	-2.419822	1.764338	-1.37	0.175	-5.946682	1.107039
3	-2.557432	1.370912	-1.87	0.067	-5.297846	.1829814
4	-2.788389	1.395259	-2.00	0.050	-5.577473	.0006939
_cons	36.23782	7.01057	5.17	0.000	22.22389	50.25175

To fit the `areg` equivalent, we type

```
. areg mpg weight gear_ratio, absorb(rep78)
Linear regression, absorbing indicators
```

Number of obs =	69
F(2, 62) =	41.64
Prob > F =	0.0000
R-squared =	0.6734
Adj R-squared =	0.6418
Root MSE =	3.5109

mpg	Coef.	Std. Err.	t	P>|t|	[95% Conf. Interval]	
weight	-.0051031	.0009206	-5.54	0.000	-.0069433	-.003263
gear_ratio	.901478	1.565552	0.58	0.567	-2.228015	4.030971
_cons	34.05889	7.056383	4.83	0.000	19.95338	48.1644

rep78	F(4, 62) =	1.117	0.356	(5 categories)

Both `regress` and `areg` display the same R^2 values, root mean squared error, and—for `weight` and `gear_ratio`—the same parameter estimates, standard errors, t statistics, significance levels, and confidence intervals. `areg`, however, does not report the coefficients for `rep78`, and, in fact, they are not even calculated. This computational trick makes the problem manageable when k is large. `areg` reports a test that the coefficients associated with `rep78` are jointly zero. Here this test has a significance level of 35.6%. This F test for `rep78` is the same that we would obtain after `regress` if we were to specify `test 1.rep78 2.rep78 3.rep78 4.rep78`; see [R] **test**.

The model F tests reported by `regress` and `areg` also differ. The `regress` command reports a test that all coefficients except that of the constant are equal to zero; thus, the dummies are included in this test. The `areg` output shows a test that all coefficients excluding the dummies and the constant are equal to zero. This is the same test that can be obtained after `regress` by typing `test weight gear_ratio`.

◁

❑ Technical note

    `areg` is designed for datasets with many groups, but not a number that grows with the sample size. Consider two different samples from the U.S. population. In the first sample, we have 10,000 individuals and we want to include an indicator for each of the 50 states, whereas in the second sample we have 3 observations on each of 10,000 individuals and we want to include an indicator for each individual. `areg` was designed for datasets similar to the first sample in which we have a fixed number of groups, the 50 states. In the second sample, the number of groups, which is the number of individuals, grows as we include more individuals in the sample. For an estimator designed to handle the case in which the number of groups grows with the sample size, see the `xtreg, fe` command in [XT] **xtreg**.

    Although the point estimates produced by `areg` and `xtreg, fe` are the same, the estimated VCEs differ when `cluster()` is specified because the commands make different assumptions about whether the number of groups increases with the sample size.

❑

❑ Technical note

    The intercept reported by `areg` deserves some explanation because, given k mutually exclusive and exhaustive dummies, it is arbitrary. `areg` identifies the model by choosing the intercept that makes the prediction calculated at the means of the independent variables equal to the mean of the dependent variable: $\overline{y} = \overline{x}\,\widehat{\beta}$.

```
. predict yhat
(option xb assumed; fitted values)

. summarize mpg yhat if rep78 != .
```

Variable	Obs	Mean	Std. Dev.	Min	Max
mpg	69	21.28986	5.866408	12	41
yhat	69	21.28986	4.383224	11.58643	28.07367

We had to include `if rep78 < .` in our `summarize` command because we have missing values in our data. `areg` automatically dropped those missing values (as it should) in forming the estimates, but `predict` with the `xb` option will make predictions for cases with missing `rep78` because it does not know that `rep78` is really part of our model.

    These predicted values do not include the absorbed effects (that is, the $\mathbf{d}_i\gamma_i$). For predicted values that include these effects, use the `xbd` option of `predict` (see [R] **areg postestimation**) or see [XT] **xtreg**.

❑

▷ Example 2

    `areg, vce(robust)` is a Huberized version of `areg`; see [P] **_robust**. Just as `areg` is equivalent to using `regress` with dummies, `areg, vce(robust)` is equivalent to using `regress, vce(robust)` with dummies. You can use `areg, vce(robust)` when you expect heteroskedastic or nonnormal errors. `areg, vce(robust)`, like ordinary regression, assumes that the observations are independent, unless the `vce(cluster clustvar)` option is specified. If the `vce(cluster clustvar)` option is specified, this independence assumption is relaxed and only the clusters identified by equal values of *clustvar* are assumed to be independent.

Assume that we were to collect data by randomly sampling 10,000 doctors (from 100 hospitals) and then sampling 10 patients of each doctor, yielding a total dataset of 100,000 patients in a cluster sample. If in some regression we wished to include effects of the hospitals to which the doctors belonged, we would want to include a dummy variable for each hospital, adding 100 variables to our model. areg could fit this model by

. areg *depvar patient_vars*, absorb(hospital) vce(cluster doctor)

◁

Saved results

areg saves the following in e():

Scalars

e(N)	number of observations
e(tss)	total sum of squares
e(df_m)	model degrees of freedom
e(rss)	residual sum of squares
e(df_r)	residual degrees of freedom
e(r2)	R-squared
e(r2_a)	adjusted R-squared
e(df_a)	degrees of freedom for absorbed effect
e(rmse)	root mean squared error
e(ll)	log likelihood
e(ll_0)	log likelihood, constant-only model
e(N_clust)	number of clusters
e(F)	F statistic
e(F_absorb)	F statistic for absorbed effect (when vce(robust) is not specified)
e(rank)	rank of e(V)

Macros

e(cmd)	areg
e(cmdline)	command as typed
e(depvar)	name of dependent variable
e(absvar)	name of absorb variable
e(wtype)	weight type
e(wexp)	weight expression
e(title)	title in estimation output
e(clustvar)	name of cluster variable
e(vce)	*vcetype* specified in vce()
e(vcetype)	title used to label Std. Err.
e(datasignature)	the checksum
e(datasignaturevars)	variables used in calculation of checksum
e(properties)	b V
e(predict)	program used to implement predict
e(marginsnotok)	predictions disallowed by margins
e(asbalanced)	factor variables fvset as asbalanced
e(asobserved)	factor variables fvset as asobserved

Matrices

e(b)	coefficient vector
e(Cns)	constraints matrix
e(V)	variance–covariance matrix of the estimators
e(V_modelbased)	model-based variance

Functions

e(sample)	marks estimation sample

Methods and formulas

`areg` is implemented as an ado-file.

`areg` begins by recalculating *depvar* and *indepvars* to have mean 0 within the groups specified by `absorb()`. The overall mean of each variable is then added back in. The adjusted *depvar* is then regressed on the adjusted *indepvars* with `regress`, yielding the coefficient estimates. The degrees of freedom of the variance–covariance matrix of the coefficients is then adjusted to account for the absorbed variables—this calculation yields the same results (up to numerical roundoff error) as if the matrix had been calculated directly by the formulas given in [R] **regress**.

`areg` with `vce(robust)` or `vce(cluster` *clustvar*`)` works similarly, calling `_robust` after `regress` to produce the Huber/White/sandwich estimator of the variance or its clustered version. See [P] **_robust**, particularly *Introduction* and *Methods and formulas*. The model F test uses the robust variance estimates. There is, however, no simple computational means of obtaining a robust test of the absorbed dummies; thus this test is not displayed when the `vce(robust)` or `vce(cluster` *clustvar*`)` option is specified.

The number of groups specified in `absorb()` are included in the degrees of freedom used in the finite-sample adjustment of the cluster–robust VCE estimator. This statement is only valid if the number of groups is small relative to the sample size. (Technically, the number of groups must remain fixed as the sample size grows.) For an estimator that allows the number of groups to grow with the sample size, see the `xtreg, fe` command in [XT] **xtreg**.

Reference

Blackwell, J. L., III. 2005. Estimation and testing of fixed-effect panel-data systems. *Stata Journal* 5: 202–207.

Also see

[R] **areg postestimation** — Postestimation tools for areg

[R] **regress** — Linear regression

[MI] **estimation** — Estimation commands for use with mi estimate

[XT] **xtreg** — Fixed-, between-, and random-effects and population-averaged linear models

[U] **20 Estimation and postestimation commands**

Title

areg postestimation — Postestimation tools for areg

Description

The following postestimation commands are available after `areg`:

Command	Description
contrast	contrasts and ANOVA-style joint tests of estimates
estat	AIC, BIC, VCE, and estimation sample summary
estimates	cataloging estimation results
lincom	point estimates, standard errors, testing, and inference for linear combinations of coefficients
linktest	link test for model specification
lrtest	likelihood-ratio test
margins	marginal means, predictive margins, marginal effects, and average marginal effects
marginsplot	graph the results from margins (profile plots, interaction plots, etc.)
nlcom	point estimates, standard errors, testing, and inference for nonlinear combinations of coefficients
predict	predictions, residuals, influence statistics, and other diagnostic measures
predictnl	point estimates, standard errors, testing, and inference for generalized predictions
pwcompare	pairwise comparisons of estimates
test	Wald tests of simple and composite linear hypotheses
testnl	Wald tests of nonlinear hypotheses

See the corresponding entries in the *Base Reference Manual* for details.

Syntax for predict

predict [*type*] *newvar* [*if*] [*in*] [, *statistic*]

where $y_j = \mathbf{x}_j \mathbf{b} + d_{\text{absorbvar}} + e_j$ and *statistic* is

statistic	Description
Main	
xb	$\mathbf{x}_j \mathbf{b}$, fitted values; the default
stdp	standard error of the prediction
<u>d</u>residuals	$d_{\text{absorbvar}} + e_j = y_j - \mathbf{x}_j \mathbf{b}$
* xbd	$\mathbf{x}_j \mathbf{b} + d_{\text{absorbvar}}$
* d	$d_{\text{absorbvar}}$
* <u>residuals</u>	residual
* <u>sco</u>re	score; equivalent to residuals

Unstarred statistics are available both in and out of sample; type predict ... if e(sample) ... if wanted only for the estimation sample. Starred statistics are calculated only for the estimation sample, even when if e(sample) is not specified.

84

Menu

Statistics > Postestimation > Predictions, residuals, etc.

Options for predict

    Main

xb, the default, calculates the prediction of $\mathbf{x}_j\mathbf{b}$, the fitted values, by using the average effect of the absorbed variable. Also see xbd below.

stdp calculates the standard error of $\mathbf{x}_j\mathbf{b}$.

dresiduals calculates $y_j - \mathbf{x}_j\mathbf{b}$, which are the residuals plus the effect of the absorbed variable.

xbd calculates $\mathbf{x}_j\mathbf{b} + d_{\text{absorbvar}}$, which are the fitted values including the individual effects of the absorbed variable.

d calculates $d_{\text{absorbvar}}$, the individual coefficients for the absorbed variable.

residuals calculates the residuals, that is, $y_j - (\mathbf{x}_j\mathbf{b} + d_{\text{absorbvar}})$.

score is a synonym for residuals.

Methods and formulas

All postestimation commands listed above are implemented as ado-files.

Also see

[R] **areg** — Linear regression with a large dummy-variable set

[U] **20 Estimation and postestimation commands**

Title

> **asclogit** — Alternative-specific conditional logit (McFadden's choice) model

Syntax

> asclogit *depvar* [*indepvars*] [*if*] [*in*] [*weight*] , case(*varname*)
>
> <u>alt</u>ernatives(*varname*) [*options*]

options	Description
Model	
* case(*varname*)	use *varname* to identify cases
* <u>alt</u>ernatives(*varname*)	use *varname* to identify the alternatives available for each case
<u>casev</u>ars(*varlist*)	case-specific variables
<u>basea</u>lternative(# | *lbl* | *str*)	alternative to normalize location
<u>noc</u>onstant	suppress alternative-specific constant terms
altwise	use alternativewise deletion instead of casewise deletion
<u>off</u>set(*varname*)	include *varname* in model with coefficient constrained to 1
<u>constr</u>aints(*constraints*)	apply specified linear constraints
<u>coll</u>inear	keep collinear variables
SE/Robust	
vce(*vcetype*)	*vcetype* may be oim, <u>r</u>obust, <u>cl</u>uster *clustvar*, <u>boot</u>strap, or <u>jack</u>knife
Reporting	
<u>level</u>(#)	set confidence level; default is level(95)
or	report odds ratios
<u>nohead</u>er	do not display the header on the coefficient table
<u>nocns</u>report	do not display constraints
display_options	control column formats and line width
Maximization	
maximize_options	control the maximization process; seldom used
<u>coefl</u>egend	display legend instead of statistics

* case(*varname*) and alternatives(*varname*) are required.

bootstrap, by, jackknife, statsby, and xi are allowed; see [U] **11.1.10 Prefix commands**.

Weights are not allowed with the bootstrap prefix; see [R] **bootstrap**.

fweights, iweights, and pweights are allowed (see [U] **11.1.6 weight**), but they are interpreted to apply to cases as a whole, not to individual observations. See *Use of weights* in [R] **clogit**.

coeflegend does not appear in the dialog box.

See [U] **20 Estimation and postestimation commands** for more capabilities of estimation commands.

Menu

Statistics > Categorical outcomes > Alternative-specific conditional logit

Description

asclogit fits McFadden's choice model, which is a specific case of the more general conditional logistic regression model (McFadden 1974). asclogit requires multiple observations for each case (individual or decision), where each observation represents an alternative that may be chosen. The cases are identified by the variable specified in the case() option, whereas the alternatives are identified by the variable specified in the alternatives() option. The outcome or chosen alternative is identified by a value of 1 in *depvar*, whereas zeros indicate the alternatives that were not chosen. There can be multiple alternatives chosen for each case.

asclogit allows two types of independent variables: alternative-specific variables and case-specific variables. Alternative-specific variables vary across both cases and alternatives and are specified in *indepvars*. Case-specific variables vary only across cases and are specified in the casevars() option.

See [R] **clogit** for a more general application of conditional logistic regression. For example, clogit would be used when you have grouped data where each observation in a group may be a different individual, but all individuals in a group have a common characteristic. You may use clogit to obtain the same estimates as asclogit by specifying the case() variable as the group() variable in clogit and generating variables that interact the casevars() in asclogit with each alternative (in the form of an indicator variable), excluding the interaction variable associated with the base alternative. asclogit takes care of this data-management burden for you. Also, for clogit, each record (row in your data) is an observation, whereas in asclogit each case, consisting of several records (the alternatives) in your data, is an observation. This last point is important because asclogit will drop observations, by default, in a casewise fashion. That is, if there is at least one missing value in any of the variables for each record of a case, the entire case is dropped from estimation. To use alternativewise deletion, specify the altwise option and only the records with missing values will be dropped from estimation.

Options

> Model

case(*varname*) specifies the numeric variable that identifies each case. case() is required and must be integer valued.

alternatives(*varname*) specifies the variable that identifies the alternatives for each case. The number of alternatives can vary with each case; the maximum number of alternatives cannot exceed the limits of tabulate oneway; see [R] **tabulate oneway**. alternatives() is required and may be a numeric or a string variable.

casevars(*varlist*) specifies the case-specific numeric variables. These are variables that are constant for each case. If there are a maximum of J alternatives, there will be $J - 1$ sets of coefficients associated with the casevars().

basealternative(# | *lbl* | *str*) specifies the alternative that normalizes the latent-variable location (the level of utility). The base alternative may be specified as a number, label, or string depending on the storage type of the variable indicating alternatives. The default is the alternative with the highest frequency.

If vce(bootstrap) or vce(jackknife) is specified, you must specify the base alternative. This is to ensure that the same model is fit with each call to asclogit.

noconstant suppresses the $J - 1$ alternative-specific constant terms.

altwise specifies that alternativewise deletion be used when marking out observations due to missing values in your variables. The default is to use casewise deletion; that is, the entire group of observations making up a case is deleted if any missing values are encountered. This option does not apply to observations that are marked out by the if or in qualifier or the by prefix.

offset(*varname*), constraints(*numlist | matname*), collinear; see [R] **estimation options**.

⎿ SE/Robust ⎼

vce(*vcetype*) specifies the type of standard error reported, which includes types that are derived from asymptotic theory, that are robust to some kinds of misspecification, that allow for intragroup correlation, and that use bootstrap or jackknife methods; see [R] *vce_option*.

⎿ Reporting ⎼

level(*#*); see [R] **estimation options**.

or reports the estimated coefficients transformed to odds ratios, that is, e^b rather than b. Standard errors and confidence intervals are similarly transformed. This option affects how results are displayed, not how they are estimated. or may be specified at estimation or when replaying previously estimated results.

noheader prevents the coefficient table header from being displayed.

nocnsreport; see [R] **estimation options**.

display_options: cformat(*%fmt*), pformat(*%fmt*), sformat(*%fmt*), and nolstretch; see [R] **estimation options**.

⎿ Maximization ⎼

maximize_options: difficult, technique(*algorithm_spec*), iterate(*#*), [no]log, trace, gradient, showstep, hessian, showtolerance, tolerance(*#*), ltolerance(*#*), nrtolerance(*#*), nonrtolerance, and from(*init_specs*); see [R] **maximize**. These options are seldom used.

technique(bhhh) is not allowed.

The initial estimates must be specified as from(*matname* [, copy]), where *matname* is the matrix containing the initial estimates and the copy option specifies that only the position of each element in *matname* is relevant. If copy is not specified, the column stripe of *matname* identifies the estimates.

The following option is available with asclogit but is not shown in the dialog box:

coeflegend; see [R] **estimation options**.

Remarks

asclogit fits McFadden's choice model (McFadden [1974]; for a brief introduction, see Greene [2012, sec. 18.2] or Cameron and Trivedi [2010, sec. 15.5]). In this model, we have a set of unordered alternatives indexed by $1, 2, \ldots, J$. Let y_{ij}, $j = 1, \ldots, J$, be an indicator variable for the alternative actually chosen by the ith individual (case). That is, $y_{ij} = 1$ if individual i chose alternative j and $y_{ij} = 0$ otherwise. The independent variables come in two forms: alternative specific and case specific. Alternative-specific variables vary among the alternatives (as well as cases), and case-specific variables vary only among cases. Assume that we have p alternative-specific variables so that for case i we have a $J \times p$ matrix, $\mathbf{X}_i$. Further, assume that we have q case-specific variables so that we have a $1 \times q$ vector $\mathbf{z}_i$ for case i. Our random-utility model can then be expressed as

$$\mathbf{u}_i = \mathbf{X}_i\boldsymbol{\beta} + (\mathbf{z}_i\mathbf{A})' + \boldsymbol{\epsilon}_i$$

Here $\boldsymbol{\beta}$ is a $p \times 1$ vector of alternative-specific regression coefficients and $\mathbf{A} = (\boldsymbol{\alpha}_1, \ldots, \boldsymbol{\alpha}_J)$ is a $q \times J$ matrix of case-specific regression coefficients. The elements of the $J \times 1$ vector $\boldsymbol{\epsilon}_i$ are independent Type I (Gumbel-type) extreme-value random variables with mean γ (the Euler–Mascheroni constant, approximately 0.577) and variance $\pi^2/6$. We must fix one of the $\boldsymbol{\alpha}_j$ to the constant vector to normalize the location. We set $\boldsymbol{\alpha}_k = 0$, where k is specified by the `basealternative()` option. The vector $\mathbf{u}_i$ quantifies the utility that the individual gains from the J alternatives. The alternative chosen by individual i is the one that maximizes utility.

> ## Example 1

We have data on 295 consumers and their choice of automobile. Each consumer chose among an American, Japanese, or European car; the variable `car` indicates the nationality of the car for each alternative. We want to explore the relationship between the choice of `car` to the consumer's sex (variable `sex`) and income (variable `income` in thousands of dollars). We also have information on the number of dealerships of each nationality in the consumer's city in the variable `dealer` that we want to include as a regressor. We assume that consumers' preferences are influenced by the number of dealerships in an area but that the number of dealerships is not influenced by consumer preferences (which we admit is a rather strong assumption). The variable `dealer` is an alternative-specific variable ($\mathbf{X}_i$ is a 3×1 vector in our previous notation), and `sex` and `income` are case-specific variables ($\mathbf{z}_i$ is a 1×2 vector). Each consumer's chosen car is indicated by the variable `choice`.

Let's list some of the data.

```
. use http://www.stata-press.com/data/r12/choice
. list id car choice dealer sex income in 1/12, sepby(id)
```

	id	car	choice	dealer	sex	income
1.	1	American	0	18	male	46.7
2.	1	Japan	0	8	male	46.7
3.	1	Europe	1	5	male	46.7
4.	2	American	1	17	male	26.1
5.	2	Japan	0	6	male	26.1
6.	2	Europe	0	2	male	26.1
7.	3	American	1	12	male	32.7
8.	3	Japan	0	6	male	32.7
9.	3	Europe	0	2	male	32.7
10.	4	American	0	18	female	49.2
11.	4	Japan	1	7	female	49.2
12.	4	Europe	0	4	female	49.2

We see, for example, that the first consumer, a male earning $46,700 per year, chose to purchase a European car even though there are more American and Japanese car dealers in his area. The fourth consumer, a female earning $49,200 per year, purchased a Japanese car.

We now fit our model.

```
. asclogit choice dealer, case(id) alternatives(car) casevars(sex income)
Iteration 0:   log likelihood = -273.55685
Iteration 1:   log likelihood = -252.75109
Iteration 2:   log likelihood = -250.78555
Iteration 3:   log likelihood =  -250.7794
Iteration 4:   log likelihood =  -250.7794
```

Alternative-specific conditional logit   Number of obs   =   885
Case variable: id   Number of cases   =   295

Alternative variable: car   Alts per case: min =   3
                                                                       avg =   3.0
                                                                       max =   3

                                                        Wald chi2(5)   =   15.86
Log likelihood = -250.7794   Prob > chi2   =   0.0072

choice	Coef.	Std. Err.	z	P>|z|	[95% Conf. Interval]	
car						
dealer	.0680938	.0344465	1.98	0.048	.00058	.1356076
American	(base alternative)					
Japan						
sex	-.5346039	.3141564	-1.70	0.089	-1.150339	.0811314
income	.0325318	.012824	2.54	0.011	.0073973	.0576663
_cons	-1.352189	.6911829	-1.96	0.050	-2.706882	.0025049
Europe						
sex	.5704109	.4540247	1.26	0.209	-.3194612	1.460283
income	.032042	.0138676	2.31	0.021	.004862	.0592219
_cons	-2.355249	.8526681	-2.76	0.006	-4.026448	-.6840501

Displaying the results as odds ratios makes interpretation easier.

```
. asclogit, or noheader
```

choice	Odds Ratio	Std. Err.	z	P>|z|	[95% Conf. Interval]	
car						
dealer	1.070466	.0368737	1.98	0.048	1.00058	1.145232
American	(base alternative)					
Japan						
sex	.5859013	.1840647	-1.70	0.089	.3165294	1.084513
income	1.033067	.013248	2.54	0.011	1.007425	1.059361
_cons	.2586735	.1787907	-1.96	0.050	.0667446	1.002508
Europe						
sex	1.768994	.8031669	1.26	0.209	.7265404	4.307178
income	1.032561	.0143191	2.31	0.021	1.004874	1.061011
_cons	.0948699	.0808925	-2.76	0.006	.0178376	.5045693

These results indicate that men (sex = 1) are less likely to pick a Japanese car over an American car than women (odds ratio 0.59) but that men are more likely to choose a European car over an American car (odds ratio 1.77). Raising a person's income increases the likelihood that he or she purchases a Japanese or European car; interestingly, the effect of higher income is about the same for these two types of cars.

◁

Daniel Little McFadden was born in 1937 in North Carolina. He studied physics, psychology, and economics at the University of Minnesota and has taught economics at Pittsburgh, Berkeley, MIT, and the University of Southern California. His contributions to logit models were triggered by a student's project on freeway routing decisions, and his work consistently links economic theory and applied problems. In 2000, he shared the Nobel Prize in Economics with James J. Heckman.

❑ Technical note

McFadden's choice model is related to multinomial logistic regression (see [R] **mlogit**). If all the independent variables are case specific, then the two models are identical. We verify this supposition by running the previous example without the alternative-specific variable, dealer.

```
. asclogit choice, case(id) alternatives(car) casevars(sex income) nolog
Alternative-specific conditional logit        Number of obs       =         885
Case variable: id                             Number of cases     =         295
Alternative variable: car                     Alts per case: min =           3
                                                             avg =         3.0
                                                             max =           3
                                              Wald chi2(4)        =       12.53
Log likelihood = -252.72012                   Prob > chi2         =      0.0138
```

choice	Coef.	Std. Err.	z	P>|z|	[95% Conf. Interval]	
American	(base alternative)					
Japan						
sex	-.4694799	.3114939	-1.51	0.132	-1.079997	.141037
income	.0276854	.0123666	2.24	0.025	.0034472	.0519236
_cons	-1.962652	.6216804	-3.16	0.002	-3.181123	-.7441807
Europe						
sex	.5388441	.4525279	1.19	0.234	-.3480942	1.425782
income	.0273669	.013787	1.98	0.047	.000345	.0543889
_cons	-3.180029	.7546837	-4.21	0.000	-4.659182	-1.700876

To run mlogit, we must rearrange the dataset. mlogit requires a dependent variable that indicates the choice—1, 2, or 3—for each individual. We will use car as our dependent variable for those observations that represent the choice actually chosen.

```
. keep if choice == 1
(590 observations deleted)

. mlogit car sex income

Iteration 0:   log likelihood =  -259.1712
Iteration 1:   log likelihood = -252.81165
Iteration 2:   log likelihood = -252.72014
Iteration 3:   log likelihood = -252.72012
```

Multinomial logistic regression

Log likelihood = -252.72012

Number of obs	=	295
LR chi2(4)	=	12.90
Prob > chi2	=	0.0118
Pseudo R2	=	0.0249

car	Coef.	Std. Err.	z	P>|z|	[95% Conf. Interval]	
American	(base outcome)					
Japan						
sex	-.4694798	.3114939	-1.51	0.132	-1.079997	.1410371
income	.0276854	.0123666	2.24	0.025	.0034472	.0519236
_cons	-1.962651	.6216803	-3.16	0.002	-3.181122	-.7441801
Europe						
sex	.5388443	.4525278	1.19	0.234	-.348094	1.425783
income	.027367	.013787	1.98	0.047	.000345	.0543889
_cons	-3.18003	.7546837	-4.21	0.000	-4.659182	-1.700877

The results are the same except for the model statistic: `asclogit` uses a Wald test and `mlogit` uses a likelihood-ratio test. If you prefer the likelihood-ratio test, you can fit the constant-only model for `asclogit` followed by the full model and use [R] **lrtest**. The following example will carry this out.

```
. use http://www.stata-press.com/data/r12/choice, clear

. asclogit choice, case(id) alternatives(car)

. estimates store null

. asclogit choice, case(id) alternatives(car) casevars(sex income)

. lrtest null .
```

❑

❑ Technical note

We force you to explicitly identify the case-specific variables in the `casevars()` option to ensure that the program behaves as you expect. For example, an `if` or `in` qualifier may drop observations in such a way that (what was expected to be) an alternative-specific variable turns into a case-specific variable. Here you would probably want `asclogit` to terminate instead of interacting the variable with the alternative indicators. This situation could also occur if `asclogit` drops cases, or observations if you use the `altwise` option, because of missing values.

❑

Saved results

asclogit saves the following in e():

Scalars

e(N)	number of observations
e(N_case)	number of cases
e(k)	number of parameters
e(k_alt)	number of alternatives
e(k_indvars)	number of alternative-specific variables
e(k_casevars)	number of case-specific variables
e(k_eq)	number of equations in e(b)
e(k_eq_model)	number of equations in overall model test
e(df_m)	model degrees of freedom
e(ll)	log likelihood
e(N_clust)	number of clusters
e(const)	constant indicator
e(i_base)	base alternative index
e(chi2)	χ^2
e(F)	F statistic
e(p)	significance
e(alt_min)	minimum number of alternatives
e(alt_avg)	average number of alternatives
e(alt_max)	maximum number of alternatives
e(rank)	rank of e(V)
e(ic)	number of iterations
e(rc)	return code
e(converged)	1 if converged, 0 otherwise

Macros

e(cmd)	asclogit
e(cmdline)	command as typed
e(depvar)	name of dependent variable
e(indvars)	alternative-specific independent variable
e(casevars)	case-specific variables
e(case)	variable defining cases
e(altvar)	variable defining alternatives
e(alteqs)	alternative equation names
e(alt#)	alternative labels
e(wtype)	weight type
e(wexp)	weight expression
e(title)	title in estimation output
e(clustvar)	name of cluster variable
e(offset)	linear offset variable
e(chi2type)	Wald, type of model χ^2 test
e(vce)	*vcetype* specified in vce()
e(vcetype)	title used to label Std. Err.
e(opt)	type of optimization
e(which)	max or min; whether optimizer is to perform maximization or minimization
e(ml_method)	type of ml method
e(user)	name of likelihood-evaluator program
e(technique)	maximization technique
e(datasignature)	the checksum
e(datasignaturevars)	variables used in calculation of checksum
e(properties)	b V
e(estat_cmd)	program used to implement estat
e(predict)	program used to implement predict
e(marginsnotok)	predictions disallowed by margins

Matrices
  e(b)   coefficient vector
  e(stats)   alternative statistics
  e(altvals)   alternative values
  e(altfreq)   alternative frequencies
  e(alt_casevars) indicators for estimated case-specific coefficients—e(k_alt)×e(k_casevars)
  e(ilog)   iteration log (up to 20 iterations)
  e(gradient)   gradient vector
  e(V)   variance–covariance matrix of the estimators
  e(V_modelbased) model-based variance

Functions
  e(sample)   marks estimation sample

Methods and formulas

asclogit is implemented as an ado-file.

In this model, we have a set of unordered alternatives indexed by $1, 2, \ldots, J$. Let $y_{ij}, j = 1, \ldots, J$, be an indicator variable for the alternative actually chosen by the ith individual (case). That is, $y_{ij} = 1$ if individual i chose alternative j and $y_{ij} = 0$ otherwise. The independent variables come in two forms: alternative specific and case specific. Alternative-specific variables vary among the alternatives (as well as cases), and case-specific variables vary only among cases. Assume that we have p alternative-specific variables so that for case i we have a $J \times p$ matrix, $\mathbf{X}_i$. Further, assume that we have q case-specific variables so that we have a $1 \times q$ vector $\mathbf{z}_i$ for case i. The deterministic component of the random-utility model can then be expressed as

$$
\begin{aligned}
\eta_i &= \mathbf{X}_i \boldsymbol{\beta} + (\mathbf{z}_i \mathbf{A})' \\
&= \mathbf{X}_i \boldsymbol{\beta} + (\mathbf{z}_i \otimes \mathbf{I}_J) \operatorname{vec}(\mathbf{A}') \\
&= (\mathbf{X}_i, \ \mathbf{z}_i \otimes \mathbf{I}_J) \begin{pmatrix} \boldsymbol{\beta} \\ \operatorname{vec}(\mathbf{A}') \end{pmatrix} \\
&= \mathbf{X}_i^* \boldsymbol{\beta}^*
\end{aligned}
$$

As before, $\boldsymbol{\beta}$ is a $p \times 1$ vector of alternative-specific regression coefficients, and $\mathbf{A} = (\boldsymbol{\alpha}_1, \ldots, \boldsymbol{\alpha}_J)$ is a $q \times J$ matrix of case-specific regression coefficients; remember that we must fix one of the $\boldsymbol{\alpha}_j$ to the constant vector to normalize the location. Here $\mathbf{I}_J$ is the $J \times J$ identity matrix, vec() is the vector function that creates a vector from a matrix by placing each column of the matrix on top of the other (see [M-5] **vec( )**), and $\otimes$ is the Kronecker product (see [M-2] **op_kronecker**).

We have rewritten the linear equation so that it is a form that can be used by clogit, namely, $\mathbf{X}_i^* \boldsymbol{\beta}^*$, where

$$\mathbf{X}_i^* = (\mathbf{X}_i, \ \mathbf{z}_i \otimes \mathbf{I}_J)$$

$$\boldsymbol{\beta}^* = \begin{pmatrix} \boldsymbol{\beta} \\ \operatorname{vec}(\mathbf{A}') \end{pmatrix}$$

With this in mind, see *Methods and formulas* in [R] **clogit** for the computational details of the conditional logit model.

This command supports the clustered version of the Huber/White/sandwich estimator of the variance using vce(robust) and vce(cluster *clustvar*). See [P] **_robust**, particularly *Maximum likelihood estimators* and *Methods and formulas*. Specifying vce(robust) is equivalent to specifying vce(cluster *casevar*), where *casevar* is the variable that identifies the cases.

References

Cameron, A. C., and P. K. Trivedi. 2010. *Microeconometrics Using Stata*. Rev. ed. College Station, TX: Stata Press.

Greene, W. H. 2012. *Econometric Analysis*. 7th ed. Upper Saddle River, NJ: Prentice Hall.

McFadden, D. L. 1974. Conditional logit analysis of qualitative choice behavior. In *Frontiers in Econometrics*, ed. P. Zarembka, 105–142. New York: Academic Press.

Also see

[R] **asclogit postestimation** — Postestimation tools for asclogit

[R] **asmprobit** — Alternative-specific multinomial probit regression

[R] **asroprobit** — Alternative-specific rank-ordered probit regression

[R] **clogit** — Conditional (fixed-effects) logistic regression

[R] **logistic** — Logistic regression, reporting odds ratios

[R] **logit** — Logistic regression, reporting coefficients

[R] **nlogit** — Nested logit regression

[R] **ologit** — Ordered logistic regression

[U] **20 Estimation and postestimation commands**

Title

asclogit postestimation — Postestimation tools for asclogit

Description

The following postestimation commands are of special interest after `asclogit`:

Commands	Description
estat alternatives	alternative summary statistics
estat mfx	marginal effects

For information about these commands, see below.

The following standard postestimation commands are also available:

Commands	Description
estat	AIC, BIC, VCE, and estimation sample summary
estimates	cataloging estimation results
hausman	Hausman's specification test
lincom	point estimates, standard errors, testing, and inference for linear combinations of coefficients
lrtest	likelihood-ratio test
nlcom	point estimates, standard errors, testing, and inference for nonlinear combinations of coefficients
predict	predicted probabilities, estimated linear predictor and its standard error
predictnl	point estimates, standard errors, testing, and inference for generalized predictions
test	Wald tests of simple and composite linear hypotheses
testnl	Wald tests of nonlinear hypotheses

See the corresponding entries in the *Base Reference Manual* for details.

Special-interest postestimation commands

`estat alternatives` displays summary statistics about the alternatives in the estimation sample.

`estat mfx` computes probability marginal effects.

Syntax for predict

> predict [*type*] *newvar* [*if*] [*in*] [, *statistic options*]

> predict [*type*] { *stub** | *newvarlist* } [*if*] [*in*], <u>sc</u>ores

statistic	Description
Main	
<u>p</u>r	probability that each alternative is chosen; the default
xb	linear prediction
stdp	standard error of the linear prediction

options	Description
Main	
*k(# | observed)	condition on # alternatives per case or on observed number of alternatives
altwise	use alternativewise deletion instead of casewise deletion when computing probabilities
<u>nooff</u>set	ignore the offset() variable specified in asclogit

*k(# | observed) may be used only with pr.

These statistics are available both in and out of sample; type predict ... if e(sample) ... if wanted only for the estimation sample.

Menu

Statistics > Postestimation > Predictions, residuals, etc.

Options for predict

⌐ Main ⌐

pr computes the probability of choosing each alternative conditioned on each case choosing k() alternatives. This is the default statistic with default k(1); one alternative per case is chosen.

xb computes the linear prediction.

stdp computes the standard error of the linear prediction.

k(# | observed) conditions the probability on # alternatives per case or on the observed number of alternatives. The default is k(1). This option may be used only with the pr option.

altwise specifies that alternativewise deletion be used when marking out observations due to missing values in your variables. The default is to use casewise deletion. The xb and stdp options always use alternativewise deletion.

nooffset is relevant only if you specified offset(*varname*) for asclogit. It modifies the calculations made by predict so that they ignore the offset variable; the linear prediction is treated as $x\beta$ rather than as $x\beta +$ offset.

scores calculates the scores for each coefficient in e(b). This option requires a new variable list of length equal to the number of columns in e(b). Otherwise, use the *stub** option to have predict generate enumerated variables with prefix *stub*.

Syntax for estat alternatives

estat alternatives

Menu

Statistics > Postestimation > Reports and statistics

Syntax for estat mfx

estat mfx [if] [in] [, options]

options	Description
Main	
varlist(varlist)	display marginal effects for varlist
at(mean [atlist] | median [atlist])	calculate marginal effects at these values
k(#)	condition on the number of alternatives chosen to be #
Options	
level(#)	set confidence interval level; default is level(95)
nodiscrete	treat indicator variables as continuous
noesample	do not restrict calculation of means and medians to the estimation sample
nowght	ignore weights when calculating means and medians

Menu

Statistics > Postestimation > Reports and statistics

Options for estat mfx

⌐ Main ⌐

varlist(varlist) specifies the variables for which to display marginal effects. The default is all variables.

at(mean [atlist] | median [atlist]) specifies the values at which the marginal effects are to be calculated. atlist is

$$[[alternative:variable = \#] [variable = \#] [alternative:offset = \#] [\dots]]$$

The default is to calculate the marginal effects at the means of the independent variables by using the estimation sample, at(mean). If offset() is used during estimation, the means of the offsets (by alternative) are computed by default.

After specifying the summary statistic, you can specify a series of specific values for variables. You can specify values for alternative-specific variables by alternative, or you can specify one value for all alternatives. You can specify only one value for case-specific variables. You specify values for the offset() variable (if present) the same way as for alternative-specific variables. For example, in the choice dataset (car choice), income is a case-specific variable, whereas dealer is an alternative-specific variable. The following would be a legal syntax for estat mfx:

```
. estat mfx, at(mean American:dealer=18 income=40)
```

When `nodiscrete` is not specified, `at(mean [atlist])` or `at(median [atlist])` has no effect on computing marginal effects for indicator variables, which are calculated as the discrete change in the simulated probability as the indicator variable changes from 0 to 1.

The mean and median computations respect any `if` or `in` qualifiers, so you can restrict the data over which the statistic is computed. You can even restrict the values to a specific case, for example,

```
. estat mfx if case==21
```

`k(#)` computes the probabilities conditioned on # alternatives chosen. The default is one alternative chosen.

⌐ Options ⌐

`level(#)` sets the confidence level; default is `level(95)`.

`nodiscrete` specifies that indicator variables be treated as continuous variables. An indicator variable is one that takes on the value 0 or 1 in the estimation sample. By default, the discrete change in the simulated probability is computed as the indicator variable changes from 0 to 1.

`noesample` specifies that the whole dataset be considered instead of only those marked in the `e(sample)` defined by the `asclogit` command.

`nowght` specifies that weights be ignored when calculating the medians.

Remarks

Remarks are presented under the following headings:

> *Predicted probabilities*
> *Obtaining estimation statistics*

Predicted probabilities

After fitting a McFadden's choice model with alternative-specific conditional logistic regression, you can use `predict` to obtain the estimated probability of alternative choices given case profiles.

▷ Example 1

In example 1 of [R] **asclogit**, we fit a model of consumer choice of automobile. The alternatives are nationality of the automobile manufacturer: American, Japanese, or European. There is one alternative-specific variable in the model, `dealer`, which contains the number of dealerships of each nationality in the consumer's city. The case-specific variables are `sex`, the consumer's sex, and `income`, the consumer's income in thousands of dollars.

```
. use http://www.stata-press.com/data/r12/choice
. asclogit choice dealer, case(id) alternatives(car) casevars(sex income)
  (output omitted)
. predict p
(option pr assumed; Pr(car))
. predict p2, k(2)
(option pr assumed; Pr(car))
. format p p2 %6.4f
```

```
. list car choice dealer sex income p p2 in 1/9, sepby(id)
```

	car	choice	dealer	sex	income	p	p2
1.	American	0	18	male	46.7	0.6025	0.8589
2.	Japan	0	8	male	46.7	0.2112	0.5974
3.	Europe	1	5	male	46.7	0.1863	0.5437
4.	American	1	17	male	26.1	0.7651	0.9293
5.	Japan	0	6	male	26.1	0.1282	0.5778
6.	Europe	0	2	male	26.1	0.1067	0.4929
7.	American	1	12	male	32.7	0.6519	0.8831
8.	Japan	0	6	male	32.7	0.1902	0.5995
9.	Europe	0	2	male	32.7	0.1579	0.5174

◁

Obtaining estimation statistics

Here we will demonstrate the specialized estat subcommands after asclogit. Use estat alternatives to obtain a table of alternative statistics. The table will contain the alternative values, labels (if any), the number of cases in which each alternative is present, the frequency that the alternative is selected, and the percent selected.

Use estat mfx to obtain marginal effects after asclogit.

▷ Example 2

We will continue with the automobile choice example, where we first list the alternative statistics and then compute the marginal effects at the mean income in our sample, assuming that there are five automobile dealers for each nationality. We will evaluate the probabilities for females because sex is coded 0 for females, and we will be obtaining the discrete change from 0 to 1.

```
. estat alternatives
```
Alternatives summary for car

index	Alternative value	label	Cases present	Frequency selected	Percent selected
1	1	American	295	192	65.08
2	2	Japan	295	64	21.69
3	3	Europe	295	39	13.22

```
. estat mfx, at(dealer=0 sex=0) varlist(sex income)
```
Pr(choice = American|1 selected) = .41964329

variable	dp/dx	Std. Err.	z	P>|z|	[95% C.I.]		X
casevars							
sex*	.026238	.068311	0.38	0.701	-.107649	.160124	0
income	-.007891	.002674	-2.95	0.003	-.013132	-.00265	42.097

(*) dp/dx is for discrete change of indicator variable from 0 to 1

Pr(choice = Japan|1 selected) = .42696187

variable	dp/dx	Std. Err.	z	P>|z|	[	95% C.I.	]	X
casevars								
sex*	-.161164	.079238	-2.03	0.042	-.316468	-.005859		0
income	.005861	.002997	1.96	0.051	-.000014	.011735		42.097

(*) dp/dx is for discrete change of indicator variable from 0 to 1

Pr(choice = Europe|1 selected) = .15339484

variable	dp/dx	Std. Err.	z	P>|z|	[	95% C.I.	]	X
casevars								
sex*	.134926	.076556	1.76	0.078	-.015122	.284973		0
income	.00203	.001785	1.14	0.255	-.001469	.00553		42.097

(*) dp/dx is for discrete change of indicator variable from 0 to 1

The marginal effect of income indicates that there is a lower chance for a consumer to buy American automobiles with an increase in income. There is an indication that men have a higher preference for European automobiles than women but a lower preference for Japanese automobiles. We did not include the marginal effects for dealer because we view these as nuisance parameters, so we adjusted the probabilities by fixing dealer to a constant, 0.

◁

Saved results

estat mfx saves the following in r():

Scalars
  r(pr_*alt*)   scalars containing the computed probability of each alternative evaluated at the value that is labeled X in the table output. Here *alt* are the labels in the macro e(alteqs).
Matrices
  r(*alt*)   matrices containing the computed marginal effects and associated statistics. There is one matrix for each alternative, where *alt* are the labels in the macro e(alteqs). Column 1 of each matrix contains the marginal effects; column 2, their standard errors; column 3, their z statistics; and columns 4 and 5, the confidence intervals. Column 6 contains the values of the independent variables used to compute the probabilities r(pr_*alt*).

Methods and formulas

All postestimation commands listed above are implemented as ado-files.

The deterministic component of the random-utility model can be expressed as

$$
\begin{aligned}
\eta &= \mathbf{X}\beta + (\mathbf{z}\mathbf{A})' \\
&= \mathbf{X}\beta + (\mathbf{z} \otimes \mathbf{I}_J)\, \mathrm{vec}(\mathbf{A}') \\
&= (\mathbf{X},\ \mathbf{z} \otimes \mathbf{I}_J) \begin{pmatrix} \beta \\ \mathrm{vec}(\mathbf{A}') \end{pmatrix} \\
&= \mathbf{X}^*\beta^*
\end{aligned}
$$

where $\mathbf{X}$ is the $J \times p$ matrix containing the alternative-specific covariates, $\mathbf{z}$ is a $1 \times q$ vector of case-specific variables, β is a $p \times 1$ vector of alternative-specific regression coefficients, and $\mathbf{A} = (\alpha_1, \ldots, \alpha_J)$ is a $q \times J$ matrix of case-specific regression coefficients (with one of the α_j fixed to the constant). Here $\mathbf{I}_J$ is the $J \times J$ identity matrix, vec() is the vector function that creates a vector from a matrix by placing each column of the matrix on top of the other (see [M-5] **vec( )**), and $\otimes$ is the Kronecker product (see [M-2] **op_kronecker**).

We have rewritten the linear equation so that it is a form that we all recognize, namely, $\eta = \mathbf{X}^*\beta^*$, where

$$
\mathbf{X}^* = (\mathbf{X},\ \mathbf{z} \otimes \mathbf{I}_J)
$$

$$
\beta^* = \begin{pmatrix} \beta \\ \mathrm{vec}(\mathbf{A}') \end{pmatrix}
$$

To compute the marginal effects, we use the derivative of the log likelihood $\partial \ell(\mathbf{y}|\eta)/\partial\eta$, where $\ell(\mathbf{y}|\eta) = \log \Pr(\mathbf{y}|\eta)$ is the log of the probability of the choice indicator vector $\mathbf{y}$ given the linear predictor vector η. Namely,

$$
\begin{aligned}
\frac{\partial \Pr(\mathbf{y}|\eta)}{\partial \mathrm{vec}(\mathbf{X}^*)'} &= \Pr(\mathbf{y}|\eta)\frac{\partial \ell(\mathbf{y}|\eta)}{\partial \eta'}\frac{\partial \eta}{\partial \mathrm{vec}(\mathbf{X}^*)'} \\
&= \Pr(\mathbf{y}|\eta)\frac{\partial \ell(\mathbf{y}|\eta)}{\partial \eta'}\left(\beta^{*\prime} \otimes \mathbf{I}_J\right)
\end{aligned}
$$

The standard errors of the marginal effects are computed using the delta method.

Also see

[R] **asclogit** — Alternative-specific conditional logit (McFadden's choice) model

[U] **20 Estimation and postestimation commands**

Title

> **asmprobit** — Alternative-specific multinomial probit regression

Syntax

asmprobit *depvar* [*indepvars*] [*if*] [*in*] [*weight*] , case(*varname*)

   alternatives(*varname*) [*options*]

options	Description
Model	
* case(*varname*)	use *varname* to identify cases
* alternatives(*varname*)	use *varname* to identify the alternatives available for each case
casevars(*varlist*)	case-specific variables
constraints(*constraints*)	apply specified linear constraints
collinear	keep collinear variables
Model 2	
correlation(*correlation*)	correlation structure of the latent-variable errors
stddev(*stddev*)	variance structure of the latent-variable errors
structural	use the structural covariance parameterization; default is the differenced covariance parameterization
factor(#)	use the factor covariance structure with dimension #
noconstant	suppress the alternative-specific constant terms
basealternative(# | *lbl* | *str*)	alternative used for normalizing location
scalealternative(# | *lbl* | *str*)	alternative used for normalizing scale
altwise	use alternativewise deletion instead of casewise deletion
SE/Robust	
vce(*vcetype*)	*vcetype* may be oim, robust, cluster *clustvar*, opg, bootstrap, or jackknife
Reporting	
level(#)	set confidence level; default is level(95)
notransform	do not transform variance–covariance estimates to the standard deviation and correlation metric
nocnsreport	do not display constraints
display_options	control column formats and line width

Integration

<u>int</u>method(*seqtype*)	type of quasi- or pseudouniform point set
<u>intp</u>oints(#)	number of points in each sequence
<u>intb</u>urn(#)	starting index in the Hammersley or Halton sequence
<u>ints</u>eed(*code* | #)	pseudouniform random-number seed
<u>anti</u>thetics	use antithetic draws
<u>nop</u>ivot	do not use integration interval pivoting
<u>initb</u>hhh(#)	use the BHHH optimization algorithm for the first # iterations
favor(<u>speed</u> | <u>space</u>)	favor speed or space when generating integration points

Maximization

maximize_options	control the maximization process
<u>coefl</u>egend	display legend instead of statistics

correlation	Description
<u>un</u>structured	one correlation parameter for each pair of alternatives; correlations with the basealternative() are zero; the default
<u>ex</u>changeable	one correlation parameter common to all pairs of alternatives; correlations with the basealternative() are zero
<u>in</u>dependent	constrain all correlation parameters to zero
<u>pa</u>ttern *matname*	user-specified matrix identifying the correlation pattern
<u>fix</u>ed *matname*	user-specified matrix identifying the fixed and free correlation parameters

stddev	Description
<u>het</u>eroskedastic	estimate standard deviation for each alternative; standard deviations for basealternative() and scalealternative() set to one
<u>hom</u>oskedastic	all standard deviations are one
<u>pa</u>ttern *matname*	user-specified matrix identifying the standard deviation pattern
<u>fix</u>ed *matname*	user-specified matrix identifying the fixed and free standard deviations

seqtype	Description
<u>ham</u>mersley	Hammersley point set
<u>hal</u>ton	Halton point set
<u>ran</u>dom	uniform pseudorandom point set

*case(*varname*) and alternatives(*varname*) are required.
bootstrap, by, jackknife, statsby, and xi are allowed; see [U] **11.1.10 Prefix commands**.
Weights are not allowed with the bootstrap prefix; see [R] **bootstrap**.
fweights, iweights, and pweights are allowed; see [U] **11.1.6 weight**.
coeflegend does not appear in the dialog box.
See [U] **20 Estimation and postestimation commands** for more capabilities of estimation commands.

Menu

Statistics > Categorical outcomes > Alternative-specific multinomial probit

Description

asmprobit fits multinomial probit (MNP) models by using maximum simulated likelihood (MSL) implemented by the Geweke–Hajivassiliou–Keane (GHK) algorithm. By estimating the variance–covariance parameters of the latent-variable errors, the model allows you to relax the independence of irrelevant alternatives (IIA) property that is characteristic of the multinomial logistic model.

asmprobit requires multiple observations for each case (decision), where each observation represents an alternative that may be chosen. The cases are identified by the variable specified in the case() option, whereas the alternatives are identified by the variable specified in the alternative() option. The outcome (chosen alternative) is identified by a value of 1 in *depvar*, with 0 indicating the alternatives that were not chosen; only one alternative may be chosen for each case.

asmprobit allows two types of independent variables: alternative-specific variables and case-specific variables. Alternative-specific variables vary across both cases and alternatives and are specified in *indepvars*. Case-specific variables vary only across cases and are specified in the casevars() option.

Options

─────┐ Model ├───

case(*varname*) specifies the variable that identifies each case. This variable identifies the individuals or entities making a choice. case() is required.

alternatives(*varname*) specifies the variable that identifies the alternatives for each case. The number of alternatives can vary with each case; the maximum number of alternatives is 20. alternatives() is required.

casevars(*varlist*) specifies the case-specific variables that are constant for each case(). If there are a maximum of J alternatives, there will be $J - 1$ sets of coefficients associated with casevars().

constraints(*constraints*), collinear; see [R] **estimation options**.

─────┐ Model 2 ├───

correlation(*correlation*) specifies the correlation structure of the latent-variable errors.

correlation(unstructured) is the most general and has $J(J - 3)/2 + 1$ unique correlation parameters. This is the default unless stdev() or structural are specified.

correlation(exchangeable) provides for one correlation coefficient common to all latent variables, except the latent variable associated with the basealternative() option.

correlation(independent) assumes that all correlations are zero.

correlation(pattern *matname*) and correlation(fixed *matname*) give you more flexibility in defining the correlation structure. See *Variance structures* later in this entry for more information.

stddev(*stddev*) specifies the variance structure of the latent-variable errors.

stddev(heteroskedastic) is the most general and has $J - 2$ estimable parameters. The standard deviations of the latent-variable errors for the alternatives specified in basealternative() and scalealternative() are fixed to one.

stddev(homoskedastic) constrains all the standard deviations to equal one.

stddev(pattern *matname*) and stddev(fixed *matname*) give you added flexibility in defining the standard deviation parameters. See *Variance structures* later in this entry for more information.

structural requests the $J \times J$ structural covariance parameterization instead of the default $J-1 \times J-1$ differenced covariance parameterization (the covariance of the latent errors differenced with that of the base alternative). The differenced covariance parameterization will achieve the same MSL regardless of the choice of basealternative() and scalealternative(). On the other hand, the structural covariance parameterization imposes more normalizations that may bound the model away from its maximum likelihood and thus prevent convergence with some datasets or choices of basealternative() and scalealternative().

factor(#) requests that the factor covariance structure of dimension # be used. The factor() option can be used with the structural option but cannot be used with stddev() or correlation(). A $\# \times J$ (or $\# \times J-1$) matrix, $\mathbf{C}$, is used to factor the covariance matrix as $I + \mathbf{C}'\mathbf{C}$, where I is the identity matrix of dimension J (or $J-1$). The column dimension of $\mathbf{C}$ depends on whether the covariance is structural or differenced. The row dimension of $\mathbf{C}$, #, must be less than or equal to floor$((J(J-1)/2-1)/(J-2))$, because there are only $J(J-1)/2-1$ identifiable variance–covariance parameters. This covariance parameterization may be useful for reducing the number of covariance parameters that need to be estimated.

If the covariance is structural, the column of $\mathbf{C}$ corresponding to the base alternative contains zeros. The column corresponding to the scale alternative has a one in the first row and zeros elsewhere. If the covariance is differenced, the column corresponding to the scale alternative (differenced with the base) has a one in the first row and zeros elsewhere.

noconstant suppresses the $J-1$ alternative-specific constant terms.

basealternative(# | *lbl* | *str*) specifies the alternative used to normalize the latent-variable location (also referred to as the level of utility). The base alternative may be specified as a number, label, or string. The standard deviation for the latent-variable error associated with the base alternative is fixed to one, and its correlations with all other latent-variable errors are set to zero. The default is the first alternative when sorted. If a fixed or pattern matrix is given in the stddev() and correlation() options, the basealternative() will be implied by the fixed standard deviations and correlations in the matrix specifications. basealternative() cannot be equal to scalealternative().

scalealternative(# | *lbl* | *str*) specifies the alternative used to normalize the latent-variable scale (also referred to as the scale of utility). The scale alternative may be specified as a number, label, or string. The default is to use the second alternative when sorted. If a fixed or pattern matrix is given in the stddev() option, the scalealternative() will be implied by the fixed standard deviations in the matrix specification. scalealternative() cannot be equal to basealternative().

If a fixed or pattern matrix is given for the stddev() option, the base alternative and scale alternative are implied by the standard deviations and correlations in the matrix specifications, and they need not be specified in the basealternative() and scalealternative() options.

altwise specifies that alternativewise deletion be used when marking out observations due to missing values in your variables. The default is to use casewise deletion; that is, the entire group of observations making up a case is deleted if any missing values are encountered. This option does not apply to observations that are marked out by the if or in qualifier or the by prefix.

SE/Robust

vce(*vcetype*) specifies the type of standard error reported, which includes types that are derived from asymptotic theory, that are robust to some kinds of misspecification, that allow for intragroup correlation, and that use bootstrap or jackknife methods; see [R] **vce_option**.

If specifying vce(bootstrap) or vce(jackknife), you must also specify basealternative() and scalealternative().

Reporting

level(*#*); see [R] **estimation options**.

notransform prevents retransforming the Cholesky-factored variance–covariance estimates to the correlation and standard deviation metric.

This option has no effect if structural is not specified because the default differenced variance–covariance estimates have no interesting interpretation as correlations and standard deviations. notransform also has no effect if the correlation() and stddev() options are specified with anything other than their default values. Here it is generally not possible to factor the variance–covariance matrix, so optimization is already performed using the standard deviation and correlation representations.

nocnsreport; see [R] **estimation options**.

display_options: cformat(% *fmt*), pformat(% *fmt*), sformat(% *fmt*), and nolstretch; see [R] **estimation options**.

Integration

intmethod(hammersley | halton | random) specifies the method of generating the point sets used in the quasi–Monte Carlo integration of the multivariate normal density. intmethod(hammersley), the default, uses the Hammersley sequence; intmethod(halton) uses the Halton sequence; and intmethod(random) uses a sequence of uniform random numbers.

intpoints(*#*) specifies the number of points to use in the quasi–Monte Carlo integration. If this option is not specified, the number of points is $50 \times J$ if intmethod(hammersley) or intmethod(halton) is used and $100 \times J$ if intmethod(random) is used. Larger values of intpoints() provide better approximations of the log likelihood, but at the cost of added computation time.

intburn(*#*) specifies where in the Hammersley or Halton sequence to start, which helps reduce the correlation between the sequences of each dimension. The default is 0. This option may not be specified with intmethod(random).

intseed(*code* | *#*) specifies the seed to use for generating the uniform pseudorandom sequence. This option may be specified only with intmethod(random). *code* refers to a string that records the state of the random-number generator runiform(); see [R] **set seed**. An integer value *#* may be used also. The default is to use the current seed value from Stata's uniform random-number generator, which can be obtained from c(seed).

antithetics specifies that antithetic draws be used. The antithetic draw for the $J - 1$ vector uniform-random variables, x, is $1 - $ x.

nopivot turns off integration interval pivoting. By default, asmprobit will pivot the wider intervals of integration to the interior of the multivariate integration. This improves the accuracy of the quadrature estimate. However, discontinuities may result in the computation of numerical second-order derivatives using finite differencing (for the Newton–Raphson optimize technique, tech(nr)) when few simulation points are used, resulting in a non–positive-definite Hessian. asmprobit uses the Broyden–Fletcher–Goldfarb–Shanno optimization algorithm, by default, which does not require computing the Hessian numerically using finite differencing.

initbhhh(#) specifies that the Berndt–Hall–Hall–Hausman (BHHH) algorithm be used for the initial # optimization steps. This option is the only way to use the BHHH algorithm along with other optimization techniques. The algorithm switching feature of ml's technique() option cannot include bhhh.

favor(speed | space) instructs asmprobit to favor either speed or space when generating the integration points. favor(speed) is the default. When favoring speed, the integration points are generated once and stored in memory, thus increasing the speed of evaluating the likelihood. This speed increase can be seen when there are many cases or when the user specifies a large number of integration points, intpoints(#). When favoring space, the integration points are generated repeatedly with each likelihood evaluation.

For unbalanced data, where the number of alternatives varies with each case, the estimates computed using intmethod(random) will vary slightly between favor(speed) and favor(space). This is because the uniform sequences will not be identical, even when initiating the sequences using the same uniform seed, intseed(*code* | #). For favor(speed), ncase blocks of intpoints(#) × $J - 2$ uniform points are generated, where J is the maximum number of alternatives. For favor(space), the column dimension of the matrices of points varies with the number of alternatives that each case has.

⎡ Maximization ⎤

maximize_options: <u>diff</u>icult, <u>tech</u>nique(*algorithm_spec*), <u>iter</u>ate(#), [<u>no</u>]<u>log</u>, <u>trac</u>e, gradient, showstep, <u>hess</u>ian, <u>showtol</u>erance, <u>tol</u>erance(#), <u>ltol</u>erance(#), <u>nrtol</u>erance(#), nonrtolerance, and from(*init_specs*); see [R] **maximize**.

The following options may be particularly useful in obtaining convergence with asmprobit: difficult, technique(*algorithm_spec*), nrtolerance(#), nonrtolerance, and from(*init_specs*).

If technique() contains more than one algorithm specification, bhhh cannot be one of them. To use the BHHH algorithm with another algorithm, use the initbhhh() option and specify the other algorithm in technique().

Setting the optimization type to technique(bhhh) resets the default *vcetype* to vce(opg).

The following option is available with asmprobit but is not shown in the dialog box:

coeflegend; see [R] **estimation options**.

Remarks

Remarks are presented under the following headings:

> *Introduction*
> *Variance structures*

Introduction

The MNP model is used with discrete dependent variables that take on more than two outcomes that do not have a natural ordering. The stochastic error terms are assumed to have a multivariate normal distribution that is heteroskedastic and correlated. Say that you have a set of J unordered alternatives that are modeled by a regression of both case-specific and alternative-specific covariates. A "case" refers to the information on one decision maker. Underlying the model is the set of J latent variables (utilities),

$$\eta_{ij} = \mathbf{x}_{ij}\boldsymbol{\beta} + \mathbf{z}_i\boldsymbol{\alpha}_j + \xi_{ij} \tag{1}$$

where i denotes cases and j denotes alternatives. $\mathbf{x}_{ij}$ is a $1 \times p$ vector of alternative-specific variables, $\boldsymbol{\beta}$ is a $p \times 1$ vector of parameters, $\mathbf{z}_i$ is a $1 \times q$ vector of case-specific variables, $\boldsymbol{\alpha}_j$ is a $q \times 1$ vector of parameters for the jth alternative, and $\xi_i = (\xi_{i1}, \ldots, \xi_{iJ})$ is distributed multivariate normal with mean zero and covariance matrix $\boldsymbol{\Omega}$. The decision maker selects the alternative whose latent variable is highest.

Because the MNP model allows for a general covariance structure in ξ_{ij}, it does not impose the IIA property inherent in multinomial logistic and conditional logistic models. That is, the MNP model permits the odds of choosing one alternative over another to depend on the remaining alternatives. For example, consider the choice of travel mode between two cities: air, train, bus, or car, as a function of the travel mode cost, travel time (alternative-specific variables), and an individual's income (a case-specific variable). The odds of choosing air travel over a bus may not be independent of the train alternative because both bus and train travel are public ground transportation. That is, the probability of choosing air travel is $\Pr(\eta_{\text{air}} > \eta_{\text{bus}}, \eta_{\text{air}} > \eta_{\text{train}}, \eta_{\text{air}} > \eta_{\text{car}})$, and the two events $\eta_{\text{air}} > \eta_{\text{bus}}$ and $\eta_{\text{air}} > \eta_{\text{train}}$ may be correlated.

An alternative to MNP that will allow a nested correlation structure in ξ_{ij} is the nested logit model (see [R] **nlogit**).

The added flexibility of the MNP model does impose a significant computation burden because of the need to evaluate probabilities from the multivariate normal distribution. These probabilities are evaluated using simulation techniques because a closed-form solution does not exist. See *Methods and formulas* for more information.

Not all the J sets of regression coefficients $\boldsymbol{\alpha}_j$ are identifiable, nor are all $J(J+1)/2$ elements of the variance–covariance matrix $\boldsymbol{\Omega}$. As described by Train (2009, sec. 2.5), the model requires normalization because both the location (level) and scale of the latent variable are irrelevant. Increasing the latent variables by a constant does not change which η_{ij} is the maximum for decision maker i, nor does multiplying them by a constant. To normalize location, we choose an alternative, indexed by k, say, and take the difference between the latent variable k and the $J-1$ others,

$$
\begin{aligned}
v_{ijk} &= \eta_{ij} - \eta_{ik} \\
&= (\mathbf{x}_{ij} - \mathbf{x}_{ik})\boldsymbol{\beta} + \mathbf{z}_i(\boldsymbol{\alpha}_j - \boldsymbol{\alpha}_k) + \xi_{ij} - \xi_{ik} \\
&= \boldsymbol{\delta}_{ij'}\boldsymbol{\beta} + \mathbf{z}_i\boldsymbol{\gamma}_{j'} + \epsilon_{ij'} \\
&= \lambda_{ij'} + \epsilon_{ij'}
\end{aligned}
\tag{2}
$$

where $j' = j$ if $j < k$ and $j' = j - 1$ if $j > k$, so that $j' = 1, \ldots, J - 1$. One can now work with the $(J-1) \times (J-1)$ covariance matrix $\boldsymbol{\Sigma}_{(k)}$ for $\boldsymbol{\epsilon}'_i = (\epsilon_{i1}, \ldots, \epsilon_{i,J-1})$. The kth alternative here is the `basealternative()` in asmprobit. From (2), the probability that decision maker i chooses alternative k, for example, is

$$
\begin{aligned}
\Pr(i \text{ chooses } k) &= \Pr(v_{i1k} \leq 0, \ldots, v_{i,J-1,k} \leq 0) \\
&= \Pr(\epsilon_{i1} \leq -\lambda_{i1}, \ldots, \epsilon_{i,J-1} \leq -\lambda_{i,J-1})
\end{aligned}
$$

To normalize for scale, one of the diagonal elements of $\boldsymbol{\Sigma}_{(k)}$ must be fixed to a constant. In asmprobit, this is the error variance for the alternative specified by `scalealternative()`. Thus there are a total of, at most, $J(J-1)/2 - 1$ identifiable variance–covariance parameters. See *Variance structures* below for more on this issue.

In fact, the model is slightly more general in that not all cases need to have faced all J alternatives. The model allows for situations in which some cases chose among all possible alternatives, whereas other cases were given a choice among a subset of them, and perhaps other cases were given a choice among a different subset. The number of observations for each case is equal to the number of alternatives faced.

The MNP model is often motivated using a random-utility consumer-choice framework. Equation (1) represents the utility that consumer i receives from good j. The consumer purchases the good for which the utility is highest. Because utility is ordinal, all that matters is the ranking of the utilities from the alternatives. Thus one must normalize for location and scale.

▷ Example 1

Application of MNP models is common in the analysis of transportation data. Greene (2012, sec. 18.2.9) uses travel-mode choice data between Sydney and Melbourne to demonstrate estimating parameters of various discrete-choice models. The data contain information on 210 individuals' choices of travel mode. The four alternatives are air, train, bus, and car, with indices 1, 2, 3, and 4, respectively. One alternative-specific variable is `travelcost`, a measure of generalized cost of travel that is equal to the sum of in-vehicle cost and a wagelike measure times the amount of time spent traveling. A second alternative-specific variable is the terminal time, `termtime`, which is zero for car transportation. Household income, `income`, is a case-specific variable.

```
. use http://www.stata-press.com/data/r12/travel
. list id mode choice travelcost termtime income in 1/12, sepby(id)
```

	id	mode	choice	travel~t	termtime	income
1.	1	air	0	70	69	35
2.	1	train	0	71	34	35
3.	1	bus	0	70	35	35
4.	1	car	1	30	0	35
5.	2	air	0	68	64	30
6.	2	train	0	84	44	30
7.	2	bus	0	85	53	30
8.	2	car	1	50	0	30
9.	3	air	0	129	69	40
10.	3	train	0	195	34	40
11.	3	bus	0	149	35	40
12.	3	car	1	101	0	40

The model of travel choice is

$$\eta_{ij} = \beta_1 \text{travelcost}_{ij} + \beta_2 \text{termtime}_{ij} + \alpha_{1j} \text{income}_i + \alpha_{0j} + \xi_{ij}$$

The alternatives can be grouped as air and ground travel. With this in mind, we set the `air` alternative to be the `basealternative()` and choose `train` as the scaling alternative. Because these are the first and second alternatives in the `mode` variable, they are also the defaults.

```
. asmprobit choice travelcost termtime, case(id) alternatives(mode)
> casevars(income)
 (output omitted )
```

Alternative-specific multinomial probit
Case variable: id

Alternative variable: mode

Integration sequence: Hammersley
Integration points: 200
Log simulated-likelihood = -190.09418

	Number of obs	=	840
Number of cases	=	210
Alts per case: min =		4
avg =		4.0
max =		4
Wald chi2(5)	=	32.05
Prob > chi2	=	0.0000

choice	Coef.	Std. Err.	z	P>|z|	[95% Conf. Interval]	
mode						
travelcost	-.00977	.0027834	-3.51	0.000	-.0152253	-.0043146
termtime	-.0377095	.0094088	-4.01	0.000	-.0561504	-.0192686
air	(base alternative)					
train						
income	-.0291971	.0089246	-3.27	0.001	-.046689	-.0117052
_cons	.5616376	.3946551	1.42	0.155	-.2118721	1.335147
bus						
income	-.0127503	.0079267	-1.61	0.108	-.0282863	.0027857
_cons	-.0571364	.4791861	-0.12	0.905	-.9963239	.882051
car						
income	-.0049086	.0077486	-0.63	0.526	-.0200957	.0102784
_cons	-1.833393	.8186156	-2.24	0.025	-3.43785	-.2289357
/lnl2_2	-.5502039	.3905204	-1.41	0.159	-1.31561	.2152021
/lnl3_3	-.6005552	.3353292	-1.79	0.073	-1.257788	.0566779
/l2_1	1.131518	.2124817	5.33	0.000	.7150612	1.547974
/l3_1	.9720669	.2352116	4.13	0.000	.5110606	1.433073
/l3_2	.5197214	.2861552	1.82	0.069	-.0411325	1.080575

```
(mode=air is the alternative normalizing location)
(mode=train is the alternative normalizing scale)
. estimates store full
```

By default, the differenced covariance parameterization is used, so the covariance matrix for this model is 3×3. There are two free variances to estimate and three correlations. To help ensure that the covariance matrix remains positive definite, asmprobit uses the square root transformation, where it optimizes on the Cholesky-factored variance–covariance. To ensure that the diagonal elements of the Cholesky estimates remain positive, we use the log transformation. The estimates labeled /lnl2_2 and /lnl3_3 in the coefficient table are the log-transformed diagonal elements of the Cholesky matrix. The estimates labeled /l2_1, /l3_1, and /l3_2 are the off-diagonal entries for elements $(2, 1)$, $(3, 1)$, and $(3, 2)$ of the Cholesky matrix.

Although the transformed parameters of the differenced covariance parameterization are difficult to interpret, you can view them untransformed by using the estat command. Typing

```
. estat correlation
```

	train	bus	car
train	1.0000		
bus	0.8909	1.0000	
car	0.7895	0.8951	1.0000

Note: correlations are for alternatives differenced with air

gives the correlations, and typing

```
. estat covariance
```

	train	bus	car
train	2		
bus	1.600208	1.613068	
car	1.37471	1.399703	1.515884

Note: covariances are for alternatives differenced with air

gives the (co)variances.

We can reduce the number of covariance parameters in the model by using the factor model by Cameron and Trivedi (2005). For large models with many alternatives, the parameter reduction can be dramatic, but for our example we will use factor(1), a one-dimension factor model, to reduce by 3 the number of parameters associated with the covariance matrix.

```
. asmprobit choice travelcost termtime, case(id) alternatives(mode)
> casevars(income) factor(1)
 (output omitted )
```

```
Alternative-specific multinomial probit      Number of obs    =        840
Case variable: id                            Number of cases  =        210

Alternative variable: mode                   Alts per case: min =          4
                                                            avg =        4.0
                                                            max =          4
Integration sequence:       Hammersley
Integration points:                200          Wald chi2(5)    =     107.85
Log simulated-likelihood = -196.85094           Prob > chi2     =     0.0000
```

choice	Coef.	Std. Err.	z	P>|z|	[95% Conf. Interval]	
mode						
travelcost	-.0093696	.0036329	-2.58	0.010	-.01649	-.0022492
termtime	-.0593173	.0064585	-9.18	0.000	-.0719757	-.0466589
air	(base alternative)					
train						
income	-.0373511	.0098219	-3.80	0.000	-.0566018	-.0181004
_cons	.1092322	.3949529	0.28	0.782	-.6648613	.8833257
bus						
income	-.0158793	.0112239	-1.41	0.157	-.0378777	.0061191
_cons	-1.082181	.4678732	-2.31	0.021	-1.999196	-.1651666
car						
income	.0042677	.0092601	0.46	0.645	-.0138817	.0224171
_cons	-3.765445	.5540636	-6.80	0.000	-4.851389	-2.6795
/c1_2	1.182805	.3060299	3.86	0.000	.5829972	1.782612
/c1_3	1.227705	.3401237	3.61	0.000	.5610747	1.894335

```
(mode=air is the alternative normalizing location)
(mode=train is the alternative normalizing scale)
```

The estimates labeled /c1_2 and /c1_3 in the coefficient table are the factor loadings. These factor loadings produce the following differenced covariance estimates:

```
. estat covariance
```

	train	bus	car
train	2		
bus	1.182805	2.399027	
car	1.227705	1.452135	2.507259

```
Note: covariances are for alternatives differenced with air
```

◁

Variance structures

The matrix Ω has $J(J+1)/2$ distinct elements because it is symmetric. Selecting a base alternative, normalizing its error variance to one, and constraining the correlations between its error and the other errors reduces the number of estimable parameters by J. Moreover, selecting a scale alternative and normalizing its error variance to one reduces the number by one, as well. Hence, there are at most $m = J(J-1)/2 - 1$ estimable parameters in Ω.

In practice, estimating all m parameters can be difficult, so one must often place more restrictions on the parameters. The `asmprobit` command provides the `correlation()` option to specify restrictions on the $J(J-3)/2+1$ correlation parameters not already restricted as a result of choosing the base alternatives, and it provides `stddev()` to specify restrictions on the $J-2$ standard deviations not already restricted as a result of choosing the base and scale alternatives.

When the `structural` option is used, `asmprobit` fits the model by assuming that all m parameters can be estimated, which is equivalent to specifying `correlation(unstructured)` and `stddev(heteroskedastic)`. The unstructured correlation structure means that all $J(J-3)/2+1$ of the remaining correlation parameters will be estimated, and the heteroskedastic specification means that all $J-2$ standard deviations will be estimated. With these default settings, the log likelihood is maximized with respect to the Cholesky decomposition of Ω, and then the parameters are transformed to the standard deviation and correlation form.

The `correlation(exchangeable)` option forces the $J(J-3)/2+1$ correlation parameters to be equal, and `correlation(independent)` forces all the correlations to be zero. Using the `stddev(homoskedastic)` option forces all J standard deviations to be one. These options may help in obtaining convergence for a model if the default options do not produce satisfactory results. In fact, when fitting a complex model, it may be advantageous to first fit a simple one and then proceed with removing the restrictions one at a time.

Advanced users may wish to specify alternative variance structures of their own choosing, and the next few paragraphs explain how to do so.

`correlation(pattern` *matname*`)` allows you to give the name of a $J \times J$ matrix that identifies a correlation structure. Sequential positive integers starting at 1 are used to identify each correlation parameter: if there are three correlation parameters, they are identified by 1, 2, and 3. The integers can be repeated to indicate that correlations with the same number should be constrained to be equal. A zero or a missing value (.) indicates that the correlation is to be set to zero. `asmprobit` considers only the elements of the matrix *below* the main diagonal.

Suppose that you have a model with four alternatives, numbered 1–4, and alternative 1 is the base. The unstructured and exchangeable correlation structures identified in the 4×4 lower triangular matrices are

$$
\begin{array}{c}
\text{unstructured} \\
\begin{array}{cccc} 1 & 2 & 3 & 4 \end{array} \\
\begin{array}{c} 1 \\ 2 \\ 3 \\ 4 \end{array}
\left(\begin{array}{cccc}
\cdot & & & \\
0 & \cdot & & \\
0 & 1 & \cdot & \\
0 & 2 & 3 & \cdot
\end{array}\right)
\end{array}
\qquad
\begin{array}{c}
\text{exchangeable} \\
\begin{array}{cccc} 1 & 2 & 3 & 4 \end{array} \\
\begin{array}{c} 1 \\ 2 \\ 3 \\ 4 \end{array}
\left(\begin{array}{cccc}
\cdot & & & \\
0 & \cdot & & \\
0 & 1 & \cdot & \\
0 & 1 & 1 & \cdot
\end{array}\right)
\end{array}
$$

`asmprobit` labels these correlation structures unstructured and exchangeable, even though the correlations corresponding to the base alternative are set to zero. More formally: these terms are appropriate when considering the $(J-1) \times (J-1)$ submatrix $\Sigma_{(k)}$ defined in the *Introduction* above.

You can also use the `correlation(fixed` *matname*`)` option to specify a matrix that specifies fixed and free parameters. Here the free parameters (those that are to be estimated) are identified by a missing value, and nonmissing values represent correlations that are to be taken as given. Below is a correlation structure that would set the correlations of alternative 1 to be 0.5:

$$
\begin{array}{c}
\begin{array}{cccc} 1 & 2 & 3 & 4 \end{array} \\
\begin{array}{c} 1 \\ 2 \\ 3 \\ 4 \end{array}
\left(\begin{array}{cccc}
\cdot & & & \\
0.5 & \cdot & & \\
0.5 & \cdot & \cdot & \\
0.5 & \cdot & \cdot & \cdot
\end{array}\right)
\end{array}
$$

The order of the elements of the `pattern` or `fixed` matrices must be the same as the numeric order of the alternative levels.

To specify the structure of the standard deviations—the diagonal elements of Ω—you can use the `stddev(pattern` *matname*`)` option, where *matname* is a $1 \times J$ matrix. Sequential positive integers starting at 1 are used to identify each standard deviation parameter. The integers can be repeated to indicate that standard deviations with the same number are to be constrained to be equal. A missing value indicates that the corresponding standard deviation is to be set to one. In the four-alternative example mentioned above, suppose that you wish to set the first and second standard deviations to one and that you wish to constrain the third and fourth standard deviations to be equal; the following `pattern` matrix will do that:

$$\begin{array}{cccc} 1 & 2 & 3 & 4 \end{array}$$
$$1 \; (\; \cdot \quad \cdot \quad 1 \quad 1 \;)$$

Using the `stddev(fixed` *matname*`)` option allows you to identify the fixed and free standard deviations. Fixed standard deviations are entered as positive real numbers, and free parameters are identified with missing values. For example, to constrain the first and second standard deviations to equal one and to allow the third and fourth to be estimated, you would use this `fixed` matrix:

$$\begin{array}{cccc} 1 & 2 & 3 & 4 \end{array}$$
$$1 \; (\; 1 \quad 1 \quad \cdot \quad \cdot \;)$$

When supplying either the `pattern` or the `fixed` matrices, you must ensure that the model is properly scaled. At least two standard deviations must be constant for the model to be scaled. A warning is issued if `asmprobit` detects that the model is not scaled.

The order of the elements of the `pattern` or `fixed` matrices must be the same as the numeric order of the alternative levels.

▷ Example 2

In example 1, we used the differenced covariance parameterization, the default. We now use the `structural` option to view the $J - 2$ standard deviation estimates and the $(J - 1)(J - 2)/2$ correlation estimates. Here we will fix the standard deviations for the `air` and `train` alternatives to 1 and the correlations between `air` and the rest of the alternatives to 0.

```
. asmprobit choice travelcost termtime, case(id) alternatives(mode)
> casevars(income) structural
```

  (*output omitted*)

```
Alternative-specific multinomial probit      Number of obs    =       840
Case variable: id                            Number of cases  =       210

Alternative variable: mode                   Alts per case: min =        4
                                                            avg =      4.0
                                                            max =        4

Integration sequence:       Hammersley
Integration points:               200        Wald chi2(5)     =     32.05
Log simulated-likelihood = -190.09418        Prob > chi2      =    0.0000
```

choice	Coef.	Std. Err.	z	P>|z|	[95% Conf. Interval]	
mode						
travelcost	-.0097703	.0027834	-3.51	0.000	-.0152257	-.0043149
termtime	-.0377103	.0094092	-4.01	0.000	-.056152	-.0192687
air	(base alternative)					
train						
income	-.0291975	.0089246	-3.27	0.001	-.0466895	-.0117055
_cons	.5616448	.3946529	1.42	0.155	-.2118607	1.33515
bus						
income	-.01275	.0079266	-1.61	0.108	-.0282858	.0027858
_cons	-.0571664	.4791996	-0.12	0.905	-.9963803	.8820476
car						
income	-.0049085	.0077486	-0.63	0.526	-.0200955	.0102785
_cons	-1.833444	.8186343	-2.24	0.025	-3.437938	-.22895
/lnsigma3	-.2447428	.4953363	-0.49	0.621	-1.215584	.7260985
/lnsigma4	-.3309429	.6494493	-0.51	0.610	-1.60384	.9419543
/atanhr3_2	1.01193	.3890994	2.60	0.009	.249309	1.774551
/atanhr4_2	.5786576	.3940461	1.47	0.142	-.1936586	1.350974
/atanhr4_3	.8885204	.5600561	1.59	0.113	-.2091693	1.98621
sigma1	1	(base alternative)				
sigma2	1	(scale alternative)				
sigma3	.7829059	.3878017			.2965368	2.067
sigma4	.7182462	.4664645			.2011227	2.564989
rho3_2	.766559	.1604596			.244269	.9441061
rho4_2	.5216891	.2868027			-.1912734	.874283
rho4_3	.7106622	.277205			-.2061713	.9630403

```
(mode=air is the alternative normalizing location)
(mode=train is the alternative normalizing scale)
```

When comparing this output to that of example 1, we see that we have achieved the same log likelihood. That is, the structural parameterization using air as the base alternative and train as the scale alternative applied no restrictions on the model. This will not always be the case. We leave it up to you to try different base and scale alternatives, and you will see that not all the different combinations will achieve the same log likelihood. This is not true for the differenced covariance parameterization: it will always achieve the same log likelihood (and the maximum possible likelihood) regardless of the base and scale alternatives. This is why it is the default parameterization.

For an exercise, we can compute the differenced covariance displayed in example 1 by using the following ado-code.

```
. estat covariance
```

	air	train	bus	car
air	1			
train	0	1		
bus	0	.6001436	.6129416	
car	0	.3747012	.399619	.5158776

```
. return list
matrices:
            r(cov) :  4 x 4
. matrix cov = r(cov)
. matrix M = (1,-1,0,0 \ 1,0,-1,0 \ 1,0,0,-1)
. matrix cov1 = M*cov*M'
. matrix list cov1
symmetric cov1[3,3]
           r1          r2          r3
r1          2
r2   1.6001436   1.6129416
r3   1.3747012    1.399619   1.5158776
```

The slight difference in the regression coefficients between the example 1 and example 2 coefficient tables reflects the accuracy of the [M-5] **ghk( )** algorithm using 200 points from the Hammersley sequence.

We now fit the model using the exchangeable correlation matrix and compare the models with a likelihood-ratio test.

```
. asmprobit choice travelcost termtime, case(id) alternatives(mode)
> casevars(income) correlation(exchangeable)
```

(*output omitted*)

Alternative-specific multinomial probit	Number of obs =	840
Case variable: id	Number of cases =	210
Alternative variable: mode	Alts per case: min =	4
	avg =	4.0
	max =	4
Integration sequence: Hammersley		
Integration points: 200	Wald chi2(5) =	53.60
Log simulated-likelihood = -190.4679	Prob > chi2 =	0.0000

choice	Coef.	Std. Err.	z	P>|z|	[95% Conf. Interval]	
mode						
travelcost	-.0084636	.0020452	-4.14	0.000	-.012472	-.0044551
termtime	-.0345394	.0072812	-4.74	0.000	-.0488103	-.0202684
air	(base alternative)					
train						
income	-.0290357	.0083226	-3.49	0.000	-.0453477	-.0127237
_cons	.5517445	.3719913	1.48	0.138	-.177345	1.280834
bus						
income	-.0132562	.0074133	-1.79	0.074	-.0277859	.0012735
_cons	-.0052517	.4337932	-0.01	0.990	-.8554708	.8449673
car						
income	-.0060878	.006638	-0.92	0.359	-.0190981	.0069224
_cons	-1.565918	.6633007	-2.36	0.018	-2.865964	-.265873
/lnsigmaP1	-.3557589	.1972809	-1.80	0.071	-.7424222	.0309045
/lnsigmaP2	-1.308596	.8872957	-1.47	0.140	-3.047663	.4304719
/atanhrP1	1.116589	.3765488	2.97	0.003	.3785667	1.854611
sigma1	1	(base alternative)				
sigma2	1	(scale alternative)				
sigma3	.7006416	.1382232			.4759596	1.031387
sigma4	.2701992	.2397466			.0474697	1.537983
rho3_2	.8063791	.131699			.3614621	.9521783
rho4_2	.8063791	.131699			.3614621	.9521783
rho4_3	.8063791	.131699			.3614621	.9521783

```
(mode=air is the alternative normalizing location)
(mode=train is the alternative normalizing scale)
```

```
. lrtest full .
```

Likelihood-ratio test	LR chi2(2) =	0.75
(Assumption: . nested in full)	Prob > chi2 =	0.6882

The likelihood-ratio test suggests that a common correlation is a plausible hypothesis, but this could be an artifact of the small sample size. The labeling of the standard deviation and correlation estimates has changed from /lnsigma and /atanhr, in the previous example, to /lnsigmaP and /atanhrP. The "P" identifies the parameter's index in the pattern matrices used by asmprobit. The pattern matrices are saved in e(stdpattern) and e(corpattern).   ◁

❑ Technical note

Another way to fit the model with the exchangeable correlation structure in example 2 is to use the `constraint` command to define the constraints on the `rho` parameters manually and then apply those.

```
. constraint 1 [atanhr3_2]_cons = [atanhr4_2]_cons
. constraint 2 [atanhr3_2]_cons = [atanhr4_3]_cons
. asmprobit choice travelcost termtime, case(id) alternatives(mode)
> casevars(income) constraints(1 2) structural
```

With this method, however, we must keep track of what parameterization of the rhos is used in estimation, and that depends on the options specified.

❑

▷ Example 3

In the last example, we used the `correlation(exchangeable)` option, reducing the number of correlation parameters from three to one. We can explore a two–correlation parameter model by specifying a `pattern` matrix in the `correlation()` option. Suppose that we wish to have the correlation between train and bus be equal to the correlation between bus and car and to have the standard deviations for the bus and car equations be equal. We will use air as the base category and train as the scale category.

```
. matrix define corpat = J(4, 4, .)
. matrix corpat[3,2] = 1
. matrix corpat[4,3] = 1
. matrix corpat[4,2] = 2
. matrix define stdpat = J(1, 4, .)
. matrix stdpat[1,3] = 1
. matrix stdpat[1,4] = 1
. asmprobit choice travelcost termtime, case(id) alternatives(mode)
> casevars(income) correlation(pattern corpat) stddev(pattern stdpat)
Iteration 0:   log simulated-likelihood = -201.33896
Iteration 1:   log simulated-likelihood = -201.00457  (backed up)
Iteration 2:   log simulated-likelihood = -200.80208  (backed up)
Iteration 3:   log simulated-likelihood = -200.79758  (backed up)
Iteration 4:   log simulated-likelihood = -200.55655  (backed up)
Iteration 5:   log simulated-likelihood =  -200.5421  (backed up)
Iteration 6:   log simulated-likelihood = -196.24925
  (output omitted )
Iteration 20:  log simulated-likelihood = -190.12874
Iteration 21:  log simulated-likelihood = -190.12871
Iteration 22:  log simulated-likelihood = -190.12871
```

```
Alternative-specific multinomial probit      Number of obs     =        840
Case variable: id                            Number of cases   =        210

Alternative variable: mode                   Alts per case: min =          4
                                                            avg =        4.0
                                                            max =          4

Integration sequence:      Hammersley
Integration points:              200         Wald chi2(5)      =      41.67
Log simulated-likelihood = -190.12871        Prob > chi2       =     0.0000
```

choice	Coef.	Std. Err.	z	P>|z|	[95% Conf. Interval]	
mode						
travelcost	-.0100335	.0026203	-3.83	0.000	-.0151692	-.0048979
termtime	-.0385731	.008608	-4.48	0.000	-.0554445	-.0217018
air	(base alternative)					
train						
income	-.029271	.0089739	-3.26	0.001	-.0468595	-.0116824
_cons	.56528	.4008037	1.41	0.158	-.2202809	1.350841
bus						
income	-.0124658	.0080043	-1.56	0.119	-.0281539	.0032223
_cons	-.0741685	.4763422	-0.16	0.876	-1.007782	.859445
car						
income	-.0046905	.0079934	-0.59	0.557	-.0203573	.0109763
_cons	-1.897931	.7912106	-2.40	0.016	-3.448675	-.3471867
/lnsigmaP1	-.197697	.2751269	-0.72	0.472	-.7369359	.3415418
/atanhrP1	.9704403	.3286981	2.95	0.003	.3262038	1.614677
/atanhrP2	.5830923	.3690419	1.58	0.114	-.1402165	1.306401
sigma1	1	(base alternative)				
sigma2	1	(scale alternative)				
sigma3	.8206185	.2257742			.4785781	1.407115
sigma4	.8206185	.2257742			.4785781	1.407115
rho3_2	.7488977	.1443485			.3151056	.9238482
rho4_2	.5249094	.2673598			-.1393048	.863362
rho4_3	.7488977	.1443485			.3151056	.9238482

(mode=air is the alternative normalizing location)
(mode=train is the alternative normalizing scale)

In the call to asmprobit, we did not need to specify the basealternative() and scalealternative() options because they are implied by the specifications of the pattern matrices.

◁

❑ Technical note

If you experience convergence problems, try specifying nopivot, increasing intpoints(), specifying antithetics, specifying technique(nr) with difficult, or specifying a switching algorithm in the technique() option. As a last resort, you can use the nrtolerance() and showtolerance options. Changing the base and scale alternative in the model specification can also affect convergence if the structural option is used.

Because simulation methods are used to obtain multivariate normal probabilities, the estimates obtained have a limited degree of precision. Moreover, the solutions are particularly sensitive to the starting values used. Experimenting with different starting values may help in obtaining convergence, and doing so is a good way to verify previous results.

If you wish to use the BHHH algorithm along with another maximization algorithm, you must specify the initbhhh(#) option, where # is the number of BHHH iterations to use before switching to the algorithm specified in technique(). The BHHH algorithm uses an outer-product-of-gradients approximation for the Hessian, and `asmprobit` must perform the gradient calculations differently than for the other algorithms.

❑

❑ Technical note

If there are no alternative-specific variables in your model, the variance–covariance matrix parameters are not identifiable. For such a model to converge, you would therefore need to use `correlation(independent)` and `stddev(homoskedastic)`. A better alternative is to use `mprobit`, which is geared specifically toward models with only case-specific variables. See [R] **mprobit**.

❑

Saved results

`asmprobit` saves the following in `e()`:

Scalars

`e(N)`	number of observations
`e(N_case)`	number of cases
`e(k)`	number of parameters
`e(k_alt)`	number of alternatives
`e(k_indvars)`	number of alternative-specific variables
`e(k_casevars)`	number of case-specific variables
`e(k_sigma)`	number of variance estimates
`e(k_rho)`	number of correlation estimates
`e(k_eq)`	number of equations in `e(b)`
`e(k_eq_model)`	number of equations in overall model test
`e(df_m)`	model degrees of freedom
`e(ll)`	log simulated-likelihood
`e(N_clust)`	number of clusters
`e(const)`	constant indicator
`e(i_base)`	base alternative index
`e(i_scale)`	scale alternative index
`e(mc_points)`	number of Monte Carlo replications
`e(mc_burn)`	starting sequence index
`e(mc_antithetics)`	antithetics indicator
`e(chi2)`	χ^2
`e(p)`	significance
`e(fullcov)`	unstructured covariance indicator
`e(structcov)`	1 if structured covariance; 0 otherwise
`e(cholesky)`	Cholesky-factored covariance indicator
`e(alt_min)`	minimum number of alternatives
`e(alt_avg)`	average number of alternatives
`e(alt_max)`	maximum number of alternatives
`e(rank)`	rank of `e(V)`
`e(ic)`	number of iterations
`e(rc)`	return code
`e(converged)`	1 if converged, 0 otherwise

Macros
e(cmd)	asmprobit
e(cmdline)	command as typed
e(depvar)	name of dependent variable
e(indvars)	alternative-specific independent variable
e(casevars)	case-specific variables
e(case)	variable defining cases
e(altvar)	variable defining alternatives
e(alteqs)	alternative equation names
e(alt#)	alternative labels
e(wtype)	weight type
e(wexp)	weight expression
e(title)	title in estimation output
e(clustvar)	name of cluster variable
e(correlation)	correlation structure
e(stddev)	variance structure
e(cov_class)	class of the covariance structure
e(chi2type)	Wald, type of model χ^2 test
e(vce)	*vcetype* specified in vce()
e(vcetype)	title used to label Std. Err.
e(opt)	type of optimization
e(which)	max or min; whether optimizer is to perform maximization or minimization
e(ml_method)	type of ml method
e(mc_method)	technique used to generate sequences
e(mc_seed)	random-number generator seed
e(user)	name of likelihood-evaluator program
e(technique)	maximization technique
e(datasignature)	the checksum
e(datasignaturevars)	variables used in calculation of checksum
e(properties)	b V
e(estat_cmd)	program used to implement estat
e(mfx_dlg)	program used to implement estat mfx dialog
e(predict)	program used to implement predict
e(marginsnotok)	predictions disallowed by margins

Matrices
e(b)	coefficient vector
e(Cns)	constraints matrix
e(stats)	alternative statistics
e(stdpattern)	variance pattern
e(stdfixed)	fixed and free standard deviations
e(altvals)	alternative values
e(altfreq)	alternative frequencies
e(alt_casevars)	indicators for estimated case-specific coefficients—e(k_alt)×e(k_casevars)
e(corpattern)	correlation structure
e(corfixed)	fixed and free correlations
e(ilog)	iteration log (up to 20 iterations)
e(gradient)	gradient vector
e(V)	variance–covariance matrix of the estimators
e(V_modelbased)	model-based variance

Functions
e(sample)	marks estimation sample

Methods and formulas

asmprobit is implemented as an ado-file.

The simulated maximum likelihood estimates for the MNP are obtained using ml; see [R] **ml**. The likelihood evaluator implements the GHK algorithm to approximate the multivariate distribution function (Geweke 1989; Hajivassiliou and McFadden 1998; Keane and Wolpin 1994). The technique is also described in detail by Genz (1992), but Genz describes a more general algorithm where both lower and upper bounds of integration are finite. We briefly describe the GHK simulator and refer you to Bolduc (1999) for the score computations.

As discussed earlier, the latent variables for a J-alternative model are $\eta_{ij} = \mathbf{x}_{ij}\boldsymbol{\beta} + \mathbf{z}_i\boldsymbol{\alpha}_j + \xi_{ij}$, for $j = 1, \ldots, J$, $i = 1, \ldots, n$, and $\boldsymbol{\xi}'_i = (\xi_{i,1}, \ldots, \xi_{i,J}) \sim \text{MVN}(\mathbf{0}, \boldsymbol{\Omega})$. The experimenter observes alternative k for the ith observation if $k = \arg \max(\eta_{ij}, j = 1, \ldots, J)$. Let

$$
\begin{aligned}
v_{ij'} &= \eta_{ij} - \eta_{ik} \\
&= (\mathbf{x}_{ij} - \mathbf{x}_{ik})\boldsymbol{\beta} + \mathbf{z}_i(\boldsymbol{\alpha}_j - \boldsymbol{\alpha}_k) + \xi_{ij} - \xi_{ik} \\
&= \boldsymbol{\delta}_{ij'}\boldsymbol{\beta} + \mathbf{z}_i\boldsymbol{\gamma}_{j'} + \epsilon_{ij'}
\end{aligned}
$$

where $j' = j$ if $j < k$ and $j' = j - 1$ if $j > k$, so that $j' = 1, \ldots, J - 1$. Further, $\epsilon_i = (\epsilon_{i1}, \ldots, \epsilon_{i,J-1}) \sim \text{MVN}(\mathbf{0}, \boldsymbol{\Sigma}_{(k)})$. $\boldsymbol{\Sigma}$ is indexed by k because it depends on the choice made. We denote the deterministic part of the model as $\lambda_{ij'} = \boldsymbol{\delta}_{ij'}\boldsymbol{\beta} + \mathbf{z}_j\boldsymbol{\gamma}_{j'}$, and the probability of this event is

$$
\begin{aligned}
\Pr(y_i = k) &= \Pr(v_{i1} \leq 0, \ldots, v_{i,J-1} \leq 0) \\
&= \Pr(\epsilon_{i1} \leq -\lambda_{i1}, \ldots, \epsilon_{i,J-1} \leq -\lambda_{i,J-1}) \\
&= (2\pi)^{-(J-1)/2} |\boldsymbol{\Sigma}_{(k)}|^{-1/2} \int_{-\infty}^{-\lambda_{i1}} \cdots \int_{-\infty}^{-\lambda_{i,J-1}} \exp\left(-\tfrac{1}{2}\mathbf{z}'\boldsymbol{\Sigma}_{(k)}^{-1}\mathbf{z}\right) d\mathbf{z}
\end{aligned}
\tag{3}
$$

Simulated likelihood

For clarity in the discussion that follows, we drop the index denoting case so that for an arbitrary observation $v' = (v_1, \ldots, v_{J-1})$, $\boldsymbol{\lambda}' = (\lambda_1, \ldots, \lambda_{J-1})$, and $\boldsymbol{\epsilon}' = (\epsilon_1, \ldots, \epsilon_{J-1})$.

The Cholesky-factored variance–covariance, $\boldsymbol{\Sigma} = \mathbf{L}\mathbf{L}'$, is lower triangular,

$$
\mathbf{L} = \begin{pmatrix}
l_{11} & 0 & \cdots & 0 \\
l_{21} & l_{22} & \cdots & 0 \\
\vdots & \vdots & & \vdots \\
l_{J-1,1} & l_{J-1,2} & \cdots & l_{J-1,J-1}
\end{pmatrix}
$$

and the correlated latent-variable errors can be expressed as linear functions of uncorrelated normal variates, $\epsilon = \mathbf{L}\boldsymbol{\zeta}$, where $\boldsymbol{\zeta}' = (\zeta_1, \ldots, \zeta_{J-1})$ and $\zeta_j \sim$ iid $N(0, 1)$. We now have $v = \lambda + \mathbf{L}\boldsymbol{\zeta}$, and by defining

$$
z_j = \begin{cases}
-\dfrac{\lambda_1}{l_{11}} & \text{for } j = 1 \\[2ex]
-\dfrac{\lambda_j + \sum_{i=1}^{j-1} l_{ji}\zeta_i}{l_{jj}} & \text{for } j = 2, \ldots, J - 1
\end{cases}
\tag{4}
$$

we can express the probability statement (3) as the product of conditional probabilities

$$
\begin{aligned}
\Pr(y_i = k) = {} &\Pr(\zeta_1 \leq z_1)\Pr(\zeta_2 \leq z_2 \mid \zeta_1 \leq z_1) \cdots \\
&\Pr(\zeta_{J-1} \leq z_{J-1} \mid \zeta_1 \leq z_1, \ldots, \zeta_{J-2} \leq z_{J-2})
\end{aligned}
$$

because

$$\Pr(v_1 \leq 0) = \Pr(\lambda_1 + l_{11}\zeta_1 \leq 0)$$

$$= \Pr\left(\zeta_1 \leq -\frac{\lambda_1}{l_{11}}\right)$$

$$\Pr(v_2 \leq 0) = \Pr(\lambda_2 + l_{21}\zeta_1 + l_{22}\zeta_2 \leq 0)$$

$$= \Pr\left(\zeta_2 \leq -\frac{\lambda_2 + l_{21}\zeta_1}{l_{22}} \mid \zeta_1 \leq -\frac{\lambda_1}{l_{11}}\right)$$

$$\cdots$$

The Monte Carlo algorithm then must make draws from the truncated standard normal distribution. It does so by generating $J - 1$ uniform variates, $\delta_j, j = 1, \ldots, J - 1$, and computing

$$\widetilde{\zeta}_j = \begin{cases} \Phi^{-1}\left\{\delta_1 \Phi\left(-\frac{\lambda_1}{l_{11}}\right)\right\} & \text{for } j = 1 \\ \Phi^{-1}\left\{\delta_j \Phi\left(\frac{-\lambda_j - \sum_{i=1}^{j-1} l_{ji}\widetilde{\zeta}_i}{l_{jj}}\right)\right\} & \text{for } j = 2, \ldots, J - 1 \end{cases}$$

Define $\widetilde{z}_j$ by replacing $\widetilde{\zeta}_i$ for ζ_i in (4) so that the simulated probability for the lth draw is

$$p_l = \prod_{j=1}^{J-1} \Phi(\widetilde{z}_j)$$

To increase accuracy, the bounds of integration, λ_j, are ordered so that the largest integration intervals are on the inside. The rows and columns of the variance–covariance matrix are pivoted accordingly (Genz 1992).

For a more detailed description of the GHK algorithm in Stata, see Gates (2006).

Repeated draws are made, say, N, and the simulated likelihood for the ith case, denoted $\widehat{L}_i$, is computed as

$$\widehat{L}_i = \frac{1}{N} \sum_{l=1}^{N} p_l$$

The overall simulated log likelihood is $\sum_i \log \widehat{L}_i$.

If the true likelihood is L_i, the error bound on the approximation can be expressed as

$$|\widehat{L}_i - L_i| \leq V(L_i) D_N\{(\delta_i)\}$$

where $V(L_i)$ is the total variation of L_i and D_N is the discrepancy, or nonuniformity, of the set of abscissas. For the uniform pseudorandom sequence, δ_i, the discrepancy is of order $O\{(\log \log N/N)^{1/2}\}$. The order of discrepancy can be improved by using quasirandom sequences.

Quasi–Monte Carlo integration is carried out by asmprobit by replacing the uniform deviates with either the Halton or the Hammersley sequences. These sequences spread the points more evenly than the uniform random sequence and have a smaller order of discrepancy, $O\left[\{(\log N)^{J-1}\}/N\right]$ and $O\left[\{(\log N)^{J-2}\}/N\right]$, respectively. The Halton sequence of dimension $J - 1$ is generated from the first $J - 1$ primes, p_k, so that on draw l we have $\mathbf{h}_l = \{r_{p_1}(l), r_{p_2}(l), \ldots, r_{p_{J-1}}(l)\}$, where

$$r_{p_k}(l) = \sum_{j=0}^{q} b_{jk}(l)p_k^{-j-1} \in (0, 1)$$

is the radical inverse function of l with base p_k so that $\sum_{j=0}^{q} b_{jk}(l)p_k^j = l$, where $p_k^q \le l < p_k^{q+1}$ (Fang and Wang 1994).

This function is demonstrated with base $p_3 = 5$ and $l = 33$, which generates $r_5(33)$. Here $q = 2$, $b_{0,3}(33) = 3$, $b_{1,5}(33) = 1$, and $b_{2,5}(33) = 1$, so that $r_5(33) = 3/5 + 1/25 + 1/625$.

The Hammersley sequence uses an evenly spaced set of points with the first $J - 2$ components of the Halton sequence

$$\mathbf{h}_l = \left\{ \frac{2l - 1}{2N}, r_{p_1}(l), r_{p_2}(l), \ldots, r_{p_{J-2}}(l) \right\}$$

for $l = 1, \ldots, N$.

For a more detailed description of the Halton and Hammersley sequences, see Drukker and Gates (2006).

Computations for the derivatives of the simulated likelihood are taken from Bolduc (1999). Bolduc gives the analytical first-order derivatives for the log of the simulated likelihood with respect to the regression coefficients and the parameters of the Cholesky-factored variance–covariance matrix. asmprobit uses these analytical first-order derivatives and numerical second-order derivatives.

This command supports the clustered version of the Huber/White/sandwich estimator of the variance using vce(robust) and vce(cluster *clustvar*). See [P] _robust, particularly *Maximum likelihood estimators* and *Methods and formulas*. Specifying vce(robust) is equivalent to specifying vce(cluster *casevar*), where *casevar* is the variable that identifies the cases.

References

Bolduc, D. 1999. A practical technique to estimate multinomial probit models in transportation. *Transportation Research Part B* 33: 63–79.

Bunch, D. S. 1991. Estimability of the multinomial probit model. *Transportation Research Part B* 25: 1–12.

Cameron, A. C., and P. K. Trivedi. 2005. *Microeconometrics: Methods and Applications*. New York: Cambridge University Press.

Cappellari, L., and S. P. Jenkins. 2003. Multivariate probit regression using simulated maximum likelihood. *Stata Journal* 3: 278–294.

Drukker, D. M., and R. Gates. 2006. Generating Halton sequences using Mata. *Stata Journal* 6: 214–228.

Fang, K.-T., and Y. Wang. 1994. *Number-theoretic Methods in Statistics*. London: Chapman & Hall.

Gates, R. 2006. A Mata Geweke–Hajivassiliou–Keane multivariate normal simulator. *Stata Journal* 6: 190–213.

Genz, A. 1992. Numerical computation of multivariate normal probabilities. *Journal of Computational and Graphical Statistics* 1: 141–149.

Geweke, J. 1989. Bayesian inference in econometric models using Monte Carlo integration. *Econometrica* 57: 1317–1339.

Geweke, J., and M. P. Keane. 2001. Computationally intensive methods for integration in econometrics. In Vol. 5 of *Handbook of Econometrics*, ed. J. Heckman and E. Leamer, 3463–3568. Amsterdam: North–Holland.

Greene, W. H. 2012. *Econometric Analysis*. 7th ed. Upper Saddle River, NJ: Prentice Hall.

Haan, P., and A. Uhlendorff. 2006. Estimation of multinomial logit models with unobserved heterogeneity using maximum simulated likelihood. *Stata Journal* 6: 229–245.

Hajivassiliou, V. A., and D. L. McFadden. 1998. The method of simulated scores for the estimation of LDV models. *Econometrica* 66: 863–896.

Hole, A. R. 2007. Fitting mixed logit models by using maximum simulated likelihood. *Stata Journal* 7: 388–401.

Keane, M. P., and K. I. Wolpin. 1994. The solution and estimation of discrete choice dynamic programming models by simulation and interpolation: Monte Carlo evidence. *Review of Economics and Statistics* 76: 648–672.

Train, K. E. 2009. *Discrete Choice Methods with Simulation.* 2nd ed. New York: Cambridge University Press.

Also see

[R] **asmprobit postestimation** — Postestimation tools for asmprobit

[R] **asclogit** — Alternative-specific conditional logit (McFadden's choice) model

[R] **asroprobit** — Alternative-specific rank-ordered probit regression

[R] **mlogit** — Multinomial (polytomous) logistic regression

[R] **mprobit** — Multinomial probit regression

[U] **20 Estimation and postestimation commands**

Title

> **asmprobit postestimation** — Postestimation tools for asmprobit

Description

The following postestimation commands are of special interest after `asmprobit`:

Command	Description
estat alternatives	alternative summary statistics
estat covariance	covariance matrix of the latent-variable errors for the alternatives
estat correlation	correlation matrix of the latent-variable errors for the alternatives
estat facweights	covariance factor weights matrix
estat mfx	marginal effects

For information about these commands, see below.

The following standard postestimation commands are also available:

Command	Description
estat	AIC, BIC, VCE, and estimation sample summary
estimates	cataloging estimation results
lincom	point estimates, standard errors, testing, and inference for linear combinations of coefficients
lrtest	likelihood-ratio test
nlcom	point estimates, standard errors, testing, and inference for nonlinear combinations of coefficients
predict	predicted probabilities, estimated linear predictor and its standard error
predictnl	point estimates, standard errors, testing, and inference for generalized predictions
test	Wald tests of simple and composite linear hypotheses
testnl	Wald tests of nonlinear hypotheses

See the corresponding entries in the *Base Reference Manual* for details.

Special-interest postestimation commands

`estat alternatives` displays summary statistics about the alternatives in the estimation sample and provides a mapping between the index numbers that label the covariance parameters of the model and their associated values and labels for the alternative variable.

`estat covariance` computes the estimated variance–covariance matrix of the latent-variable errors for the alternatives. The estimates are displayed, and the variance–covariance matrix is stored in `r(cov)`.

`estat correlation` computes the estimated correlation matrix of the latent-variable errors for the alternatives. The estimates are displayed, and the correlation matrix is stored in `r(cor)`.

estat facweights displays the covariance factor weights matrix and stores it in r(C).

estat mfx computes the simulated probability marginal effects.

Syntax for predict

predict [*type*] *newvar* [*if*] [*in*] [, *statistic* altwise]

predict [*type*] { *stub* | *newvarlist* } [*if*] [*in*], <u>sc</u>ores

statistic	Description
Main	
<u>pr</u>	probability alternative is chosen; the default
xb	linear prediction
stdp	standard error of the linear prediction

These statistics are available both in and out of sample; type predict ... if e(sample) ... if wanted only for the estimation sample.

Menu

Statistics > Postestimation > Predictions, residuals, etc.

Options for predict

⌐ Main ⌐

pr, the default, calculates the probability that alternative j is chosen in case i.

xb calculates the linear prediction $\mathbf{x}_{ij}\boldsymbol{\beta} + \mathbf{z}_i\boldsymbol{\alpha}_j$ for alternative j and case i.

stdp calculates the standard error of the linear predictor.

altwise specifies that alternativewise deletion be used when marking out observations due to missing values in your variables. The default is to use casewise deletion. The xb and stdp options always use alternativewise deletion.

scores calculates the scores for each coefficient in e(b). This option requires a new variable list of length equal to the number of columns in e(b). Otherwise, use the *stub** option to have predict generate enumerated variables with prefix *stub*.

Syntax for estat alternatives

estat <u>alternatives</u>

Menu

Statistics > Postestimation > Reports and statistics

Syntax for estat covariance

estat <u>cov</u>ariance [, <u>form</u>at(%*fmt*) <u>bor</u>der(*bspec*) left(#)]

Menu

Statistics > Postestimation > Reports and statistics

Options for estat covariance

format(%*fmt*) sets the matrix display format. The default is format(%9.0g).

border(*bspec*) sets the matrix display border style. The default is border(all). See [P] **matlist**.

left(#) sets the matrix display left indent. The default is left(2). See [P] **matlist**.

Syntax for estat correlation

estat <u>cor</u>relation [, <u>form</u>at(%*fmt*) <u>bor</u>der(*bspec*) left(#)]

Menu

Statistics > Postestimation > Reports and statistics

Options for estat correlation

format(%*fmt*) sets the matrix display format. The default is format(%9.4f).

border(*bspec*) sets the matrix display border style. The default is border(all). See [P] **matlist**.

left(#) sets the matrix display left indent. The default is left(2). See [P] **matlist**.

Syntax for estat facweights

estat <u>fac</u>weights [, <u>form</u>at(%*fmt*) <u>bor</u>der(*bspec*) left(#)]

Menu

Statistics > Postestimation > Reports and statistics

Options for estat facweights

format(%*fmt*) sets the matrix display format. The default is format(%9.0f).

border(*bspec*) sets the matrix display border style. The default is border(all). See [P] **matlist**.

left(#) sets the matrix display left indent. The default is left(2). See [P] **matlist**.

Syntax for estat mfx

estat mfx $\begin{bmatrix} if \end{bmatrix}$ $\begin{bmatrix} in \end{bmatrix}$ $\begin{bmatrix} , & options \end{bmatrix}$

options	Description
Main	
<u>var</u>list(*varlist*)	display marginal effects for *varlist*
at(mean $\begin{bmatrix} atlist \end{bmatrix}$ | median $\begin{bmatrix} atlist \end{bmatrix}$)	calculate marginal effects at these values
Options	
<u>l</u>evel(#)	set confidence interval level; default is level(95)
<u>nod</u>iscrete	treat indicator variables as continuous
<u>noe</u>sample	do not restrict calculation of means and medians to the estimation sample
<u>now</u>ght	ignore weights when calculating means and medians

Menu

Statistics > Postestimation > Reports and statistics

Options for estat mfx

⌐ Main ⌐

<u>var</u>list(*varlist*) specifies the variables for which to display marginal effects. The default is all variables.

at(mean $\begin{bmatrix} atlist \end{bmatrix}$ \| median $\begin{bmatrix} atlist \end{bmatrix}$) specifies the values at which the marginal effects are to be calculated. *atlist* is

$$\begin{bmatrix} \begin{bmatrix} alternative:variable \ = \ \# \end{bmatrix} \ \begin{bmatrix} variable \ = \ \# \end{bmatrix} \ \begin{bmatrix} \dots \end{bmatrix} \end{bmatrix}$$

The default is to calculate the marginal effects at the means of the independent variables at the estimation sample, at(mean).

After specifying the summary statistic, you can specify a series of specific values for variables. You can specify values for alternative-specific variables by alternative, or you can specify one value for all alternatives. You can specify only one value for case-specific variables. For example, in the travel dataset, income is a case-specific variable, whereas termtime and travelcost are alternative-specific variables. The following would be a legal syntax for estat mfx:

. estat mfx, at(mean air:termtime=50 travelcost=100 income=60)

When nodiscrete is not specified, at(mean $\begin{bmatrix} atlist \end{bmatrix}$) or at(median $\begin{bmatrix} atlist \end{bmatrix}$) has no effect on computing marginal effects for indicator variables, which are calculated as the discrete change in the simulated probability as the indicator variable changes from 0 to 1.

The mean and median computations respect any if and in qualifiers, so you can restrict the data over which the means or medians are computed. You can even restrict the values to a specific case; for example,

. estat mfx if case==21

⌐ Options ⌐

<u>l</u>evel(#) specifies the confidence level, as a percentage, for confidence intervals. The default is level(95) or as set by set level; see [U] **20.7 Specifying the width of confidence intervals**.

nodiscrete specifies that indicator variables be treated as continuous variables. An indicator variable is one that takes on the value 0 or 1 in the estimation sample. By default, the discrete change in the simulated probability is computed as the indicator variable changes from 0 to 1.

noesample specifies that the whole dataset be considered instead of only those marked in the e(sample) defined by the asmprobit command.

nowght specifies that weights be ignored when calculating the means or medians.

Remarks

Remarks are presented under the following headings:

> *Predicted probabilities*
> *Obtaining estimation statistics*
> *Obtaining marginal effects*

Predicted probabilities

After fitting an alternative-specific multinomial probit model, you can use predict to obtain the simulated probabilities that an individual will choose each of the alternatives. When evaluating the multivariate normal probabilities via Monte Carlo simulation, predict uses the same method to generate the random sequence of numbers as the previous call to asmprobit. For example, if you specified intmethod(Halton) when fitting the model, predict also uses the Halton sequence.

▷ Example 1

In example 1 of [R] **asmprobit**, we fit a model of individuals' travel-mode choices. We can obtain the simulated probabilities that an individual chooses each alternative by using predict:

```
. use http://www.stata-press.com/data/r12/travel
. asmprobit choice travelcost termtime, case(id) alternatives(mode)
> casevars(income)
  (output omitted )
. predict prob
(option pr assumed; Pr(mode))
. list id mode prob choice in 1/12, sepby(id)
```

	id	mode	prob	choice
1.	1	air	.1494137	0
2.	1	train	.329167	0
3.	1	bus	.1320298	0
4.	1	car	.3898562	1
5.	2	air	.2565875	0
6.	2	train	.2761054	0
7.	2	bus	.0116135	0
8.	2	car	.4556921	1
9.	3	air	.2098406	0
10.	3	train	.1081824	0
11.	3	bus	.1671841	0
12.	3	car	.5147822	1

◁

Obtaining estimation statistics

Once you have fit a multinomial probit model, you can obtain the estimated variance or correlation matrices for the model alternatives by using the estat command.

▷ Example 2

To display the correlations of the errors in the latent-variable equations, we type

. estat correlation

	train	bus	car
train	1.0000		
bus	0.8909	1.0000	
car	0.7895	0.8951	1.0000

Note: correlations are for alternatives differenced with air

The covariance matrix can be displayed by typing

. estat covariance

	train	bus	car
train	2		
bus	1.600208	1.613068	
car	1.37471	1.399703	1.515884

Note: covariances are for alternatives differenced with air

◁

Obtaining marginal effects

The marginal effects are computed as the derivative of the simulated probability for an alternative with respect to an independent variable. A table of marginal effects is displayed for each alternative, with the table containing the marginal effect for each case-specific variable and the alternative for each alternative-specific variable.

By default, the marginal effects are computed at the means of each continuous independent variable over the estimation sample. For indicator variables, the difference in the simulated probability evaluated at 0 and 1 is computed by default. Indicator variables will be treated as continuous variables if the nodiscrete option is used.

▷ Example 3

Continuing with our model from example 1, we obtain the marginal effects for alternatives air, train, bus, and car evaluated at the mean values of each independent variable. Recall that the travelcost and termtime variables are alternative specific, taking on different values for each alternative, so they have a separate marginal effect for each alternative.

```
. estat mfx
```

Pr(choice = air) = .29434926

variable	dp/dx	Std. Err.	z	P>|z|	[95% C.I.]		X
travelcost							
air	-.002688	.000677	-3.97	0.000	-.004015	-.001362	102.65
train	.0009	.000436	2.07	0.039	.000046	.001755	130.2
bus	.000376	.000271	1.39	0.166	-.000155	.000908	115.26
car	.001412	.00051	2.77	0.006	.000412	.002412	95.414
termtime							
air	-.010376	.002711	-3.83	0.000	-.015689	-.005063	61.01
train	.003475	.001639	2.12	0.034	.000264	.006687	35.69
bus	.001452	.001008	1.44	0.150	-.000523	.003427	41.657
car	.005449	.002164	2.52	0.012	.001209	.00969	0
casevars							
income	.003891	.001847	2.11	0.035	.000271	.007511	34.548

Pr(choice = train) = .29531182

variable	dp/dx	Std. Err.	z	P>|z|	[95% C.I.]		X
travelcost							
air	.000899	.000436	2.06	0.039	.000045	.001753	102.65
train	-.004081	.001466	-2.78	0.005	-.006953	-.001208	130.2
bus	.001278	.00063	2.03	0.042	.000043	.002513	115.26
car	.001904	.000887	2.15	0.032	.000166	.003641	95.414
termtime							
air	.003469	.001638	2.12	0.034	.000258	.00668	61.01
train	-.01575	.00247	-6.38	0.000	-.020591	-.010909	35.69
bus	.004934	.001593	3.10	0.002	.001812	.008056	41.657
car	.007348	.002228	3.30	0.001	.00298	.011715	0
casevars							
income	-.00957	.002223	-4.31	0.000	-.013927	-.005214	34.548

Pr(choice = bus) = .08880039

variable	dp/dx	Std. Err.	z	P>|z|	[95% C.I.]		X
travelcost							
air	.00038	.000274	1.39	0.165	-.000157	.000916	102.65
train	.001279	.00063	2.03	0.042	.000044	.002514	130.2
bus	-.003182	.001175	-2.71	0.007	-.005485	-.00088	115.26
car	.001523	.000675	2.26	0.024	.0002	.002847	95.414
termtime							
air	.001466	.001017	1.44	0.149	-.000526	.003459	61.01
train	.004937	.001591	3.10	0.002	.001819	.008055	35.69
bus	-.012283	.002804	-4.38	0.000	-.017778	-.006788	41.657
car	.00588	.002255	2.61	0.009	.001461	.010299	0
casevars							
income	.000435	.001461	0.30	0.766	-.002428	.003298	34.548

```
Pr(choice = car) = .32168607
```

variable	dp/dx	Std. Err.	z	P>|z|	[95% C.I.]		X
travelcost							
air	.00141	.000509	2.77	0.006	.000411	.002408	102.65
train	.001903	.000886	2.15	0.032	.000166	.003641	130.2
bus	.001523	.000675	2.25	0.024	.000199	.002847	115.26
car	-.004836	.001539	-3.14	0.002	-.007853	-.001819	95.414
termtime							
air	.005441	.002161	2.52	0.012	.001205	.009677	61.01
train	.007346	.002228	3.30	0.001	.00298	.011713	35.69
bus	.005879	.002256	2.61	0.009	.001456	.010301	41.657
car	-.018666	.003938	-4.74	0.000	-.026385	-.010948	0
casevars							
income	.005246	.002166	2.42	0.015	.001002	.00949	34.548

First, we note that there is a separate marginal effects table for each alternative and that table begins by reporting the overall probability of choosing the alternative, for example, 0.2944 for air travel. We see in the first table that a unit increase in terminal time for air travel from 61.01 minutes will result in a decrease in probability of choosing air travel (when the probability is evaluated at the mean of all variables) by approximately 0.01, with a 95% confidence interval of about -0.016 to -0.005. Travel cost has a less negative effect of choosing air travel (at the average cost of 102.65). Alternatively, an increase in terminal time and travel cost for train, bus, or car from these mean values will increase the chance for air travel to be chosen. Also, with an increase in income from 34.5, it would appear that an individual would be more likely to choose air or automobile travel over bus or train. (While the marginal effect for bus travel is positive, it is not significant.)

◁

▷ Example 4

Plotting the simulated probability marginal effect evaluated over a range of values for an independent variable may be more revealing than a table of values. Below are the commands for generating the simulated probability marginal effect of air travel for increasing air travel terminal time. We fix all other independent variables at their medians.

```
. qui gen meff = .
. qui gen tt = .
. qui gen lb = .
. qui gen ub = .
. forvalues i=0/19 {
  2.          local termtime = 5+5*'i'
  3.          qui replace tt = 'termtime' if _n == 'i'+1
  4.          qui estat mfx, at(median air:termtime='termtime') var(termtime)
  5.          mat air = r(air)
  6.          qui replace meff = air[1,1] if _n == 'i'+1
  7.          qui replace lb = air[1,5] if _n == 'i'+1
  8.          qui replace ub = air[1,6] if _n == 'i'+1
  9.          qui replace prob = r(pr_air) if _n == 'i'+1
 10. }
. label variable tt "terminal time"
```

```
. twoway (rarea lb ub tt, pstyle(ci)) (line meff tt, lpattern(solid)), name(meff)
> legend(off) title("  marginal effect of air travel" "terminal time and"
> "95% confidence interval", position(3))

. twoway line prob tt, name(prob) title("  probability of choosing" "air travel",
> position(3)) graphregion(margin(r+9)) ytitle("") xtitle("")

. graph combine prob meff, cols(1) graphregion(margin(l+5 r+5))
```

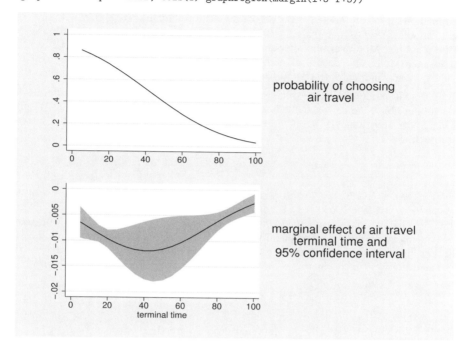

From the graphs, we see that the simulated probability of choosing air travel decreases in an sigmoid fashion. The marginal effects display the rate of change in the simulated probability as a function of the air travel terminal time. The rate of change in the probability of choosing air travel decreases until the air travel terminal time reaches about 45; thereafter, it increases.

◁

Saved results

estat mfx saves the following in r():

Scalars
    r(pr_*alt*)   scalars containing the computed probability of each alternative evaluated at the value that is labeled X in the table output. Here *alt* are the labels in the macro e(alteqs).

Matrices
    r(*alt*)   matrices containing the computed marginal effects and associated statistics. There is one matrix for each alternative, where *alt* are the labels in the macro e(alteqs). Column 1 of each matrix contains the marginal effects; column 2, their standard errors; columns 3 and 4, their z statistics and the p-values for the z statistics; and columns 5 and 6, the confidence intervals. Column 7 contains the values of the independent variables used to compute the probabilities r(pr_*alt*).

Methods and formulas

All postestimation commands listed above are implemented as ado-files.

Marginal effects

The marginal effects are computed as the derivative of the simulated probability with respect to each independent variable. A set of marginal effects is computed for each alternative; thus, for J alternatives, there will be J tables. Moreover, the alternative-specific variables will have J entries, one for each alternative in each table. The details of computing the effects are different for alternative-specific variables and case-specific variables, as well as for continuous and indicator variables.

We use the latent-variable notation of asmprobit (see [R] **asmprobit**) for a J-alternative model and, for notational convenience, we will drop any subscripts involving observations. We then have the following linear functions $\eta_j = \mathbf{x}_j\boldsymbol{\beta} + \mathbf{z}\boldsymbol{\alpha}_j$, for $j = 1, \ldots, J$. Let k index the alternative of interest, and then

$$v_{j'} = \eta_j - \eta_k$$
$$= (\mathbf{x}_j - \mathbf{x}_k)\boldsymbol{\beta} + \mathbf{z}(\boldsymbol{\alpha}_j - \boldsymbol{\alpha}_k) + \epsilon_{j'}$$

where $j' = j$ if $j < k$ and $j' = j - 1$ if $j > k$, so that $j' = 1, \ldots, J - 1$ and $\epsilon_{j'} \sim \text{MVN}(\mathbf{0}, \boldsymbol{\Sigma})$. Denote $p_k = \Pr(v_1 \le 0, \ldots, v_{J-1} \le 0)$ as the simulated probability of choosing alternative k given profile $\mathbf{x}_k$ and $\mathbf{z}$. The marginal effects are then $\partial p_k/\partial \mathbf{x}_k$, $\partial p_k/\partial \mathbf{x}_j$, and $\partial p_k/\partial \mathbf{z}$, where $k = 1, \ldots, J$, $j \ne k$. asmprobit analytically computes the first-order derivatives of the simulated probability with respect to the v's, and the marginal effects for $\mathbf{x}$'s and $\mathbf{z}$ are obtained via the chain rule. The standard errors for the marginal effects are computed using the delta method.

Also see

[R] **asmprobit** — Alternative-specific multinomial probit regression

[U] **20 Estimation and postestimation commands**

Title

> **asroprobit** — Alternative-specific rank-ordered probit regression

Syntax

asroprobit *depvar* [*indepvars*] [*if*] [*in*] [*weight*] , case(*varname*)

alternatives(*varname*) [*options*]

options	Description
Model	
* case(*varname*)	use *varname* to identify cases
* alternatives(*varname*)	use *varname* to identify the alternatives available for each case
casevars(*varlist*)	case-specific variables
constraints(*constraints*)	apply specified linear constraints
collinear	keep collinear variables
Model 2	
correlation(*correlation*)	correlation structure of the latent-variable errors
stddev(*stddev*)	variance structure of the latent-variable errors
structural	use the structural covariance parameterization; default is the differenced covariance parameterization
factor(*#*)	use the factor covariance structure with dimension *#*
noconstant	suppress the alternative-specific constant terms
basealternative(*# | lbl | str*)	alternative used for normalizing location
scalealternative(*# | lbl | str*)	alternative used for normalizing scale
altwise	use alternativewise deletion instead of casewise deletion
reverse	interpret the lowest rank in *depvar* as the best; the default is the highest rank is the best
SE/Robust	
vce(*vcetype*)	*vcetype* may be oim, robust, cluster *clustvar*, opg, bootstrap, or jackknife
Reporting	
level(*#*)	set confidence level; default is level(95)
notransform	do not transform variance–covariance estimates to the standard deviation and correlation metric
nocnsreport	do not display constraints
display_options	control column formats and line width

Integration

<u>intmethod</u>(*seqtype*)	type of quasi- or pseudouniform sequence
<u>intpoints</u>(#)	number of points in each sequence
<u>intburn</u>(#)	starting index in the Hammersley or Halton sequence
<u>intseed</u>(*code* | #)	pseudouniform random-number seed
<u>antithetics</u>	use antithetic draws
<u>nopivot</u>	do not use integration interval pivoting
<u>initbhhh</u>(#)	use the BHHH optimization algorithm for the first # iterations
favor(<u>speed</u> | <u>space</u>)	favor speed or space when generating integration points

Maximization

maximize_options	control the maximization process
<u>coefl</u>egend	display legend instead of statistics

correlation	Description
<u>uns</u>tructured	one correlation parameter for each pair of alternatives; correlations with the basealternative() are zero; the default
<u>exc</u>hangeable	one correlation parameter common to all pairs of alternatives; correlations with the basealternative() are zero
<u>ind</u>ependent	constrain all correlation parameters to zero
<u>pat</u>tern *matname*	user-specified matrix identifying the correlation pattern
<u>fix</u>ed *matname*	user-specified matrix identifying the fixed and free correlation parameters

stddev	Description
<u>het</u>eroskedastic	estimate standard deviation for each alternative; standard deviations for basealternative() and scalealternative() set to one
<u>hom</u>oskedastic	all standard deviations are one
<u>pat</u>tern *matname*	user-specified matrix identifying the standard deviation pattern
<u>fix</u>ed *matname*	user-specified matrix identifying the fixed and free standard deviations

seqtype	Description
<u>ham</u>mersley	Hammersley point set
<u>hal</u>ton	Halton point set
<u>ran</u>dom	uniform pseudorandom point set

[*] case(*varname*) and alternatives(*varname*) are required.
bootstrap, by, jackknife, statsby, and xi are allowed; see [U] **11.1.10 Prefix commands**.
Weights are not allowed with the bootstrap prefix; see [R] **bootstrap**.
fweights, iweights, and pweights are allowed; see [U] **11.1.6 weight**.
coeflegend does not appear in the dialog box.
See [U] **20 Estimation and postestimation commands** for more capabilities of estimation commands.

Menu

Statistics > Ordinal outcomes > Rank-ordered probit regression

Description

asroprobit fits rank-ordered probit (ROP) models by using maximum simulated likelihood (MSL). The model allows you to relax the independence of irrelevant alternatives (IIA) property that is characteristic of the rank-ordered logistic model by estimating the variance–covariance parameters of the latent-variable errors. Each unique identifier in the case() variable has multiple alternatives identified in the alternatives() variable, and *depvar* contains the ranked alternatives made by each case. Only the order in the ranks, not the magnitude of their differences, is assumed to be relevant. By default, the largest rank indicates the more desirable alternative. Use the reverse option if the lowest rank should be interpreted as the more desirable alternative. Tied ranks are allowed, but they increase the computation time because all permutations of the tied ranks are used in computing the likelihood for each case. asroprobit allows two types of independent variables: alternative-specific variables, in which the values of each variable vary with each alternative, and case-specific variables, which vary with each case.

The estimation technique of asroprobit is nearly identical to that of asmprobit, and the two routines share many of the same options; see [R] **asmprobit**.

Options

```
                Model
```

case(*varname*) specifies the variable that identifies each case. This variable identifies the individuals or entities making a choice. case() is required.

alternatives(*varname*) specifies the variable that identifies the alternatives available for each case. The number of alternatives can vary with each case; the maximum number of alternatives is 20. alternatives() is required.

casevars(*varlist*) specifies the case-specific variables that are constant for each case(). If there are a maximum of J alternatives, there will be $J - 1$ sets of coefficients associated with casevars().

constraints(*constraints*), collinear; see [R] **estimation options**.

```
                Model 2
```

correlation(*correlation*) specifies the correlation structure of the latent-variable errors.

correlation(unstructured) is the most general and has $J(J - 3)/2 + 1$ unique correlation parameters. This is the default unless stddev() or structural are specified.

correlation(exchangeable) provides for one correlation coefficient common to all latent variables, except the latent variable associated with the basealternative().

correlation(independent) assumes that all correlations are zero.

correlation(pattern *matname*) and correlation(fixed *matname*) give you more flexibility in defining the correlation structure. See *Variance structures* in [R] **asmprobit** for more information.

stddev(*stddev*) specifies the variance structure of the latent-variable errors.

stddev(heteroskedastic) is the most general and has $J - 2$ estimable parameters. The standard deviations of the latent-variable errors for the alternatives specified in basealternative() and scalealternative() are fixed to one.

`stddev(homoskedastic)` constrains all the standard deviations to equal one.

`stddev(pattern` *matname*`)` and `stddev(fixed` *matname*`)` give you added flexibility in defining the standard deviation parameters. See *Variance structures* in [R] **asmprobit** for more information.

`structural` requests the $J \times J$ structural covariance parameterization instead of the default $J-1 \times J-1$ differenced covariance parameterization (the covariance of the latent errors differenced with that of the base alternative). The differenced covariance parameterization will achieve the same maximum simulated likelihood regardless of the choice of `basealternative()` and `scalealternative()`. On the other hand, the structural covariance parameterization imposes more normalizations that may bound the model away from its maximum likelihood and thus prevent convergence with some datasets or choices of `basealternative()` and `scalealternative()`.

`factor(#)` requests that the factor covariance structure of dimension # be used. The `factor()` option can be used with the `structural` option but cannot be used with `stddev()` or `correlation()`. A $\# \times J$ (or $\# \times J - 1$) matrix, $\mathbf{C}$, is used to factor the covariance matrix as $I + \mathbf{C}'\mathbf{C}$, where I is the identity matrix of dimension J (or $J - 1$). The column dimension of $\mathbf{C}$ depends on whether the covariance is structural or differenced. The row dimension of $\mathbf{C}$, #, must be less than or equal to `floor`$((J(J-1)/2-1)/(J-2))$, because there are only $J(J-1)/2-1$ identifiable variance–covariance parameters. This covariance parameterization may be useful for reducing the number of covariance parameters that need to be estimated.

If the covariance is structural, the column of $\mathbf{C}$ corresponding to the base alternative contains zeros. The column corresponding to the scale alternative has a one in the first row and zeros elsewhere. If the covariance is differenced, the column corresponding to the scale alternative (differenced with the base) has a one in the first row and zeros elsewhere.

`noconstant` suppresses the $J - 1$ alternative-specific constant terms.

`basealternative(# | lbl | str)` specifies the alternative used to normalize the latent-variable location (also referred to as the level of utility). The base alternative may be specified as a number, label, or string. The standard deviation for the latent-variable error associated with the base alternative is fixed to one, and its correlations with all other latent-variable errors are set to zero. The default is the first alternative when sorted. If a `fixed` or `pattern` matrix is given in the `stddev()` and `correlation()` options, the `basealternative()` will be implied by the fixed standard deviations and correlations in the matrix specifications. `basealternative()` cannot be equal to `scalealternative()`.

`scalealternative(# | lbl | str)` specifies the alternative used to normalize the latent-variable scale (also referred to as the scale of utility). The scale alternative may be specified as a number, label, or string. The default is to use the second alternative when sorted. If a `fixed` or `pattern` matrix is given in the `stddev()` option, the `scalealternative()` will be implied by the fixed standard deviations in the matrix specification. `scalealternative()` cannot be equal to `basealternative()`.

If a `fixed` or `pattern` matrix is given for the `stddev()` option, the base alternative and scale alternative are implied by the standard deviations and correlations in the matrix specifications, and they need not be specified in the `basealternative()` and `scalealternative()` options.

`altwise` specifies that alternativewise deletion be used when marking out observations due to missing values in your variables. The default is to use casewise deletion; that is, the entire group of observations making up a case is deleted if any missing values are encountered. This option does not apply to observations that are marked out by the `if` or `in` qualifier or the `by` prefix.

`reverse` directs `asroprobit` to interpret the rank in *depvar* that is smallest in value as the preferred alternative. By default, the rank that is largest in value is the favored alternative.

SE/Robust

vce(*vcetype*) specifies the type of standard error reported, which includes types that are derived from asymptotic theory, that are robust to some kinds of misspecification, that allow for intragroup correlation, and that use bootstrap or jackknife methods; see [R] *vce_option*.

If specifying vce(bootstrap) or vce(jackknife), you must also specify basealternative() and scalealternative().

Reporting

level(*#*); see [R] **estimation options**.

notransform prevents retransforming the Cholesky-factored variance–covariance estimates to the correlation and standard deviation metric.

This option has no effect if structural is not specified because the default differenced variance–covariance estimates have no interesting interpretation as correlations and standard deviations. notransform also has no effect if the correlation() and stddev() options are specified with anything other than their default values. Here it is generally not possible to factor the variance–covariance matrix, so optimization is already performed using the standard deviation and correlation representations.

nocnsreport; see [R] **estimation options**.

display_options: cformat(%*fmt*), pformat(%*fmt*), sformat(%*fmt*), and nolstretch; see [R] **estimation options**.

Integration

intmethod(hammersley | halton | random) specifies the method of generating the point sets used in the quasi–Monte Carlo integration of the multivariate normal density. intmethod(hammersley), the default, uses the Hammersley sequence; intmethod(halton) uses the Halton sequence; and intmethod(random) uses a sequence of uniform random numbers.

intpoints(*#*) specifies the number of points to use in the quasi–Monte Carlo integration. If this option is not specified, the number of points is $50 \times J$ if intmethod(hammersley) or intmethod(halton) is used and $100 \times J$ if intmethod(random) is used. Larger values of intpoints() provide better approximations of the log likelihood, but at the cost of added computation time.

intburn(*#*) specifies where in the Hammersley or Halton sequence to start, which helps reduce the correlation between the sequences of each dimension. The default is 0. This option may not be specified with intmethod(random).

intseed(*code* | *#*) specifies the seed to use for generating the uniform pseudorandom sequence. This option may be specified only with intmethod(random). *code* refers to a string that records the state of the random-number generator runiform(); see [R] **set seed**. An integer value *#* may be used also. The default is to use the current seed value from Stata's uniform random-number generator, which can be obtained from c(seed).

antithetics specifies that antithetic draws be used. The antithetic draw for the $J - 1$ vector uniform-random variables, $\mathbf{x}$, is $1 - \mathbf{x}$.

nopivot turns off integration interval pivoting. By default, asroprobit will pivot the wider intervals of integration to the interior of the multivariate integration. This improves the accuracy of the quadrature estimate. However, discontinuities may result in the computation of numerical second-order derivatives using finite differencing (for the Newton–Raphson optimize technique, tech(nr)) when few simulation points are used, resulting in a non–positive-definite Hessian. asroprobit uses the Broyden–Fletcher–Goldfarb–Shanno optimization algorithm, by default, which does not require computing the Hessian numerically using finite differencing.

initbhhh(#) specifies that the Berndt–Hall–Hall–Hausman (BHHH) algorithm be used for the initial # optimization steps. This option is the only way to use the BHHH algorithm along with other optimization techniques. The algorithm switching feature of ml's technique() option cannot include bhhh.

favor(speed | space) instructs asroprobit to favor either speed or space when generating the integration points. favor(speed) is the default. When favoring speed, the integration points are generated once and stored in memory, thus increasing the speed of evaluating the likelihood. This speed increase can be seen when there are many cases or when the user specifies a large number of integration points, intpoints(#). When favoring space, the integration points are generated repeatedly with each likelihood evaluation.

For unbalanced data, where the number of alternatives varies with each case, the estimates computed using intmethod(random) will vary slightly between favor(speed) and favor(space). This is because the uniform sequences will not be identical, even when initiating the sequences using the same uniform seed, intseed(*code* | #). For favor(speed), ncase blocks of intpoints(#) $\times$ $J - 2$ uniform points are generated, where J is the maximum number of alternatives. For favor(space), the column dimension of the matrices of points varies with the number of alternatives that each case has.

> ___Maximization___
>
> *maximize_options*: difficult, technique(*algorithm_spec*), iterate(#), [no]log, trace, gradient, showstep, hessian, showtolerance, tolerance(#), ltolerance(#), nrtolerance(#), nonrtolerance, and from(*init_specs*); see [R] **maximize**.

The following options may be particularly useful in obtaining convergence with asroprobit: difficult, technique(*algorithm_spec*), nrtolerance(#), nonrtolerance, and from(*init_specs*).

If technique() contains more than one algorithm specification, bhhh cannot be one of them. To use the BHHH algorithm with another algorithm, use the initbhhh() option and specify the other algorithm in technique().

Setting the optimization type to technique(bhhh) resets the default *vcetype* to vce(opg).

When specifying from(*matname* [, copy]), the values in *matname* associated with the latent-variable error variances must be for the log-transformed standard deviations and inverse-hyperbolic tangent-transformed correlations. This option makes using the coefficient vector from a previously fitted asroprobit model convenient as a starting point.

The following option is available with asroprobit but is not shown in the dialog box:

coeflegend; see [R] **estimation options**.

Remarks

The mathematical description and numerical computations of the rank-ordered probit model are similar to that of the multinomial probit model. The only difference is that the dependent variable of the rank-ordered probit model is ordinal, showing preferences among alternatives, as opposed to the binary dependent variable of the multinomial probit model, indicating a chosen alternative. We will describe how the likelihood of a ranking is computed using the latent-variable framework here, but for details of the latent-variable parameterization of these models and the method of maximum simulated likelihood, see [R] **asmprobit**.

Consider the latent-variable parameterization of a J alternative rank-ordered probit model. Using the notation from `asmprobit`, we have variables η_{ij}, $j = 1, \ldots, J$, such that

$$\eta_{ij} = \mathbf{x}_{ij}\boldsymbol{\beta} + \mathbf{z}_i\boldsymbol{\alpha}_j + \xi_{ij}$$

Here the $\mathbf{x}_{ij}$ are the alternative-specific independent variables, the $\mathbf{z}_i$ are the case-specific variables, and the ξ_{ij} are multivariate normal with mean zero and covariance $\boldsymbol{\Omega}$. Without loss of generality, assume that individual i ranks the alternatives in order of the alternative indices $j = 1, 2, \ldots, J$, so the alternative J is the preferred alternative and alternative 1 is the least preferred alternative. The probability of this ranking given $\boldsymbol{\beta}$ and $\boldsymbol{\alpha}_j$ is the probability that $\eta_{i,J-1} - \eta_{i,J} \leq 0$ and $\eta_{i,J-2} - \eta_{i,J-1} \leq 0$, $\ldots$, and $\eta_{i,1} - \eta_{i,2} \leq 0$.

▷ Example 1

Long and Freese (2006) provide an example of a rank-ordered logit model with alternative-specific variables. We use this dataset to demonstrate `asroprobit`. The data come from the Wisconsin Longitudinal Study. This is a study of 1957 Wisconsin high school graduates that were asked to rate their relative preference of four job characteristics: esteem, a job other people regard highly; variety, a job that is not repetitive and allows you to do a variety of things; autonomy, a job where your supervisor does not check on you frequently; and security, a job with a low risk of being laid off. The case-specific covariates are gender, `female`, an indicator variable for females, and `score`, a score on a general mental ability test measured in standard deviations. The alternative-specific variables are `high` and `low`, which indicate whether the respondent's current job is high or low in esteem, variety, autonomy, or security. This approach provides three states for a respondent's current job status for each alternative, $(1, 0)$, $(0, 1)$, and $(0, 0)$, using the notation (`high`, `low`). The score $(1, 1)$ is omitted because the respondent's current job cannot be considered both high and low in one of the job characteristics. The $(0, 0)$ score would indicate that the respondent's current job does not rank high or low (is neutral) in a job characteristic. The alternatives are ranked such that 1 is the preferred alternative and 4 is the least preferred.

```
. use http://www.stata-press.com/data/r12/wlsrank
(1992 Wisconsin Longitudinal Study data on job values)
. list id jobchar rank female score high low in 1/12, sepby(id)
```

	id	jobchar	rank	female	score	high	low
1.	1	security	1	1	.0492111	0	0
2.	1	autonomy	4	1	.0492111	0	0
3.	1	variety	1	1	.0492111	0	0
4.	1	esteem	3	1	.0492111	0	0
5.	5	security	2	1	2.115012	1	0
6.	5	variety	2	1	2.115012	1	0
7.	5	esteem	2	1	2.115012	1	0
8.	5	autonomy	1	1	2.115012	0	0
9.	7	autonomy	1	0	1.701852	1	0
10.	7	variety	1	0	1.701852	0	1
11.	7	esteem	4	0	1.701852	0	0
12.	7	security	1	0	1.701852	0	0

The three cases listed have tied ranks. `asroprobit` will allow ties, but at the cost of increased computation time. To evaluate the likelihood of the first observation, `asroprobit` must compute

$$\Pr(\text{esteem} = 3, \text{ variety} = 1, \text{ autonomy} = 4, \text{ security} = 2)+$$

$$\Pr(\text{esteem} = 3, \text{ variety} = 2, \text{ autonomy} = 4, \text{ security} = 1)$$

and both of these probabilities are estimated using simulation. In fact, the full dataset contains 7,237 tied ranks and asroprobit takes a great deal of time to estimate the parameters. For exposition, we estimate the rank-ordered probit model by using the cases without ties. These cases are marked in the variable noties.

The model of job preference is

$$\eta_{ij} = \beta_1 \text{high}_{ij} + \beta_2 \text{low}_{ij} + \alpha_{1j} \text{female}_i + \alpha_{2j} \text{score}_i + \alpha_{0j} + \xi_{ij}$$

for $j = 1, 2, 3, 4$. The base alternative will be esteem, so $\alpha_{01} = \alpha_{11} = \alpha_{21} = 0$.

```
. asroprobit rank high low if noties, case(id) alternatives(jobchar)
> casevars(female score) reverse
note: variable high has 107 cases that are not alternative-specific: there is
        no within-case variability
note: variable low has 193 cases that are not alternative-specific: there is
        no within-case variability
Iteration 0:   log simulated-likelihood = -1103.2768
Iteration 1:   log simulated-likelihood = -1089.3361  (backed up)
  (output omitted )
Alternative-specific rank-ordered probit      Number of obs    =        1660
Case variable: id                             Number of cases  =         415

Alternative variable: jobchar                 Alts per case: min =          4
                                                             avg =        4.0
                                                             max =          4

Integration sequence:       Hammersley
Integration points:            200            Wald chi2(8)     =       34.01
Log simulated-likelihood = -1080.2206         Prob > chi2      =      0.0000
```

rank	Coef.	Std. Err.	z	P>|z|	[95% Conf. Interval]	
jobchar						
high	.3741029	.0925685	4.04	0.000	.192672	.5555337
low	-.0697443	.1093317	-0.64	0.524	-.2840305	.1445419
esteem	(base alternative)					
variety						
female	.1351487	.1843088	0.73	0.463	-.2260899	.4963873
score	.1405482	.0977567	1.44	0.151	-.0510515	.3321479
_cons	1.735016	.1451343	11.95	0.000	1.450558	2.019474
autonomy						
female	.2561828	.1679565	1.53	0.127	-.0730059	.5853715
score	.1898853	.0875668	2.17	0.030	.0182575	.361513
_cons	.7009797	.1227336	5.71	0.000	.4604262	.9415333
security						
female	.232622	.2057547	1.13	0.258	-.1706497	.6358938
score	-.1780076	.1102115	-1.62	0.106	-.3940181	.038003
_cons	1.343766	.1600059	8.40	0.000	1.030161	1.657372
/lnl2_2	.1805151	.0757296	2.38	0.017	.0320878	.3289424
/lnl3_3	.4843091	.0793343	6.10	0.000	.3288168	.6398014
/l2_1	.6062037	.1169368	5.18	0.000	.3770117	.8353957
/l3_1	.4509217	.1431183	3.15	0.002	.1704151	.7314283
/l3_2	.2289447	.1226081	1.87	0.062	-.0113627	.4692521

```
(jobchar=esteem is the alternative normalizing location)
(jobchar=variety is the alternative normalizing scale)
```

We specified the `reverse` option because a rank of 1 is the highest preference. The variance–covariance estimates are for the Cholesky-factored variance–covariance for the latent-variable errors differenced with that of alternative `esteem`. We can view the estimated correlations by entering

    . estat correlation

	variety	autonomy	security
variety	1.0000		
autonomy	0.4516	1.0000	
security	0.2652	0.2399	1.0000

Note: correlations are for alternatives differenced with esteem

and typing

    . estat covariance

	variety	autonomy	security
variety	2		
autonomy	.8573015	1.80229	
security	.6376996	.5475882	2.890048

Note: covariances are for alternatives differenced with esteem

gives the (co)variances. [R] **mprobit** explains that if the latent-variable errors are independent, then the correlations in the differenced parameterization should be ~0.5 and the variances should be ~2.0, which seems to be the case here.

The coefficient estimates for the probit models can be difficult to interpret because of the normalization for location and scale. The regression estimates for the case-specific variables will be relative to the base alternative and the regression estimates for both the case-specific and alternative-specific variables are affected by the scale normalization. The more pronounced the heteroskedasticity and correlations, the more pronounced the resulting estimate differences when choosing alternatives to normalize for location and scale. However, when using the differenced covariance structure, you will obtain the same model likelihood regardless of which alternatives you choose as the base and scale alternatives. For model interpretation, you can examine the estimated probabilities and marginal effects by using postestimation routines `predict` and `estat mfx`. See [R] **asroprobit postestimation**.

◁

Saved results

asroprobit saves the following in e():

Scalars

e(N)	number of observations
e(N_case)	number of cases
e(N_ties)	number of ties
e(k)	number of parameters
e(k_alt)	number of alternatives
e(k_indvars)	number of alternative-specific variables
e(k_casevars)	number of case-specific variables
e(k_sigma)	number of variance estimates
e(k_rho)	number of correlation estimates
e(k_eq)	number of equations in e(b)
e(k_eq_model)	number of equations in overall model test
e(df_m)	model degrees of freedom
e(ll)	log simulated-likelihood
e(N_clust)	number of clusters
e(const)	constant indicator
e(i_base)	base alternative index
e(i_scale)	scale alternative index
e(mc_points)	number of Monte Carlo replications
e(mc_burn)	starting sequence index
e(mc_antithetics)	antithetics indicator
e(reverse)	1 if minimum rank is best, 0 if maximum rank is best
e(chi2)	χ^2
e(p)	significance
e(fullcov)	unstructured covariance indicator
e(structcov)	1 if structured covariance; 0 otherwise
e(cholesky)	Cholesky-factored covariance indicator
e(alt_min)	minimum number of alternatives
e(alt_avg)	average number of alternatives
e(alt_max)	maximum number of alternatives
e(rank)	rank of e(V)
e(ic)	number of iterations
e(rc)	return code
e(converged)	1 if converged, 0 otherwise

Macros

e(cmd)	asroprobit
e(cmdline)	command as typed
e(depvar)	name of dependent variable
e(indvars)	alternative-specific independent variable
e(casevars)	case-specific variables
e(case)	variable defining cases
e(altvar)	variable defining alternatives
e(alteqs)	alternative equation names
e(alt#)	alternative labels
e(wtype)	weight type
e(wexp)	weight expression
e(title)	title in estimation output
e(clustvar)	name of cluster variable
e(correlation)	correlation structure
e(stddev)	variance structure
e(chi2type)	Wald, type of model χ^2 test
e(vce)	*vcetype* specified in vce()
e(vcetype)	title used to label Std. Err.
e(opt)	type of optimization
e(which)	max or min; whether optimizer is to perform maximization or minimization
e(ml_method)	type of ml method
e(mc_method)	Hammersley, Halton, or uniform random; technique to generate sequences
e(mc_seed)	random-number generator seed
e(user)	name of likelihood-evaluator program
e(technique)	maximization technique
e(datasignature)	the checksum
e(datasignaturevars)	variables used in calculation of checksum
e(properties)	b V
e(estat_cmd)	program used to implement estat
e(mfx_dlg)	program used to implement estat mfx dialog
e(predict)	program used to implement predict
e(marginsnotok)	predictions disallowed by margins

Matrices

e(b)	coefficient vector
e(Cns)	constraints matrix
e(stats)	alternative statistics
e(stdpattern)	variance pattern
e(stdfixed)	fixed and free standard deviations
e(altvals)	alternative values
e(altfreq)	alternative frequencies
e(alt_casevars)	indicators for estimated case-specific coefficients—e(k_alt)×e(k_casevars)
e(corpattern)	correlation structure
e(corfixed)	fixed and free correlations
e(ilog)	iteration log (up to 20 iterations)
e(gradient)	gradient vector
e(V)	variance–covariance matrix of the estimators
e(V_modelbased)	model-based variance

Functions

e(sample)	marks estimation sample

Methods and formulas

asroprobit is implemented as an ado-file.

From a computational perspective, asroprobit is similar to asmprobit and the two programs share many numerical tools. Therefore, we will use the notation from *Methods and formulas* in [R] **asmprobit** to discuss the rank-ordered probit probability model.

The latent variables for a J-alternative model are $\eta_{ij} = \mathbf{x}_{ij}\boldsymbol{\beta} + \mathbf{z}_i\boldsymbol{\alpha}_j + \xi_{ij}$, for $j = 1, \ldots, J$, $i = 1, \ldots, n$, and $\boldsymbol{\xi}_i' = (\xi_{i,1}, \ldots, \xi_{i,J}) \sim \text{MVN}(\mathbf{0}, \boldsymbol{\Omega})$. Without loss of generality, assume for the ith observation that an individual ranks the alternatives in the order of their numeric indices, $\mathbf{y}_i = (J, J - 1, \ldots, 1)$, so the first alternative is the most preferred and the last alternative is the least preferred. We can then difference the latent variables such that

$$
\begin{aligned}
v_{ik} &= \eta_{i,k+1} - \eta_{i,k} \\
&= (\mathbf{x}_{i,k+1} - \mathbf{x}_{i,k})\boldsymbol{\beta} + \mathbf{z}_i(\boldsymbol{\alpha}_{k+1} - \boldsymbol{\alpha}_k) + \xi_{i,k+1} - \xi_{ik} \\
&= \boldsymbol{\delta}_{ik}\boldsymbol{\beta} + \mathbf{z}_i\boldsymbol{\gamma}_k + \epsilon_{ik}
\end{aligned}
$$

for $k = 1, \ldots, J - 1$ and where $\boldsymbol{\epsilon}_i = (\epsilon_{i1}, \ldots, \epsilon_{i,J-1}) \sim \text{MVN}(\mathbf{0}, \boldsymbol{\Sigma}_{(i)})$. $\boldsymbol{\Sigma}$ is indexed by i because it is specific to the ranking of individual i. We denote the deterministic part of the model as $\lambda_{ik} = \boldsymbol{\delta}_{ik}\boldsymbol{\beta} + \mathbf{z}_j\boldsymbol{\gamma}_k$, and the probability of this event is

$$
\begin{aligned}
\Pr(\mathbf{y}_i) &= \Pr(v_{i1} \leq 0, \ldots, v_{i,J-1} \leq 0) \\
&= \Pr(\epsilon_{i1} \leq -\lambda_{i1}, \ldots, \epsilon_{i,J-1} \leq -\lambda_{i,J-1}) \\
&= (2\pi)^{-(J-1)/2} |\boldsymbol{\Sigma}_{(i)}|^{-1/2} \int_{-\infty}^{-\lambda_{i1}} \cdots \int_{-\infty}^{-\lambda_{i,J-1}} \exp\left(-\tfrac{1}{2}\mathbf{z}'\boldsymbol{\Sigma}_{(i)}^{-1}\mathbf{z}\right) d\mathbf{z}
\end{aligned}
$$

The integral has the same form as (3) of *Methods and formulas* in [R] **asmprobit**. See [R] **asmprobit** for details on evaluating this integral numerically by using simulation.

asroprobit handles tied ranks by enumeration. For k tied ranks, it will generate $k!$ rankings, where ! is the factorial operator $k! = k(k-1)(k-2)\cdots(2)(1)$. For two sets of tied ranks of size k_1 and k_2, asroprobit will generate $k_1!k_2!$ rankings. The total probability is the sum of the probability of each ranking. For example, if there are two tied ranks such that $\mathbf{y}_i = (J, J, J - 2, \ldots, 1)$, then asroprobit will evaluate $\Pr(\mathbf{y}_i) = \Pr(\mathbf{y}_i^{(1)}) + \Pr(\mathbf{y}_i^{(2)})$, where $\mathbf{y}_i^{(1)} = (J, J - 1, J - 2, \ldots, 1)$ and $\mathbf{y}_i^{(2)} = (J - 1, J, J - 2, \ldots, 1)$.

This command supports the clustered version of the Huber/White/sandwich estimator of the variance using vce(robust) and vce(cluster *clustvar*). See [P] **_robust**, particularly *Maximum likelihood estimators* and *Methods and formulas*. Specifying vce(robust) is equivalent to specifying vce(cluster *casevar*), where *casevar* is the variable that identifies the cases.

Reference

Long, J. S., and J. Freese. 2006. *Regression Models for Categorical Dependent Variables Using Stata*. 2nd ed. College Station, TX: Stata Press.

Also see

[R] **asroprobit postestimation** — Postestimation tools for asroprobit

[R] **asmprobit** — Alternative-specific multinomial probit regression

[R] **mlogit** — Multinomial (polytomous) logistic regression

[R] **mprobit** — Multinomial probit regression

[R] **oprobit** — Ordered probit regression

[U] **20 Estimation and postestimation commands**

Title

asroprobit postestimation — Postestimation tools for asroprobit

Description

The following postestimation commands are of special interest after asroprobit:

Command	Description
estat alternatives	alternative summary statistics
estat covariance	covariance matrix of the latent-variable errors for the alternatives
estat correlation	correlation matrix of the latent-variable errors for the alternatives
estat facweights	covariance factor weights matrix
estat mfx	marginal effects

For information about these commands, see below.

The following standard postestimation commands are also available:

Command	Description
estat	AIC, BIC, VCE, and estimation sample summary
estimates	cataloging estimation results
lincom	point estimates, standard errors, testing, and inference for linear combinations of coefficients
lrtest	likelihood-ratio test
nlcom	point estimates, standard errors, testing, and inference for nonlinear combinations of coefficients
predict	predicted probabilities, estimated linear predictor and its standard error
predictnl	point estimates, standard errors, testing, and inference for generalized predictions
test	Wald tests of simple and composite linear hypotheses
testnl	Wald tests of nonlinear hypotheses

See the corresponding entries in the *Base Reference Manual* for details.

Special-interest postestimation commands

estat alternatives displays summary statistics about the alternatives in the estimation sample. The command also provides a mapping between the index numbers that label the covariance parameters of the model and their associated values and labels for the alternative variable.

estat covariance computes the estimated variance–covariance matrix of the latent-variable errors for the alternatives. The estimates are displayed, and the variance–covariance matrix is stored in r(cov).

estat correlation computes the estimated correlation matrix of the latent-variable errors for the alternatives. The estimates are displayed, and the correlation matrix is stored in r(cor).

estat facweights displays the covariance factor weights matrix and stores it in r(C).

149

estat mfx computes marginal effects of a simulated probability of a set of ranked alternatives. The probability is stored in r(pr), the matrix of rankings is stored in r(ranks), and the matrix of marginal-effect statistics is stored in r(mfx).

Syntax for predict

predict [*type*] *newvar* [*if*] [*in*] [, *statistic* altwise]

predict [*type*] { *stub** | *newvarlist* } [*if*] [*in*], scores

statistic	Description
Main	
pr	probability of each ranking, by case; the default
pr1	probability that each alternative is preferred
xb	linear prediction
stdp	standard error of the linear prediction

These statistics are available both in and out of sample; type predict ... if e(sample) ... if wanted only for the estimation sample.

Menu

Statistics > Postestimation > Predictions, residuals, etc.

Options for predict

⌐ Main ⌐

pr, the default, calculates the probability of each ranking. For each case, one probability is computed for the ranks in e(depvar).

pr1 calculates the probability that each alternative is preferred.

xb calculates the linear prediction $\mathbf{x}_{ij}\boldsymbol{\beta} + \mathbf{z}_i\boldsymbol{\alpha}_j$ for alternative j and case i.

stdp calculates the standard error of the linear predictor.

altwise specifies that alternativewise deletion be used when marking out observations due to missing values in your variables. The default is to use casewise deletion. The xb and stdp options always use alternativewise deletion.

scores calculates the scores for each coefficient in e(b). This option requires a new variable list of length equal to the number of columns in e(b). Otherwise, use the *stub** option to have predict generate enumerated variables with prefix *stub*.

Syntax for estat alternatives

estat alternatives

Menu

Statistics > Postestimation > Reports and statistics

Syntax for estat covariance

estat <u>cov</u>ariance $\left[\ ,\ \underline{\text{format}}(\%\,fmt)\ \underline{\text{bor}}\text{der}(bspec)\ \text{left}(\#)\ \right]$

Menu

Statistics > Postestimation > Reports and statistics

Options for estat covariance

format(%*fmt*) sets the matrix display format. The default is format(%9.0g).

border(*bspec*) sets the matrix display border style. The default is border(all). See [P] **matlist**.

left(*#*) sets the matrix display left indent. The default is left(2). See [P] **matlist**.

Syntax for estat correlation

estat <u>cor</u>relation $\left[\ ,\ \underline{\text{format}}(\%\,fmt)\ \underline{\text{bor}}\text{der}(bspec)\ \text{left}(\#)\ \right]$

Menu

Statistics > Postestimation > Reports and statistics

Options for estat correlation

format(%*fmt*) sets the matrix display format. The default is format(%9.4f).

border(*bspec*) sets the matrix display border style. The default is border(all). See [P] **matlist**.

left(*#*) sets the matrix display left indent. The default is left(2). See [P] **matlist**.

Syntax for estat facweights

estat <u>fac</u>weights $\left[\ ,\ \underline{\text{format}}(\%\,fmt)\ \underline{\text{bor}}\text{der}(bspec)\ \text{left}(\#)\ \right]$

Menu

Statistics > Postestimation > Reports and statistics

Options for estat facweights

format(%*fmt*) sets the matrix display format. The default is format(%9.0f).

border(*bspec*) sets the matrix display border style. The default is border(all). See [P] **matlist**.

left(*#*) sets the matrix display left indent. The default is left(2). See [P] **matlist**.

Syntax for estat mfx

estat mfx [*if*] [*in*] [, *options*]

options	Description
Main	
varlist(*varlist*)	display marginal effects for *varlist*
at(median [*atlist*])	calculate marginal effects at these values
rank(*ranklist*)	calculate marginal effects for the simulated probability of these ranked alternatives
Options	
level(*#*)	set confidence interval level; default is level(95)
nodiscrete	treat indicator variables as continuous
noesample	do not restrict calculation of the medians to the estimation sample
nowght	ignore weights when calculating medians

Menu

Statistics > Postestimation > Reports and statistics

Options for estat mfx

⌐ Main ⌐

varlist(*varlist*) specifies the variables for which to display marginal effects. The default is all variables.

at(median [*atlist*]) specifies the values at which the marginal effects are to be calculated. *atlist* is

[[*alternative*:*variable* = #] [*variable* = #] [...]])

The marginal effects are calculated at the medians of the independent variables.

After specifying the summary statistic, you can specify specific values for variables. You can specify values for alternative-specific variables by alternative, or you can specify one value for all alternatives. You can specify only one value for case-specific variables. For example, in the wlsrank dataset, female and score are case-specific variables, whereas high and low are alternative-specific variables. The following would be a legal syntax for estat mfx:

. estat mfx, at(median high=0 esteem:high=1 low=0 security:low=1 female=1)

When nodiscrete is not specified, at(median [*atlist*]) has no effect on computing marginal effects for indicator variables, which are calculated as the discrete change in the simulated probability as the indicator variable changes from 0 to 1.

The median computations respect any if or in qualifiers, so you can restrict the data over which the medians are computed. You can even restrict the values to a specific case, for example,

. estat mfx if case==13

rank(*ranklist*) specifies the ranks for the alternatives. *ranklist* is

alternative = # *alternative* = # [...]])

The default is to rank the calculated latent variables. Alternatives excluded from rank() are omitted from the analysis. You must therefore specify at least two alternatives in rank(). You may have tied ranks in the rank specification. Only the order in the ranks is relevant.

___Options___

level(*#*) specifies the confidence level, as a percentage, for confidence intervals. The default is level(95) or as set by set level; see [U] **20.7 Specifying the width of confidence intervals**.

nodiscrete specifies that indicator variables be treated as continuous variables. An indicator variable is one that takes on the value 0 or 1 in the estimation sample. By default, the discrete change in the simulated probability is computed as the indicator variable changes from 0 to 1.

noesample specifies that the whole dataset be considered instead of only those marked in the e(sample) defined by the asroprobit command.

nowght specifies that weights be ignored when calculating the medians.

Remarks

Remarks are presented under the following headings:

> *Predicted probabilities*
> *Obtaining estimation statistics*

Predicted probabilities

After fitting an alternative-specific rank-ordered probit model, you can use predict to obtain the probabilities of alternative rankings or the probabilities of each alternative being preferred. When evaluating the multivariate normal probabilities via (quasi) Monte Carlo, predict uses the same method to generate the (quasi) random sequence of numbers as the previous call to asroprobit. For example, if you specified intmethod(halton) when fitting the model, predict also uses Halton sequences.

▷ Example 1

In example 1 of [R] **asroprobit**, we fit a model of job characteristic preferences. This is a study of 1957 Wisconsin high school graduates that were asked to rate their relative preference of four job characteristics: esteem, a job other people regard highly; variety, a job that is not repetitive and allows you to do a variety of things; autonomy, a job where your supervisor does not check on you frequently; and security, a job with a low risk of being laid off. The case-specific covariates are gender, female, an indicator variable for females, and score, a score on a general mental ability test measured in standard deviations. The alternative-specific variables are high and low, which indicate whether the respondent's current job is high or low in esteem, variety, autonomy, or security. This approach provides three states for a respondent's current job status for each alternative, $(1, 0)$, $(0, 1)$, and $(0, 0)$, using the notation (high, low). The score $(1, 1)$ is omitted because the respondent's current job cannot be considered both high and low in one of the job characteristics. The $(0, 0)$ score would indicate that the respondent's current job does not rank high or low (is neutral) in a job characteristic. The alternatives are ranked such that 1 is the preferred alternative and 4 is the least preferred.

We can obtain the probabilities of the observed alternative rankings, the pr option, and the probability of each alternative being preferred, the pr1 option, by using predict:

```
. use http://www.stata-press.com/data/r12/wlsrank
(1992 Wisconsin Longitudinal Study data on job values)
. asroprobit rank high low if noties, case(id) alternatives(jobchar)
> casevars(female score) reverse
  (output omitted )
. keep if e(sample)
(11244 observations deleted)
. predict prob, pr
. predict prob1, pr1
. list id jobchar prob prob1 rank female score high low in 1/12
```

	id	jobchar	prob	prob1	rank	female	score	high	low
1.	13	security	.0421807	.2784269	3	0	.3246512	0	1
2.	13	autonomy	.0421807	.1029036	1	0	.3246512	0	0
3.	13	variety	.0421807	.6026725	2	0	.3246512	1	0
4.	13	esteem	.0421807	.0160111	4	0	.3246512	0	1
5.	19	autonomy	.0942025	.1232488	4	1	.0492111	0	0
6.	19	esteem	.0942025	.0140261	3	1	.0492111	0	0
7.	19	security	.0942025	.4601368	1	1	.0492111	1	0
8.	19	variety	.0942025	.4025715	2	1	.0492111	0	0
9.	22	esteem	.1414177	.0255264	4	1	1.426412	1	0
10.	22	variety	.1414177	.4549441	1	1	1.426412	0	0
11.	22	security	.1414177	.2629494	3	1	1.426412	0	0
12.	22	autonomy	.1414177	.2566032	2	1	1.426412	1	0

The prob variable is constant for each case because it contains the probability of the ranking in the rank variable. On the other hand, the prob1 variable contains the estimated probability of each alternative being preferred. For each case, the sum of the values in prob1 will be approximately 1.0. They do not add up to exactly 1.0 because of approximations due to the GHK algorithm.

◁

Obtaining estimation statistics

For examples of the specialized estat subcommands covariance and correlation, see [R] **asmprobit postestimation**. The entry also has a good example of computing marginal effects after asmprobit that is applicable to asroprobit. Below we will elaborate further on marginal effects after asroprobit where we manipulate the rank() option.

▷ Example 2

We will continue with the preferred job characteristics example where we first compute the marginal effects for case id = 13.

```
. estat mfx if id==13, rank(security=3 autonomy=1 variety=2 esteem=4)

Pr(esteem=4 variety=2 autonomy=1 security=3) = .04218068
```

variable	dp/dx	Std. Err.	z	P>|z|	[95% C.I.]	X
high*						
esteem	-.008713	.001964	-4.44	0.000	-.012562 -.004864	0
variety	-.009102	.003127	-2.91	0.004	-.015231 -.002973	1
autonomy	.025535	.007029	3.63	0.000	.011758 .039313	0
security	-.003745	.001394	-2.69	0.007	-.006477 -.001013	0
low*						
esteem	.001614	.002646	0.61	0.542	-.003572 .0068	1
variety	.001809	.003012	0.60	0.548	-.004094 .007712	0
autonomy	-.003849	.006104	-0.63	0.528	-.015813 .008115	0
security	.000582	.000985	0.59	0.554	-.001348 .002513	1
casevars						
female*	.009767	.009064	1.08	0.281	-.007998 .027533	0
score	.008587	.004488	1.91	0.056	-.00021 .017384	.32465

(*) dp/dx is for discrete change of indicator variable from 0 to 1

Next we compute the marginal effects for the probability that autonomy is preferred given the profile of case id = 13.

```
. estat mfx if id==13, rank(security=2 autonomy=1 variety=2 esteem=2)

Pr(esteem=3 variety=4 autonomy=1 security=2) +
Pr(esteem=4 variety=3 autonomy=1 security=2) +
Pr(esteem=2 variety=4 autonomy=1 security=3) +
Pr(esteem=4 variety=2 autonomy=1 security=3) +
Pr(esteem=2 variety=3 autonomy=1 security=4) +
Pr(esteem=3 variety=2 autonomy=1 security=4) = .10276103
```

variable	dp/dx	Std. Err.	z	P>|z|	[95% C.I.]	X
high*						
esteem	-.003524	.001258	-2.80	0.005	-.005989 -.001059	0
variety	-.036203	.00894	-4.05	0.000	-.053724 -.018681	1
autonomy	.057279	.013801	4.15	0.000	.030231 .084328	0
security	-.0128	.002665	-4.80	0.000	-.018024 -.007576	0
low*						
esteem	.000518	.000833	0.62	0.534	-.001116 .002151	1
variety	.006409	.010588	0.61	0.545	-.014343 .027161	0
autonomy	-.008818	.013766	-0.64	0.522	-.035799 .018163	0
security	.002314	.003697	0.63	0.531	-.004932 .009561	1
casevars						
female*	.013839	.021607	0.64	0.522	-.028509 .056188	0
score	.017917	.011062	1.62	0.105	-.003764 .039598	.32465

(*) dp/dx is for discrete change of indicator variable from 0 to 1

The probability computed by estat mfx matches the probability computed by predict, pr1 only within three digits. This outcome is because of how the computation is carried out and the numeric inaccuracy of the GHK simulator using a Hammersley point set of length 200. The computation carried out by estat mfx literally computes all six probabilities listed in the header of the MFX table and sums them. The computation by predict, pr1 is the same as predict after asmprobit (multinomial probit): it computes the probability that autonomy is chosen, thus requiring only one call to the GHK simulator. Hence, there is a difference in the reported values even though the two probability statements are equivalent.

◁

Saved results

estat mfx saves the following in r():

Scalars
  r(pr)   scalar containing the computed probability of the ranked alternatives.

Matrices
  r(ranks)   column vector containing the alternative ranks. The rownames identify the alternatives.

  r(mfx)   matrix containing the computed marginal effects and associated statistics. Column 1 of the matrix contains the marginal effects; column 2, their standard errors; column 3, their z statistics; and columns 4 and 5, the confidence intervals. Column 6 contains the values of the independent variables used to compute the probabilities r(pr).

Methods and formulas

All postestimation commands listed above are implemented as ado-files.

Also see

[R] **asroprobit** — Alternative-specific rank-ordered probit regression

[R] **asmprobit** — Alternative-specific multinomial probit regression

[U] **20 Estimation and postestimation commands**

Title

> **BIC note** — Calculating and interpreting BIC

Description

This entry discusses a statistical issue that arises when using the Bayesian information criterion (BIC) to compare models.

Stata calculates BIC, assuming $N = $ e(N)—we will explain—but sometimes it would be better if a different N were used. Commands that calculate BIC have an n() option, allowing you to specify the N to be used.

In summary,

1. If you are comparing results estimated by the same estimation command, using the default BIC calculation is probably fine. There is an issue, but most researchers would ignore it.

2. If you are comparing results estimated by different estimation commands, you need to be on your guard.

   a. If the different estimation commands share the same definitions of observations, independence, and the like, you are back in case 1.

   b. If they differ in these regards, you need to think about the value of N that should be used. For example, logit and xtlogit differ in that the former assumes independent observations and the latter, independent panels.

   c. If estimation commands differ in the events being used over which the likelihood function is calculated, the information criteria may not be comparable at all. We say information *criteria* because this would apply equally to the Akaike information criterion (AIC), as well as to BIC. For instance, streg and stcox produce such incomparable results. The events used by streg are the actual survival times, whereas the events used by stcox are failures within risk pools, conditional on the times at which failures occurred.

Remarks

Remarks are presented under the following headings:

> *Background*
> *The problem of determining N*
> *The problem of conformable likelihoods*
> *The first problem does not arise with AIC; the second problem does*
> *Calculating BIC correctly*

Background

The AIC and the BIC are two popular measures for comparing maximum likelihood models. AIC and BIC are defined as

$$\text{AIC} = -2 \times \ln(\text{likelihood}) + 2 \times k$$

$$\text{BIC} = -2 \times \ln(\text{likelihood}) + \ln(N) \times k$$

where

$$k = \text{number of parameters estimated}$$

$$N = \text{number of observations}$$

We are going to discuss AIC along with BIC because AIC has some of the problems that BIC has, but not all.

AIC and BIC can be viewed as measures that combine fit and complexity. Fit is measured negatively by $-2 \times \ln(\text{likelihood})$; the larger the value, the worse the fit. Complexity is measured positively, either by $2 \times k$ (AIC) or $\ln(N) \times k$ (BIC).

Given two models fit on the same data, the model with the smaller value of the information criterion is considered to be better.

There is substantial literature on these measures: see Akaike (1974); Raftery (1995); Sakamoto, Ishiguro, and Kitagawa (1986); and Schwarz (1978).

When Stata calculates the above measures, it uses the rank of e(V) for k and it uses e(N) for N. e(V) and e(N) are Stata notation for results stored by the estimation command. e(V) is the variance–covariance matrix of the estimated parameters, and e(N) is the number of observations in the dataset used in calculating the result.

The problem of determining N

The difference between AIC and BIC is that AIC uses the constant 2 to weight k, whereas BIC uses $\ln(N)$.

Determining what value of N should be used is problematic. Despite appearances, the definition "N is the number of observations" is not easy to make operational. N does not appear in the likelihood function itself, N is not the output of a standard statistical formula, and what is an observation is often subjective.

▷ Example 1

Often what is meant by N is obvious. Consider a simple logit model. What is meant by N is the number of observations that are statistically independent and that corresponds to M, the number of observations in the dataset used in the calculation. We will write $N = M$.

But now assume that the same dataset has a grouping variable and the data are thought to be clustered within group. To keep the problem simple, let's pretend that there are G groups and m observations within group, so that $M = G \times m$. Because you are worried about intragroup correlation, you fit your model with xtlogit, grouping on the grouping variable. Now you wish to calculate BIC. What is the N that should be used? $N = M$ or $N = G$?

That is a deep question. If the observations really are independent, then you should use $N = M$. If the observations within group are not just correlated but are duplicates of one another, and they had to be so, then you should use $M = G$. Between those two extremes, you should probably use a number between N and G, but determining what that number should be from measured correlations is difficult. Using $N = M$ is conservative in that, if anything, it overweights complexity. Conservativeness, however, is subjective, too: using $N = G$ could be considered more conservative in that fewer constraints are being placed on the data.

When the estimated correlation is high, our reaction would be that using $N = G$ is probably more reasonable. Our first reaction, however, would be that using BIC to compare models is probably a misuse of the measure.

Stata uses $N = M$. An informal survey of web-based literature suggests that $N = M$ is the popular choice.

There is another reason, not so good, to choose $N = M$. It makes across-model comparisons more likely to be valid when performed without thinking about the issue. Say that you wish to compare the `logit` and `xtlogit` results. Thus you need to calculate

$$\text{BIC}_p = -2 \times \ln(\text{likelihood}_p) + \ln(N_p) \times k$$

$$\text{BIC}_x = -2 \times \ln(\text{likelihood}_x) + \ln(N_x) \times k$$

Whatever N you use, you must use the same N in both formulas. Stata's choice of $N = M$ at least meets that test.

◁

▷ Example 2

In the above example, using $N = M$ is reasonable. Now let's look at when using $N = M$ is wrong, even if popular.

Consider a model fit by `stcox`. Using $N = M$ is certainly wrong if for no other reason than M is not even a well-defined number. The same data can be represented by different datasets with different numbers of observations. For example, in one dataset, there might be 1 observation per subject. In another, the same subjects could have two records each, the first recording the first half of the time at risk and the second recording the remaining part. All statistics calculated by Stata on either dataset would be the same, but M would be different.

Deciding on the right definition, however, is difficult. Viewed one way, N in the Cox regression case should be the number of risk pools, R, because the Cox regression calculation is made on the basis of the independent risk pools. Viewed another way, N should be the number of subjects, N_{subj}, because, even though the likelihood function is based on risk pools, the parameters estimated are at the subject level.

You can decide which argument you prefer.

For parametric survival models, in single-record data, $N = M$ is unambiguously correct. For multirecord data, there is an argument for $N = M$ and for $N = N_{\text{subj}}$.

◁

The problem of conformable likelihoods

The problem of conformable likelihoods does not concern N. Researchers sometimes use information criteria such as BIC and AIC to make comparisons across models. For that to be valid, the likelihoods must be conformable; that is, the likelihoods must all measure the same thing.

It is common to think of the likelihood function as the $\Pr(\text{data} \mid \text{parameters})$, but in fact, the likelihood is

$$\Pr(\text{particular events in the data} \mid \text{parameters})$$

You must ensure that the events are the same.

For instance, they are not the same in the semiparametric Cox regression and the various parametric survival models. In Cox regression, the events are, at each failure time, that the subjects observed to fail in fact failed, given that failures occurred at those times. In the parametric models, the events are that each subject failed exactly when the subject was observed to fail.

The formula for AIC and BIC is

$$\text{measure} = -2 \times \ln(\text{likelihood}) + \text{complexity}$$

When you are comparing models, if the likelihoods are measuring different events, even if the models obtain estimates of the same parameters, differences in the information measures are irrelevant.

The first problem does not arise with AIC; the second problem does

Regardless of model, the problem of defining N never arises with AIC because N is not used in the AIC calculation. AIC uses a constant 2 to weight complexity as measured by k, rather than $\ln(N)$.

For both AIC and BIC, however, the likelihood functions must be conformable; that is, they must be measuring the same event.

Calculating BIC correctly

When using BIC to compare results, and especially when using BIC to compare results from different models, you should think carefully about how N should be defined. Then specify that number by using the n() option:

```
. estimates stats full sub, n(74)
```

Model	Obs	ll(null)	ll(model)	df	AIC	BIC
full	102	-45.03321	-20.59083	4	49.18167	58.39793
sub	102	-45.03321	-27.17516	3	60.35031	67.26251

```
Note:  N = 74 used in calculating BIC
```

Both estimates stats and estat ic allow the n() option; see [R] **estimates stats** and [R] **estat**.

Methods and formulas

AIC and BIC are defined as

$$\text{AIC} = -2 \times \ln(\text{likelihood}) + 2 \times k$$

$$\text{BIC} = -2 \times \ln(\text{likelihood}) + \ln(N) \times k$$

where k is the model degrees of freedom calculated as the rank of variance–covariance matrix of the parameters e(V) and N is the number of observations used in estimation or, more precisely, the number of independent terms in the likelihood. Operationally, N is defined as e(N) unless the n() option is specified.

References

Akaike, H. 1974. A new look at the statistical model identification. *IEEE Transactions on Automatic Control* 19: 716–723.

Raftery, A. 1995. Bayesian model selection in social research. In Vol. 25 of *Sociological Methodology*, ed. P. V. Marsden, 111–163. Oxford: Blackwell.

Sakamoto, Y., M. Ishiguro, and G. Kitagawa. 1986. *Akaike Information Criterion Statistics*. Dordrecht, The Netherlands: Reidel.

Schwarz, G. 1978. Estimating the dimension of a model. *Annals of Statistics* 6: 461–464.

Also see

[R] **estat** — Postestimation statistics

[R] **estimates stats** — Model statistics

Title

> **binreg** — Generalized linear models: Extensions to the binomial family

Syntax

binreg *depvar* [*indepvars*] [*if*] [*in*] [*weight*] [, *options*]

options	Description
Model	
<u>nocon</u>stant	suppress constant term
or	use logit link and report odds ratios
rr	use log link and report risk ratios
hr	use log-complement link and report health ratios
rd	use identity link and report risk differences
n(*#* | *varname*)	use *#* or *varname* for number of trials
exposure(*varname*)	include ln(*varname*) in model with coefficient constrained to 1
<u>off</u>set(*varname*)	include *varname* in model with coefficient constrained to 1
<u>constraints</u>(*constraints*)	apply specified linear constraints
<u>collinear</u>	keep collinear variables
mu(*varname*)	use *varname* as the initial estimate for the mean of *depvar*
<u>init</u>(*varname*)	synonym for mu(*varname*)
SE/Robust	
vce(*vcetype*)	*vcetype* may be eim, <u>r</u>obust, <u>c</u>luster *clustvar*, oim, opg, <u>boot</u>strap, jackknife, hac *kernel*, jackknife1, or <u>un</u>biased
t(*varname*)	variable name corresponding to time
<u>vf</u>actor(*#*)	multiply variance matrix by scalar *#*
disp(*#*)	quasi-likelihood multiplier
<u>scale</u>(x2 | dev | *#*)	set the scale parameter; default is scale(1)
Reporting	
<u>level</u>(*#*)	set confidence level; default is level(95)
<u>coef</u>ficients	report nonexponentiated coefficients
<u>nocns</u>report	do not display constraints
display_options	control column formats, row spacing, line width, and display of omitted variables and base and empty cells
Maximization	
irls	use iterated, reweighted least-squares optimization; the default
ml	use maximum likelihood optimization
maximize_options	control the maximization process; seldom used
fisher(*#*)	Fisher scoring steps
search	search for good starting values
<u>coef</u>legend	display legend instead of statistics

indepvars may contain factor variables; see [U] **11.4.3 Factor variables**.

depvar and *indepvars* may contain time-series operators; see [U] **11.4.4 Time-series varlists**.

bootstrap, by, jackknife, mi estimate, rolling, and statsby are allowed; see [U] **11.1.10 Prefix commands**.

vce(bootstrap), vce(jackknife), and vce(jackknife1) are not allowed with the mi estimate prefix; see [MI] **mi estimate**.

Weights are not allowed with the bootstrap prefix; see [R] **bootstrap**.

aweights are not allowed with the jackknife prefix; see [R] **jackknife**.

fweights, aweights, iweights, and pweights are allowed; see [U] **11.1.6 weight**.

coeflegend does not appear in the dialog box.

See [U] **20 Estimation and postestimation commands** for more capabilities of estimation commands.

Menu

Statistics > Generalized linear models > GLM for the binomial family

Description

binreg fits generalized linear models for the binomial family. It estimates odds ratios, risk ratios, health ratios, and risk differences. The available links are

Option	Implied link	Parameter
or	logit	odds ratios = $\exp(\beta)$
rr	log	risk ratios = $\exp(\beta)$
hr	log complement	health ratios = $\exp(\beta)$
rd	identity	risk differences = β

Estimates of odds, risk, and health ratios are obtained by exponentiating the appropriate coefficients. The or option produces the same results as Stata's logistic command, and or coefficients yields the same results as the logit command. When no link is specified, or is assumed.

Options

    ┌ Model ┐

noconstant; see [R] **estimation options**.

or requests the logit link and results in odds ratios if coefficients is not specified.

rr requests the log link and results in risk ratios if coefficients is not specified.

hr requests the log-complement link and results in health ratios if coefficients is not specified.

rd requests the identity link and results in risk differences.

n(# | *varname*) specifies either a constant integer to use as the denominator for the binomial family or a variable that holds the denominator for each observation.

exposure(*varname*), offset(*varname*), constraints(*constraints*), collinear; see [R] **estimation options**. constraints(*constraints*) and collinear are not allowed with irls.

mu(*varname*) specifies *varname* containing an initial estimate for the mean of *depvar*. This option can be useful if you encounter convergence difficulties. init(*varname*) is a synonym.

vce(*vcetype*) specifies the type of standard error reported, which includes types that are robust to some kinds of misspecification, that allow for intragroup correlation, that are derived from asymptotic theory, and that use bootstrap or jackknife methods; see [R] *vce_option*.

vce(eim), the default, uses the expected information matrix (EIM) for the variance estimator.

binreg also allows the following:

vce(hac *kernel* [*#*]) specifies that a heteroskedasticity- and autocorrelation-consistent (HAC) variance estimate be used. HAC refers to the general form for combining weighted matrices to form the variance estimate. There are three kernels built into binreg. *kernel* is a user-written program or one of

nwest | gallant | anderson

If *#* is not specified, $N - 2$ is assumed.

vce(jackknife1) specifies that the one-step jackknife estimate of variance be used.

vce(unbiased) specifies that the unbiased sandwich estimate of variance be used.

t(*varname*) specifies the variable name corresponding to time; see [TS] **tsset**. binreg does not always need to know t(), though it does if vce(hac ...) is specified. Then you can either specify the time variable with t(), or you can tsset your data before calling binreg. When the time variable is required, binreg assumes that the observations are spaced equally over time.

vfactor(*#*) specifies a scalar by which to multiply the resulting variance matrix. This option allows users to match output with other packages, which may apply degrees of freedom or other small-sample corrections to estimates of variance.

disp(*#*) multiplies the variance of *depvar* by *#* and divides the deviance by *#*. The resulting distributions are members of the quasilikelihood family.

scale(x2 | dev | *#*) overrides the default scale parameter. This option is allowed only with Hessian (information matrix) variance estimates.

By default, scale(1) is assumed for the discrete distributions (binomial, Poisson, and negative binomial), and scale(x2) is assumed for the continuous distributions (Gaussian, gamma, and inverse Gaussian).

scale(x2) specifies that the scale parameter be set to the Pearson chi-squared (or generalized chi-squared) statistic divided by the residual degrees of freedom, which was recommended by McCullagh and Nelder (1989) as a good general choice for continuous distributions.

scale(dev) sets the scale parameter to the deviance divided by the residual degrees of freedom. This option provides an alternative to scale(x2) for continuous distributions and overdispersed or underdispersed discrete distributions.

scale(*#*) sets the scale parameter to *#*.

level(*#*), noconstant; see [R] **estimation options**.

coefficients displays the nonexponentiated coefficients and corresponding standard errors and confidence intervals. This option has no effect when the rd option is specified, because it always presents the nonexponentiated coefficients.

nocnsreport; see [R] **estimation options**.

display_options: <u>noomit</u>ted, vsquish, <u>noempty</u>cells, <u>base</u>levels, <u>allbase</u>levels, cformat(*%fmt*), pformat(*%fmt*), sformat(*%fmt*), and nolstretch; see [R] **estimation options**.

⌐──────┌ Maximization ┌──┐

irls requests iterated, reweighted least-squares (IRLS) optimization of the deviance instead of Newton–Raphson optimization of the log likelihood. This option is the default.

ml requests that optimization be carried out by using Stata's ml command; see [R] **ml**.

maximize_options: <u>techn</u>ique(*algorithm_spec*), [no]log, <u>trace</u>, gradient, showstep, <u>hessian</u>, showtolerance, <u>diff</u>icult, <u>iter</u>ate(*#*), <u>tol</u>erance(*#*), <u>ltol</u>erance(*#*), <u>nrtol</u>erance(*#*), nonrtolerance, and from(*init_specs*); see [R] **maximize**. These options are seldom used.

Setting the optimization method to ml, with technique() set to something other than BHHH, changes the *vcetype* to vce(oim). Specifying technique(bhhh) changes *vcetype* to vce(opg).

fisher(*#*) specifies the number of Newton–Raphson steps that should use the Fisher scoring Hessian or EIM before switching to the observed information matrix (OIM). This option is available only if ml is specified and is useful only for Newton–Raphson optimization.

search specifies that the command search for good starting values. This option is available only if ml is specified and is useful only for Newton–Raphson optimization.

The following option is available with binreg but is not shown in the dialog box:

coeflegend; see [R] **estimation options**.

Remarks

Wacholder (1986) suggests methods for estimating risk ratios and risk differences from prospective binomial data. These estimates are obtained by selecting the proper link functions in the generalized linear-model framework. (See *Methods and formulas* for details; also see [R] **glm**.)

▷ Example 1

Wacholder (1986) presents an example, using data from Wright et al. (1983), of an investigation of the relationship between alcohol consumption and the risk of a low-birthweight baby. Covariates examined included whether the mother smoked (yes or no), mother's social class (three levels), and drinking frequency (light, moderate, or heavy). The data for the 18 possible categories determined by the covariates are illustrated below.

Let's first describe the data and list a few observations.

```
. use http://www.stata-press.com/data/r12/binreg
. list
```

	cat	d	n	alc	smo	soc
1.	1	11	84	3	1	1
2.	2	5	79	2	1	1
3.	3	11	169	1	1	1
4.	4	6	28	3	2	1
5.	5	3	13	2	2	1
6.	6	1	26	1	2	1
7.	7	4	22	3	1	2
8.	8	3	25	2	1	2
9.	9	12	162	1	1	2
10.	10	4	17	3	2	2
11.	11	2	7	2	2	2
12.	12	6	38	1	2	2
13.	13	0	14	3	1	3
14.	14	1	18	2	1	3
15.	15	12	91	1	1	3
16.	16	7	19	3	2	3
17.	17	2	18	2	2	3
18.	18	8	70	1	2	3

Each observation corresponds to one of the 18 covariate structures. The number of low-birthweight babies from n in each category is given by the d variable.

We begin by estimating risk ratios:

```
. binreg d i.soc i.alc i.smo, n(n) rr
Iteration 1:   deviance =   14.2879
Iteration 2:   deviance =    13.607
Iteration 3:   deviance = 13.60503
Iteration 4:   deviance = 13.60503
```

```
Generalized linear models                     No. of obs       =        18
Optimization      : MQL Fisher scoring        Residual df      =        12
                    (IRLS EIM)                Scale parameter  =         1
Deviance          =    13.6050268            (1/df) Deviance   =  1.133752
Pearson           =    11.51517095           (1/df) Pearson    =  .9595976
Variance function: V(u) = u*(1-u/n)           [Binomial]
Link function     : g(u) = ln(u/n)            [Log]
                                              BIC              = -21.07943
```

d	Risk Ratio	EIM Std. Err.	z	P>|z|	[95% Conf. Interval]	
soc						
2	1.340001	.3127382	1.25	0.210	.848098	2.11721
3	1.349487	.3291488	1.23	0.219	.8366715	2.176619
alc						
2	1.191157	.3265354	0.64	0.523	.6960276	2.038503
3	1.974078	.4261751	3.15	0.002	1.293011	3.013884
2.smo	1.648444	.332875	2.48	0.013	1.109657	2.448836
_cons	.0630341	.0128061	-13.61	0.000	.0423297	.0938656

By default, Stata reports the risk ratios (the exponentiated regression coefficients) estimated by the model. We can see that the risk ratio comparing heavy drinkers with light drinkers, after adjusting for smoking and social class, is 1.974078. That is, mothers who drink heavily during their pregnancy have approximately twice the risk of delivering low-birthweight babies as mothers who are light drinkers.

The nonexponentiated coefficients can be obtained with the `coefficients` option:

```
. binreg d i.soc i.alc i.smo, n(n) rr coefficients
Iteration 1:   deviance =   14.2879
Iteration 2:   deviance =   13.607
Iteration 3:   deviance =  13.60503
Iteration 4:   deviance =  13.60503
```

Generalized linear models		No. of obs	=	18
Optimization : MQL Fisher scoring		Residual df	=	12
(IRLS EIM)		Scale parameter	=	1
Deviance = 13.6050268		(1/df) Deviance	=	1.133752
Pearson = 11.51517095		(1/df) Pearson	=	.9595976
Variance function: V(u) = u*(1−u/n)		[Binomial]		
Link function : g(u) = ln(u/n)		[Log]		
		BIC	=	−21.07943

d	Coef.	EIM Std. Err.	z	P>|z|	[95% Conf. Interval]	
soc						
2	.2926702	.2333866	1.25	0.210	−.1647591	.7500994
3	.2997244	.2439066	1.23	0.219	−.1783238	.7777726
alc						
2	.1749248	.274133	0.64	0.523	−.362366	.7122156
3	.6801017	.2158856	3.15	0.002	.2569737	1.10323
2.smo	.4998317	.2019329	2.48	0.013	.1040505	.8956129
_cons	−2.764079	.2031606	−13.61	0.000	−3.162266	−2.365891

Risk differences are obtained with the rd option:

```
. binreg d i.soc i.alc i.smo, n(n) rd
Iteration 1:   deviance =  18.67277
Iteration 2:   deviance =  14.94364
Iteration 3:   deviance =   14.9185
Iteration 4:   deviance =  14.91762
Iteration 5:   deviance =  14.91758
Iteration 6:   deviance =  14.91758
Iteration 7:   deviance =  14.91758
```

Generalized linear models		No. of obs	=	18
Optimization : MQL Fisher scoring		Residual df	=	12
(IRLS EIM)		Scale parameter	=	1
Deviance = 14.91758277		(1/df) Deviance	=	1.243132
Pearson = 12.60353235		(1/df) Pearson	=	1.050294
Variance function: V(u) = u*(1-u/n)		[Binomial]		
Link function : g(u) = u/n		[Identity]		
		BIC	=	-19.76688

d	Risk Diff.	EIM Std. Err.	z	P>|z|	[95% Conf. Interval]	
soc						
2	.0263817	.0232124	1.14	0.256	-.0191137	.0718771
3	.0365553	.0268668	1.36	0.174	-.0161026	.0892132
alc						
2	.0122539	.0257713	0.48	0.634	-.0382569	.0627647
3	.0801291	.0302878	2.65	0.008	.020766	.1394921
2.smo	.0542415	.0270838	2.00	0.045	.0011582	.1073248
_cons	.059028	.0160693	3.67	0.000	.0275327	.0905232

The risk difference between heavy drinkers and light drinkers is simply the value of the coefficient for 3.alc = 0.0801291. Because the risk differences are obtained directly from the coefficients estimated by using the identity link, the coefficients option has no effect here.

Health ratios are obtained with the hr option. The health ratios (exponentiated coefficients for the log-complement link) are reported directly.

```
. binreg d i.soc i.alc i.smo, n(n) hr
Iteration 1:    deviance =  21.15233
Iteration 2:    deviance =  15.16467
Iteration 3:    deviance =  15.13205
Iteration 4:    deviance =  15.13114
Iteration 5:    deviance =  15.13111
Iteration 6:    deviance =  15.13111
Iteration 7:    deviance =  15.13111
```

Generalized linear models				No. of obs	=	18
Optimization	: MQL Fisher scoring			Residual df	=	12
	(IRLS EIM)			Scale parameter =		1
Deviance	= 15.13110545			(1/df) Deviance =		1.260925
Pearson	= 12.84203917			(1/df) Pearson =		1.07017
Variance function: $V(u) = u*(1-u/n)$				[Binomial]		
Link function	: $g(u) = \ln(1-u/n)$			[Log complement]		
				BIC		= −19.55336

d	HR	EIM Std. Err.	z	P>|z|	[95% Conf. Interval]	
soc						
2	.9720541	.024858	−1.11	0.268	.9245342	1.022017
3	.9597182	.0290412	−1.36	0.174	.9044535	1.01836
alc						
2	.9871517	.0278852	−0.46	0.647	.9339831	1.043347
3	.9134243	.0325726	−2.54	0.011	.8517631	.9795493
2.smo	.9409983	.0296125	−1.93	0.053	.8847125	1.000865
_cons	.9409945	.0163084	−3.51	0.000	.9095674	.9735075

(HR) Health ratios

To see the nonexponentiated coefficients, we can specify the coefficients option.

◁

Saved results

binreg, irls saves the following in e():

Scalars

e(N)	number of observations
e(k)	number of parameters
e(k_eq_model)	number of equations in overall model test
e(df_m)	model degrees of freedom
e(df)	residual degrees of freedom
e(phi)	model scale parameter
e(disp)	dispersion parameter
e(bic)	model BIC
e(N_clust)	number of clusters
e(deviance)	deviance
e(deviance_s)	scaled deviance
e(deviance_p)	Pearson deviance
e(deviance_ps)	scaled Pearson deviance
e(dispers)	dispersion
e(dispers_s)	scaled dispersion
e(dispers_p)	Pearson dispersion
e(dispers_ps)	scaled Pearson dispersion
e(vf)	factor set by vfactor(), 1 if not set
e(rank)	rank of e(V)
e(rc)	return code

Macros

e(cmd)	binreg
e(cmdline)	command as typed
e(depvar)	name of dependent variable
e(eform)	eform() option implied by or, rr, hr, or rd
e(varfunc)	program to calculate variance function
e(varfunct)	variance title
e(varfuncf)	variance function
e(link)	program to calculate link function
e(linkt)	link title
e(linkf)	link function
e(m)	number of binomial trials
e(wtype)	weight type
e(wexp)	weight expression
e(title_fl)	family–link title
e(clustvar)	name of cluster variable
e(offset)	linear offset variable
e(cons)	noconstant or not set
e(hac_kernel)	HAC kernel
e(hac_lag)	HAC lag
e(vce)	vcetype specified in vce()
e(vcetype)	title used to label Std. Err.
e(opt)	type of optimization
e(opt1)	optimization title, line 1
e(opt2)	optimization title, line 2
e(properties)	b V
e(predict)	program used to implement predict
e(marginsok)	predictions allowed by margins
e(marginsnotok)	predictions disallowed by margins
e(asbalanced)	factor variables fvset as asbalanced
e(asobserved)	factor variables fvset as asobserved

Matrices

e(b)	coefficient vector
e(Cns)	constraints matrix
e(V)	variance–covariance matrix of the estimators
e(V_modelbased)	model-based variance

Functions

e(sample)	marks estimation sample

binreg, ml saves the following in e():

Scalars

e(N)	number of observations
e(k)	number of parameters
e(k_eq)	number of equations in e(b)
e(k_eq_model)	number of equations in overall model test
e(k_dv)	number of dependent variables
e(df_m)	model degrees of freedom
e(df)	residual degrees of freedom
e(phi)	model scale parameter
e(aic)	model AIC, if ml
e(bic)	model BIC
e(ll)	log likelihood, if ml
e(N_clust)	number of clusters
e(chi2)	χ^2
e(p)	significance of model test
e(deviance)	deviance
e(deviance_s)	scaled deviance
e(deviance_p)	Pearson deviance
e(deviance_ps)	scaled Pearson deviance
e(dispers)	dispersion
e(dispers_s)	scaled dispersion
e(dispers_p)	Pearson dispersion
e(dispers_ps)	scaled Pearson dispersion
e(vf)	factor set by vfactor(), 1 if not set
e(rank)	rank of e(V)
e(ic)	number of iterations
e(rc)	return code
e(converged)	1 if converged, 0 otherwise

Macros

e(cmd)	binreg
e(cmdline)	command as typed
e(depvar)	name of dependent variable
e(eform)	eform() option implied by or, rr, hr, or rd
e(varfunc)	program to calculate variance function
e(varfunct)	variance title
e(varfuncf)	variance function
e(link)	program to calculate link function
e(linkt)	link title
e(linkf)	link function
e(m)	number of binomial trials
e(wtype)	weight type
e(wexp)	weight expression
e(title)	title in estimation output
e(title_fl)	family–link title
e(clustvar)	name of cluster variable
e(offset)	linear offset variable
e(cons)	noconstant or not set
e(hac_kernel)	HAC kernel
e(hac_lag)	HAC lag
e(chi2type)	Wald; type of model χ^2 test
e(vce)	vcetype specified in vce()
e(vcetype)	title used to label Std. Err.
e(opt)	type of optimization
e(opt1)	optimization title, line 1
e(which)	max or min; whether optimizer is to perform maximization or minimization
e(ml_method)	type of ml method
e(user)	name of likelihood-evaluator program
e(technique)	maximization technique
e(properties)	b V
e(predict)	program used to implement predict
e(marginsok)	predictions allowed by margins
e(marginsnotok)	predictions disallowed by margins
e(asbalanced)	factor variables fvset as asbalanced
e(asobserved)	factor variables fvset as asobserved

Matrices

e(b)	coefficient vector
e(Cns)	constraints matrix
e(ilog)	iteration log (up to 20 iterations)
e(gradient)	gradient vector
e(V)	variance–covariance matrix of the estimators
e(V_modelbased)	model-based variance

Functions

e(sample)	marks estimation sample

Methods and formulas

binreg is implemented as an ado-file.

Let π_i be the probability of success for the ith observation, $i = 1, \ldots, N$, and let $X\beta$ be the linear predictor. The link function relates the covariates of each observation to its respective probability through the linear predictor.

In logistic regression, the logit link is used:

$$\ln\left(\frac{\pi}{1-\pi}\right) = X\beta$$

The regression coefficient β_k represents the change in the logarithm of the odds associated with a one-unit change in the value of the X_k covariate; thus $\exp(\beta_k)$ is the ratio of the odds associated with a change of one unit in X_k.

For risk differences, the identity link $\pi = X\beta$ is used. The regression coefficient β_k represents the risk difference associated with a change of one unit in X_k. When using the identity link, you can obtain fitted probabilities outside the interval $(0, 1)$. As suggested by Wacholder, at each iteration, fitted probabilities are checked for range conditions (and put back in range if necessary). For example, if the identity link results in a fitted probability that is smaller than 1e–4, the probability is replaced with 1e–4 before the link function is calculated.

A similar adjustment is made for the logarithmic link, which is used for estimating the risk ratio, $\ln(\pi) = X\beta$, where $\exp(\beta_k)$ is the risk ratio associated with a change of one unit in X_k, and for the log-complement link used to estimate the probability of no disease or health, where $\exp(\beta_k)$ represents the "health ratio" associated with a change of one unit in X_k.

This command supports the Huber/White/sandwich estimator of the variance and its clustered version using vce(robust) and vce(cluster *clustvar*), respectively. See [P] _robust, particularly *Maximum likelihood estimators* and *Methods and formulas*.

References

Cummings, P. 2009. Methods for estimating adjusted risk ratios. *Stata Journal* 9: 175–196.

Hardin, J. W., and M. A. Cleves. 1999. sbe29: Generalized linear models: Extensions to the binomial family. *Stata Technical Bulletin* 50: 21–25. Reprinted in *Stata Technical Bulletin Reprints*, vol. 9, pp. 140–146. College Station, TX: Stata Press.

Kleinbaum, D. G., and M. Klein. 2010. *Logistic Regression: A Self-Learning Text*. 3rd ed. New York: Springer.

McCullagh, P., and J. A. Nelder. 1989. *Generalized Linear Models*. 2nd ed. London: Chapman & Hall/CRC.

Wacholder, S. 1986. Binomial regression in GLIM: Estimating risk ratios and risk differences. *American Journal of Epidemiology* 123: 174–184.

Wright, J. T., I. G. Barrison, I. G. Lewis, K. D. MacRae, E. J. Waterson, P. J. Toplis, M. G. Gordon, N. F. Morris, and I. M. Murray-Lyon. 1983. Alcohol consumption, pregnancy and low birthweight. *Lancet* 1: 663–665.

Also see

[R] **binreg postestimation** — Postestimation tools for binreg

[R] **glm** — Generalized linear models

[MI] **estimation** — Estimation commands for use with mi estimate

[U] **20 Estimation and postestimation commands**

Title

binreg postestimation — Postestimation tools for binreg

Description

The following postestimation commands are available after `binreg`:

Command	Description
contrast	contrasts and ANOVA-style joint tests of estimates
estat	AIC, BIC, VCE, and estimation sample summary
estimates	cataloging estimation results
lincom	point estimates, standard errors, testing, and inference for linear combinations of coefficients
linktest	link test for model specification
margins	marginal means, predictive margins, marginal effects, and average marginal effects
marginsplot	graph the results from margins (profile plots, interaction plots, etc.)
nlcom	point estimates, standard errors, testing, and inference for nonlinear combinations of coefficients
predict	predictions, residuals, influence statistics, and other diagnostic measures
predictnl	point estimates, standard errors, testing, and inference for generalized predictions
pwcompare	pairwise comparisons of estimates
test	Wald tests of simple and composite linear hypotheses
testnl	Wald tests of nonlinear hypotheses

See the corresponding entries in the *Base Reference Manual* for details.

Syntax for predict

predict [*type*] *newvar* [*if*] [*in*] [, *statistic options*]

statistic	Description
Main	
mu	expected value of y; the default
xb	linear prediction $\eta = \mathbf{x}\widehat{\beta}$
eta	synonym for xb
stdp	standard error of the linear prediction
anscombe	Anscombe (1953) residuals
cooksd	Cook's distance
deviance	deviance residuals
hat	diagonals of the "hat" matrix as an analog to simple linear regression
likelihood	weighted average of the standardized deviance and standard Pearson residuals
pearson	Pearson residuals
response	differences between the observed and fitted outcomes
score	first derivative of the log likelihood with respect to $\mathbf{x}_j\beta$
working	working residuals

options	Description
Options	
nooffset	modify calculations to ignore the offset variable
adjusted	adjust deviance residual to speed up convergence
standardized	multiply residual by the factor $(1 - h)^{1/2}$
studentized	multiply residual by one over the square root of the estimated scale parameter
modified	modify denominator of residual to be a reasonable estimate of the variance of *depvar*

These statistics are available both in and out of sample; type predict ... if e(sample) ... if wanted only for the estimation sample.

Menu

Statistics > Postestimation > Predictions, residuals, etc.

Options for predict

> Main

mu, the default, specifies that predict calculate the expected value of y, equal to $g^{-1}(\mathbf{x}\widehat{\beta})$ [$ng^{-1}(\mathbf{x}\widehat{\beta})$ for the binomial family].

xb calculates the linear prediction $\eta = \mathbf{x}\widehat{\beta}$.

eta is a synonym for xb.

stdp calculates the standard error of the linear prediction.

anscombe calculates the Anscombe (1953) residuals to produce residuals that closely follow a normal distribution.

cooksd calculates Cook's distance, which measures the aggregate change in the estimated coefficients when each observation is left out of the estimation.

deviance calculates the deviance residuals, which are recommended by McCullagh and Nelder (1989) and others as having the best properties for examining goodness of fit of a GLM. They are approximately normally distributed if the model is correct and may be plotted against the fitted values or against a covariate to inspect the model's fit. Also see the pearson option below.

hat calculates the diagonals of the "hat" matrix as an analog to simple linear regression.

likelihood calculates a weighted average of the standardized deviance and standardized Pearson (described below) residuals.

pearson calculates the Pearson residuals, which often have markedly skewed distributions for nonnormal family distributions. Also see the deviance option above.

response calculates the differences between the observed and fitted outcomes.

score calculates the equation-level score, $\partial \ln L / \partial (\mathbf{x}_j \beta)$.

working calculates the working residuals, which are response residuals weighted according to the derivative of the link function.

⌐Options⌐

nooffset is relevant only if you specified offset(*varname*) for binreg. It modifies the calculations made by predict so that they ignore the offset variable; the linear prediction is treated as $\mathbf{x}_j \mathbf{b}$ rather than as $\mathbf{x}_j \mathbf{b} + \text{offset}_j$.

adjusted adjusts the deviance residual to make the convergence to the limiting normal distribution faster. The adjustment deals with adding to the deviance residual a higher-order term depending on the variance function family. This option is allowed only when deviance is specified.

standardized requests that the residual be multiplied by the factor $(1 - h)^{-1/2}$, where h is the diagonal of the hat matrix. This step is done to take into account the correlation between *depvar* and its predicted value.

studentized requests that the residual be multiplied by one over the square root of the estimated scale parameter.

modified requests that the denominator of the residual be modified to be a reasonable estimate of the variance of *depvar*. The base residual is multiplied by the factor $(k/w)^{-1/2}$, where k is either one or the user-specified dispersion parameter and w is the specified weight (or one if left unspecified).

Methods and formulas

All postestimation commands listed above are implemented as ado-files.

References

Anscombe, F. J. 1953. Contribution of discussion paper by H. Hotelling "New light on the correlation coefficient and its transforms". *Journal of the Royal Statistical Society, Series B* 15: 229–230.

McCullagh, P., and J. A. Nelder. 1989. *Generalized Linear Models*. 2nd ed. London: Chapman & Hall/CRC.

Also see

[R] **binreg** — Generalized linear models: Extensions to the binomial family

[U] **20 Estimation and postestimation commands**

Title

> **biprobit** — Bivariate probit regression

Syntax

Bivariate probit regression

> biprobit *depvar₁* *depvar₂* [*indepvars*] [*if*] [*in*] [*weight*] [, *options*]

Seemingly unrelated bivariate probit regression

> biprobit *equation₁* *equation₂* [*if*] [*in*] [*weight*] [, *su_options*]

where *equation₁* and *equation₂* are specified as

> ([*eqname*:] *depvar* [=] [*indepvars*] [, <u>noco</u>nstant <u>off</u>set(*varname*)])

options	Description
Model	
<u>noco</u>nstant	suppress constant term
<u>partial</u>	fit partial observability model
offset1(*varname*)	offset variable for first equation
offset2(*varname*)	offset variable for second equation
<u>cons</u>traints(*constraints*)	apply specified linear constraints
<u>col</u>linear	keep collinear variables
SE/Robust	
vce(*vcetype*)	*vcetype* may be oim, <u>r</u>obust, <u>c</u>luster *clustvar*, opg, <u>boot</u>strap, or <u>jack</u>knife
Reporting	
<u>l</u>evel(#)	set confidence level; default is level(95)
noskip	perform likelihood-ratio test
<u>nocns</u>report	do not display constraints
display_options	control column formats, row spacing, line width, and display of omitted variables and base and empty cells
Maximization	
maximize_options	control the maximization process; seldom used
<u>coefl</u>egend	display legend instead of statistics

su_options	Description
Model	
<u>partial</u>	fit partial observability model
<u>constraints</u>(*constraints*)	apply specified linear constraints
<u>collinear</u>	keep collinear variables
SE/Robust	
vce(*vcetype*)	*vcetype* may be oim, <u>r</u>obust, <u>cl</u>uster *clustvar*, opg, <u>boot</u>strap, or <u>jackknife</u>
Reporting	
<u>level</u>(*#*)	set confidence level; default is level(95)
noskip	perform likelihood-ratio test
<u>nocns</u>report	do not display constraints
display_options	control column formats, row spacing, line width, and display of omitted variables and base and empty cells
Maximization	
maximize_options	control the maximization process; seldom used
<u>coefl</u>egend	display legend instead of statistics

indepvars may contain factor variables; see [U] **11.4.3 Factor variables**.

depvar$_1$, *depvar*$_2$, *indepvars*, and *depvar* may contain time-series operators; see [U] **11.4.4 Time-series varlists**.

bootstrap, by, jackknife, rolling, statsby, and svy are allowed; see [U] **11.1.10 Prefix commands**.

Weights are not allowed with the bootstrap prefix; see [R] **bootstrap**.

vce(), noskip, and weights are not allowed with the svy prefix; see [SVY] **svy**.

pweights, fweights, and iweights are allowed; see [U] **11.1.6 weight**.

coeflegend does not appear in the dialog box.

See [U] **20 Estimation and postestimation commands** for more capabilities of estimation commands.

Menu

biprobit

Statistics > Binary outcomes > Bivariate probit regression

seemingly unrelated biprobit

Statistics > Binary outcomes > Seemingly unrelated bivariate probit regression

Description

biprobit fits maximum-likelihood two-equation probit models—either a bivariate probit or a seemingly unrelated probit (limited to two equations).

Options

    ___| Model |___

noconstant; see [R] **estimation options**.

partial specifies that the partial observability model be fit. This particular model commonly has poor convergence properties, so we recommend that you use the difficult option if you want to fit the Poirier partial observability model; see [R] **maximize**.

This model computes the product of the two dependent variables so that you do not have to replace each with the product.

offset1(*varname*), offset2(*varname*), constraints(*constraints*), collinear; see [R] **estimation options**.

    ___| SE/Robust |___

vce(*vcetype*) specifies the type of standard error reported, which includes types that are derived from asymptotic theory, that are robust to some kinds of misspecification, that allow for intragroup correlation, and that use bootstrap or jackknife methods; see [R] ***vce_option***.

    ___| Reporting |___

level(*#*); see [R] **estimation options**.

noskip specifies that a full maximum-likelihood model with only a constant for the regression equation be fit. This model is not displayed but is used as the base model to compute a likelihood-ratio test for the model test statistic displayed in the estimation header. By default, the overall model test statistic is an asymptotically equivalent Wald test of all the parameters in the regression equation being zero (except the constant). For many models, this option can substantially increase estimation time.

nocnsreport; see [R] **estimation options**.

display_options: noomitted, vsquish, noemptycells, baselevels, allbaselevels, cformat(% *fmt*), pformat(% *fmt*), sformat(% *fmt*), and nolstretch; see [R] **estimation options**.

    ___| Maximization |___

maximize_options: difficult, technique(*algorithm_spec*), iterate(*#*), [no]log, trace, gradient, showstep, hessian, showtolerance, tolerance(*#*), ltolerance(*#*), nrtolerance(*#*), nonrtolerance, and from(*init_specs*); see [R] **maximize**. These options are seldom used.

Setting the optimization type to technique(bhhh) resets the default *vcetype* to vce(opg).

The following option is available with biprobit but is not shown in the dialog box:

coeflegend; see [R] **estimation options**.

Remarks

For a good introduction to the bivariate probit models, see Greene (2012, 738–752) and Pindyck and Rubinfeld (1998). Poirier (1980) explains the partial observability model. Van de Ven and Van Pragg (1981) explain the probit model with sample selection; see [R] **heckprob** for details.

▷ Example 1

We use the data from Pindyck and Rubinfeld (1998, 332). In this dataset, the variables are whether children attend private school (`private`), number of years the family has been at the present residence (`years`), log of property tax (`logptax`), log of income (`loginc`), and whether the head of the household voted for an increase in property taxes (`vote`).

We wish to model the bivariate outcomes of whether children attend private school and whether the head of the household voted for an increase in property tax based on the other covariates.

```
. use http://www.stata-press.com/data/r12/school
. biprobit private vote years logptax loginc
Fitting comparison equation 1:
Iteration 0:    log likelihood = -31.967097
Iteration 1:    log likelihood = -31.452424
Iteration 2:    log likelihood = -31.448958
Iteration 3:    log likelihood = -31.448958
Fitting comparison equation 2:
Iteration 0:    log likelihood = -63.036914
Iteration 1:    log likelihood = -58.534843
Iteration 2:    log likelihood = -58.497292
Iteration 3:    log likelihood = -58.497288
Comparison:     log likelihood = -89.946246
Fitting full model:
Iteration 0:    log likelihood = -89.946246
Iteration 1:    log likelihood = -89.258897
Iteration 2:    log likelihood = -89.254028
Iteration 3:    log likelihood = -89.254028
```

Bivariate probit regression

Log likelihood = -89.254028

	Number of obs	=	95
	Wald chi2(6)	=	9.59
	Prob > chi2	=	0.1431

	Coef.	Std. Err.	z	P>|z|	[95% Conf. Interval]	
private						
years	-.0118884	.0256778	-0.46	0.643	-.0622159	.0384391
logptax	-.1066962	.6669782	-0.16	0.873	-1.413949	1.200557
loginc	.3762037	.5306484	0.71	0.478	-.663848	1.416255
_cons	-4.184694	4.837817	-0.86	0.387	-13.66664	5.297253
vote						
years	-.0168561	.0147834	-1.14	0.254	-.0458309	.0121188
logptax	-1.288707	.5752266	-2.24	0.025	-2.416131	-.1612839
loginc	.998286	.4403565	2.27	0.023	.1352031	1.861369
_cons	-.5360573	4.068509	-0.13	0.895	-8.510188	7.438073
/athrho	-.2764525	.2412099	-1.15	0.252	-.7492153	.1963102
rho	-.2696186	.2236753			-.6346806	.1938267

Likelihood-ratio test of rho=0:   chi2(1) = 1.38444   Prob > chi2 = 0.2393

The output shows several iteration logs. The first iteration log corresponds to running the univariate probit model for the first equation, and the second log corresponds to running the univariate probit for the second model. If $\rho = 0$, the sum of the log likelihoods from these two models will equal the log likelihood of the bivariate probit model; this sum is printed in the iteration log as the comparison log likelihood.

The final iteration log is for fitting the full bivariate probit model. A likelihood-ratio test of the log likelihood for this model and the comparison log likelihood is presented at the end of the output. If we had specified the vce(robust) option, this test would be presented as a Wald test instead of as a likelihood-ratio test.

We could have fit the same model by using the seemingly unrelated syntax as

```
. biprobit (private=years logptax loginc) (vote=years logptax loginc)
```
◁

Saved results

biprobit saves the following in e():

Scalars
e(N)	number of observations
e(k)	number of parameters
e(k_eq)	number of equations in e(b)
e(k_aux)	number of auxiliary parameters
e(k_eq_model)	number of equations in overall model test
e(k_dv)	number of dependent variables
e(df_m)	model degrees of freedom
e(ll)	log likelihood
e(ll_0)	log likelihood, constant-only model (noskip only)
e(ll_c)	log likelihood, comparison model
e(N_clust)	number of clusters
e(chi2)	χ^2
e(chi2_c)	χ^2 for comparison test
e(p)	significance
e(rho)	ρ
e(rank)	rank of e(V)
e(rank0)	rank of e(V) for constant-only model
e(ic)	number of iterations
e(rc)	return code
e(converged)	1 if converged, 0 otherwise

Macros
e(cmd)	biprobit
e(cmdline)	command as typed
e(depvar)	names of dependent variables
e(wtype)	weight type
e(wexp)	weight expression
e(title)	title in estimation output
e(clustvar)	name of cluster variable
e(offset1)	offset for first equation
e(offset2)	offset for second equation
e(chi2type)	Wald or LR; type of model χ^2 test
e(chi2_ct)	Wald or LR; type of model χ^2 test corresponding to e(chi2_c)
e(vce)	*vcetype* specified in vce()
e(vcetype)	title used to label Std. Err.
e(opt)	type of optimization
e(which)	max or min; whether optimizer is to perform maximization or minimization
e(ml_method)	type of ml method
e(user)	name of likelihood-evaluator program
e(technique)	maximization technique
e(properties)	b V
d(predict)	program used to implement predict
e(asbalanced)	factor variables fvset as asbalanced
e(asobserved)	factor variables fvset as asobserved

Matrices
     e(b)   coefficient vector
     e(Cns)   constraints matrix
     e(ilog)   iteration log (up to 20 iterations)
     e(gradient)   gradient vector
     e(V)   variance–covariance matrix of the estimators
     e(V_modelbased)   model-based variance
Functions
     e(sample)   marks estimation sample

Methods and formulas

biprobit is implemented as an ado-file.

The log likelihood, $\ln L$, is given by

$$\xi_j^\beta = x_j\beta + \text{offset}_j^\beta$$
$$\xi_j^\gamma = z_j\gamma + \text{offset}_j^\gamma$$
$$q_{1j} = \begin{cases} 1 & \text{if } y_{1j} \neq 0 \\ -1 & \text{otherwise} \end{cases}$$
$$q_{2j} = \begin{cases} 1 & \text{if } y_{2j} \neq 0 \\ -1 & \text{otherwise} \end{cases}$$
$$\rho_j^* = q_{1j}q_{2j}\rho$$
$$\ln L = \sum_{j=1}^{n} w_j \ln\Phi_2\left(q_{1j}\xi_j^\beta, q_{2j}\xi_j^\gamma, \rho_j^*\right)$$

where $\Phi_2()$ is the cumulative bivariate normal distribution function (with mean $[0\ 0]'$) and w_j is an optional weight for observation j. This derivation assumes that

$$y_{1j}^* = x_j\beta + \epsilon_{1j} + \text{offset}_j^\beta$$
$$y_{2j}^* = z_j\gamma + \epsilon_{2j} + \text{offset}_j^\gamma$$
$$E(\epsilon_1) = E(\epsilon_2) = 0$$
$$\text{Var}(\epsilon_1) = \text{Var}(\epsilon_2) = 1$$
$$\text{Cov}(\epsilon_1, \epsilon_2) = \rho$$

where y_{1j}^* and y_{2j}^* are the unobserved latent variables; instead, we observe only $y_{ij} = 1$ if $y_{ij}^* > 0$ and $y_{ij} = 0$ otherwise (for $i = 1, 2$).

In the maximum likelihood estimation, ρ is not directly estimated, but atanh ρ is

$$\text{atanh}\,\rho = \frac{1}{2}\ln\left(\frac{1+\rho}{1-\rho}\right)$$

From the form of the likelihood, if $\rho = 0$, then the log likelihood for the bivariate probit models is equal to the sum of the log likelihoods of the two univariate probit models. A likelihood-ratio test may therefore be performed by comparing the likelihood of the full bivariate model with the sum of the log likelihoods for the univariate probit models.

This command supports the Huber/White/sandwich estimator of the variance and its clustered version using vce(robust) and vce(cluster *clustvar*), respectively. See [P] **_robust**, particularly *Maximum likelihood estimators* and *Methods and formulas*.

biprobit also supports estimation with survey data. For details on VCEs with survey data, see [SVY] **variance estimation**.

References

De Luca, G. 2008. SNP and SML estimation of univariate and bivariate binary-choice models. *Stata Journal* 8: 190–220.

Greene, W. H. 2012. *Econometric Analysis*. 7th ed. Upper Saddle River, NJ: Prentice Hall.

Hardin, J. W. 1996. sg61: Bivariate probit models. *Stata Technical Bulletin* 33: 15–20. Reprinted in *Stata Technical Bulletin Reprints*, vol. 6, pp. 152–158. College Station, TX: Stata Press.

Heckman, J. 1979. Sample selection bias as a specification error. *Econometrica* 47: 153–161.

Pindyck, R. S., and D. L. Rubinfeld. 1998. *Econometric Models and Economic Forecasts*. 4th ed. New York: McGraw–Hill.

Poirier, D. J. 1980. Partial observability in bivariate probit models. *Journal of Econometrics* 12: 209–217.

Van de Ven, W. P. M. M., and B. M. S. Van Pragg. 1981. The demand for deductibles in private health insurance: A probit model with sample selection. *Journal of Econometrics* 17: 229–252.

Also see

[R] **biprobit postestimation** — Postestimation tools for biprobit

[R] **mprobit** — Multinomial probit regression

[R] **probit** — Probit regression

[SVY] **svy estimation** — Estimation commands for survey data

[U] **20 Estimation and postestimation commands**

Title

biprobit postestimation — Postestimation tools for biprobit

Description

The following postestimation commands are available after `biprobit`:

Command	Description
contrast	contrasts and ANOVA-style joint tests of estimates
estat	AIC, BIC, VCE, and estimation sample summary
estat (svy)	postestimation statistics for survey data
estimates	cataloging estimation results
lincom	point estimates, standard errors, testing, and inference for linear combinations of coefficients
lrtest[1]	likelihood-ratio test
margins	marginal means, predictive margins, marginal effects, and average marginal effects
marginsplot	graph the results from margins (profile plots, interaction plots, etc.)
nlcom	point estimates, standard errors, testing, and inference for nonlinear combinations of coefficients
predict	predictions, residuals, influence statistics, and other diagnostic measures
predictnl	point estimates, standard errors, testing, and inference for generalized predictions
pwcompare	pairwise comparisons of estimates
suest	seemingly unrelated estimation
test	Wald tests of simple and composite linear hypotheses
testnl	Wald tests of nonlinear hypotheses

[1] `lrtest` is not appropriate with svy estimation results.

See the corresponding entries in the *Base Reference Manual* for details, but see [SVY] **estat** for details about `estat` (svy).

Syntax for predict

predict [*type*] *newvar* [*if*] [*in*] [, *statistic* <u>nooff</u>set]

predict [*type*] { *stub** | *newvar*$_{eq1}$ *newvar*$_{eq2}$ *newvar*$_{athrho}$ } [*if*] [*in*] , <u>sc</u>ores

statistic	Description
Main	
<u>p</u>11	$\Phi_2(\mathbf{x}_j\mathbf{b}, \mathbf{z}_j\mathbf{g}, \rho)$, predicted probability $\Pr(y_{1j}=1, y_{2j}=1)$; the default
p10	$\Phi_2(\mathbf{x}_j\mathbf{b}, -\mathbf{z}_j\mathbf{g}, -\rho)$, predicted probability $\Pr(y_{1j}=1, y_{2j}=0)$
p01	$\Phi_2(-\mathbf{x}_j\mathbf{b}, \mathbf{z}_j\mathbf{g}, -\rho)$, predicted probability $\Pr(y_{1j}=0, y_{2j}=1)$
p00	$\Phi_2(-\mathbf{x}_j\mathbf{b}, -\mathbf{z}_j\mathbf{g}, \rho)$, predicted probability $\Pr(y_{1j}=0, y_{2j}=0)$
pmarg1	$\Phi(\mathbf{x}_j\mathbf{b})$, marginal success probability for equation 1
pmarg2	$\Phi(\mathbf{z}_j\mathbf{g})$, marginal success probability for equation 2
pcond1	$\Phi_2(\mathbf{x}_j\mathbf{b}, \mathbf{z}_j\mathbf{g}, \rho)/\Phi(\mathbf{z}_j\mathbf{g})$, conditional probability of success for equation 1
pcond2	$\Phi_2(\mathbf{x}_j\mathbf{b}, \mathbf{z}_j\mathbf{g}, \rho)/\Phi(\mathbf{x}_j\mathbf{b})$, conditional probability of success for equation 2
xb1	$\mathbf{x}_j\mathbf{b}$, linear prediction for equation 1
xb2	$\mathbf{z}_j\mathbf{g}$, linear prediction for equation 2
stdp1	standard error of the linear prediction for equation 1
stdp2	standard error of the linear prediction for equation 2

where $\Phi()$ is the standard normal-distribution function and $\Phi_2()$ is the bivariate standard normal-distribution function.

These statistics are available both in and out of sample; type predict ... if e(sample) ... if wanted only for the estimation sample.

Menu

Statistics > Postestimation > Predictions, residuals, etc.

Options for predict

    ⌐ Main ⌐

p11, the default, calculates the bivariate predicted probability $\Pr(y_{1j}=1, y_{2j}=1)$.

p10 calculates the bivariate predicted probability $\Pr(y_{1j}=1, y_{2j}=0)$.

p01 calculates the bivariate predicted probability $\Pr(y_{1j}=0, y_{2j}=1)$.

p00 calculates the bivariate predicted probability $\Pr(y_{1j}=0, y_{2j}=0)$.

pmarg1 calculates the univariate (marginal) predicted probability of success $\Pr(y_{1j}=1)$.

pmarg2 calculates the univariate (marginal) predicted probability of success $\Pr(y_{2j}=1)$.

pcond1 calculates the conditional (on success in equation 2) predicted probability of success $\Pr(y_{1j}=1, y_{2j}=1)/\Pr(y_{2j}=1)$.

pcond2 calculates the conditional (on success in equation 1) predicted probability of success $\Pr(y_{1j}=1, y_{2j}=1)/\Pr(y_{1j}=1)$.

xb1 calculates the probit linear prediction $\mathbf{x}_j\mathbf{b}$.

xb2 calculates the probit linear prediction $\mathbf{z}_j\mathbf{g}$.

stdp1 calculates the standard error of the linear prediction for equation 1.

stdp2 calculates the standard error of the linear prediction for equation 2.

nooffset is relevant only if you specified offset1(*varname*) or offset2(*varname*) for biprobit. It modifies the calculations made by predict so that they ignore the offset variables; the linear predictions are treated as $\mathbf{x}_j\mathbf{b}$ rather than as $\mathbf{x}_j\mathbf{b} + \text{offset}_{1j}$ and $\mathbf{z}_j\boldsymbol{\gamma}$ rather than as $\mathbf{z}_j\boldsymbol{\gamma} + \text{offset}_{2j}$.

scores calculates equation-level score variables.

The first new variable will contain $\partial \ln L / \partial(\mathbf{x}_j\boldsymbol{\beta})$.

The second new variable will contain $\partial \ln L / \partial(\mathbf{z}_j\boldsymbol{\gamma})$.

The third new variable will contain $\partial \ln L / \partial(\text{atanh}\,\rho)$.

Methods and formulas

All postestimation commands listed above are implemented as ado-files.

Also see

[R] **biprobit** — Bivariate probit regression

[U] **20 Estimation and postestimation commands**

Title

> **bitest** — Binomial probability test

Syntax

Binomial probability test

> bitest *varname*== $\#_p$ $\left[\,if\,\right]$ $\left[\,in\,\right]$ $\left[\,weight\,\right]$ $\left[\,, \underline{d}etail\,\right]$

Immediate form of binomial probability test

> bitesti $\#_N$ $\#_{\mathrm{succ}}$ $\#_p$ $\left[\,, \underline{d}etail\,\right]$

by is allowed with bitest; see [D] **by**.

fweights are allowed with bitest; see [U] **11.1.6 weight**.

Menu

bitest

Statistics > Summaries, tables, and tests > Classical tests of hypotheses > Binomial probability test

bitesti

Statistics > Summaries, tables, and tests > Classical tests of hypotheses > Binomial probability test calculator

Description

bitest performs exact hypothesis tests for binomial random variables. The null hypothesis is that the probability of a success on a trial is $\#_p$. The total number of trials is the number of nonmissing values of *varname* (in bitest) or $\#_N$ (in bitesti). The number of observed successes is the number of 1s in *varname* (in bitest) or $\#_{\mathrm{succ}}$ (in bitesti). *varname* must contain only 0s, 1s, and missing.

bitesti is the immediate form of bitest; see [U] **19 Immediate commands** for a general introduction to immediate commands.

Option

> ┌─ Advanced ┐
>
> detail shows the probability of the observed number of successes, k_{obs}; the probability of the number of successes on the opposite tail of the distribution that is used to compute the two-sided p-value, k_{opp}; and the probability of the point next to k_{opp}. This information can be safely ignored. See the technical note below for details.

188

Remarks

Remarks are presented under the following headings:

> bitest
> bitesti

bitest

▷ Example 1

We test 15 university students for high levels of one measure of visual quickness which, from other evidence, we believe is present in 30% of the nonuniversity population. Included in our data is quick, taking on the values 1 ("success") or 0 ("failure") depending on the outcome of the test.

```
. use http://www.stata-press.com/data/r12/quick
. bitest quick == 0.3
```

Variable	N	Observed k	Expected k	Assumed p	Observed p
quick	15	7	4.5	0.30000	0.46667

```
Pr(k >= 7)            = 0.131143  (one-sided test)
Pr(k <= 7)            = 0.949987  (one-sided test)
Pr(k <= 1 or k >= 7) = 0.166410  (two-sided test)
```

The first part of the output reveals that, assuming a true probability of success of 0.3, the expected number of successes is 4.5 and that we observed seven. Said differently, the assumed frequency under the null hypothesis H_0 is 0.3, and the observed frequency is 0.47.

The first line under the table is a one-sided test; it is the probability of observing seven or more successes conditional on $p = 0.3$. It is a test of $H_0: p = 0.3$ versus the alternative hypothesis $H_A: p > 0.3$. Said in English, the alternative hypothesis is that more than 30% of university students score at high levels on this test of visual quickness. The p-value for this hypothesis test is 0.13.

The second line under the table is a one-sided test of H_0 versus the opposite alternative hypothesis $H_A: p < 0.3$.

The third line is the two-sided test. It is a test of H_0 versus the alternative hypothesis $H_A: p \neq 0.3$.

◁

❑ Technical note

The p-value of a hypothesis test is the probability (calculated assuming H_0 is true) of observing any outcome as extreme or more extreme than the observed outcome, with extreme meaning in the direction of the alternative hypothesis. In example 1, the outcomes $k = 8$, 9, ..., 15 are clearly "more extreme" than the observed outcome $k_{obs} = 7$ when considering the alternative hypothesis $H_A: p \neq 0.3$. However, outcomes with only a few successes are also in the direction of this alternative hypothesis. For two-sided hypotheses, outcomes with k successes are considered "as extreme or more extreme" than the observed outcome k_{obs} if $\Pr(k) \leq \Pr(k_{obs})$. Here $\Pr(k = 0)$ and $\Pr(k = 1)$ are both less than $\Pr(k = 7)$, so they are included in the two-sided p-value.

The detail option allows you to see the probability (assuming that H_0 is true) of the observed successes ($k = 7$) and the probability of the boundary point ($k = 1$) of the opposite tail used for the two-sided p-value.

```
. bitest quick == 0.3, detail
        Variable │        N   Observed k   Expected k   Assumed p   Observed p
        ─────────┼──────────────────────────────────────────────────────────────
           quick │       15            7          4.5     0.30000      0.46667
   Pr(k >= 7)                  = 0.131143  (one-sided test)
   Pr(k <= 7)                  = 0.949987  (one-sided test)
   Pr(k <= 1 or k >= 7)        = 0.166410  (two-sided test)
   Pr(k == 7)                  = 0.081130  (observed)
   Pr(k == 2)                  = 0.091560
   Pr(k == 1)                  = 0.030520  (opposite extreme)
```

Also shown is the probability of the point next to the boundary point. This probability, namely, $\Pr(k = 2) = 0.092$, is certainly close to the probability of the observed outcome $\Pr(k = 7) = 0.081$, so some people might argue that $k = 2$ should be included in the two-sided p-value. Statisticians (at least some we know) would reply that the p-value is a precisely defined concept and that this is an arbitrary "fuzzification" of its definition. When you compute exact p-values according to the precise definition of a p-value, your type I error is never more than what you say it is—so no one can criticize you for being anticonservative. Including the point $k = 2$ is being overly conservative because it makes the p-value larger yet. But it is your choice; being overly conservative, at least in statistics, is always safe. Know that bitest and bitesti always keep to the precise definition of a p-value, so if you wish to include this extra point, you must do so by hand or by using the r() saved results; see *Saved results* below.

❏

bitesti

▷ Example 2

The binomial test is a function of two statistics and one parameter: N, the number of observations; k_{obs}, the number of observed successes; and p, the assumed probability of a success on a trial. For instance, in a city of $N = 2{,}500{,}000$, we observe $k_{\mathrm{obs}} = 36$ cases of a particular disease when the population rate for the disease is $p = 0.00001$.

```
. bitesti 2500000 36 .00001
         N   Observed k   Expected k   Assumed p   Observed p
   ──────────────────────────────────────────────────────────────
   2500000           36           25     0.00001      0.00001
   Pr(k >= 36)                = 0.022458  (one-sided test)
   Pr(k <= 36)                = 0.985448  (one-sided test)
   Pr(k <= 14 or k >= 36)     = 0.034859  (two-sided test)
```

◁

▷ Example 3

Boice and Monson (1977) present data on breast cancer cases and person-years of observations for women with tuberculosis who were repeatedly exposed to multiple x-ray fluoroscopies and for women with tuberculosis who were not. The data are

	Exposed	Not exposed	Total
Breast cancer	41	15	56
Person-years	28,010	19,017	47,027

We can thus test whether x-ray fluoroscopic examinations are associated with breast cancer; the assumed rate of exposure is $p = 28010/47027$.

```
. bitesti 56 41 28010/47027
           N   Observed k   Expected k   Assumed p   Observed p

          56           41     33.35446     0.59562     0.73214
    Pr(k >= 41)              = 0.023830   (one-sided test)
    Pr(k <= 41)              = 0.988373   (one-sided test)
    Pr(k <= 25 or k >= 41)  = 0.040852   (two-sided test)
```

◁

Saved results

`bitest` and `bitesti` save the following in `r()`:

Scalars

`r(N)`	number N of trials	`r(k_opp)`	opposite extreme k
`r(P_p)`	assumed probability p of success	`r(P_k)`	probability of observed k (detail only)
`r(k)`	observed number k of successes	`r(P_oppk)`	probability of opposite extreme k (detail only)
`r(p_l)`	lower one-sided p-value	`r(k_nopp)`	k next to opposite extreme (detail only)
`r(p_u)`	upper one-sided p-value	`r(P_noppk)`	probability of k next to opposite extreme (detail only)
`r(p)`	two-sided p-value		

Methods and formulas

`bitest` and `bitesti` are implemented as ado-files.

Let N, k_{obs}, and p be, respectively, the number of observations, the observed number of successes, and the assumed probability of success on a trial. The expected number of successes is Np, and the observed probability of success on a trial is k_{obs}/N.

`bitest` and `bitesti` compute exact p-values based on the binomial distribution. The upper one-sided p-value is

$$\Pr(k \geq k_{\text{obs}}) = \sum_{m=k_{\text{obs}}}^{N} \binom{N}{m} p^m (1-p)^{N-m}$$

The lower one-sided p-value is

$$\Pr(k \leq k_{\text{obs}}) = \sum_{m=0}^{k_{\text{obs}}} \binom{N}{m} p^m (1-p)^{N-m}$$

If $k_{\text{obs}} \geq Np$, the two-sided p-value is

$$\Pr(k \leq k_{\text{opp}} \text{ or } k \geq k_{\text{obs}})$$

where k_{opp} is the largest number $\leq Np$ such that $\Pr(k = k_{\text{opp}}) \leq \Pr(k = k_{\text{obs}})$. If $k_{\text{obs}} < Np$, the two-sided p-value is

$$\Pr(k \leq k_{\text{obs}} \text{ or } k \geq k_{\text{opp}})$$

where k_{opp} is the smallest number $\geq Np$ such that $\Pr(k = k_{\text{opp}}) \leq \Pr(k = k_{\text{obs}})$.

References

Boice, J. D., Jr., and R. R. Monson. 1977. Breast cancer in women after repeated fluoroscopic examinations of the chest. *Journal of the National Cancer Institute* 59: 823–832.

Hoel, P. G. 1984. *Introduction to Mathematical Statistics.* 5th ed. New York: Wiley.

Also see

[R] **ci** — Confidence intervals for means, proportions, and counts

[R] **prtest** — One- and two-sample tests of proportions

Title

bootstrap — Bootstrap sampling and estimation

Syntax

bootstrap *exp_list* [, *options eform_option*] : *command*

options	Description
Main	
<u>reps</u>(#)	perform # bootstrap replications; default is reps(50)
Options	
<u>stra</u>ta(*varlist*)	variables identifying strata
<u>si</u>ze(#)	draw samples of size #; default is _N
<u>cl</u>uster(*varlist*)	variables identifying resampling clusters
<u>idc</u>luster(*newvar*)	create new cluster ID variable
<u>sa</u>ving(*filename*, ...)	save results to *filename*; save statistics in double precision; save results to *filename* every # replications
bca	compute acceleration for BC_a confidence intervals
mse	use MSE formula for variance estimation
Reporting	
<u>l</u>evel(#)	set confidence level; default is level(95)
notable	suppress table of results
<u>noh</u>eader	suppress table header
<u>nol</u>egend	suppress table legend
<u>v</u>erbose	display the full table legend
nodots	suppress replication dots
<u>noi</u>sily	display any output from *command*
<u>trace</u>	trace *command*
<u>ti</u>tle(*text*)	use *text* as title for bootstrap results
display_options	control column formats, row spacing, line width, and display of omitted variables and base and empty cells
eform_option	display coefficient table in exponentiated form
Advanced	
nodrop	do not drop observations
nowarn	do not warn when e(sample) is not set
force	do not check for *weights* or svy commands; seldom used
reject(*exp*)	identify invalid results
seed(#)	set random-number seed to #
group(*varname*)	ID variable for groups within cluster()
<u>jackknife</u>opts(*jkopts*)	options for jackknife; see [R] **jackknife**
<u>coefl</u>egend	display legend instead of statistics

weights are not allowed in *command*.

group(), jackknifeopts(), and coeflegend do not appear in the dialog box.

See [U] **20 Estimation and postestimation commands** for more capabilities of estimation commands.

exp_list contains	(*name*: *elist*)
	elist
	eexp
elist contains	*newvar* = (*exp*)
	(*exp*)
eexp is	*specname*
	[*eqno*]*specname*
specname is	_b
	_b[]
	_se
	_se[]
eqno is	# #
	name

exp is a standard Stata expression; see [U] **13 Functions and expressions**.

Distinguish between [ ], which are to be typed, and [ ], which indicate optional arguments.

Menu

Statistics > Resampling > Bootstrap estimation

Description

bootstrap performs bootstrap estimation. Typing

    . bootstrap *exp_list*, reps(*#*): *command*

executes *command* multiple times, bootstrapping the statistics in *exp_list* by resampling observations (with replacement) from the data in memory *#* times. This method is commonly referred to as the nonparametric bootstrap.

command defines the statistical command to be executed. Most Stata commands and user-written programs can be used with bootstrap, as long as they follow standard Stata syntax; see [U] **11 Language syntax**. If the bca option is supplied, *command* must also work with jackknife; see [R] **jackknife**. The by prefix may not be part of *command*.

exp_list specifies the statistics to be collected from the execution of *command*. If *command* changes the contents in e(b), *exp_list* is optional and defaults to _b.

Because bootstrapping is a random process, if you want to be able to reproduce results, set the random-number seed by specifying the seed(*#*) option or by typing

    . set seed *#*

where *#* is a seed of your choosing, before running bootstrap; see [R] **set seed**.

Many estimation commands allow the `vce(bootstrap)` option. For those commands, we recommend using `vce(bootstrap)` over `bootstrap` because the estimation command already handles clustering and other model-specific details for you. The `bootstrap` prefix command is intended for use with nonestimation commands, such as `summarize`, user-written commands, or functions of coefficients.

`bs` and `bstrap` are synonyms for `bootstrap`.

Options

___ Main ___

`reps(#)` specifies the number of bootstrap replications to be performed. The default is 50. A total of 50–200 replications are generally adequate for estimates of standard error and thus are adequate for normal-approximation confidence intervals; see Mooney and Duval (1993, 11). Estimates of confidence intervals using the percentile or bias-corrected methods typically require 1,000 or more replications.

___ Options ___

`strata(varlist)` specifies the variables that identify strata. If this option is specified, bootstrap samples are taken independently within each stratum.

`size(#)` specifies the size of the samples to be drawn. The default is _N, meaning to draw samples of the same size as the data. If specified, # must be less than or equal to the number of observations within `strata()`.

If `cluster()` is specified, the default size is the number of clusters in the original dataset. For unbalanced clusters, resulting sample sizes will differ from replication to replication. For cluster sampling, # must be less than or equal to the number of clusters within `strata()`.

`cluster(varlist)` specifies the variables that identify resampling clusters. If this option is specified, the sample drawn during each replication is a bootstrap sample of clusters.

`idcluster(newvar)` creates a new variable containing a unique identifier for each resampled cluster. This option requires that `cluster()` also be specified.

`saving(filename[, suboptions])` creates a Stata data file (`.dta` file) consisting of (for each statistic in *exp_list*) a variable containing the replicates.

   `double` specifies that the results for each replication be stored as `doubles`, meaning 8-byte reals. By default, they are stored as `floats`, meaning 4-byte reals. This option may be used without the `saving()` option to compute the variance estimates by using double precision.

   `every(#)` specifies that results be written to disk every #th replication. `every()` should be specified only in conjunction with `saving()` when *command* takes a long time for each replication. This option will allow recovery of partial results should some other software crash your computer. See [P] **postfile**.

   `replace` specifies that *filename* be overwritten if it exists. This option does not appear in the dialog box.

`bca` specifies that `bootstrap` estimate the acceleration of each statistic in *exp_list*. This estimate is used to construct BC_a confidence intervals. Type `estat bootstrap, bca` to display the BC_a confidence interval generated by the `bootstrap` command.

`mse` specifies that `bootstrap` compute the variance by using deviations of the replicates from the observed value of the statistics based on the entire dataset. By default, `bootstrap` computes the variance by using deviations from the average of the replicates.

level(*#*); see [R] **estimation options**.

notable suppresses the display of the table of results.

noheader suppresses the display of the table header. This option implies nolegend. This option may also be specified when replaying estimation results.

nolegend suppresses the display of the table legend. This option may also be specified when replaying estimation results.

verbose specifies that the full table legend be displayed. By default, coefficients and standard errors are not displayed. This option may also be specified when replaying estimation results.

nodots suppresses display of the replication dots. By default, one dot character is displayed for each successful replication. A red 'x' is displayed if *command* returns an error or if one of the values in *exp_list* is missing.

noisily specifies that any output from *command* be displayed. This option implies the nodots option.

trace causes a trace of the execution of *command* to be displayed. This option implies the noisily option.

title(*text*) specifies a title to be displayed above the table of bootstrap results. The default title is the title saved in e(title) by an estimation command, or if e(title) is not filled in, Bootstrap results is used. title() may also be specified when replaying estimation results.

display_options: noomitted, vsquish, noemptycells, baselevels, allbaselevels, cformat(%*fmt*), pformat(%*fmt*), sformat(%*fmt*), and nolstretch; see [R] **estimation options**.

eform_option causes the coefficient table to be displayed in exponentiated form; see [R] ***eform_option***.

command determines which of the following are allowed (eform(*string*) and eform are always allowed):

eform_option	Description
eform(*string*)	use *string* for the column title
eform	exponentiated coefficient, *string* is exp(b)
hr	hazard ratio, *string* is Haz. Ratio
shr	subhazard ratio, *string* is SHR
irr	incidence-rate ratio, *string* is IRR
or	odds ratio, *string* is Odds Ratio
rrr	relative-risk ratio, *string* is RRR

nodrop prevents observations outside e(sample) and the if and in qualifiers from being dropped before the data are resampled.

nowarn suppresses the display of a warning message when *command* does not set e(sample).

force suppresses the restriction that *command* not specify weights or be a svy command. This is a rarely used option. Use it only if you know what you are doing.

reject(*exp*) identifies an expression that indicates when results should be rejected. When *exp* is true, the resulting values are reset to missing values.

seed(*#*) sets the random-number seed. Specifying this option is equivalent to typing the following command prior to calling bootstrap:

. set seed *#*

The following options are available with bootstrap but are not shown in the dialog box:

group(*varname*) re-creates *varname* containing a unique identifier for each group across the resampled clusters. This option requires that idcluster() also be specified.

This option is useful for maintaining unique group identifiers when sampling clusters with replacement. Suppose that cluster 1 contains 3 groups. If the idcluster(newclid) option is specified and cluster 1 is sampled multiple times, newclid uniquely identifies each copy of cluster 1. If group(newgroupid) is also specified, newgroupid uniquely identifies each copy of each group.

jackknifeopts(*jkopts*) identifies options that are to be passed to jackknife when it computes the acceleration values for the BC_a confidence intervals; see [R] **jackknife**. This option requires the bca option and is mostly used for passing the eclass, rclass, or n(*#*) option to jackknife.

coeflegend; see [R] **estimation options**.

Remarks

Remarks are presented under the following headings:

Introduction
Regression coefficients
Expressions
Combining bootstrap datasets
A note about macros
Achieved significance level
Bootstrapping a ratio
Warning messages and e(sample)
Bootstrapping statistics from data with a complex structure

Introduction

With few assumptions, bootstrapping provides a way of estimating standard errors and other measures of statistical precision (Efron 1979; Efron and Stein 1981; Efron 1982; Efron and Tibshirani 1986; Efron and Tibshirani 1993; also see Davison and Hinkley [1997]; Guan [2003]; Mooney and Duval [1993]; Poi [2004]; and Stine [1990]). It provides a way to obtain such measures when no formula is otherwise available or when available formulas make inappropriate assumptions. Cameron and Trivedi (2010, chap. 13) discuss many bootstrapping topics and demonstrate how to do them in Stata.

To illustrate bootstrapping, suppose that you have a dataset containing N observations and an estimator that, when applied to the data, produces certain statistics. You draw, with replacement, N observations from the N-observation dataset. In this random drawing, some of the original observations will appear once, some more than once, and some not at all. Using the resampled dataset, you apply the estimator and collect the statistics. This process is repeated many times; each time, a new random sample is drawn and the statistics are recalculated.

This process builds a dataset of replicated statistics. From these data, you can calculate the standard error by using the standard formula for the sample standard deviation

$$\widehat{se} = \left\{ \frac{1}{k-1} \sum (\widehat{\theta}_i - \overline{\theta})^2 \right\}^{1/2}$$

where $\widehat{\theta}_i$ is the statistic calculated using the ith bootstrap sample and k is the number of replications. This formula gives an estimate of the standard error of the statistic, according to Hall and Wilson (1991). Although the average, $\overline{\theta}$, of the bootstrapped estimates is used in calculating the standard deviation, it is not used as the estimated value of the statistic itself. Instead, the original observed value of the statistic, $\widehat{\theta}$, is used, meaning the value of the statistic computed using the original N observations.

You might think that $\overline{\theta}$ is a better estimate of the parameter than $\widehat{\theta}$, but it is not. If the statistic is biased, bootstrapping exaggerates the bias. In fact, the bias can be estimated as $\overline{\theta} - \widehat{\theta}$ (Efron 1982, 33). Knowing this, you might be tempted to subtract this estimate of bias from $\widehat{\theta}$ to produce an unbiased statistic. The bootstrap bias estimate has an indeterminate amount of random error, so this unbiased estimator may have greater mean squared error than the biased estimator (Mooney and Duval 1993; Hinkley 1978). Thus $\widehat{\theta}$ is the best point estimate of the statistic.

The logic behind the bootstrap is that all measures of precision come from a statistic's sampling distribution. When the statistic is estimated on a sample of size N from some population, the sampling distribution tells you the relative frequencies of the values of the statistic. The sampling distribution, in turn, is determined by the distribution of the population and the formula used to estimate the statistic.

Sometimes the sampling distribution can be derived analytically. For instance, if the underlying population is distributed normally and you calculate means, the sampling distribution for the mean is also normal but has a smaller variance than that of the population. In other cases, deriving the sampling distribution is difficult, as when means are calculated from nonnormal populations. Sometimes, as in the case of means, it is not too difficult to derive the sampling distribution as the sample size goes to infinity ($N \rightarrow \infty$). However, such asymptotic distributions may not perform well when applied to finite samples.

If you knew the population distribution, you could obtain the sampling distribution by simulation: you could draw random samples of size N, calculate the statistic, and make a tally. Bootstrapping does precisely this, but it uses the observed distribution of the sample in place of the true population distribution. Thus the bootstrap procedure hinges on the assumption that the observed distribution is a good estimate of the underlying population distribution. In return, the bootstrap produces an estimate, called the bootstrap distribution, of the sampling distribution. From this, you can estimate the standard error of the statistic, produce confidence intervals, etc.

The accuracy with which the bootstrap distribution estimates the sampling distribution depends on the number of observations in the original sample and the number of replications in the bootstrap. A crudely estimated sampling distribution is adequate if you are only going to extract, say, a standard error. A better estimate is needed if you want to use the 2.5th and 97.5th percentiles of the distribution to produce a 95% confidence interval. To extract many features simultaneously about the distribution, an even better estimate is needed. Generally, replications on the order of 1,000 produce very good estimates, but only 50–200 replications are needed for estimates of standard errors. See Poi (2004) for a method to choose the number of bootstrap replications.

Regression coefficients

▷ Example 1

Let's say that we wish to compute bootstrap estimates for the standard errors of the coefficients from the following regression:

```
. use http://www.stata-press.com/data/r12/auto
(1978 Automobile Data)

. regress mpg weight gear foreign
```

Source	SS	df	MS
Model	1629.67805	3	543.226016
Residual	813.781411	70	11.6254487
Total	2443.45946	73	33.4720474

Number of obs =	74
F(3, 70) =	46.73
Prob > F =	0.0000
R-squared =	0.6670
Adj R-squared =	0.6527
Root MSE =	3.4096

mpg	Coef.	Std. Err.	t	P>|t|	[95% Conf. Interval]
weight	-.006139	.0007949	-7.72	0.000	-.0077245 -.0045536
gear_ratio	1.457113	1.541286	0.95	0.348	-1.616884 4.53111
foreign	-2.221682	1.234961	-1.80	0.076	-4.684735 .2413715
_cons	36.10135	6.285984	5.74	0.000	23.56435 48.63835

To run the bootstrap, we simply prefix the above regression command with the bootstrap command (specifying its options before the colon separator). We must set the random-number seed before calling bootstrap.

```
. bootstrap, reps(100) seed(1): regress mpg weight gear foreign
(running regress on estimation sample)
```

Bootstrap replications (100)

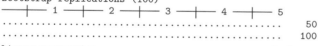

Linear regression

Number of obs =	74
Replications =	100
Wald chi2(3) =	111.96
Prob > chi2 =	0.0000
R-squared =	0.6670
Adj R-squared =	0.6527
Root MSE =	3.4096

mpg	Observed Coef.	Bootstrap Std. Err.	z	P>|z|	Normal-based [95% Conf. Interval]
weight	-.006139	.0006498	-9.45	0.000	-.0074127 -.0048654
gear_ratio	1.457113	1.297786	1.12	0.262	-1.086501 4.000727
foreign	-2.221682	1.162728	-1.91	0.056	-4.500587 .0572236
_cons	36.10135	4.71779	7.65	0.000	26.85465 45.34805

The displayed confidence interval is based on the assumption that the sampling (and hence bootstrap) distribution is approximately normal (see *Methods and formulas* below). Because this confidence interval is based on the standard error, it is a reasonable estimate if normality is approximately true, even for a few replications. Other types of confidence intervals are available after bootstrap; see [R] **bootstrap postestimation**.

We could instead supply names to our expressions when we run bootstrap. For example,

```
. bootstrap diff=(_b[weight]-_b[gear]): regress mpg weight gear foreign
```

would bootstrap a statistic, named `diff`, equal to the difference between the coefficients on `weight` and `gear_ratio`.

◁

Expressions

▷ Example 2

When we use bootstrap, the list of statistics can contain complex expressions, as long as each expression is enclosed in parentheses. For example, to bootstrap the range of a variable x, we could type

```
. bootstrap range=(r(max)-r(min)), reps(1000): summarize x
```

Of course, we could also bootstrap the minimum and maximum and later compute the range.

```
. bootstrap max=r(max) min=r(min), reps(1000) saving(mybs): summarize x
. use mybs, clear
(bootstrap: summarize)
. generate range = max - min
. bstat range, stat(19.5637501)
```

The difference between the maximum and minimum of x in the sample is 19.5637501.

The `stat()` option to bstat specifies the observed value of the statistic (`range`) to be summarized. This option is useful when, as shown above, the statistic of ultimate interest is not specified directly to bootstrap but instead is calculated by other means.

Here the observed values of `r(max)` and `r(min)` are saved as characteristics of the dataset created by bootstrap and are thus available for retrieval by bstat; see [R] **bstat**. The observed range, however, is unknown to bstat, so it must be specified.

◁

Combining bootstrap datasets

You can combine two datasets from separate runs of bootstrap by using append (see [D] **append**) and then get the bootstrap statistics for the combined datasets by running bstat. The runs must have been performed independently (having different starting random-number seeds), and the original dataset, command, and bootstrap statistics must have been all the same.

A note about macros

In the previous example, we executed the command

```
. bootstrap max=r(max) min=r(min), reps(1000) saving(mybs): summarize x
```

We did not enclose r(max) and r(min) in single quotes, as we would in most other contexts, because it would not produce what was intended:

```
. bootstrap 'r(max)' 'r(min)', reps(1000) saving(mybs): summarize x
```

To understand why, note that 'r(max)', like any reference to a local macro, will evaluate to a literal string containing the contents of r(max) *before* bootstrap is even executed. Typing the command above would appear to Stata as if we had typed

```
. bootstrap 14.5441234 33.4393293, reps(1000) saving(mybs): summarize x
```

Even worse, the current contents of r(min) and r(max) could be empty, producing an even more confusing result. To avoid this outcome, refer to statistics by name (for example, r(max)) and not by value (for example, 'r(max)').

Achieved significance level

▷ Example 3

Suppose that we wish to estimate the *achieved significance level* (ASL) of a test statistic by using the bootstrap. ASL is another name for *p*-value. An example is

$$ \text{ASL} = \Pr\left(\widehat{\theta}^* \geq \widehat{\theta} | H_0\right) $$

for an upper-tailed, alternative hypothesis, where H_0 denotes the null hypothesis, $\widehat{\theta}$ is the observed value of the test statistic, and $\widehat{\theta}^*$ is the random variable corresponding to the test statistic, assuming that H_0 is true.

Here we will compare the mean miles per gallon (mpg) between foreign and domestic cars by using the two-sample *t* test with unequal variances. The following results indicate the *p*-value to be 0.0034 for the two-sided test using Satterthwaite's approximation. Thus assuming that mean mpg is the same for foreign and domestic cars, we would expect to observe a *t* statistic more extreme (in absolute value) than 3.1797 in about 0.3% of all possible samples of the type that we observed. Thus we have evidence to reject the null hypothesis that the means are equal.

```
. use http://www.stata-press.com/data/r12/auto
(1978 Automobile Data)

. ttest mpg, by(foreign) unequal

Two-sample t test with unequal variances
```

Group	Obs	Mean	Std. Err.	Std. Dev.	[95% Conf. Interval]	
Domestic	52	19.82692	.657777	4.743297	18.50638	21.14747
Foreign	22	24.77273	1.40951	6.611187	21.84149	27.70396
combined	74	21.2973	.6725511	5.785503	19.9569	22.63769
diff		-4.945804	1.555438		-8.120053	-1.771556

```
      diff = mean(Domestic) - mean(Foreign)                         t =  -3.1797
  Ho: diff = 0                     Satterthwaite's degrees of freedom =   30.5463

     Ha: diff < 0                   Ha: diff != 0                  Ha: diff > 0
  Pr(T < t) = 0.0017         Pr(|T| > |t|) = 0.0034         Pr(T > t) = 0.9983
```

We also place the value of the test statistic in a scalar for later use.

```
. scalar tobs = r(t)
```

Efron and Tibshirani (1993, 224) describe an alternative to Satterthwaite's approximation that estimates the ASL by bootstrapping the statistic from the test of equal means. Their idea is to recenter the two samples to the combined sample mean so that the data now conform to the null hypothesis but that the variances within the samples remain unchanged.

```
. summarize mpg, meanonly

. scalar omean = r(mean)

. summarize mpg if foreign==0, meanonly

. replace mpg = mpg - r(mean) + scalar(omean) if foreign==0
mpg was int now float
(52 real changes made)

. summarize mpg if foreign==1, meanonly

. replace mpg = mpg - r(mean) + scalar(omean) if foreign==1
(22 real changes made)

. sort foreign

. by foreign: summarize mpg
```

```
-> foreign = Domestic
    Variable |      Obs       Mean    Std. Dev.      Min        Max
-------------+--------------------------------------------------------
         mpg |       52    21.2973    4.743297    13.47037   35.47038

-> foreign = Foreign
    Variable |      Obs       Mean    Std. Dev.      Min        Max
-------------+--------------------------------------------------------
         mpg |       22    21.2973    6.611187    10.52457   37.52457
```

Each sample (foreign and domestic) is a stratum, so the bootstrapped samples must have the same number of foreign and domestic cars as the original dataset. This requirement is facilitated by the `strata()` option to `bootstrap`. By typing the following, we bootstrap the test statistic using the modified dataset and save the values in `bsauto2.dta`:

```
. keep mpg foreign

. set seed 1

. bootstrap t=r(t), rep(1000) strata(foreign) saving(bsauto2) nodots: ttest mpg,
> by(foreign) unequal
```

Warning: Because **ttest** is not an estimation command or does not set
          e(sample), **bootstrap** has no way to determine which observations are
          used in calculating the statistics and so assumes that all
          observations are used. This means that no observations will be
          excluded from the resampling because of missing values or other
          reasons.

          If the assumption is not true, press Break, save the data, and drop
          the observations that are to be excluded. Be sure that the dataset
          in memory contains only the relevant data.

Bootstrap results

| Number of strata | = | 2 | | Number of obs | = | 74 |
| | | | | Replications | = | 1000 |

```
        command:  ttest mpg, by(foreign) unequal
              t:  r(t)
```

	Observed Coef.	Bootstrap Std. Err.	z	P>|z|	Normal-based [95% Conf. Interval]	
t	1.75e-07	1.036437	0.00	1.000	-2.031379	2.031379

We can use the data in bsauto2.dta to estimate ASL via the fraction of bootstrap test statistics that are more extreme than 3.1797.

```
. use bsauto2, clear
(bootstrap: ttest)

. generate indicator = abs(t)>=abs(scalar(tobs))

. summarize indicator, meanonly

. display "ASLboot = " r(mean)
ASLboot = .005
```

The result is $\text{ASL}_{\text{boot}} = 0.005$. Assuming that the mean mpg is the same between foreign and domestic cars, we would expect to observe a t statistic more extreme (in absolute value) than 3.1797 in about 0.5% of all possible samples of the type we observed. This finding is still strong evidence to reject the hypothesis that the means are equal.

◁

Bootstrapping a ratio

▷ Example 4

Suppose that we wish to produce a bootstrap estimate of the ratio of two means. Because summarize saves results for only one variable, we must call summarize twice to compute the means. Actually, we could use collapse to compute the means in one call, but calling summarize twice is much faster. Thus we will have to write a small program that will return the results we want.

We write the program below and save it to a file called `ratio.ado` (see [U] **17 Ado-files**). Our program takes two variable names as input and saves them in the `local` macros `y` (first variable) and `x` (second variable). It then computes one statistic: the mean of '`y`' divided by the mean of '`x`'. This value is returned as a scalar in `r(ratio)`. `ratio` also returns the ratio of the number of observations used to the mean for each variable.

```
program myratio, rclass
        version 12
        args y x
        confirm var `y'
        confirm var `x'
        tempname ymean yn
        summarize `y', meanonly
        scalar `ymean' = r(mean)
        return scalar n_`y' = r(N)
        summarize `x', meanonly
        return scalar n_`x' = r(N)
        return scalar ratio = `ymean'/r(mean)
end
```

Remember to test any newly written commands before using them with `bootstrap`.

```
. use http://www.stata-press.com/data/r12/auto, clear
(1978 Automobile Data)
. summarize price
```

Variable	Obs	Mean	Std. Dev.	Min	Max
price	74	6165.257	2949.496	3291	15906

```
. scalar mean1=r(mean)
. summarize weight
```

Variable	Obs	Mean	Std. Dev.	Min	Max
weight	74	3019.459	777.1936	1760	4840

```
. scalar mean2=r(mean)
. di scalar(mean1)/scalar(mean2)
2.0418412
. myratio price weight
. return list
scalars:
             r(ratio) =  2.041841210168278
          r(n_weight) =  74
           r(n_price) =  74
```

The results of running `bootstrap` on our program are

```
. use http://www.stata-press.com/data/r12/auto
(1978 Automobile Data)
. set seed 1
. bootstrap ratio=r(ratio), reps(1000) nowarn nodots: myratio price weight
Bootstrap results                            Number of obs    =         74
                                             Replications     =       1000
        command:  myratio price weight
          ratio:  r(ratio)
```

	Observed Coef.	Bootstrap Std. Err.	z	P>|z|	Normal-based [95% Conf. Interval]	
ratio	2.041841	.0942932	21.65	0.000	1.85703	2.226652

As mentioned previously, we should specify the `saving()` option if we wish to save the bootstrap dataset.

◁

Warning messages and e(sample)

`bootstrap` is not meant to be used with weighted calculations. `bootstrap` determines the presence of weights by parsing the prefixed command with standard syntax. However, commands like `stcox` and `streg` require that weights be specified in `stset`, and some user commands may allow weights to be specified by using an option instead of the standard syntax. Both cases pose a problem for `bootstrap` because it cannot determine the presence of weights under these circumstances. In these cases, we can only assume that you know what you are doing.

`bootstrap` does not know which variables of the dataset in memory matter to the calculation at hand. You can speed their execution by dropping unnecessary variables because, otherwise, they are included in each bootstrap sample.

You should thus drop observations with missing values. Leaving in missing values causes no problem in one sense because all Stata commands deal with missing values gracefully. It does, however, cause a statistical problem. Bootstrap sampling is defined as drawing, with replacement, samples of size N from a set of N observations. `bootstrap` determines N by counting the number of observations in memory, not counting the number of nonmissing values on the relevant variables. The result is that too many observations are resampled; the resulting bootstrap samples, because they are drawn from a population with missing values, are of unequal sizes.

If the number of missing values relative to the sample size is small, this will make little difference. If you have many missing values, however, you should first drop the observations that contain them.

▷ Example 5

To illustrate, we use the previous example but replace some of the values of `price` with missing values. The number of values of `price` used to compute the mean for each bootstrap is not constant. This is the purpose of the `Warning` message.

```
. use http://www.stata-press.com/data/r12/auto
(1978 Automobile Data)
. replace price = . if inlist(_n,1,3,5,7)
(4 real changes made, 4 to missing)
. set seed 1
```

```
. bootstrap ratio=r(ratio) np=r(n_price) nw=r(n_weight), reps(100) nodots:
> myratio price weight
```

Warning: Because **myratio** is not an estimation command or does not set
            e(sample), **bootstrap** has no way to determine which observations are
            used in calculating the statistics and so assumes that all
            observations are used. This means that no observations will be
            excluded from the resampling because of missing values or other
            reasons.

            If the assumption is not true, press Break, save the data, and drop
            the observations that are to be excluded. Be sure that the dataset
            in memory contains only the relevant data.

Bootstrap results   Number of obs   =   74
                                               Replications   =   100

```
      command:  myratio price weight
        ratio:  r(ratio)
           np:  r(n_price)
           nw:  r(n_weight)
```

	Observed Coef.	Bootstrap Std. Err.	z	P>|z|	Normal-based [95% Conf. Interval]	
ratio	2.063051	.0893669	23.09	0.000	1.887896	2.238207
np	70	1.872178	37.39	0.000	66.3306	73.6694
nw	74	.	.	.	.	.

◁

Bootstrapping statistics from data with a complex structure

Here we describe how to bootstrap statistics from data with a complex structure, for example, longitudinal or panel data, or matched data. bootstrap, however, is not designed to work with complex survey data. It is important to include all necessary information about the structure of the data in the bootstrap syntax to obtain correct bootstrap estimates for standard errors and confidence intervals.

bootstrap offers several options identifying the specifics of the data. These options are strata(), cluster(), idcluster(), and group(). The usage of strata() was described in example 3 above. Below we demonstrate several examples that require specifying the other three options.

▷ Example 6

Suppose that the auto data in example 1 above are clustered by rep78. We want to obtain bootstrap estimates for the standard errors of the difference between the coefficients on weight and gear_ratio, taking into account clustering.

We supply the cluster(rep78) option to bootstrap to request resampling from clusters rather than from observations in the dataset.

```
. use http://www.stata-press.com/data/r12/auto, clear
(1978 Automobile Data)

. keep if rep78 < .
(5 observations deleted)

. bootstrap diff=(_b[weight]-_b[gear]), seed(1) cluster(rep78): regress mpg weight
> gear foreign
(running regress on estimation sample)

Bootstrap replications (50)
──────┼─── 1 ───┼─── 2 ───┼─── 3 ───┼─── 4 ───┼─── 5
..................................................   50
```

Linear regression

| | | Number of obs | = | 69 |
| | | Replications | = | 50 |

      command: regress mpg weight gear foreign
         diff: _b[weight]-_b[gear]

(Replications based on 5 clusters in rep78)

	Observed Coef.	Bootstrap Std. Err.	z	P>|z|	Normal-based [95% Conf. Interval]	
diff	-1.910396	1.876778	-1.02	0.309	-5.588812	1.768021

We drop missing values in rep78 before issuing the command because bootstrap does not allow missing values in cluster(). See the section above about using bootstrap when variables contain missing values.

We can also obtain these same results by using the following syntax:

```
. bootstrap diff=(_b[weight]-_b[gear]), seed(1): regress mpg weight gear foreign,
> vce(cluster rep78)
```

When only clustered information is provided to the command, bootstrap can pick up the vce(cluster *clustvar*) option from the main command and use it to resample from clusters.

                                                           ◁

▷ Example 7

Suppose now that we have matched data and want to use bootstrap to obtain estimates of the standard errors of the exponentiated difference between two coefficients (or, equivalently, the ratio of two odds ratios) estimated by clogit. Consider the example of matched case–control data on birthweight of infants described in example 2 of [R] **clogit**.

The infants are paired by being matched on mother's age. All groups, defined by the pairid variable, have 1:2 matching. clogit requires that the matching information, pairid, be supplied to the group() (or, equivalently, strata()) option to be used in computing the parameter estimates. Because the data are matched, we need to resample from groups rather than from the whole dataset. However, simply supplying the grouping variable pairid in cluster() is not enough with bootstrap, as it is with clustered data.

```
. use http://www.stata-press.com/data/r12/lowbirth2, clear
(Applied Logistic Regression, Hosmer & Lemeshow)

. bootstrap ratio=exp(_b[smoke]-_b[ptd]), seed(1) cluster(pairid): clogit low lwt
> smoke ptd ht ui i.race, group(pairid)
(running clogit on estimation sample)

Bootstrap replications (50)
———+— 1 ——+— 2 ——+— 3 ——+— 4 ——+— 5
.................................................. 50
Bootstrap results                      Number of obs      =      112
                                       Replications       =       50

     command:  clogit low lwt smoke ptd ht ui i.race, group(pairid)
       ratio:  exp(_b[smoke]-_b[ptd])

                       (Replications based on 56 clusters in pairid)
```

	Observed Coef.	Bootstrap Std. Err.	z	P>|z|	Normal-based [95% Conf. Interval]	
ratio	.6654095	17.71791	0.04	0.970	-34.06106	35.39187

For the syntax above, imagine that the first pair was sampled twice during a replication. Then the bootstrap sample has four subjects with pairid equal to one, which clearly violates the original 1:2 matching design. As a result, the estimates of the coefficients obtained from this bootstrap sample will be incorrect.

Therefore, in addition to resampling from groups, we need to ensure that resampled groups are uniquely identified in each of the bootstrap samples. The idcluster(*newcluster*) option is designed for this. It requests that at each replication bootstrap create the new variable, *newcluster*, containing unique identifiers for all resampled groups. Thus, to make sure that the correct matching is preserved during each replication, we need to specify the grouping variable in cluster(), supply a variable name to idcluster(), and use this variable as the grouping variable with clogit, as we demonstrate below.

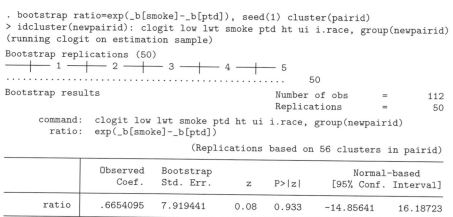

```
. bootstrap ratio=exp(_b[smoke]-_b[ptd]), seed(1) cluster(pairid)
> idcluster(newpairid): clogit low lwt smoke ptd ht ui i.race, group(newpairid)
(running clogit on estimation sample)

Bootstrap replications (50)
———+— 1 ——+— 2 ——+— 3 ——+— 4 ——+— 5
.................................................. 50
Bootstrap results                      Number of obs      =      112
                                       Replications       =       50

     command:  clogit low lwt smoke ptd ht ui i.race, group(newpairid)
       ratio:  exp(_b[smoke]-_b[ptd])

                       (Replications based on 56 clusters in pairid)
```

	Observed Coef.	Bootstrap Std. Err.	z	P>|z|	Normal-based [95% Conf. Interval]	
ratio	.6654095	7.919441	0.08	0.933	-14.85641	16.18723

Note the difference between the estimates of the bootstrap standard error for the two specifications of the bootstrap syntax.

◁

❑ Technical note

Similarly, when you have panel (longitudinal) data, all resampled panels must be unique in each of the bootstrap samples to obtain correct bootstrap estimates of statistics. Therefore, both cluster(*panelvar*) and idcluster(*newpanelvar*) must be specified with bootstrap, and i(*newpanelvar*) must be used with the main command. Moreover, you must clear the current xtset settings by typing xtset, clear before calling bootstrap.

❑

▷ Example 8

Continuing with our birthweight data, suppose that we have more information about doctors supervising women's pregnancies. We believe that the data on the pairs of infants from the same doctor may be correlated and want to adjust standard errors for possible correlation among the pairs. clogit offers the vce(cluster *clustvar*) option to do this.

Let's add a cluster variable to our dataset. One thing to keep in mind is that to use vce(cluster *clustvar*), groups in group() must be nested within clusters.

```
. use http://www.stata-press.com/data/r12/lowbirth2, clear
(Applied Logistic Regression, Hosmer & Lemeshow)
. set seed 12345
. by pairid, sort: egen byte doctor = total(int(2*runiform()+1)*(_n == 1))
. clogit low lwt smoke ptd ht ui i.race, group(pairid) vce(cluster doctor)
Iteration 0:   log pseudolikelihood = -26.768693
Iteration 1:   log pseudolikelihood = -25.810476
Iteration 2:   log pseudolikelihood = -25.794296
Iteration 3:   log pseudolikelihood = -25.794271
Iteration 4:   log pseudolikelihood = -25.794271
Conditional (fixed-effects) logistic regression   Number of obs   =        112
                                                   Wald chi2(1)    =          .
                                                   Prob > chi2     =          .
Log pseudolikelihood = -25.794271                  Pseudo R2       =     0.3355
```

(Std. Err. adjusted for 2 clusters in doctor)

low	Coef.	Robust Std. Err.	z	P>|z|	[95% Conf. Interval]	
lwt	-.0183757	.0217802	-0.84	0.399	-.0610641	.0243128
smoke	1.400656	.0085545	163.73	0.000	1.38389	1.417423
ptd	1.808009	.938173	1.93	0.054	-.0307765	3.646794
ht	2.361152	1.587013	1.49	0.137	-.7493362	5.47164
ui	1.401929	.8568119	1.64	0.102	-.2773913	3.08125
race						
2	.5713643	.0672593	8.49	0.000	.4395385	.7031902
3	-.0253148	.9149785	-0.03	0.978	-1.81864	1.76801

To obtain correct bootstrap standard errors of the exponentiated difference between the two coefficients in this example, we need to make sure that both resampled clusters and groups within resampled clusters are unique in each of the bootstrap samples. To achieve this, bootstrap needs the information about clusters in cluster(), the variable name of the new identifier for clusters in idcluster(), and the information about groups in group(). We demonstrate the corresponding syntax of bootstrap below.

```
. bootstrap ratio=exp(_b[smoke]-_b[ptd]), seed(1) cluster(doctor)
> idcluster(uidoctor) group(pairid): clogit low lwt smoke ptd ht ui i.race,
> group(pairid)
(running clogit on estimation sample)

Bootstrap replications (50)
————+— 1 ——+— 2 ——+— 3 ——+— 4 ——+— 5
.................................................. 50

Bootstrap results                         Number of obs    =      112
                                          Replications     =       50

     command:  clogit low lwt smoke ptd ht ui i.race, group(pairid)
       ratio:  exp(_b[smoke]-_b[ptd])
                        (Replications based on 2 clusters in doctor)
```

	Observed Coef.	Bootstrap Std. Err.	z	P>|z|	Normal-based [95% Conf. Interval]	
ratio	.6654095	.3156251	2.11	0.035	.0467956	1.284023

In the above syntax, although we specify group(pairid) with clogit, it is not the group identifiers of the original pairid variable that are used to compute parameter estimates from bootstrap samples. The way bootstrap works is that, at each replication, the clusters defined by doctor are resampled and the new variable, uidoctor, uniquely identifying resampled clusters is created. After that, another new variable uniquely identifying the (uidoctor, group) combination is created and renamed to have the same name as the grouping variable, pairid. This newly defined grouping variable is then used by clogit to obtain the parameter estimates from this bootstrap sample of clusters. After all replications are performed, the original values of the grouping variable are restored.

◁

❏ Technical note

The same logic must be used when running bootstrap with commands designed for panel (longitudinal) data that allow specifying the cluster(*clustervar*) option. To ensure that the combination of (*clustervar, panelvar*) values are unique in each of the bootstrap samples, cluster(*clustervar*), idcluster(*newclustervar*), and group(*panelvar*) must be specified with bootstrap, and i(*panelvar*) must be used with the main command.

❏

Bradley Efron was born in 1938 in Minnesota and studied mathematics and statistics at Caltech and Stanford; he has lived in northern California since 1960. He has worked on empirical Bayes, survival analysis, exponential families, bootstrap and jackknife methods, and confidence intervals, in conjunction with applied work in biostatistics, astronomy, and physics.

Saved results

bootstrap saves the following in e():

Scalars
e(N)	sample size
e(N_reps)	number of complete replications
e(N_misreps)	number of incomplete replications
e(N_strata)	number of strata
e(N_clust)	number of clusters
e(k_eq)	number of equations in e(b)
e(k_exp)	number of standard expressions
e(k_eexp)	number of extended expressions (i.e., _b)
e(k_extra)	number of extra equations beyond the original ones from e(b)
e(level)	confidence level for bootstrap CIs
e(bs_version)	version for bootstrap results
e(rank)	rank of e(V)

Macros
e(cmdname)	command name from *command*
e(cmd)	same as e(cmdname) or bootstrap
e(command)	*command*
e(cmdline)	command as typed
e(prefix)	bootstrap
e(title)	title in estimation output
e(strata)	strata variables
e(cluster)	cluster variables
e(seed)	initial random-number seed
e(size)	from the size(#) option
e(exp#)	expression for the #th statistic
e(mse)	mse, if specified
e(vce)	bootstrap
e(vcetype)	title used to label Std. Err.
e(properties)	b V

Matrices
e(b)	observed statistics
e(b_bs)	bootstrap estimates
e(reps)	number of nonmissing results
e(bias)	estimated biases
e(se)	estimated standard errors
e(z0)	median biases
e(accel)	estimated accelerations
e(ci_normal)	normal-approximation CIs
e(ci_percentile)	percentile CIs
e(ci_bc)	bias-corrected CIs
e(ci_bca)	bias-corrected and accelerated CIs
e(V)	bootstrap variance–covariance matrix
e(V_modelbased)	model-based variance

When *exp_list* is _b, bootstrap will also carry forward most of the results already in e() from *command*.

Methods and formulas

bootstrap is implemented as an ado-file.

Let $\widehat{\theta}$ be the observed value of the statistic, that is, the value of the statistic calculated with the original dataset. Let $i = 1, 2, \ldots, k$ denote the bootstrap samples, and let $\widehat{\theta}_i$ be the value of the statistic from the ith bootstrap sample.

When the `mse` option is specified, the standard error is estimated as

$$
\widehat{se}_{\text{MSE}} = \left\{ \frac{1}{k} \sum_{i=1}^{k} (\widehat{\theta}_i - \widehat{\theta})^2 \right\}^{1/2}
$$

Otherwise, the standard error is estimated as

$$
\widehat{se} = \left\{ \frac{1}{k-1} \sum_{i=1}^{k} (\widehat{\theta}_i - \overline{\theta})^2 \right\}^{1/2}
$$

where

$$
\overline{\theta} = \frac{1}{k} \sum_{i=1}^{k} \widehat{\theta}_i
$$

The variance–covariance matrix is similarly computed. The bias is estimated as

$$
\widehat{\text{bias}} = \overline{\theta} - \widehat{\theta}
$$

Confidence intervals with nominal coverage rates $1 - \alpha$ are calculated according to the following formulas. The normal-approximation method yields the confidence intervals

$$
\left[\widehat{\theta} - z_{1-\alpha/2}\, \widehat{se}, \ \widehat{\theta} + z_{1-\alpha/2}\, \widehat{se} \right]
$$

where $z_{1-\alpha/2}$ is the $(1 - \alpha/2)$th quantile of the standard normal distribution. If the `mse` option is specified, `bootstrap` will report the normal confidence interval using $\widehat{se}_{\text{MSE}}$ instead of $\widehat{se}$. `estat bootstrap` only uses $\widehat{se}$ in the normal confidence interval.

The percentile method yields the confidence intervals

$$
\left[\theta^*_{\alpha/2}, \ \theta^*_{1-\alpha/2} \right]
$$

where θ^*_p is the pth quantile (the $100p$th percentile) of the bootstrap distribution $(\widehat{\theta}_1, \dots, \widehat{\theta}_k)$.

Let

$$
z_0 = \Phi^{-1} \{ \#(\widehat{\theta}_i \leq \widehat{\theta})/k \}
$$

where $\#(\widehat{\theta}_i \leq \widehat{\theta})$ is the number of elements of the bootstrap distribution that are less than or equal to the observed statistic and Φ is the standard cumulative normal. z_0 is known as the median bias of $\widehat{\theta}$. Let

$$
a = \frac{\sum_{i=1}^{n} (\overline{\theta}_{(\cdot)} - \widehat{\theta}_{(i)})^3}{6 \{ \sum_{i=1}^{n} (\overline{\theta}_{(\cdot)} - \widehat{\theta}_{(i)})^2 \}^{3/2}}
$$

where $\widehat{\theta}_{(i)}$ are the leave-one-out (jackknife) estimates of $\widehat{\theta}$ and $\overline{\theta}_{(\cdot)}$ is their mean. This expression is known as the jackknife estimate of acceleration for $\widehat{\theta}$. Let

$$
p_1 = \Phi \left\{ z_0 + \frac{z_0 - z_{1-\alpha/2}}{1 - a(z_0 - z_{1-\alpha/2})} \right\}
$$
$$
p_2 = \Phi \left\{ z_0 + \frac{z_0 + z_{1-\alpha/2}}{1 - a(z_0 + z_{1-\alpha/2})} \right\}
$$

where $z_{1-\alpha/2}$ is the $(1-\alpha/2)$th quantile of the normal distribution. The bias-corrected and accelerated (BC$_a$) method yields confidence intervals

$$\left[\, \theta^*_{p_1},\, \theta^*_{p_2} \,\right]$$

where θ^*_p is the pth quantile of the bootstrap distribution as defined previously. The bias-corrected (but not accelerated) method is a special case of BC$_a$ with $a = 0$.

References

Ängquist, L. 2010. Stata tip 92: Manual implementation of permutations and bootstraps. *Stata Journal* 10: 686–688.

Cameron, A. C., and P. K. Trivedi. 2010. *Microeconometrics Using Stata*. Rev. ed. College Station, TX: Stata Press.

Davison, A. C., and D. V. Hinkley. 1997. *Bootstrap Methods and Their Application*. Cambridge: Cambridge University Press.

Efron, B. 1979. Bootstrap methods: Another look at the jackknife. *Annals of Statistics* 7: 1–26.

———. 1982. *The Jackknife, the Bootstrap and Other Resampling Plans*. Philadelphia: Society for Industrial and Applied Mathematics.

Efron, B., and C. Stein. 1981. The jackknife estimate of variance. *Annals of Statistics* 9: 586–596.

Efron, B., and R. J. Tibshirani. 1986. Bootstrap methods for standard errors, confidence intervals, and other measures of statistical accuracy. *Statistical Science* 1: 54–77.

———. 1993. *An Introduction to the Bootstrap*. New York: Chapman & Hall/CRC.

Gleason, J. R. 1997. ip18: A command for randomly resampling a dataset. *Stata Technical Bulletin* 37: 17–22. Reprinted in *Stata Technical Bulletin Reprints*, vol. 7, pp. 77–83. College Station, TX: Stata Press.

———. 1999. ip18.1: Update to resample. *Stata Technical Bulletin* 52: 9–10. Reprinted in *Stata Technical Bulletin Reprints*, vol. 9, p. 119. College Station, TX: Stata Press.

Gould, W. W. 1994. ssi6.2: Faster and easier bootstrap estimation. *Stata Technical Bulletin* 21: 24–33. Reprinted in *Stata Technical Bulletin Reprints*, vol. 4, pp. 211–223. College Station, TX: Stata Press.

Guan, W. 2003. From the help desk: Bootstrapped standard errors. *Stata Journal* 3: 71–80.

Hall, P., and S. R. Wilson. 1991. Two guidelines for bootstrap hypothesis testing. *Biometrics* 47: 757–762.

Hamilton, L. C. 1991. ssi2: Bootstrap programming. *Stata Technical Bulletin* 4: 18–27. Reprinted in *Stata Technical Bulletin Reprints*, vol. 1, pp. 208–220. College Station, TX: Stata Press.

———. 1992. *Regression with Graphics: A Second Course in Applied Statistics*. Belmont, CA: Duxbury.

———. 2009. *Statistics with Stata (Updated for Version 10)*. Belmont, CA: Brooks/Cole.

Hinkley, D. V. 1978. Improving the jackknife with special reference to correlation estimation. *Biometrika* 65: 13–22.

Holmes, S., C. Morris, and R. J. Tibshirani. 2003. Bradley Efron: A conversation with good friends. *Statistical Science* 18: 268–281.

Mooney, C. Z., and R. D. Duval. 1993. *Bootstrapping: A Nonparametric Approach to Statistical Inference*. Newbury Park, CA: Sage.

Poi, B. P. 2004. From the help desk: Some bootstrapping techniques. *Stata Journal* 4: 312–328.

Royston, P., and W. Sauerbrei. 2009. Bootstrap assessment of the stability of multivariable models. *Stata Journal* 9: 547–570.

Stine, R. 1990. An introduction to bootstrap methods: Examples and ideas. In *Modern Methods of Data Analysis*, ed. J. Fox and J. S. Long, 353–373. Newbury Park, CA: Sage.

Also see

[R] **bootstrap postestimation** — Postestimation tools for bootstrap

[R] **jackknife** — Jackknife estimation

[R] **permute** — Monte Carlo permutation tests

[R] **simulate** — Monte Carlo simulations

[SVY] **svy bootstrap** — Bootstrap for survey data

[U] **13.5 Accessing coefficients and standard errors**

[U] **13.6 Accessing results from Stata commands**

[U] **20 Estimation and postestimation commands**

Title

bootstrap postestimation — Postestimation tools for bootstrap

Description

The following postestimation command is of special interest after `bootstrap`:

Command	Description
estat bootstrap	percentile-based and bias-corrected CI tables

For information about `estat bootstrap`, see below.

The following standard postestimation commands are also available:

Command	Description
*contrast	contrasts and ANOVA-style joint tests of estimates
estat	AIC, BIC, VCE, and estimation sample summary
estimates	cataloging estimation results
*hausman	Hausman's specification test
*lincom	point estimates, standard errors, testing, and inference for linear combinations of coefficients
*margins	marginal means, predictive margins, marginal effects, and average marginal effects
*marginsplot	graph the results from margins (profile plots, interaction plots, etc.)
*nlcom	point estimates, standard errors, testing, and inference for nonlinear combinations of coefficients
*predict	predictions, residuals, influence statistics, and other diagnostic measures
*predictnl	point estimates, standard errors, testing, and inference for generalized predictions
*pwcompare	pairwise comparisons of estimates
*test	Wald tests of simple and composite linear hypotheses
*testnl	Wald tests of nonlinear hypotheses

*This postestimation command is allowed if it may be used after *command*.

See the corresponding entries in the *Stata Base Reference Manual* for details.

Special-interest postestimation command

`estat bootstrap` displays a table of confidence intervals for each statistic from a bootstrap analysis.

Syntax for predict

The syntax of `predict` (and even if `predict` is allowed) following `bootstrap` depends upon the *command* used with `bootstrap`. If `predict` is not allowed, neither is `predictnl`.

Syntax for estat bootstrap

estat <u>boot</u>strap [, *options*]

options	Description
bc	bias-corrected CIs; the default
bca	bias-corrected and accelerated (BC_a) CIs
<u>n</u>ormal	normal-based CIs
<u>p</u>ercentile	percentile CIs
all	all available CIs
<u>noh</u>eader	suppress table header
<u>nol</u>egend	suppress table legend
<u>v</u>erbose	display the full table legend

bc, bca, normal, and percentile may be used together.

Menu

Statistics > Postestimation > Reports and statistics

Options for estat bootstrap

bc is the default and displays bias-corrected confidence intervals.

bca displays bias-corrected and accelerated confidence intervals. This option assumes that you also specified the bca option on the bootstrap prefix command.

normal displays normal approximation confidence intervals.

percentile displays percentile confidence intervals.

all displays all available confidence intervals.

noheader suppresses display of the table header. This option implies nolegend.

nolegend suppresses display of the table legend, which identifies the rows of the table with the expressions they represent.

verbose requests that the full table legend be displayed.

Remarks

▷ Example 1

The estat bootstrap postestimation command produces a table containing the observed value of the statistic, an estimate of its bias, the bootstrap standard error, and up to four different confidence intervals.

If we were interested merely in getting bootstrap standard errors for the model coefficients, we could use the bootstrap prefix with our estimation command. If we were interested in performing a thorough bootstrap analysis of the model coefficients, we could use the estat bootstrap postestimation command after fitting the model with the bootstrap prefix.

Using example 1 from [R] **bootstrap**, we need many more replications for the confidence interval types other than the normal based, so let's rerun the estimation command. We will reset the random-number seed—in case we wish to reproduce the results—increase the number of replications, and save the bootstrap distribution as a dataset called bsauto.dta.

```
. use http://www.stata-press.com/data/r12/auto
(1978 Automobile Data)

. set seed 1

. bootstrap _b, reps(1000) saving(bsauto) bca: regress mpg weight gear foreign
  (output omitted )

. estat bootstrap, all
```

Linear regression   Number of obs   =   74
                                                   Replications   =   1000

mpg	Observed Coef.	Bias	Bootstrap Std. Err.	[95% Conf. Interval]	
weight	-.00613903	.0000567	.000628	-.0073699	-.0049082	(N)
				-.0073044	-.0048548	(P)
				-.0074355	-.004928	(BC)
				-.0075282	-.0050258	(BCa)
gear_ratio	1.4571134	.1051696	1.4554785	-1.395572	4.309799	(N)
				-1.262111	4.585372	(P)
				-1.523927	4.174376	(BC)
				-1.492223	4.231356	(BCa)
foreign	-2.2216815	-.0196361	1.2023286	-4.578202	.1348393	(N)
				-4.442199	.2677989	(P)
				-4.155504	.6170642	(BC)
				-4.216531	.5743973	(BCa)
_cons	36.101353	-.502281	5.4089441	25.50002	46.70269	(N)
				24.48569	46.07086	(P)
				25.59799	46.63227	(BC)
				25.85658	47.02108	(BCa)

```
(N)    normal confidence interval
(P)    percentile confidence interval
(BC)   bias-corrected confidence interval
(BCa)  bias-corrected and accelerated confidence interval
```

The estimated standard errors here differ from our previous estimates using only 100 replications by, respectively, 8%, 3%, 11%, and 6%; see example 1 of [R] **bootstrap**. So much for our advice that 50–200 replications are good enough to estimate standard errors. Well, the more replications the better—that advice you should believe.

Which of the methods to compute confidence intervals should we use? If the statistic is unbiased, the percentile (P) and bias-corrected (BC) methods should give similar results. The bias-corrected confidence interval will be the same as the percentile confidence interval when the observed value of the statistic is equal to the median of the bootstrap distribution. Thus, for unbiased statistics, the two methods should give similar results as the number of replications becomes large. For biased statistics, the bias-corrected method should yield confidence intervals with better coverage probability (closer to the nominal value of 95% or whatever was specified) than the percentile method. For statistics with variances that vary as a function of the parameter of interest, the bias-corrected and accelerated method (BC_a) will typically have better coverage probability than the others.

When the bootstrap distribution is approximately normal, all these methods should give similar confidence intervals as the number of replications becomes large. If we examine the normality of these bootstrap distributions using, say, the pnorm command (see [R] **diagnostic plots**), we see that they closely follow a normal distribution. Thus here, the normal approximation would also be a valid choice. The chief advantage of the normal-approximation method is that it (supposedly) requires fewer replications than the other methods. Of course, it should be used only when the bootstrap distribution exhibits normality.

We can load `bsauto.dta` containing the bootstrap distributions for these coefficients:

```
. use bsauto
(bootstrap: regress)

. describe *

              storage   display    value
variable name   type    format     label      variable label

_b_weight       float    %9.0g                 _b[weight]
_b_gear_ratio   float    %9.0g                 _b[gear_ratio]
_b_foreign      float    %9.0g                 _b[foreign]
_b_cons         float    %9.0g                 _b[_cons]
```

We can now run other commands, such as `pnorm`, on the bootstrap distributions. As with all standard estimation commands, we can use the `bootstrap` command to replay its output table. The default variable names assigned to the statistics in *exp_list* are `_bs_1`, `_bs_2`, ..., and each variable is labeled with the associated expression. The naming convention for the extended expressions `_b` and `_se` is to prepend `_b_` and `_se_`, respectively, onto the name of each element of the coefficient vector. Here the first coefficient is `_b[weight]`, so `bootstrap` named it `_b_weight`.

◁

Methods and formulas

All postestimation commands listed above are implemented as ado-files.

Also see

[R] **bootstrap** — Bootstrap sampling and estimation

[U] **20 Estimation and postestimation commands**

Title

> **boxcox** — Box–Cox regression models

Syntax

> boxcox *depvar* [*indepvars*] [*if*] [*in*] [*weight*] [, *options*]

options	Description
Model	
noconstant	suppress constant term
model(lhsonly)	left-hand-side Box–Cox model; the default
model(rhsonly)	right-hand-side Box–Cox model
model(lambda)	both sides Box–Cox model with same parameter
model(theta)	both sides Box–Cox model with different parameters
notrans(*varlist*)	nontransformed independent variables
Reporting	
level(#)	set confidence level; default is level(95)
lrtest	perform likelihood-ratio test
Maximization	
nolog	suppress full-model iteration log
nologlr	suppress restricted-model lrtest iteration log
maximize_options	control the maximization process; seldom used

depvar and *indepvars* may contain time-series operators; see [U] **11.4.4 Time-series varlists**.
bootstrap, by, jackknife, rolling, statsby, and xi are allowed; see [U] **11.1.10 Prefix commands**.
Weights are not allowed with the bootstrap prefix; see [R] **bootstrap**.
fweights and iweights are allowed; see [U] **11.1.6 weight**.
See [U] **20 Estimation and postestimation commands** for more capabilities of estimation commands.

Menu

Statistics > Linear models and related > Box-Cox regression

Description

boxcox finds the maximum likelihood estimates of the parameters of the Box–Cox transform, the coefficients on the independent variables, and the standard deviation of the normally distributed errors for a model in which *depvar* is regressed on *indepvars*. You can fit the following models:

Option	Estimates
lhsonly	$y_j^{(\theta)} = \beta_1 x_{1j} + \beta_2 x_{2j} + \cdots + \beta_k x_{kj} + \epsilon_j$
rhsonly	$y_j = \beta_1 x_{1j}^{(\lambda)} + \beta_2 x_{2j}^{(\lambda)} + \cdots + \beta_k x_{kj}^{(\lambda)} + \epsilon_j$
rhsonly notrans()	$y_j = \beta_1 x_{1j}^{(\lambda)} + \beta_2 x_{2j}^{(\lambda)} + \cdots + \beta_k x_{kj}^{(\lambda)} + \gamma_1 z_{1j} + \cdots + \gamma_l z_{lj} + \epsilon_j$
lambda	$y_j^{(\lambda)} = \beta_1 x_{1j}^{(\lambda)} + \beta_2 x_{2j}^{(\lambda)} + \cdots + \beta_k x_{kj}^{(\lambda)} + \epsilon_j$
lambda notrans()	$y_j^{(\lambda)} = \beta_1 x_{1j}^{(\lambda)} + \beta_2 x_{2j}^{(\lambda)} + \cdots + \beta_k x_{kj}^{(\lambda)} + \gamma_1 z_{1j} + \cdots + \gamma_l z_{lj} + \epsilon_j$
theta	$y_j^{(\theta)} = \beta_1 x_{1j}^{(\lambda)} + \beta_2 x_{2j}^{(\lambda)} + \cdots + \beta_k x_{kj}^{(\lambda)} + \epsilon_j$
theta notrans()	$y_j^{(\theta)} = \beta_1 x_{1j}^{(\lambda)} + \beta_2 x_{2j}^{(\lambda)} + \cdots + \beta_k x_{kj}^{(\lambda)} + \gamma_1 z_{1j} + \cdots + \gamma_l z_{lj} + \epsilon_j$

Any variable to be transformed must be strictly positive.

Options

Model

noconstant; see [R] **estimation options**.

model(lhsonly | rhsonly | lambda | theta) specifies which of the four models to fit.

   model(lhsonly) applies the Box–Cox transform to *depvar* only. model(lhsonly) is the default.

   model(rhsonly) applies the transform to the *indepvars* only.

   model(lambda) applies the transform to both *depvar* and *indepvars*, and they are transformed by the same parameter.

   model(theta) applies the transform to both *depvar* and *indepvars*, but this time, each side is transformed by a separate parameter.

notrans(*varlist*) specifies that the variables in *varlist* be included as nontransformed independent variables.

Reporting

level(*#*); see [R] **estimation options**.

lrtest specifies that a likelihood-ratio test of significance be performed and reported for each independent variable.

Maximization

nolog suppresses the iteration log when fitting the full model.

nologlr suppresses the iteration log when fitting the restricted models required by the lrtest option.

maximize_options: <u>iter</u>ate(*#*) and from(*init_specs*); see [R] **maximize**.

Model	Initial value specification
lhsonly	from(θ_0, copy)
rhsonly	from(λ_0, copy)
lambda	from(λ_0, copy)
theta	from($\lambda_0\ \theta_0$, copy)

Remarks

Remarks are presented under the following headings:

Introduction
Theta model
Lambda model
Left-hand-side-only model
Right-hand-side-only model

Introduction

The Box–Cox transform

$$y^{(\lambda)} = \frac{y^\lambda - 1}{\lambda}$$

has been widely used in applied data analysis. Box and Cox (1964) developed the transformation and argued that the transformation could make the residuals more closely normal and less heteroskedastic. Cook and Weisberg (1982) discuss the transform in this light. Because the transform embeds several popular functional forms, it has received some attention as a method for testing functional forms, in particular,

$$y^{(\lambda)} = \begin{cases} y - 1 & \text{if } \lambda = 1 \\ \ln(y) & \text{if } \lambda = 0 \\ 1 - 1/y & \text{if } \lambda = -1 \end{cases}$$

Davidson and MacKinnon (1993) discuss this use of the transform. Atkinson (1985) also gives a good general treatment.

Theta model

boxcox obtains the maximum likelihood estimates of the parameters for four different models. The most general of the models, the theta model, is

$$y_j^{(\theta)} = \beta_0 + \beta_1 x_{1j}^{(\lambda)} + \beta_2 x_{2j}^{(\lambda)} + \cdots + \beta_k x_{kj}^{(\lambda)} + \gamma_1 z_{1j} + \gamma_2 z_{2j} + \cdots + \gamma_l z_{lj} + \epsilon_j$$

where $\epsilon \sim N(0, \sigma^2)$. Here the dependent variable, y, is subject to a Box–Cox transform with parameter θ. Each of the *indepvars*, $x_1, x_2, \ldots, x_k$, is transformed by a Box–Cox transform with parameter λ. The $z_1, z_2, \ldots, z_l$ specified in the notrans() option are independent variables that are not transformed.

Box and Cox (1964) argued that this transformation would leave behind residuals that more closely follow a normal distribution than those produced by a simple linear regression model. Bear in mind that the normality of ϵ is assumed and that boxcox obtains maximum likelihood estimates of the $k + l + 4$ parameters under this assumption. boxcox does not choose λ and θ so that the residuals are approximately normally distributed. If you are interested in this type of transformation to normality, see the official Stata commands lnskew0 and bcskew0 in [R] **lnskew0**. However, those commands work on a more restrictive model in which none of the independent variables is transformed.

▷ Example 1

Consider an example using the `auto` data.

```
. use http://www.stata-press.com/data/r12/auto
(1978 Automobile Data)
. boxcox mpg weight price, notrans(foreign) model(theta) lrtest
Fitting comparison model

Iteration 0:   log likelihood = -234.39434
Iteration 1:   log likelihood = -228.26891
Iteration 2:   log likelihood = -228.26777
Iteration 3:   log likelihood = -228.26777

Fitting full model

Iteration 0:   log likelihood = -194.13727
  (output omitted )

Fitting comparison models for LR tests

Iteration 0:   log likelihood = -179.58214
Iteration 1:   log likelihood = -177.59036
Iteration 2:   log likelihood = -177.58739
Iteration 3:   log likelihood = -177.58739

Iteration 0:   log likelihood = -203.92855
Iteration 1:   log likelihood = -201.30202
Iteration 2:   log likelihood = -201.18257
Iteration 3:   log likelihood = -201.18233
Iteration 4:   log likelihood = -201.18233

Iteration 0:   log likelihood = -178.83799
Iteration 1:   log likelihood = -175.98405
Iteration 2:   log likelihood = -175.97931
Iteration 3:   log likelihood = -175.97931
```

				Number of obs	=	74
				LR chi2(4)	=	105.19
Log likelihood = -175.67343				Prob > chi2	=	0.000

mpg	Coef.	Std. Err.	z	P>|z|	[95% Conf. Interval]
/lambda	.7601691	.6289991	1.21	0.227	-.4726465 1.992985
/theta	-.7189315	.3244439	-2.22	0.027	-1.35483 -.0830332

Estimates of scale-variant parameters

	Coef.	chi2(df)	P>chi2(df)	df of chi2
Notrans				
foreign	-.0114338	3.828	0.050	1
_cons	1.377399			
Trans				
weight	-.000239	51.018	0.000	1
price	-6.18e-06	0.612	0.434	1
/sigma	.0143509			

Test H0:	Restricted log likelihood	chi2	Prob > chi2
theta=lambda = -1	-181.64479	11.94	0.001
theta=lambda = 0	-178.2406	5.13	0.023
theta=lambda = 1	-194.13727	36.93	0.000

The output is composed of the iteration logs and three distinct tables. The first table contains a standard header for a maximum likelihood estimator and a standard output table for the Box–Cox transform parameters. The second table contains the estimates of the scale-variant parameters. The third table contains the output from likelihood-ratio tests on three standard functional form specifications.

If we were to interpret this output, the right-hand-side transformation would not significantly add to the regression, whereas the left-hand-side transformation would make the 5% but not the 1% cutoff. price is certainly not significant, and foreign lies right on the 5% cutoff. weight is clearly significant. The output also shows that the linear and multiplicative inverse specifications are both strongly rejected. A natural log specification can be rejected at the 5% level but not at the 1% level.

◁

❏ Technical note

Spitzer (1984) showed that the Wald tests of the joint significance of the coefficients of the right-hand-side variables, either transformed or untransformed, are not invariant to changes in the scale of the transformed dependent variable. Davidson and MacKinnon (1993) also discuss this point. This problem demonstrates that Wald statistics can be manipulated in nonlinear models. Lafontaine and White (1986) analyze this problem numerically, and Phillips and Park (1988) analyze it by using Edgeworth expansions. See Drukker (2000b) for a more detailed discussion of this issue. Because the parameter estimates and their Wald tests are not scale invariant, no Wald tests or confidence intervals are reported for these parameters. However, when the lrtest option is specified, likelihood-ratio tests are performed and reported. Schlesselman (1971) showed that, if a constant is included in the model, the parameter estimates of the Box–Cox transforms are scale invariant. For this reason, we strongly recommend that you not use the noconstant option.

The lrtest option does not perform a likelihood-ratio test on the constant, so no value for this statistic is reported. Unless the data are properly scaled, the restricted model does not often converge. For this reason, no likelihood-ratio test on the constant is performed by the lrtest option. However, if you have a special interest in performing this test, you can do so by fitting the constrained model separately. If problems with convergence are encountered, rescaling the data by their means may help.

❏

Lambda model

A less general model than the one above is called the lambda model. It specifies that the same parameter be used in both the left-hand-side and right-hand-side transformations. Specifically,

$$y_j^{(\lambda)} = \beta_0 + \beta_1 x_{1j}^{(\lambda)} + \beta_2 x_{2j}^{(\lambda)} + \cdots + \beta_k x_{kj}^{(\lambda)} + \gamma_1 z_{1j} + \gamma_2 z_{2j} + \cdots + \gamma_l z_{lj} + \epsilon_j$$

where $\epsilon \sim N(0, \sigma^2)$. Here the *depvar* variable, y, and each of the *indepvars*, $x_1, x_2, \ldots, x_k$, is transformed by a Box–Cox transform with the common parameter λ. Again the $z_1, z_2, \ldots, z_l$ are independent variables that are not transformed.

Left-hand-side-only model

Even more restrictive than a common transformation parameter is transforming the dependent variable only. Because the dependent variable is on the left-hand side of the equation, this model is known as the lhsonly model. Here you are estimating the parameters of the model

$$y_j^{(\theta)} = \beta_0 + \beta_1 x_{1j} + \beta_2 x_{2j} + \cdots + \beta_k x_{kj} + \epsilon_j$$

where $\epsilon \sim N(0, \sigma^2)$. Here only the *depvar*, y, is transformed by a Box–Cox transform with the parameter θ.

▷ Example 2

We again hypothesize mpg to be a function of weight, price, and foreign in a Box–Cox model in which only mpg is subject to the transform:

```
. boxcox mpg weight price foreign, model(lhs) lrtest nolog nologlr
Fitting comparison model
Fitting full model
Fitting comparison models for LR tests
```

Number of obs	=	74
LR chi2(3)	=	105.04
Log likelihood = -175.74705 · Prob > chi2	=	0.000

mpg	Coef.	Std. Err.	z	P>|z|	[95% Conf. Interval]
/theta	-.7826999	.281954	-2.78	0.006	-1.33532 -.2300802

Estimates of scale-variant parameters

	Coef.	chi2(df)	P>chi2(df)	df of chi2
Notrans				
weight	-.0000294	58.056	0.000	1
price	-4.66e-07	0.469	0.493	1
foreign	-.0097564	4.644	0.031	1
_cons	1.249845			
/sigma	.0118454			

Test HO:	Restricted log likelihood	LR statistic chi2	P-value Prob > chi2
theta = -1	-176.04312	0.59	0.442
theta = 0	-179.54104	7.59	0.006
theta = 1	-194.13727	36.78	0.000

This model rejects both linear and log specifications of mpg but fails to reject the hypothesis that 1/mpg is linear in the independent variables. These findings are in line with what an engineer would have expected. In engineering terms, gallons per mile represents actual energy consumption, and energy consumption should be approximately linear in weight.

◁

Right-hand-side-only model

The fourth model leaves the *depvar* alone and transforms a subset of the *indepvars* using the parameter λ. This is the rhsonly model. In this model, the *depvar*, y, is given by

$$y_j = \beta_0 + \beta_1 x_{1j}^{(\lambda)} + \beta_2 x_{2j}^{(\lambda)} + \cdots + \beta_k x_{kj}^{(\lambda)} + \gamma_1 z_{1j} + \gamma_2 z_{2j} + \cdots + \gamma_l z_{lj} + \epsilon_j$$

where $\epsilon \sim N(0, \sigma^2)$. Here each of the *indepvars*, $x_1, x_2, \ldots, x_k$, is transformed by a Box–Cox transform with the parameter λ. Again the $z_1, z_2, \ldots, z_l$ are independent variables that are not transformed.

▷ Example 3

Here is an example with the rhsonly model. price and foreign are not included in the list of covariates. (You are invited to use the auto data and check that they fare no better here than above.)

```
. boxcox mpg weight, model(rhs) lrtest nolog nologlr
Fitting full model
Fitting comparison models for LR tests
Comparison model for LR test on weight is a linear regression.
Lambda is not identified in the restricted model.
```

			Number of obs	=	74
			LR chi2(2)	=	82.90
Log likelihood = -192.94368			Prob > chi2	=	0.000

	Coef.	Std. Err.	z	P>|z|	[95% Conf. Interval]
mpg					
/lambda	-.4460916	.6551107	-0.68	0.496	-1.730085 .8379018

Estimates of scale-variant parameters

	Coef.	chi2(df)	P>chi2(df)	df of chi2
Notrans				
_cons	1359.092			
Trans				
weight	-614.3876	82.901	0.000	1
/sigma	3.281854			

Test HO:	Restricted log likelihood	LR statistic chi2	P-value Prob > chi2
lambda = -1	-193.2893	0.69	0.406
lambda = 0	-193.17892	0.47	0.493
lambda = 1	-195.38869	4.89	0.027

The interpretation of the output is similar to that in all the cases above, with one caveat. As requested, a likelihood-ratio test was performed on the lone independent variable. However, when it is dropped to form the constrained model, the comparison model is not a right-hand-side-only Box–Cox model but rather a simple linear regression on a constant model. When weight is dropped, there are no longer any transformed variables. Hence, λ is not identified, and it must also be dropped. This process leaves a linear regression on a constant as the "comparison model". It also implies that the test statistic has 2 degrees of freedom instead of 1. At the top of the output, a more concise warning informs you of this point.

A similar identification issue can also arise in the `lambda` and `theta` models when only one independent variable is specified. In these cases, warnings also appear on the output.

◁

Saved results

boxcox saves the following in e():

Scalars
e(N)	number of observations
e(ll)	log likelihood
e(chi2)	LR statistic of full vs. comparison
e(df_m)	full model degrees of freedom
e(ll0)	log likelihood of the restricted model
e(df_r)	restricted model degrees of freedom
e(ll_t1)	log likelihood of model $\lambda=\theta=1$
e(chi2_t1)	LR of $\lambda=\theta=1$ vs. full model
e(p_t1)	p-value of $\lambda=\theta=1$ vs. full model
e(ll_tm1)	log likelihood of model $\lambda=\theta=-1$
e(chi2_tm1)	LR of $\lambda=\theta=-1$ vs. full model
e(p_tm1)	p-value of $\lambda=\theta=-1$ vs. full model
e(ll_t0)	log likelihood of model $\lambda=\theta=0$
e(chi2_t0)	LR of $\lambda=\theta=0$ vs. full model
e(p_t0)	p-value of $\lambda=\theta=0$ vs. full model
e(rank)	rank of e(V)
e(ic)	number of iterations
e(rc)	return code

Macros
e(cmd)	boxcox
e(cmdline)	command as typed
e(depvar)	name of dependent variable
e(model)	lhsonly, rhsonly, lambda, or theta
e(wtype)	weight type
e(wexp)	weight expression
e(ntrans)	yes if nontransformed *indepvars*
e(chi2type)	LR; type of model χ^2 test
e(lrtest)	lrtest, if requested
e(properties)	b V
e(predict)	program used to implement predict
e(marginsnotok)	predictions disallowed by margins

Matrices
e(b)	coefficient vector
e(V)	variance–covariance matrix of the estimators (see note below)
e(pm)	p-values for LR tests on *indepvars*
e(df)	degrees of freedom of LR tests on *indepvars*
e(chi2m)	LR statistics for tests on *indepvars*

Functions
e(sample)	marks estimation sample

e(V) contains all zeros, except for the elements that correspond to the parameters of the Box–Cox transform.

Methods and formulas

boxcox is implemented as an ado-file.

In the internal computations,

$$y^{(\lambda)} = \begin{cases} \frac{y^{\lambda}-1}{\lambda} & \text{if } |\lambda| > 10^{-10} \\ \\ \ln(y) & \text{otherwise} \end{cases}$$

The unconcentrated log likelihood for the `theta` model is

$$\ln L = \left(\frac{-N}{2}\right)\left\{ \ln(2\pi) + \ln(\sigma^2) \right\} + (\theta - 1)\sum_{i=1}^{N}\ln(y_i) - \left(\frac{1}{2\sigma^2}\right)\text{SSR}$$

where

$$\text{SSR} = \sum_{i=1}^{N}(y_i^{(\theta)} - \beta_0 + \beta_1 x_{i1}^{(\lambda)} + \beta_2 x_{i2}^{(\lambda)} + \cdots + \beta_k x_{ik}^{(\lambda)} + \gamma_1 z_{i1} + \gamma_2 z_{i2} + \cdots + \gamma_l z_{il})^2$$

Writing the SSR in matrix form,

$$\text{SSR} = (\mathbf{Y}^{(\theta)} - \mathbf{X}^{(\lambda)}\mathbf{b}' - \mathbf{Z}\mathbf{g}')'(\mathbf{Y}^{(\theta)} - \mathbf{X}^{(\lambda)}\mathbf{b}' - \mathbf{Z}\mathbf{g}')$$

where $\mathbf{Y}^{(\theta)}$ is an $N \times 1$ vector of elementwise transformed data, $\mathbf{X}^{(\lambda)}$ is an $N \times k$ matrix of elementwise transformed data, $\mathbf{Z}$ is an $N \times l$ matrix of untransformed data, $\mathbf{b}$ is a $1 \times k$ vector of coefficients, and $\mathbf{g}$ is a $1 \times l$ vector of coefficients. Letting

$$\mathbf{W}_\lambda = \left(\mathbf{X}^{(\lambda)}\ \mathbf{Z}\right)$$

be the horizontal concatenation of $\mathbf{X}^{(\lambda)}$ and $\mathbf{Z}$ and

$$\mathbf{d}' = \begin{pmatrix} \mathbf{b}' \\ \mathbf{g}' \end{pmatrix}$$

be the vertical concatenation of the coefficients yields

$$\text{SSR} = (\mathbf{Y}^{(\theta)} - \mathbf{W}_\lambda \mathbf{d}')'(\mathbf{Y}^{(\theta)} - \mathbf{W}_\lambda \mathbf{d}')$$

For given values of λ and θ, the solutions for $\mathbf{d}'$ and σ^2 are

$$\widehat{\mathbf{d}}' = (\mathbf{W}_\lambda'\mathbf{W}_\lambda)^{-1}\mathbf{W}_\lambda'\mathbf{Y}^{(\theta)}$$

and

$$\widehat{\sigma}^2 = \frac{1}{N}\left(\mathbf{Y}^{(\theta)} - \mathbf{W}_\lambda\widehat{\mathbf{d}}'\right)'\left(\mathbf{Y}^{(\theta)} - \mathbf{W}_\lambda\widehat{\mathbf{d}}'\right)$$

Substituting these solutions into the log-likelihood function yields the concentrated log-likelihood function

$$\ln L_c = \left(-\frac{N}{2}\right)\left\{ \ln(2\pi) + 1 + \ln(\widehat{\sigma}^2) \right\} + (\theta - 1)\sum_{i=1}^{N}\ln(y_i)$$

Similar calculations yield the concentrated log-likelihood function for the `lambda` model,

$$\ln L_c = \left(-\frac{N}{2}\right)\left\{\ln(2\pi) + 1 + \ln(\widehat{\sigma}^{\,2})\right\} + (\lambda - 1)\sum_{i=1}^{N}\ln(y_i)$$

the `lhsonly` model,

$$\ln L_c = \left(-\frac{N}{2}\right)\left\{\ln(2\pi) + 1 + \ln(\widehat{\sigma}^{\,2})\right\} + (\theta - 1)\sum_{i=1}^{N}\ln(y_i)$$

and the `rhsonly` model,

$$\ln L_c = \left(-\frac{N}{2}\right)\left\{\ln(2\pi) + 1 + \ln(\widehat{\sigma}^{\,2})\right\}$$

where $\widehat{\sigma}^{\,2}$ is specific to each model and is defined analogously to that in the `theta` model.

References

Atkinson, A. C. 1985. *Plots, Transformations, and Regression: An Introduction to Graphical Methods of Diagnostic Regression Analysis.* Oxford: Oxford University Press.

Box, G. E. P., and D. R. Cox. 1964. An analysis of transformations. *Journal of the Royal Statistical Society, Series B* 26: 211–252.

Carroll, R. J., and D. Ruppert. 1988. *Transformation and Weighting in Regression.* New York: Chapman & Hall.

Cook, R. D., and S. Weisberg. 1982. *Residuals and Influence in Regression.* New York: Chapman & Hall/CRC.

Davidson, R., and J. G. MacKinnon. 1993. *Estimation and Inference in Econometrics.* New York: Oxford University Press.

Drukker, D. M. 2000a. sg130: Box–Cox regression models. *Stata Technical Bulletin* 54: 27–36. Reprinted in *Stata Technical Bulletin Reprints*, vol. 9, pp. 307–319. College Station, TX: Stata Press.

———. 2000b. sg131: On the manipulability of Wald tests in Box–Cox regression models. *Stata Technical Bulletin* 54: 36–42. Reprinted in *Stata Technical Bulletin Reprints*, vol. 9, pp. 319–327. College Station, TX: Stata Press.

Lafontaine, F., and K. J. White. 1986. Obtaining any Wald statistic you want. *Economics Letters* 21: 35–40.

Lindsey, C., and S. J. Sheather. 2010a. Power transformation via multivariate Box–Cox. *Stata Journal* 10: 69–81.

———. 2010b. Optimal power transformation via inverse response plots. *Stata Journal* 10: 200–214.

Phillips, P. C. B., and J. Y. Park. 1988. On the formulation of Wald tests of nonlinear restrictions. *Econometrica* 56: 1065–1083.

Schlesselman, J. J. 1971. Power families: A note on the Box and Cox transformation. *Journal of the Royal Statistical Society, Series B* 33: 307–311.

Spitzer, J. J. 1984. Variance estimates in models with the Box–Cox transformation: Implications for estimation and hypothesis testing. *Review of Economics and Statistics* 66: 645–652.

Also see

[R] **boxcox postestimation** — Postestimation tools for boxcox

[R] **regress** — Linear regression

[R] **lnskew0** — Find zero-skewness log or Box–Cox transform

[U] **20 Estimation and postestimation commands**

Title

boxcox postestimation — Postestimation tools for boxcox

Description

The following postestimation commands are available after boxcox:

Command	Description
estat	AIC, BIC, VCE, and estimation sample summary
estimates	cataloging estimation results
*lincom	point estimates, standard errors, testing, and inference for linear combinations of coefficients
*nlcom	point estimates, standard errors, testing, and inference for nonlinear combinations of coefficients
predict	predictions, residuals, influence statistics, and other diagnostic measures
*test	Wald tests of simple and composite linear hypotheses
*testnl	Wald tests of nonlinear hypotheses

*Inference is valid only for hypotheses concerning λ and θ.

See the corresponding entries in the *Base Reference Manual* for details.

Syntax for predict

predict [*type*] *newvar* [*if*] [*in*] [, *statistic*]

statistic	Description
Main	
xbt	transformed linear prediction; the default
yhat	predicted value of y
residuals	residuals

These statistics are available both in and out of sample; type predict ... if e(sample) ... if wanted only for the estimation sample.

Menu

Statistics > Postestimation > Predictions, residuals, etc.

Options for predict

Main

xbt, the default, calculates the "linear" prediction. For all the models except model(lhsonly), all the *indepvars* except those specified in the notrans() option of boxcox are transformed.

229

yhat calculates the predicted value of y.

residuals calculates the residuals after the predicted value of y has been subtracted from the actual value.

Remarks

boxcox estimates variances only for the λ and θ parameters (see the technical note in [R] **boxcox**), so the extent to which postestimation commands can be used following boxcox is limited. Formulas used in lincom, nlcom, test, and testnl are dependent on the estimated variances. Therefore, the use of these commands is limited and generally applicable only to inferences on the λ and θ coefficients.

Methods and formulas

All postestimation commands listed above are implemented as ado-files.

Also see

[R] **boxcox** — Box–Cox regression models

[R] **lnskew0** — Find zero-skewness log or Box–Cox transform

[U] **20 Estimation and postestimation commands**

Title

brier — Brier score decomposition

Syntax

`brier` *outcomevar forecastvar* $\begin{bmatrix} if \end{bmatrix}$ $\begin{bmatrix} in \end{bmatrix}$ $\begin{bmatrix} , \underline{g}roup(\#) \end{bmatrix}$

by is allowed; see [D] **by**.

Menu

Statistics > Epidemiology and related > Other > Brier score decomposition

Description

`brier` computes the Yates, Sanders, and Murphy decompositions of the Brier Mean Probability Score. *outcomevar* contains 0/1 values reflecting the actual outcome of the experiment, and *forecastvar* contains the corresponding probabilities as predicted by, say, logit, probit, or a human forecaster.

Option

⌐ Main ⌐

`group(#)` specifies the number of groups that will be used to compute the decomposition. `group(10)` is the default.

Remarks

You have a binary (0/1) response and a formula that predicts the corresponding probabilities of having observed a positive outcome (1). If the probabilities were obtained from logistic regression, there are many methods that assess goodness of fit (see, for instance, `estat gof` in [R] **logistic**). However, the probabilities might be computed from a published formula or from a model fit on another sample, both completely unrelated to the data at hand, or perhaps the forecasts are not from a formula at all. In any case, you now have a *test dataset* consisting of the forecast probabilities and observed outcomes. Your test dataset might, for instance, record predictions made by a meteorologist on the probability of rain along with a variable recording whether it actually rained.

The Brier score is an aggregate measure of disagreement between the observed outcome and a prediction—the average squared error difference. The Brier score decomposition is a partition of the Brier score into components that suggest reasons for discrepancy. These reasons fall roughly into three groups: 1) lack of overall calibration between the average predicted probability and the actual probability of the event in your data, 2) misfit of the data in groups defined within your sample, and 3) inability to match actual 0 and 1 responses.

Problem 1 refers to simply overstating or understating the probabilities.

Problem 2 refers to what is standardly called a goodness-of-fit test: the data are grouped, and the predictions for the group are compared with the outcomes.

Problem 3 refers to an individual-level measure of fit. Imagine that the grouped outcomes are predicted on average correctly but that within the group, the outcomes are poorly predicted.

Using logit or probit analysis to fit your data will guarantee that there is no lack of fit due to problem 1, and a good model fitter will be able to avoid problem 2. Problem 3 is inherent in any prediction exercise.

▷ Example 1

We have data on the outcomes of 20 basketball games (win) and the probability of victory predicted by a local pundit (for).

```
. use http://www.stata-press.com/data/r12/bball
. summarize win for
    Variable |        Obs        Mean    Std. Dev.       Min        Max
-------------+--------------------------------------------------------
         win |         20         .65    .4893605          0          1
         for |         20       .4785    .2147526        .15         .9

. brier win for, group(5)
Mean probability of outcome    0.6500
                of forecast    0.4785

Correlation                    0.5907
ROC area                       0.8791   p = 0.0030

Brier score                    0.1828
Spiegelhalter's z-statistic   -0.6339   p = 0.7369
Sanders-modified Brier score   0.1861
Sanders resolution             0.1400
Outcome index variance         0.2275
Murphy resolution              0.0875
Reliability-in-the-small       0.0461
Forecast variance              0.0438
Excess forecast variance       0.0285
Minimum forecast variance      0.0153
Reliability-in-the-large       0.0294
2*Forecast-Outcome-Covar       0.1179
```

The mean probabilities of forecast and outcome are simply the mean of the predicted probabilities and the actual outcomes (wins/losses). The correlation is the product-moment correlation between them.

The Brier score measures the total difference between the event (winning) and the forecast probability of that event as an average squared difference. As a benchmark, a perfect forecaster would have a Brier score of 0, a perfect misforecaster (predicts probability of win is 1 when loses and 0 when wins) would have a Brier score of 1, and a fence-sitter (forecasts every game as 50/50) would have a Brier score of 0.25. Our pundit is doing reasonably well.

Spiegelhalter's z statistic is a standard normal test statistic for testing whether an individual Brier score is extreme. The ROC area is the area under the receiver operating curve, and the associated test is a test of whether it is greater than 0.5. The more accurate the forecast probabilities, the larger the ROC area.

The Sanders-modified Brier score measures the difference between a grouped forecast measure and the event, where the data are grouped by sorting the sample on the forecast and dividing it into approximately equally sized groups. The difference between the modified and the unmodified score is typically minimal. For this and the other statistics that require grouping—the Sanders and Murphy resolutions and reliability-in-the-small—to be well-defined, group boundaries are chosen so as not to allocate observations with the same forecast probability to different groups. This task is done by grouping on the forecast using xtile, n(#), with # being the number of groups; see [D] **pctile**.

The Sanders resolution measures error that arises from statistical considerations in evaluating the forecast for a group. A group with all positive or all negative outcomes would have a Sanders resolution of 0; it would most certainly be feasible to predict exactly what happened to each member of the group. If the group had 40% positive responses, on the other hand, a forecast that assigned $p = 0.4$ to each member of the group would be a good one, and yet, there would be "errors" in the squared difference sense. The "error" would be $(1 - 0.4)^2$ or $(0 - 0.4)^2$ for each member. The Sanders resolution is the average across groups of such "expected" errors. The 0.1400 value in our data from an overall Brier score of 0.1828 or 0.1861 suggests that a substantial portion of the "error" in our data is inherent.

Outcome index variance is just the variance of the outcome variable. This is the expected value of the Brier score if all the forecast probabilities were merely the average observed outcome. Remember that a fence-sitter has an expected Brier score of 0.25; a smarter fence sitter (who would guess $p = 0.65$ for these data) would have a Brier score of 0.2275.

The Murphy resolution measures the variation in the average outcomes across groups. If all groups have the same frequency of positive outcomes, little information in any forecast is possible, and the Murphy resolution is 0. If groups differ markedly, the Murphy resolution is as large as 0.25. The 0.0875 means that there is some variation but not a lot, and 0.0875 is probably higher than in most real cases. If you had groups in your data that varied between 40% and 60% positive outcomes, the Murphy resolution would be 0.01; between 30% and 70%, it would be 0.04.

Reliability-in-the-small measures the error that comes from the average forecast within group not measuring the average outcome within group—a classical goodness-of-fit measure, with 0 meaning a perfect fit and 1 meaning a complete lack of fit. The calculated value of 0.0461 shows some amount of lack of fit. Remember, the number is squared, and we are saying that probabilities could be just more than $\sqrt{0.0461} = 0.215$ or 21.5% off.

Forecast variance measures the amount of discrimination being attempted—that is, the variation in the forecasted probabilities. A small number indicates a fence-sitter making constant predictions. If the forecasts were from a logistic regression model, forecast variance would tend to increase with the amount of information available. Our pundit shows considerable forecast variance of 0.0438 (standard deviation $\sqrt{0.0438} = 0.2093$), which is in line with the reliability-in-the-small, suggesting that the forecaster is attempting as much variation as is available in these data.

Excess forecast variance is the amount of actual forecast variance over a theoretical minimum. The theoretical minimum—called the minimum forecast variance—corresponds to forecasts of p_0 for observations ultimately observed to be negative responses and p_1 for observations ultimately observed to be positive outcomes. Moreover, p_0 and p_1 are set to the average forecasts made for the ultimate negative and positive outcomes. These predictions would be just as good as the predictions the forecaster did make, and any variation in the actual forecast probabilities above this is useless. If this number is large, above 1%–2%, then the forecaster may be attempting more than is possible. The 0.0285 in our data suggests this possibility.

Reliability-in-the-large measures the discrepancy between the mean forecast and the observed fraction of positive outcomes. This discrepancy will be 0 for forecasts made by most statistical models—at least when measured on the same sample used for estimation—because they, by design, reproduce sample means. For our human pundit, the 0.0294 says that there is a $\sqrt{0.0294}$, or 17-percentage-point, difference. (This difference can also be found by calculating the difference in the averages of the observed outcomes and forecast probabilities: $0.65 - 0.4785 = 0.17$.) That difference, however, is not significant, as we would see if we typed `ttest win=for`; see [R] **ttest**. If these data were larger and the bias persisted, this difference would be a critical shortcoming of the forecast.

Twice the forecast-outcome covariance is a measure of how accurately the forecast corresponds to the outcome. The concept is similar to that of R-squared in linear regression.

◁

Saved results

brier saves the following in r():

Scalars

r(p_roc)	significance of ROC area	r(murphy)	Murphy resolution
r(roc_area)	ROC area	r(relinsm)	reliability-in-the-small
r(z)	Spiegelhalter's z statistic	r(Var_f)	forecast variance
r(p)	significance of z statistic	r(Var_fex)	excess forecast variance
r(brier)	Brier score	r(Var_fmin)	minimum forecast variance
r(brier_s)	Sanders-modified Brier score	r(relinla)	reliability-in-the-large
r(sanders)	Sanders resolution	r(cov_2f)	2×forecast-outcome-covariance
r(oiv)	outcome index variance		

Methods and formulas

brier is implemented as an ado-file.

See Wilks (2006, 284–287, 289–292, 298–299) or Schmidt and Griffith (2005) for a discussion of the Brier score.

Let d_j, $j = 1, \ldots, N$, be the observed outcomes with $d_j = 0$ or $d_j = 1$, and let f_j be the corresponding forecasted probabilities that d_j is 1, $0 \le f_j \le 1$. Assume that the data are ordered so that $f_{j+1} \ge f_j$ (brier sorts the data to obtain this order). Divide the data into K nearly equally sized groups, with group 1 containing observations 1 through $j_2 - 1$, group 2 containing observations j_2 through $j_3 - 1$, and so on.

Define

$$\overline{f}_0 = \text{average } f_j \text{ among } d_j = 0$$
$$\overline{f}_1 = \text{average } f_j \text{ among } d_j = 1$$
$$\overline{f} = \text{average } f_j$$
$$\overline{d} = \text{average } d_j$$
$$\widetilde{f}_k = \text{average } f_j \text{ in group } k$$
$$\widetilde{d}_k = \text{average } d_j \text{ in group } k$$
$$\widetilde{n}_k = \text{number of observations in group } k$$

The Brier score is $\sum_j (d_j - f_j)^2 / N$.

The Sanders-modified Brier score is $\sum_j (d_j - \widetilde{f}_{k(j)})^2 / N$.

Let p_j denote the true but unknown probability that $d_j = 1$. Under the null hypothesis that $p_j = f_j$ for all j, Spiegelhalter (1986) determined that the expectation and variance of the Brier score is given by the following:

$$E(\text{Brier}) = \frac{1}{N} \sum_{j=1}^{N} f_j(1 - f_j)$$

$$\text{Var}(\text{Brier}) = \frac{1}{N^2} \sum_{j=1}^{N} f_j(1 - f_j)(1 - 2f_j)^2$$

Denoting the observed value of the Brier score by $O(\text{Brier})$, Spiegelhalter's z statistic is given by

$$Z = \frac{O(\text{Brier}) - E(\text{Brier})}{\sqrt{\text{Var}(\text{Brier})}}$$

The corresponding p-value is given by the upper-tail probability of Z under the standard normal distribution.

The area under the ROC curve is estimated by applying the trapezoidal rule to the empirical ROC curve. This area is Wilcoxon's test statistic, so the corresponding p-value is just that of a one-sided Wilcoxon test of the null hypothesis that the distribution of predictions is constant across the two outcomes.

The Sanders resolution is $\sum_k \tilde{n}_k \{\tilde{d}_k(1 - \tilde{d}_k)\}/N$.

The outcome index variance is $\bar{d}(1 - \bar{d})$.

The Murphy resolution is $\sum_k \tilde{n}_k(\tilde{d}_k - \bar{d})^2/N$.

Reliability-in-the-small is $\sum_k \tilde{n}_k(\tilde{d}_k - \tilde{f}_k)^2/N$.

The forecast variance is $\sum_j (f_j - \bar{f})^2/N$.

The minimum forecast variance is $\left\{ \sum_{j \in F}(f_j - \bar{f}_0)^2 + \sum_{j \in S}(f_j - \bar{f}_1)^2 \right\}/N$, where F is the set of observations for which $d_j = 0$ and S is the complement.

The excess forecast variance is the difference between the forecast variance and the minimum forecast variance.

Reliability-in-the-large is $(\bar{f} - \bar{d})^2$.

Twice the outcome covariance is $2(\bar{f}_1 - \bar{f}_0)\bar{d}(1 - \bar{d})$.

> Glenn Wilson Brier (1913–1998) was an American meteorological statistician who, after obtaining degrees in physics and statistics, was for many years head of meteorological statistics at the U.S. Weather Bureau in Washington, DC. In the latter part of his career, he was associated with Colorado State University. Brier worked especially on verification and evaluation of predictions and forecasts, statistical decision making, the statistical theory of turbulence, the analysis of weather modification experiments, and the application of permutation techniques.

Acknowledgment

We thank Richard Goldstein for his contributions to this improved version of `brier`.

References

Brier, G. W. 1950. Verification of forecasts expressed in terms of probability. *Monthly Weather Review* 78: 1–3.

Goldstein, R. 1996. sg55: Extensions to the brier command. *Stata Technical Bulletin* 32: 21–22. Reprinted in *Stata Technical Bulletin Reprints*, vol. 6, pp. 133–134. College Station, TX: Stata Press.

Hadorn, D. C., E. B. Keeler, W. H. Rogers, and R. H. Brook. 1993. *Assessing the Performance of Mortality Prediction Models*. Santa Monica, CA: Rand.

Holloway, L., and P. Mielke. 1998. Glenn Wilson Brier 1913–1998. *Bulletin of the American Meteorological Society* 79: 1438–1439.

Jolliffe, I. T., and D. B. Stephenson, ed. 2003. *Forecast Verification: A Practitioner's Guide in Atmospheric Science*. Chichester, UK: Wiley.

Murphy, A. H. 1973. A new vector partition of the probability score. *Journal of Applied Meteorology* 12: 595–600.

——. 1997. Forecast verification. In *Economic Value of Weather and Climate Forecasts*, ed. R. W. Katz and A. H. Murphy, 19–74. Cambridge: Cambridge University Press.

Redelmeier, D. A., D. A. Bloch, and D. H. Hickam. 1991. Assessing predictive accuracy: How to compare Brier scores. *Journal of Clinical Epidemiology* 44: 1141–1146.

Rogers, W. H. 1992. sbe9: Brier score decomposition. *Stata Technical Bulletin* 10: 20–22. Reprinted in *Stata Technical Bulletin Reprints*, vol. 2, pp. 92–94. College Station, TX: Stata Press.

Sanders, F. 1963. On subjective probability forecasting. *Journal of Applied Meteorology* 2: 191–201.

Schmidt, C. H., and J. L. Griffith. 2005. Multivariate classification rules: Calibration and discrimination. In Vol. 2 of *Encyclopedia of Biostatistics*, ed. P. Armitage and T. Colton, 3492–3494. Chichester, UK: Wiley.

Spiegelhalter, D. J. 1986. Probabilistic prediction in patient management and clinical trials. *Statistics in Medicine* 5: 421–433.

Von Storch, H., and F. W. Zwiers. 1999. *Statistical Analysis in Climate Research*. Cambridge: Cambridge University Press.

Wilks, D. S. 2006. *Statistical Methods in the Atmospheric Sciences*. 2nd ed. Burlington, MA: Academic Press.

Yates, J. F. 1982. External correspondence: Decompositions of the mean probability score. *Organizational Behavior and Human Performance* 30: 132–156.

Also see

[R] **logistic** — Logistic regression, reporting odds ratios

[R] **logit** — Logistic regression, reporting coefficients

[R] **probit** — Probit regression

Title

bsample — Sampling with replacement

Syntax

bsample [*exp*] [*if*] [*in*] [, *options*]

where *exp* is a standard Stata expression; see [U] **13 Functions and expressions**.

options	Description
<u>str</u>ata(*varlist*)	variables identifying strata
<u>cl</u>uster(*varlist*)	variables identifying resampling clusters
idcluster(*newvar*)	create new cluster ID variable
<u>wei</u>ght(*varname*)	replace *varname* with frequency weights

Menu

Statistics > Resampling > Draw bootstrap sample

Description

bsample draws bootstrap samples (random samples with replacement) from the data in memory.

exp specifies the size of the sample, which must be less than or equal to the number of sampling units in the data. The observed number of units is the default when *exp* is not specified.

For bootstrap sampling of the observations, *exp* must be less than or equal to _N (the number of observations in the data; see [U] **13.4 System variables (_variables)**).

For stratified bootstrap sampling, *exp* must be less than or equal to _N within the strata identified by the strata() option.

For clustered bootstrap sampling, *exp* must be less than or equal to N_c (the number of clusters identified by the cluster() option).

For stratified bootstrap sampling of clusters, *exp* must be less than or equal to N_c within the strata identified by the strata() option.

Observations that do not meet the optional if and in criteria are dropped (not sampled).

Options

strata(*varlist*) specifies the variables identifying strata. If strata() is specified, bootstrap samples are selected within each stratum.

cluster(*varlist*) specifies the variables identifying resampling clusters. If cluster() is specified, the sample drawn during each replication is a bootstrap sample of clusters.

idcluster(*newvar*) creates a new variable containing a unique identifier for each resampled cluster.

weight(*varname*) specifies a variable in which the sampling frequencies will be placed. *varname* must be an existing variable, which will be replaced. After bsample, *varname* can be used as an fweight in any Stata command that accepts fweights, which can speed up resampling for commands like regress and summarize. This option cannot be combined with idcluster().

By default, bsample replaces the data in memory with the sampled observations; however, specifying the weight() option causes only the specified *varname* to be changed.

Remarks

Below is a series of examples illustrating how bsample is used with various sampling schemes.

▷ Example 1: Bootstrap sampling

We have data on the characteristics of hospital patients and wish to draw a bootstrap sample of 200 patients. We type

```
. use http://www.stata-press.com/data/r12/bsample1
. bsample 200
. count
  200
```

◁

▷ Example 2: Stratified samples with equal sizes

Among the variables in our dataset is female, an indicator for the female patients. To get a bootstrap sample of 200 female patients and 200 male patients, we type

```
. use http://www.stata-press.com/data/r12/bsample1, clear
. bsample 200, strata(female)
. tab female
```

female	Freq.	Percent	Cum.
male	200	50.00	50.00
female	200	50.00	100.00
Total	400	100.00	

◁

▷ Example 3: Stratified samples with unequal sizes

To sample 300 females and 200 males, we must generate a variable that is 300 for females and 200 for males and then use this variable in *exp* when we call bsample.

```
. use http://www.stata-press.com/data/r12/bsample1, clear
. gen nsamp = cond(female,300,200)
. bsample nsamp, strata(female)
. tab female
```

female	Freq.	Percent	Cum.
male	200	40.00	40.00
female	300	60.00	100.00
Total	500	100.00	

◁

▷ Example 4: Samples satisfying a condition

For a bootstrap sample of 200 female patients, we type

```
. use http://www.stata-press.com/data/r12/bsample1, clear
. bsample 200 if female
. tab female
```

female	Freq.	Percent	Cum.
female	200	100.00	100.00
Total	200	100.00	

◁

▷ Example 5: Generating frequency weights

To identify the sampled observations using frequency weights instead of dropping unsampled observations, we use the `weight()` option (we will need to supply it an existing variable name) and type

```
. use http://www.stata-press.com/data/r12/bsample1, clear
. set seed 1234
. gen fw = .
(5810 missing values generated)
. bsample 200 if female, weight(fw)
. tabulate fw female
```

fw	female male	female	Total
0	2,392	3,221	5,613
1	0	194	194
2	0	3	3
Total	2,392	3,418	5,810

Note that $(194 \times 1) + (3 \times 2) = 200$.

◁

▷ Example 6: Oversampling observations

`bsample` requires the expression in *exp* to evaluate to a number that is less than or equal to the number of observations. To sample twice as many male and female patients as there are already in memory, we must expand the data before using `bsample`. For example,

```
. use http://www.stata-press.com/data/r12/bsample1, clear
. set seed 1234
. expand 2
(5810 observations created)
. bsample, strata(female)
```

```
. tab female
```

female	Freq.	Percent	Cum.
male	4,784	41.17	41.17
female	6,836	58.83	100.00
Total	11,620	100.00	

◁

▷ Example 7: Stratified oversampling with unequal sizes

To sample twice as many female patients as male patients, we must expand the records for the female patients because there are less than twice as many of them as there are male patients, but first put the number of observed male patients in a local macro. After expanding the female records, we generate a variable that contains the number of observations to sample within the two groups.

```
. use http://www.stata-press.com/data/r12/bsample1, clear
. set seed 1234
. count if !female
 2392
. local nmale = r(N)
. expand 2 if female
(3418 observations created)
. gen nsamp = cond(female,2*`nmale',`nmale')
. bsample nsamp, strata(female)
. tab female
```

female	Freq.	Percent	Cum.
male	2,392	33.33	33.33
female	4,784	66.67	100.00
Total	7,176	100.00	

◁

▷ Example 8: Oversampling of clusters

For clustered data, sampling more clusters than are present in the original dataset requires more than just expanding the data. To illustrate, suppose we wanted a bootstrap sample of eight clusters from a dataset consisting of five clusters of observations.

```
. use http://www.stata-press.com/data/r12/bsample2, clear
. tabstat x, stat(n mean) by(group)
Summary for variables: x
     by categories of: group
```

group	N	mean
A	15	-.3073028
B	10	-.00984
C	11	.0810985
D	11	-.1989179
E	29	-.095203
Total	76	-.1153269

bsample will complain if we simply expand the dataset.

```
. use http://www.stata-press.com/data/r12/bsample2
. expand 3
(152 observations created)
. bsample 8, cluster(group)
resample size must not be greater than number of clusters
r(498);
```

Expanding the data will only partly solve the problem. We also need a new variable that uniquely identifies the copied clusters. We use the expandcl command to accomplish both these tasks; see [D] **expandcl**.

```
. use http://www.stata-press.com/data/r12/bsample2, clear
. set seed 1234
. expandcl 2, generate(expgroup) cluster(group)
(76 observations created)
. tabstat x, stat(n mean) by(expgroup)
Summary for variables: x
     by categories of: expgroup
```

expgroup	N	mean
1	15	-.3073028
2	15	-.3073028
3	10	-.00984
4	10	-.00984
5	11	.0810985
6	11	.0810985
7	11	-.1989179
8	11	-.1989179
9	29	-.095203
10	29	-.095203
Total	152	-.1153269

```
. gen fw = .
(152 missing values generated)
. bsample 8, cluster(expgroup) weight(fw)
. tabulate fw group
```

fw	group A	B	C	D	E	Total
0	15	10	0	0	29	54
1	15	10	22	22	0	69
2	0	0	0	0	29	29
Total	30	20	22	22	58	152

The results from tabulate on the generated frequency weight variable versus the original cluster ID (group) show us that the bootstrap sample contains one copy of cluster A, one copy of cluster B, two copies of cluster C, two copies of cluster D, and two copies of cluster E $(1 + 1 + 2 + 2 + 2 = 8)$.

◁

▷ Example 9: Stratified oversampling of clusters

Suppose that we have a dataset containing two strata with five clusters in each stratum, but the cluster identifiers are not unique between the strata. To get a stratified bootstrap sample with eight clusters in each stratum, we first use expandcl to expand the data and get a new cluster ID variable. We use cluster(strid group) in the call to expandcl; this action will uniquely identify the $2 * 5 = 10$ clusters across the strata.

```
. use http://www.stata-press.com/data/r12/bsample2, clear
. set seed 1234
. tab group strid
```

group	strid 1	2	Total
A	7	8	15
B	5	5	10
C	5	6	11
D	5	6	11
E	14	15	29
Total	36	40	76

```
. expandcl 2, generate(expgroup) cluster(strid group)
(76 observations created)
```

Now we can use bsample with the expanded data, stratum ID variable, and new cluster ID variable.

```
. gen fw = .
(152 missing values generated)
. bsample 8, cluster(expgroup) str(strid) weight(fw)
. by strid, sort: tabulate fw group
```

-> strid = 1

fw	group A	B	C	D	E	Total
0	0	5	0	5	14	24
1	14	5	10	5	0	34
2	0	0	0	0	14	14
Total	14	10	10	10	28	72

-> strid = 2

fw	group A	B	C	D	E	Total
0	8	10	0	6	0	24
1	8	0	6	6	15	35
2	0	0	6	0	15	21
Total	16	10	12	12	30	80

The results from by strid: tabulate on the generated frequency weight variable versus the original cluster ID (group) show us how many times each cluster was sampled for each stratum. For stratum 1, the bootstrap sample contains two copies of cluster A, one copy of cluster B, two copies of cluster C, one copy of cluster D, and two copies of cluster E $(2 + 1 + 2 + 1 + 2 = 8)$. For stratum 2, the bootstrap sample contains one copy of cluster A, zero copies of cluster B, three copies of cluster C, one copy of cluster D, and three copies of cluster E $(1 + 0 + 3 + 1 + 3 = 8)$.

◁

Methods and formulas

bsample is implemented as an ado-file.

Also see

[R] **bootstrap** — Bootstrap sampling and estimation

[R] **bstat** — Report bootstrap results

[R] **simulate** — Monte Carlo simulations

[D] **sample** — Draw random sample

Title

> **bstat** — Report bootstrap results

Syntax

Bootstrap statistics from variables

> bstat [*varlist*] [*if*] [*in*] [, *options*]

Bootstrap statistics from file

> bstat [*namelist*] [using *filename*] [*if*] [*in*] [, *options*]

options	Description
Main	
<u>stat</u>(*vector*)	observed values for each statistic
accel(*vector*)	acceleration values for each statistic
mse	use MSE formula for variance estimation
Reporting	
<u>level</u>(#)	set confidence level; default is level(95)
n(#)	# of observations from which bootstrap samples were taken
notable	suppress table of results
<u>nohe</u>ader	suppress table header
<u>nol</u>egend	suppress table legend
<u>ver</u>bose	display the full table legend
<u>title</u>(*text*)	use *text* as title for bootstrap results
display_options	control column formats and line width

See [U] **20 Estimation and postestimation commands** for more capabilities of estimation commands.

Menu

Statistics > Resampling > Report bootstrap results

Description

bstat is a programmer's command that computes and displays estimation results from bootstrap statistics.

For each variable in *varlist* (the default is all variables), then bstat computes a covariance matrix, estimates bias, and constructs several different confidence intervals (CIs). The following CIs are constructed by bstat:

1. Normal CIs (using the normal approximation)

2. Percentile CIs

3. Bias-corrected (BC) CIs

4. Bias-corrected and accelerated (BC_a) CIs (optional)

244

estat bootstrap displays a table of one or more of the above confidence intervals; see [R] **bootstrap postestimation**.

If there are bootstrap estimation results in e(), bstat replays them. If given the using modifier, bstat uses the data in *filename* to compute the bootstrap statistics while preserving the data currently in memory. Otherwise, bstat uses the data in memory to compute the bootstrap statistics.

The following options may be used to replay estimation results from bstat:

level(*#*) notable noheader nolegend verbose title(*text*)

For all other options and the qualifiers using, if, and in, bstat requires a bootstrap dataset.

Options

_____ Main _____

stat(*vector*) specifies the observed value of each statistic (that is, the value of the statistic using the original dataset).

accel(*vector*) specifies the acceleration of each statistic, which is used to construct BC_a CIs.

mse specifies that bstat compute the variance by using deviations of the replicates from the observed value of the statistics. By default, bstat computes the variance by using deviations from the average of the replicates.

_____ Reporting _____

level(*#*); see [R] **estimation options**.

n(*#*) specifies the number of observations from which bootstrap samples were taken. This value is used in no calculations but improves the table header when this information is not saved in the bootstrap dataset.

notable suppresses the display of the output table.

noheader suppresses the display of the table header. This option implies nolegend.

nolegend suppresses the display of the table legend.

verbose specifies that the full table legend be displayed. By default, coefficients and standard errors are not displayed.

title(*text*) specifies a title to be displayed above the table of bootstrap results; the default title is Bootstrap results.

display_options: cformat(*% fmt*), pformat(*% fmt*), sformat(*% fmt*), and nolstretch; see [R] **estimation options**.

Remarks

Remarks are presented under the following headings:

> *Bootstrap datasets*
> *Creating a bootstrap dataset*

Bootstrap datasets

Although bstat allows you to specify the observed value and acceleration of each bootstrap statistic via the stat() and accel() options, programmers may be interested in what bstat uses when these options are not supplied.

When working from a bootstrap dataset, bstat first checks the data characteristics (see [P] **char**) that it understands:

_dta[bs_version] identifies the version of the bootstrap dataset. This characteristic may be empty (not defined), 2, or 3; otherwise, bstat will quit and display an error message. This version tells bstat which other characteristics to look for in the bootstrap dataset.

> bstat uses the following characteristics from version 3 bootstrap datasets:
> > _dta[N]
> > _dta[N_strata]
> > _dta[N_cluster]
> > _dta[command]
> > *varname*[observed]
> > *varname*[acceleration]
> > *varname*[expression]
>
> bstat uses the following characteristics from version 2 bootstrap datasets:
> > _dta[N]
> > _dta[N_strata]
> > _dta[N_cluster]
> > *varname*[observed]
> > *varname*[acceleration]

> An empty bootstrap dataset version implies that the dataset was created by the bstrap command in a version of Stata earlier than Stata 8. Here bstat expects *varname*[bstrap] to contain the observed value of the statistic identified by *varname* (*varname*[observed] in version 2). All other characteristics are ignored.

_dta[N] is the number of observations in the observed dataset. This characteristic may be overruled by specifying the n() option.

_dta[N_strata] is the number of strata in the observed dataset.

_dta[N_cluster] is the number of clusters in the observed dataset.

_dta[command] is the command used to compute the observed values of the statistics.

varname[observed] is the observed value of the statistic identified by *varname*. To specify a different value, use the stat() option.

varname[acceleration] is the estimate of acceleration for the statistic identified by *varname*. To specify a different value, use the accel() option.

varname[expression] is the expression or label that describes the statistic identified by *varname*.

Creating a bootstrap dataset

Suppose that we are interested in obtaining bootstrap statistics by resampling the residuals from a regression (which is not possible with the bootstrap command). After loading some data, we run a regression, save some results relevant to the bstat command, and save the residuals in a new variable, res.

```
. use http://www.stata-press.com/data/r12/auto
(1978 Automobile Data)

. regress mpg weight length
```

Source	SS	df	MS
Model	1616.08062	2	808.040312
Residual	827.378835	71	11.653223
Total	2443.45946	73	33.4720474

```
      Number of obs =      74
      F(  2,    71) =   69.34
      Prob > F      =  0.0000
      R-squared     =  0.6614
      Adj R-squared =  0.6519
      Root MSE      =  3.4137
```

mpg	Coef.	Std. Err.	t	P>|t|	[95% Conf. Interval]	
weight	-.0038515	.001586	-2.43	0.018	-.0070138	-.0006891
length	-.0795935	.0553577	-1.44	0.155	-.1899736	.0307867
_cons	47.88487	6.08787	7.87	0.000	35.746	60.02374

```
. matrix b = e(b)

. local n = e(N)

. predict res, residuals
```

We can resample the residual values in res by generating a random observation ID (rid), generate a new response variable (y), and run the original regression with the new response variables.

```
. set seed 54321

. gen rid = int(_N*runiform())+1

. matrix score double y = b

. replace y = y + res[rid]
(74 real changes made)

. regress y weight length
```

Source	SS	df	MS
Model	1773.23548	2	886.617741
Residual	608.747732	71	8.57391172
Total	2381.98321	73	32.629907

```
      Number of obs =      74
      F(  2,    71) =  103.41
      Prob > F      =  0.0000
      R-squared     =  0.7444
      Adj R-squared =  0.7372
      Root MSE      =  2.9281
```

y	Coef.	Std. Err.	t	P>|t|	[95% Conf. Interval]	
weight	-.0059938	.0013604	-4.41	0.000	-.0087064	-.0032813
length	-.0127875	.0474837	-0.27	0.788	-.1074673	.0818924
_cons	42.23195	5.22194	8.09	0.000	31.8197	52.6442

Instead of programming this resampling inside a loop, it is much more convenient to write a short program and use the simulate command; see [R] **simulate**. In the following, mysim_r requires the user to specify a coefficient vector and a residual variable. mysim_r then retrieves the list of predictor variables (removing _cons from the list), generates a new temporary response variable with the resampled residuals, and regresses the new response variable on the predictors.

```
program mysim_r
        version 12
        syntax name(name=bvector), res(varname)
        tempvar y rid
        local xvars : colnames `bvector'
        local cons _cons
        local xvars : list xvars - cons
        matrix score double `y' = `bvector'
        gen long `rid' = int(_N*runiform()) + 1
        replace `y' = `y' + `res'[`rid']
        regress `y' `xvars'
end
```

We can now give mysim_r a test run, but we first set the random-number seed (to reproduce results).

```
. set seed 54321

. mysim_r b, res(res)
(74 real changes made)
```

Source	SS	df	MS		Number of obs =	74
					F(2, 71) =	103.41
Model	1773.23548	2	886.617741		Prob > F =	0.0000
Residual	608.747732	71	8.57391172		R-squared =	0.7444
					Adj R-squared =	0.7372
Total	2381.98321	73	32.629907		Root MSE =	2.9281

__000000	Coef.	Std. Err.	t	P>|t|	[95% Conf. Interval]	
weight	-.0059938	.0013604	-4.41	0.000	-.0087064	-.0032813
length	-.0127875	.0474837	-0.27	0.788	-.1074673	.0818924
_cons	42.23195	5.22194	8.09	0.000	31.8197	52.6442

Now that we have a program that will compute the results we want, we can use simulate to generate a bootstrap dataset and bstat to display the results.

```
. set seed 54321

. simulate, reps(200) nodots: mysim_r b, res(res)
        command:  mysim_r b, res(res)

. bstat, stat(b) n(`n')
```

Bootstrap results

Number of obs =   74
Replications   =   200

	Observed Coef.	Bootstrap Std. Err.	z	P>|z|	Normal-based [95% Conf. Interval]	
_b_weight	-.0038515	.0015715	-2.45	0.014	-.0069316	-.0007713
_b_length	-.0795935	.0552415	-1.44	0.150	-.1878649	.0286779
_b_cons	47.88487	6.150069	7.79	0.000	35.83096	59.93879

Finally, we see that simulate created some of the data characteristics recognized by bstat. All we need to do is correctly specify the version of the bootstrap dataset, and bstat will automatically use the relevant data characteristics.

```
. char list
_dta[seed]:                      X681014b5c43f462544a474abacbdd93d12a1
_dta[command]:                   mysim_r b, res(res)
_b_weight[is_eexp]:              1
_b_weight[colname]:              weight
_b_weight[coleq]:                _
_b_weight[expression]:           _b[weight]
_b_length[is_eexp]:              1
_b_length[colname]:              length
_b_length[coleq]:                _
_b_length[expression]:           _b[length]
_b_cons[is_eexp]:                1
_b_cons[colname]:                _cons
_b_cons[coleq]:                  _
_b_cons[expression]:             _b[_cons]

. char _dta[bs_version] 3

. bstat, stat(b) n('n')
```

```
Bootstrap results                          Number of obs      =          74
                                           Replications       =         200

        command:  mysim_r b, res(res)
```

	Observed Coef.	Bootstrap Std. Err.	z	P>|z|	Normal-based [95% Conf. Interval]	
weight	-.0038515	.0015715	-2.45	0.014	-.0069316	-.0007713
length	-.0795935	.0552415	-1.44	0.150	-.1878649	.0286779
_cons	47.88487	6.150069	7.79	0.000	35.83096	59.93879

See Poi (2004) for another example of residual resampling.

Saved results

bstat saves the following in e():

Scalars
e(N)	sample size
e(N_reps)	number of complete replications
e(N_misreps)	number of incomplete replications
e(N_strata)	number of strata
e(N_clust)	number of clusters
e(k_aux)	number of auxiliary parameters
e(k_eq)	number of equations in e(b)
e(k_exp)	number of standard expressions
e(k_eexp)	number of extended expressions (i.e., _b)
e(k_extra)	number of extra equations beyond the original ones from e(b)
e(level)	confidence level for bootstrap CIs
e(bs_version)	version for bootstrap results
e(rank)	rank of e(V)

Macros
e(cmd)	bstat
e(command)	from _dta[command]
e(cmdline)	command as typed
e(title)	title in estimation output
e(exp#)	expression for the #th statistic
e(prefix)	bootstrap
e(mse)	mse if specified
e(vce)	bootstrap
e(vcetype)	title used to label Std. Err.
e(properties)	b V

Matrices
e(b)	observed statistics
e(b_bs)	bootstrap estimates
e(reps)	number of nonmissing results
e(bias)	estimated biases
e(se)	estimated standard errors
e(z0)	median biases
e(accel)	estimated accelerations
e(ci_normal)	normal-approximation CIs
e(ci_percentile)	percentile CIs
e(ci_bc)	bias-corrected CIs
e(ci_bca)	bias-corrected and accelerated CIs
e(V)	bootstrap variance–covariance matrix

Methods and formulas

bstat is implemented as an ado-file.

Reference

Poi, B. P. 2004. From the help desk: Some bootstrapping techniques. *Stata Journal* 4: 312–328.

Also see

[R] **bootstrap** — Bootstrap sampling and estimation

[R] **bsample** — Sampling with replacement

Title

> **centile** — Report centile and confidence interval

Syntax

centile $\left[\,varlist\,\right]$ $\left[\,if\,\right]$ $\left[\,in\,\right]$ $\left[\,,\,options\,\right]$

options	Description
Main	
<u>c</u>entile(*numlist*)	report specified centiles; default is centile(50)
Options	
<u>c</u>ci	binomial exact; conservative confidence interval
<u>n</u>ormal	normal, based on observed centiles
<u>m</u>eansd	normal, based on mean and standard deviation
<u>l</u>evel(*#*)	set confidence level; default is level(95)

by is allowed; see [D] **by**.

Menu

Statistics > Summaries, tables, and tests > Summary and descriptive statistics > Centiles with CIs

Description

centile estimates specified centiles and calculates confidence intervals. If no *varlist* is specified, centile calculates centiles for all the variables in the dataset. If centile() is not specified, medians (centile(50)) are reported.

Options

> Main

centile(*numlist*) specifies the centiles to be reported. The default is to display the 50th centile. Specifying centile(5) requests that the fifth centile be reported. Specifying centile(5 50 95) requests that the 5th, 50th, and 95th centiles be reported. Specifying centile(10(10)90) requests that the 10th, 20th, ..., 90th centiles be reported; see [U] **11.1.8 numlist**.

> Options

cci (conservative confidence interval) forces the confidence limits to fall exactly on sample values. Confidence intervals displayed with the cci option are slightly wider than those with the default (nocci) option.

normal causes the confidence interval to be calculated by using a formula for the standard error of a normal-distribution quantile given by Kendall and Stuart (1969, 237). The normal option is useful when you want empirical centiles—that is, centiles based on sample order statistics rather than on the mean and standard deviation—and are willing to assume normality.

251

meansd causes the centile and confidence interval to be calculated based on the sample mean and standard deviation, and it assumes normality.

level(#) specifies the confidence level, as a percentage, for confidence intervals. The default is level(95) or as set by set level; see [R] **level**.

Remarks

The qth centile of a continuous random variable, X, is defined as the value of C_q, which fulfills the condition $\Pr(X \le C_q) = q/100$. The value of q must be in the range $0 < q < 100$, though q is not necessarily an integer. By default, centile estimates C_q for the variables in *varlist* and for the values of q given in centile(*numlist*). It makes no assumptions about the distribution of X, and, if necessary, uses linear interpolation between neighboring sample values. Extreme centiles (for example, the 99th centile in samples smaller than 100) are fixed at the minimum or maximum sample value. An "exact" confidence interval for C_q is also given, using the binomial-based method described below in *Methods and formulas* and in Conover (1999, 143–148). Again linear interpolation is used to improve the accuracy of the estimated confidence limits, but extremes are fixed at the minimum or maximum sample value.

You can prevent centile from interpolating when calculating binomial-based confidence intervals by specifying cci. The resulting intervals are generally wider than with the default; that is, the coverage (confidence level) tends to be greater than the nominal value (given as usual by level(#), by default 95%).

If the data are believed to be normally distributed (a common case), there are two alternative methods for estimating centiles. If normal is specified, C_q is calculated, as just described, but its confidence interval is based on a formula for the standard error (se) of a normal-distribution quantile given by Kendall and Stuart (1969, 237). If meansd is alternatively specified, C_q is estimated as $\bar{x} + z_q \times s$, where $\bar{x}$ and s are the sample mean and standard deviation, and z_q is the qth centile of the standard normal distribution (for example, $z_{95} = 1.645$). The confidence interval is derived from the se of the estimate of C_q.

▷ Example 1

Using auto.dta, we estimate the 5th, 50th, and 95th centiles of the price variable:

```
. use http://www.stata-press.com/data/r12/auto
(1978 Automobile Data)
. format price %8.2fc
. centile price, centile(5 50 95)
```

Variable	Obs	Percentile	Centile	— Binom. Interp. — [95% Conf. Interval]	
price	74	5	3,727.75	3,291.23	3,914.16
		50	5,006.50	4,593.57	5,717.90
		95	13,498.00	11,061.53	15,865.30

summarize produces somewhat different results from centile; see *Methods and formulas*.

```
. summarize price, detail
```

                                Price

	Percentiles	Smallest		
1%	3291	3291		
5%	3748	3299		
10%	3895	3667	Obs	74
25%	4195	3748	Sum of Wgt.	74
50%	5006.5		Mean	6165.257
		Largest	Std. Dev.	2949.496
75%	6342	13466		
90%	11385	13594	Variance	8699526
95%	13466	14500	Skewness	1.653434
99%	15906	15906	Kurtosis	4.819188

The confidence limits produced by using the cci option are slightly wider than those produced without this option:

```
. centile price, c(5 50 95) cci
```

Variable	Obs	Percentile	Centile	— Binomial Exact — [95% Conf. Interval]	
price	74	5	3,727.75	3,291.00	3,955.00
		50	5,006.50	4,589.00	5,719.00
		95	13,498.00	10,372.00	15,906.00

If we are willing to assume that price is normally distributed, we could include either the normal or the meansd option:

```
. centile price, c(5 50 95) normal
```

Variable	Obs	Percentile	Centile	— Normal, based on observed centiles — [95% Conf. Interval]	
price	74	5	3,727.75	3,211.19	4,244.31
		50	5,006.50	4,096.68	5,916.32
		95	13,498.00	5,426.81	21,569.19

```
. centile price, c(5 50 95) meansd
```

Variable	Obs	Percentile	Centile	— Normal, based on mean and std. dev.— [95% Conf. Interval]	
price	74	5	1,313.77	278.93	2,348.61
		50	6,165.26	5,493.24	6,837.27
		95	11,016.75	9,981.90	12,051.59

With the normal option, the centile estimates are, by definition, the same as before. The confidence intervals for the 5th and 50th centiles are similar to the previous ones, but the interval for the 95th centile is different. The results using the meansd option also differ from both previous sets of estimates.

We can use sktest (see [R] **sktest**) to check the correctness of the normality assumption:

```
. sktest price
```

                    Skewness/Kurtosis tests for Normality

Variable	Obs	Pr(Skewness)	Pr(Kurtosis)	joint adj chi2(2)	Prob>chi2
price	74	0.0000	0.0127	21.77	0.0000

`sktest` reveals that `price` is definitely not normally distributed, so the normal assumption is not reasonable, and the `normal` and `meansd` options are not appropriate for these data. We should rely on the results from the default choice, which does not assume normality. If the data are normally distributed, however, the precision of the estimated centiles and their confidence intervals will be ordered (best) `meansd` > `normal` > [default] (worst). The `normal` option is useful when we really do want empirical centiles (that is, centiles based on sample order statistics rather than on the mean and standard deviation) but are willing to assume normality.

◁

Saved results

centile saves the following in `r()`:

Scalars
`r(N)`	number of observations
`r(n_cent)`	number of centiles requested
`r(c_#)`	value of # centile
`r(lb_#)`	#-requested centile lower confidence bound
`r(ub_#)`	#-requested centile upper confidence bound

Macros
`r(centiles)`	centiles requested

Methods and formulas

centile is implemented as an ado-file.

Methods and formulas are presented under the following headings:

> *Default case*
> *Normal case*
> *meansd case*

Default case

The calculation is based on the method of Mood and Graybill (1963, 408). Let $x_1 \leq x_2 \leq \cdots \leq x_n$ be a sample of size n arranged in ascending order. Denote the estimated qth centile of the x's as c_q. We require that $0 < q < 100$. Let $R = (n+1)q/100$ have integer part r and fractional part f; that is, $r = \text{int}(R)$ and $f = R - r$. (If R is itself an integer, then $r = R$ and $f = 0$.) Note that $0 \leq r \leq n$. For convenience, define $x_0 = x_1$ and $x_{n+1} = x_n$. C_q is estimated by

$$c_q = x_r + f \times (x_{r+1} - x_r)$$

that is, c_q is a weighted average of x_r and x_{r+1}. Loosely speaking, a (conservative) $p\%$ confidence interval for C_q involves finding the observations ranked t and u, which correspond, respectively, to the $\alpha = (100 - p)/200$ and $1 - \alpha$ quantiles of a binomial distribution with parameters n and $q/100$, that is, $\text{B}(n, q/100)$. More precisely, define the ith value $(i = 0, \ldots, n)$ of the cumulative binomial distribution function as $F_i = \Pr(S \leq i)$, where S has distribution $\text{B}(n, q/100)$. For convenience, let $F_{-1} = 0$ and $F_{n+1} = 1$. t is found such that $F_t \leq \alpha$ and $F_{t+1} > \alpha$, and u is found such that $1 - F_u \leq \alpha$ and $1 - F_{u-1} > \alpha$.

With the `cci` option in force, the (conservative) confidence interval is (x_{t+1}, x_{u+1}), and its actual coverage probability is $F_u - F_t$.

The default case uses linear interpolation on the F_i as follows. Let

$$g = (\alpha - F_t)/(F_{t+1} - F_t)$$
$$h = \{\alpha - (1 - F_u)\}/\{(1 - F_{u-1}) - (1 - F_u)\}$$
$$= (\alpha - 1 + F_u)/(F_u - F_{u-1})$$

The interpolated lower and upper confidence limits (c_{qL}, c_{qU}) for C_q are

$$c_{qL} = x_{t+1} + g \times (x_{t+2} - x_{t+1})$$
$$c_{qU} = x_{u+1} - h \times (x_{u+1} - x_u)$$

Suppose that we want a 95% confidence interval for the median of a sample of size 13. $n = 13$, $q = 50$, $p = 95$, $\alpha = 0.025$, $R = 14 \times 50/100 = 7$, and $f = 0$. Therefore, the median is the 7th observation. Some example data, x_i, and the values of F_i are as follows:

i	F_i	$1 - F_i$	x_i	i	F_i	$1 - F_i$	x_i
0	0.0001	0.9999	–	7	0.7095	0.2905	33
1	0.0017	0.9983	5	8	0.8666	0.1334	37
2	0.0112	0.9888	7	9	0.9539	0.0461	45
3	0.0461	0.9539	10	10	0.9888	0.0112	59
4	0.1334	0.8666	15	11	0.9983	0.0017	77
5	0.2905	0.7095	23	12	0.9999	0.0001	104
6	0.5000	0.5000	28	13	1.0000	0.0000	211

The median is $x_7 = 33$. Also, $F_2 \leq 0.025$ and $F_3 > 0.025$, so $t = 2$; $1 - F_{10} \leq 0.025$ and $1 - F_9 > 0.025$, so $u = 10$. The conservative confidence interval is therefore

$$(c_{50L}, c_{50U}) = (x_{t+1}, x_{u+1}) = (x_3, x_{11}) = (10, 77)$$

with actual coverage $F_{10} - F_2 = 0.9888 - 0.0112 = 0.9776$ (97.8% confidence). For the interpolation calculation, we have

$$g = (0.025 - 0.0112)/(0.0461 - 0.0112) = 0.395$$
$$h = (0.025 - 1 + 0.9888)/(0.9888 - 0.9539) = 0.395$$

So,

$$c_{50L} = x_3 + 0.395 \times (x_4 - x_3) = 10 + 0.395 \times 5 = 11.98$$
$$c_{50U} = x_{11} - 0.395 \times (x_{11} - x_{10}) = 77 - 0.395 \times 18 = 69.89$$

Normal case

The value of c_q is as above. Its se is given by the formula

$$s_q = \sqrt{q(100 - q)} \Big/ \left\{ 100\sqrt{n} Z(c_q; \overline{x}, s) \right\}$$

where $\overline{x}$ and s are the mean and standard deviation of the x_i, and

$$Z(Y; \mu, \sigma) = \left(1/\sqrt{2\pi\sigma^2}\right) e^{-(Y-\mu)^2/2\sigma^2}$$

is the density function of a normally distributed variable Y with mean μ and standard deviation σ. The confidence interval for C_q is $(c_q - z_{100(1-\alpha)} s_q, c_q + z_{100(1-\alpha)} s_q)$.

meansd case

The value of c_q is $\overline{x} + z_q \times s$. Its se is given by the formula

$$s_q^\star = s\sqrt{1/n + z_q^2/(2n - 2)}$$

The confidence interval for C_q is $(c_q - z_{100(1-\alpha)} \times s_q^\star, c_q + z_{100(1-\alpha)} \times s_q^\star)$.

Acknowledgment

centile was written by Patrick Royston, MRC Clinical Trials Unit, London.

References

Conover, W. J. 1999. *Practical Nonparametric Statistics*. 3rd ed. New York: Wiley.

Kendall, M. G., and A. Stuart. 1969. *The Advanced Theory of Statistics, Vol. 1: Distribution Theory*. 3rd ed. London: Griffin.

Mood, A. M., and F. A. Graybill. 1963. *Introduction to the Theory of Statistics*. 2nd ed. New York: McGraw–Hill.

Newson, R. 2000. snp16: Robust confidence intervals for median and other percentile differences between two groups. *Stata Technical Bulletin* 58: 30–35. Reprinted in *Stata Technical Bulletin Reprints*, vol. 10, pp. 324–331. College Station, TX: Stata Press.

Royston, P. 1992. sg7: Centile estimation command. *Stata Technical Bulletin* 8: 12–15. Reprinted in *Stata Technical Bulletin Reprints*, vol. 2, pp. 122–125. College Station, TX: Stata Press.

Stuart, A., and J. K. Ord. 1994. *Kendall's Advanced Theory of Statistics: Distribution Theory, Vol I*. 6th ed. London: Arnold.

Also see

[R] **ci** — Confidence intervals for means, proportions, and counts

[R] **summarize** — Summary statistics

[D] **pctile** — Create variable containing percentiles

Title

> **ci** — Confidence intervals for means, proportions, and counts

Syntax

Syntax for ci

ci [*varlist*] [*if*] [*in*] [*weight*] [, *options*]

Immediate command for variable distributed as normal

cii $\#_{obs}$ $\#_{mean}$ $\#_{sd}$ [, *ciin_option*]

Immediate command for variable distributed as binomial

cii $\#_{obs}$ $\#_{succ}$ [, *ciib_options*]

Immediate command for variable distributed as Poisson

cii $\#_{exposure}$ $\#_{events}$, <u>poi</u>sson [*ciip_options*]

options	Description
Main	
<u>b</u>inomial	binomial 0/1 variables; compute exact confidence intervals
<u>po</u>isson	Poisson variables; compute exact confidence intervals
<u>e</u>xposure(*varname*)	exposure variable; implies poisson
<u>exa</u>ct	calculate exact confidence intervals; the default
<u>wa</u>ld	calculate Wald confidence intervals
<u>wi</u>lson	calculate Wilson confidence intervals
agresti	calculate Agresti–Coull confidence intervals
jeffreys	calculate Jeffreys confidence intervals
<u>to</u>tal	add output for all groups combined (for use with by only)
separator(#)	draw separator line after every # variables; default is separator(5)
<u>l</u>evel(#)	set confidence level; default is level(95)

by is allowed with ci; see [D] **by**.

aweights and fweights are allowed, but aweights may not be specified with the binomial or poisson options; see [U] **11.1.6 weight**.

ciin_option	Description
<u>l</u>evel(#)	set confidence level; default is level(95)

257

ciib_options	Description
level(#)	set confidence level; default is level(95)
exact	calculate exact confidence intervals; the default
wald	calculate Wald confidence intervals
wilson	calculate Wilson confidence intervals
agresti	calculate Agresti–Coull confidence intervals
jeffreys	calculate Jeffreys confidence intervals

ciip_options	Description
*poisson	numbers are Poisson-distributed counts
level(#)	set confidence level; default is level(95)

*poisson is required.

Menu

ci

Statistics > Summaries, tables, and tests > Summary and descriptive statistics > Confidence intervals

cii for variable distributed as normal

Statistics > Summaries, tables, and tests > Summary and descriptive statistics > Normal CI calculator

cii for variable distributed as binomial

Statistics > Summaries, tables, and tests > Summary and descriptive statistics > Binomial CI calculator

cii for variable distributed as Poisson

Statistics > Summaries, tables, and tests > Summary and descriptive statistics > Poisson CI calculator

Description

ci computes standard errors and confidence intervals for each of the variables in *varlist*.

cii is the immediate form of ci; see [U] **19 Immediate commands** for a general discussion of immediate commands.

In the binomial and Poisson variants of cii, the second number specified ($\#_{succ}$ or $\#_{events}$) must be an integer or between 0 and 1. If the number is between 0 and 1, Stata interprets it as the fraction of successes or events and converts it to an integer number representing the number of successes or events. The computation then proceeds as if two integers had been specified.

Options

⌐ Main ⌐

binomial tells ci that the variables are 0/1 variables and that binomial confidence intervals will be calculated. (cii produces binomial confidence intervals when only two numbers are specified.)

poisson specifies that the variables (or numbers for cii) are Poisson-distributed counts; exact Poisson confidence intervals will be calculated.

exposure(*varname*) is used only with poisson. You do not need to specify poisson if you specify exposure(); poisson is assumed. *varname* contains the total exposure (typically a time or an area) during which the number of events recorded in *varlist* were observed.

exact, wald, wilson, agresti, and jeffreys specify that variables are 0/1 and specify how binomial confidence intervals are to be calculated.

exact is the default and specifies exact (also known in the literature as Clopper–Pearson [1934]) binomial confidence intervals.

wald specifies calculation of Wald confidence intervals.

wilson specifies calculation of Wilson confidence intervals.

agresti specifies calculation of Agresti–Coull confidence intervals.

jeffreys specifies calculation of Jeffreys confidence intervals.

See Brown, Cai, and DasGupta (2001) for a discussion and comparison of the different binomial confidence intervals.

total is for use with the by prefix. It requests that, in addition to output for each by-group, output be added for all groups combined.

separator(*#*) specifies how often separation lines should be inserted into the output. The default is separator(5), meaning that a line is drawn after every five variables. separator(10) would draw the line after every 10 variables. separator(0) suppresses the separation line.

level(*#*) specifies the confidence level, as a percentage, for confidence intervals. The default is level(95) or as set by set level; see [R] **level**.

Remarks

Remarks are presented under the following headings:

> *Ordinary confidence intervals*
> *Binomial confidence intervals*
> *Poisson confidence intervals*
> *Immediate form*

Ordinary confidence intervals

▷ Example 1

Without the binomial or poisson options, ci produces "ordinary" confidence intervals, meaning those that are correct if the variable is distributed normally, and *asymptotically* correct for all other distributions satisfying the conditions of the central limit theorem.

```
. use http://www.stata-press.com/data/r12/auto
(1978 Automobile Data)

. ci mpg price
```

Variable	Obs	Mean	Std. Err.	[95% Conf. Interval]	
mpg	74	21.2973	.6725511	19.9569	22.63769
price	74	6165.257	342.8719	5481.914	6848.6

The standard error of the mean of mpg is 0.67, and the 95% confidence interval is $[19.96, 22.64]$. We can obtain wider confidence intervals, 99%, by typing

```
. ci mpg price, level(99)
```

Variable	Obs	Mean	Std. Err.	[99% Conf. Interval]
mpg	74	21.2973	.6725511	19.51849 23.07611
price	74	6165.257	342.8719	5258.405 7072.108

◁

▷ Example 2

by() breaks out the confidence intervals according to by-group; total adds an overall summary. For instance,

```
. ci mpg, by(foreign) total
```

```
-> foreign = Domestic
```

Variable	Obs	Mean	Std. Err.	[95% Conf. Interval]
mpg	52	19.82692	.657777	18.50638 21.14747

```
-> foreign = Foreign
```

Variable	Obs	Mean	Std. Err.	[95% Conf. Interval]
mpg	22	24.77273	1.40951	21.84149 27.70396

```
-> Total
```

Variable	Obs	Mean	Std. Err.	[95% Conf. Interval]
mpg	74	21.2973	.6725511	19.9569 22.63769

◁

❏ Technical note

You can control the formatting of the numbers in the output by specifying a display format for the variable; see [U] **12.5 Formats: Controlling how data are displayed**. For instance,

```
. format mpg %9.2f
. ci mpg
```

Variable	Obs	Mean	Std. Err.	[95% Conf. Interval]
mpg	74	21.30	0.67	19.96 22.64

❏

Binomial confidence intervals

▷ Example 3

We have data on employees, including a variable marking whether the employee was promoted last year.

```
. use http://www.stata-press.com/data/r12/promo
. ci promoted, binomial
```

Variable	Obs	Mean	Std. Err.	— Binomial Exact — [95% Conf. Interval]	
promoted	20	.1	.067082	.0123485	.3169827

The above interval is the default for binomial data, known equivalently as both the exact binomial and the Clopper–Pearson interval.

Nominally, the interpretation of a 95% confidence interval is that under repeated samples or experiments, 95% of the resultant intervals would contain the unknown parameter in question. However, for binomial data, the actual coverage probability, regardless of method, usually differs from that interpretation. This result occurs because of the discreteness of the binomial distribution, which produces only a finite set of outcomes, meaning that coverage probabilities are subject to discrete jumps and the exact nominal level cannot always be achieved. Therefore, the term *exact confidence interval* refers to its being derived from the binomial distribution, the distribution exactly generating the data, rather than resulting in exactly the nominal coverage.

For the Clopper–Pearson interval, the actual coverage probability is guaranteed to be greater than or equal to the nominal confidence level, here 95%. Because of the way it is calculated—see *Methods and formulas*—it may also be interpreted as follows: If the true probability of being promoted were 0.012, the chances of observing a result as extreme or more extreme than the result observed ($20 \times 0.1 = 2$ or more promotions) would be 2.5%. If the true probability of being promoted were 0.317, the chances of observing a result as extreme or more extreme than the result observed (two or fewer promotions) would be 2.5%.

◁

▷ Example 4

The Clopper–Pearson interval is desirable because it guarantees nominal coverage; however, by dropping this restriction, you may obtain accurate intervals that are not as conservative. In this vein, you might opt for the Wilson (1927) interval,

```
. ci promoted, binomial wilson
```

Variable	Obs	Mean	Std. Err.	——— Wilson ——— [95% Conf. Interval]	
promoted	20	.1	.067082	.0278665	.3010336

the Agresti–Coull (1998) interval,

```
. ci promoted, binomial agresti
```

Variable	Obs	Mean	Std. Err.	— Agresti-Coull — [95% Conf. Interval]	
promoted	20	.1	.067082	.0156562	.3132439

or the Bayesian-derived Jeffreys interval (Brown, Cai, and DasGupta 2001),

```
. ci promoted, binomial jeffreys
```

Variable	Obs	Mean	Std. Err.	——— Jeffreys ——— [95% Conf. Interval]	
promoted	20	.1	.067082	.0213725	.2838533

Picking the best interval is a matter of balancing accuracy (coverage) against precision (average interval length) and depends on sample size and success probability. Brown, Cai, and DasGupta (2001) recommend the Wilson or Jeffreys interval for small sample sizes ($\leq$40) yet favor the Agresti–Coull interval for its simplicity, decent performance for sample sizes less than or equal to 40, and performance comparable to Wilson/Jeffreys for sample sizes greater than 40. They also deem the Clopper–Pearson interval to be "wastefully conservative and [. . .] not a good choice for practical use", unless of course one requires, at a minimum, the nominal coverage level.

◁

Finally, the binomial Wald confidence interval is obtained by specifying the `binomial` and `wald` options. The Wald interval is the one taught in most introductory statistics courses and for the above is simply, for level $1 - \alpha$, Mean$\pm z_\alpha$(Std. Err.), where z_α is the $1 - \alpha/2$ quantile of the standard normal. Because its overall poor performance makes it impractical, the Wald interval is available mainly for pedagogical purposes. The binomial Wald interval is also similar to the interval produced by treating binary data as normal data and using `ci` without the `binomial` option, with two exceptions. First, when `binomial` is specified, the calculation of the standard error uses denominator n rather than $n - 1$, used for normal data. Second, confidence intervals for normal data are based on the t distribution rather than the standard normal. Of course, both discrepancies vanish as sample size increases.

❏ Technical note

Let's repeat example 3, but this time with data in which there are no promotions over the observed period:

```
. use http://www.stata-press.com/data/r12/promonone
. ci promoted, binomial
```

Variable	Obs	Mean	Std. Err.	— Binomial Exact — [95% Conf. Interval]	
promoted	20	0	0	0	.1684335*

(*) one-sided, 97.5% confidence interval

The confidence interval is $[0, 0.168]$, and this is the confidence interval that most books publish. It is not, however, a true 95% confidence interval because the lower tail has vanished. As Stata notes, it is a one-sided, 97.5% confidence interval. If you wanted to put 5% in the right tail, you could type `ci promoted, binomial level(90)`.

❏

❏ Technical note

`ci` with the `binomial` option ignores any variables that do not take on the values 0 and 1 exclusively. For instance, with our automobile dataset,

```
. use http://www.stata-press.com/data/r12/auto
(1978 Automobile Data)
. ci mpg foreign, binomial
```

Variable	Obs	Mean	Std. Err.	— Binomial Exact — [95% Conf. Interval]	
foreign	74	.2972973	.0531331	.196584	.4148353

We also requested the confidence interval for `mpg`, but Stata ignored us. It does that so you can type `ci, binomial` and obtain correct confidence intervals for all the variables that are 0/1 in your data.

❏

Poisson confidence intervals

▷ Example 5

We have data on the number of bacterial colonies on a Petri dish. The dish has been divided into 36 small squares, and the number of colonies in each square has been counted. Each observation in our dataset represents a square on the dish. The variable count records the number of colonies in each square counted, which varies from 0 to 5.

```
. use http://www.stata-press.com/data/r12/petri
. ci count, poisson
```

Variable	Exposure	Mean	Std. Err.	— Poisson Exact — [95% Conf. Interval]	
count	36	2.333333	.2545875	1.861158	2.888825

ci reports that the average number of colonies per square is 2.33. If the expected number of colonies per square were as low as 1.86, the probability of observing 2.33 or more colonies per square would be 2.5%. If the expected number were as large as 2.89, the probability of observing 2.33 or fewer colonies per square would be 2.5%.

◁

❑ Technical note

The number of "observations"—how finely the Petri dish is divided—makes no difference. The Poisson distribution is a function only of the count. In example 4, we observed a total of $2.33 \times 36 = 84$ colonies and a confidence interval of $[1.86 \times 36, 2.89 \times 36] = [67, 104]$. We would obtain the same $[67, 104]$ confidence interval if our dish were divided into, say, 49 squares, rather than 36.

For the counts, it is not even important that all the squares be of the same size. For *rates*, however, such differences do matter, but in an easy-to-calculate way. Rates are obtained from counts by dividing by exposure, which is typically a number multiplied by either time or an area. For our Petri dishes, we divide by an area to obtain a rate, but if our example were cast in terms of being infected by a disease, we might divide by person-years to obtain the rate. Rates are convenient because they are easier to compare: we might have 2.3 colonies per square inch or 0.0005 infections per person-year.

So, let's assume that we wish to obtain the number of colonies per square inch, and, moreover, that not all the "squares" on our dish are of equal size. We have a variable called area that records the area of each "square":

```
. ci count, exposure(area)
```

Variable	Exposure	Mean	Std. Err.	— Poisson Exact — [95% Conf. Interval]	
count	3	28	3.055051	22.3339	34.66591

The rates are now in more familiar terms. In our sample, there are 28 colonies per square inch and the 95% confidence interval is $[22.3, 34.7]$. When we did not specify exposure(), ci assumed that each observation contributed 1 to exposure.

❑

❑ Technical note

As with the `binomial` option, if there were no colonies on our dish, `ci` would calculate a one-sided confidence interval:

```
. use http://www.stata-press.com/data/r12/petrinone
. ci count, poisson
```

Variable	Exposure	Mean	Std. Err.	— Poisson Exact — [95% Conf. Interval]	
count	36	0	0	0	.1024689*

(*) one-sided, 97.5% confidence interval

❑

Immediate form

▷ Example 6

We are reading a soon-to-be-published paper by a colleague. In it is a table showing the number of observations, mean, and standard deviation of 1980 median family income for the Northeast and West. We correctly think that the paper would be much improved if it included the confidence intervals. The paper claims that for 166 cities in the Northeast, the average of median family income is $19,509 with a standard deviation of $4,379:

For the Northeast:

```
. cii 166 19509 4379
```

Variable	Obs	Mean	Std. Err.	[95% Conf. Interval]	
	166	19509	339.8763	18837.93	20180.07

For the West:

```
. cii 256 22557 5003
```

Variable	Obs	Mean	Std. Err.	[95% Conf. Interval]	
	256	22557	312.6875	21941.22	23172.78

◁

▷ Example 7

We flip a coin 10 times, and it comes up heads only once. We are shocked and decide to obtain a 99% confidence interval for this coin:

```
. cii 10 1, level(99)
```

Variable	Obs	Mean	Std. Err.	— Binomial Exact — [99% Conf. Interval]	
	10	.1	.0948683	.0005011	.5442871

◁

▷ Example 8

The number of reported traffic accidents in Santa Monica over a 24-hour period is 27. We need know nothing else:

. cii 1 27, poisson

Variable	Exposure	Mean	Std. Err.	— Poisson Exact — [95% Conf. Interval]	
	1	27	5.196152	17.79317	39.28358

◁

Saved results

ci and cii saves the following in r():

Scalars

r(N)	number of observations or exposure	r(lb)	lower bound of confidence interval
r(mean)	mean	r(ub)	upper bound of confidence interval
r(se)	estimate of standard error		

Methods and formulas

ci and cii are implemented as ado-files.

Methods and formulas are presented under the following headings:

Ordinary
Binomial
Poisson

Ordinary

Define n, $\overline{x}$, and s^2 as, respectively, the number of observations, (weighted) average, and (unbiased) estimated variance of the variable in question; see [R] **summarize**.

The standard error of the mean, s_μ, is defined as $\sqrt{s^2/n}$.

Let α be $1 - l/100$, where l is the significance level specified by the user. Define t_α as the two-sided t statistic corresponding to a significance level of α with $n - 1$ degrees of freedom; t_α is obtained from Stata as invttail(n-1,0.5*α). The lower and upper confidence bounds are, respectively, $\overline{x} - s_\mu t_\alpha$ and $\overline{x} + s_\mu t_\alpha$.

Binomial

Given k successes of n trials, the estimated probability is $\widehat{p} = k/n$ with standard error $\sqrt{\widehat{p}(1 - \widehat{p})/n}$. ci calculates the exact (Clopper–Pearson) confidence interval $[p_1, p_2]$ such that

$$\Pr(K \geq k | p = p_1) = \alpha/2$$

and

$$\Pr(K \leq k | p = p_2) = \alpha/2$$

where K is distributed as binomial(n, p). The endpoints may be obtained directly by using Stata's invbinomial() function. If $k = 0$ or $k = n$, the calculation of the appropriate tail is skipped.

The Wald interval is $\widehat{p} \pm z_\alpha \sqrt{\widehat{p}(1 - \widehat{p})/n}$, where z_α is the $1 - \alpha/2$ quantile of the standard normal. The interval is obtained by inverting the acceptance region of the large-sample Wald test of $H_0 \colon p = p_0$ versus the two-sided alternative. That is, the confidence interval is the set of all p_0 such that

$$\left| \frac{\widehat{p} - p_0}{\sqrt{n^{-1}\widehat{p}(1 - \widehat{p})}} \right| \leq z_\alpha$$

The Wilson interval is a variation on the Wald interval, using the null standard error $\sqrt{n^{-1}p_0(1 - p_0)}$ in place of the estimated standard error $\sqrt{n^{-1}\widehat{p}(1 - \widehat{p})}$ in the above expression. Inverting this acceptance region is more complicated yet results in the closed form

$$\frac{k + z_\alpha^2/2}{n + z_\alpha^2} \pm \frac{z_\alpha n^{1/2}}{n + z_\alpha^2/2} \left\{ \widehat{p}(1 - \widehat{p}) + \frac{z_\alpha^2}{4n} \right\}^{1/2}$$

The Agresti–Coull interval is basically a Wald interval that borrows its center from the Wilson interval. Defining $\widetilde{k} = k + z_\alpha^2/2$, $\widetilde{n} = n + z_\alpha^2$, and (hence) $\widetilde{p} = \widetilde{k}/\widetilde{n}$, the Agresti–Coull interval is

$$\widetilde{p} \pm z_\alpha \sqrt{\widetilde{p}(1 - \widetilde{p})/\widetilde{n}}$$

When $\alpha = 0.05$, z_α is near enough to 2 that $\widetilde{p}$ can be thought of as a typical estimate of proportion where two successes and two failures have been added to the sample (Agresti and Coull 1998). This typical estimate of proportion makes the Agresti–Coull interval an easy-to-present alternative for introductory statistics students.

The Jeffreys interval is a Bayesian interval and is based on the Jeffreys prior, which is the Beta$(1/2, 1/2)$ distribution. Assigning this prior to p results in a posterior distribution for p that is Beta with parameters $k + 1/2$ and $n - k + 1/2$. The Jeffreys interval is then taken to be the $1 - \alpha$ central posterior probability interval, namely, the $\alpha/2$ and $1 - \alpha/2$ quantiles of the Beta$(k + 1/2, n - k + 1/2)$ distribution. These quantiles may be obtained directly by using Stata's invibeta() function.

Poisson

Given the total cases, k, the estimate of the expected count λ is k, and its standard error is $\sqrt{k}$. ci calculates the exact confidence interval $[\lambda_1, \lambda_2]$ such that

$$\Pr(K \geq k | \lambda = \lambda_1) = \alpha/2$$

and

$$\Pr(K \leq k | \lambda = \lambda_2) = \alpha/2$$

where K is Poisson with mean λ. Solution is by Newton's method. If $k = 0$, the calculation of λ_1 is skipped. All values are then reported as rates, which are the above numbers divided by the total exposure.

Harold Jeffreys (1891–1989) was born near Durham, England, and spent more than 75 years studying and working at the University of Cambridge, principally on theoretical and observational problems in geophysics, astronomy, mathematics, and statistics. He developed a systematic Bayesian approach to inference in his monograph *Theory of Probability*.

Edwin Bidwell (E. B.) Wilson (1879–1964) majored in mathematics at Harvard and studied and taught at Yale and MIT before returning to Harvard in 1922. He worked in mathematics, physics, and statistics. His method for binomial intervals can be considered a precursor, for a particular problem, of Neyman's concept of confidence intervals.

Jerzy Neyman (1894–1981) was born in Bendery, Russia, now Moldavia. He studied and then taught at Kharkov University, moving from physics to mathematics. In 1921, Neyman moved to Poland, where he worked in statistics at Bydgoszcz and then Warsaw. Neyman received a Rockefeller Fellowship to work with Karl Pearson at University College London. There, he collaborated with Egon Pearson, Karl's son, on the theory of hypothesis testing. Life in Poland became progressively more difficult, and Neyman returned to UCL to work there from 1934 to 1938. At this time, he published on the theory of confidence intervals. He then was offered a post in California at Berkeley, where he settled. Neyman established an outstanding statistics department and remained highly active in research, including applications in astronomy, meteorology, and medicine. He was one of the great statisticians of the 20th century.

Acknowledgment

We thank Nicholas J. Cox of Durham University for his assistance with the `jeffreys` and `wilson` options.

References

Agresti, A., and B. A. Coull. 1998. Approximate is better than "exact" for interval estimation of binomial proportions. *American Statistician* 52: 119–126.

Brown, L. D., T. T. Cai, and A. DasGupta. 2001. Interval estimation for a binomial proportion. *Statistical Science* 16: 101–133.

Campbell, M. J., D. Machin, and S. J. Walters. 2007. *Medical Statistics: A Textbook for the Health Sciences*. 4th ed. Chichester, UK: Wiley.

Clopper, C. J., and E. S. Pearson. 1934. The use of confidence or fiducial limits illustrated in the case of the binomial. *Biometrika* 26: 404–413.

Cook, A. 1990. Sir Harold Jeffreys. 2 April 1891–18 March 1989. *Biographical Memoirs of Fellows of the Royal Society* 36: 303–333.

Gleason, J. R. 1999. sg119: Improved confidence intervals for binomial proportions. *Stata Technical Bulletin* 52: 16–18. Reprinted in *Stata Technical Bulletin Reprints*, vol. 9, pp. 208–211. College Station, TX: Stata Press.

Jeffreys, H. 1946. An invariant form for the prior probability in estimation problems. *Proceedings of the Royal Society of London, Series A* 186: 453–461.

Lindley, D. V. 2001. Harold Jeffreys. In *Statisticians of the Centuries*, ed. C. C. Heyde and E. Seneta, 402–405. New York: Springer.

Reid, C. 1982. *Neyman—from Life*. New York: Springer.

Rothman, K. J., S. Greenland, and T. L. Lash. 2008. *Modern Epidemiology*. 3rd ed. Philadelphia: Lippincott Williams & Wilkins.

Seed, P. T. 2001. sg159: Confidence intervals for correlations. *Stata Technical Bulletin* 59: 27–28. Reprinted in *Stata Technical Bulletin Reprints*, vol. 10, pp. 267–269. College Station, TX: Stata Press.

Stigler, S. M. 1997. Wilson, Edwin Bidwell. In *Leading Personalities in Statistical Sciences: From the Seventeenth Century to the Present*, ed. N. L. Johnson and S. Kotz, 344–346. New York: Wiley.

Utts, J. M. 2005. *Seeing Through Statistics*. 3rd ed. Belmont, CA: Brooks/Cole.

Wang, D. 2000. sg154: Confidence intervals for the ratio of two binomial proportions by Koopman's method. *Stata Technical Bulletin* 58: 16–19. Reprinted in *Stata Technical Bulletin Reprints*, vol. 10, pp. 244–247. College Station, TX: Stata Press.

Wilson, E. B. 1927. Probable inference, the law of succession, and statistical inference. *Journal of the American Statistical Association* 22: 209–212.

Also see

[R] **bitest** — Binomial probability test

[R] **prtest** — One- and two-sample tests of proportions

[R] **ttest** — Mean-comparison tests

[R] **ameans** — Arithmetic, geometric, and harmonic means

[R] **centile** — Report centile and confidence interval

[R] **summarize** — Summary statistics

[D] **pctile** — Create variable containing percentiles

Title

> **clogit** — Conditional (fixed-effects) logistic regression

Syntax

$$\underline{\text{clogit}} \;\; depvar \;\; \big[\,indepvars\,\big] \;\; \big[\,if\,\big] \;\; \big[\,in\,\big] \;\; \big[\,weight\,\big] \;, \;\; \underline{\text{gr}}\text{oup}(varname) \;\; \big[\,options\,\big]$$

options	Description
Model	
* <u>group</u>(*varname*)	matched group variable
<u>off</u>set(*varname*)	include *varname* in model with coefficient constrained to 1
<u>constraints</u>(*constraints*)	apply specified linear constraints
<u>coll</u>inear	keep collinear variables
SE/Robust	
vce(*vcetype*)	*vcetype* may be oim, <u>r</u>obust, <u>cl</u>uster *clustvar*, opg, <u>boot</u>strap, or <u>jackknife</u>
nonest	do not check that panels are nested within clusters
Reporting	
<u>level</u>(#)	set confidence level; default is level(95)
or	report odds ratios
<u>nocns</u>report	do not display constraints
display_options	control column formats, row spacing, line width, and display of omitted variables and base and empty cells
Maximization	
maximize_options	control the maximization process; seldom used
<u>coefl</u>egend	display legend instead of statistics

* group(*varname*) is required.

indepvars may contain factor variables; see [U] **11.4.3 Factor variables**.

bootstrap, by, fracpoly, jackknife, mfp, mi estimate, nestreg, rolling, statsby, stepwise, and svy are allowed; see [U] **11.1.10 Prefix commands**.

vce(bootstrap) and vce(jackknife) are not allowed with the mi estimate prefix; see [MI] **mi estimate**.

Weights are not allowed with the bootstrap prefix; see [R] **bootstrap**.

vce(), nonest, and weights are not allowed with the svy prefix; see [SVY] **svy**.

fweights, iweights, and pweights are allowed (see [U] **11.1.6 weight**), but they are interpreted to apply to groups as a whole, not to individual observations. See *Use of weights* below.

coeflegend does not appear in the dialog box.

See [U] **20 Estimation and postestimation commands** for more capabilities of estimation commands.

Menu

Statistics > Categorical outcomes > Conditional logistic regression

Description

clogit fits what biostatisticians and epidemiologists call conditional logistic regression for matched case–control groups (see, for example, Hosmer and Lemeshow [2000, chap. 7]) and what economists and other social scientists call fixed-effects logit for panel data (see, for example, Chamberlain [1980]). Computationally, these models are the same. *depvar* equal to nonzero and nonmissing (typically *depvar* equal to one) indicates a positive outcome, whereas *depvar* equal to zero indicates a negative outcome.

See [R] **asclogit** if you want to fit McFadden's choice model (McFadden 1974). Also see [R] **logistic** for a list of related estimation commands.

Options

Model

group(*varname*) is required; it specifies an identifier variable (numeric or string) for the matched groups. strata(*varname*) is a synonym for group().

offset(*varname*), constraints(*constraints*), collinear; see [R] **estimation options**.

SE/Robust

vce(*vcetype*) specifies the type of standard error reported, which includes types that are derived from asymptotic theory, that are robust to some kinds of misspecification, that allow for intragroup correlation, and that use bootstrap or jackknife methods; see [R] *vce_option*.

nonest, available only with vce(cluster *clustvar*), prevents checking that matched groups are nested within clusters. It is the user's responsibility to verify that the standard errors are theoretically correct.

Reporting

level(#); see [R] **estimation options**.

or reports the estimated coefficients transformed to odds ratios, that is, e^b rather than b. Standard errors and confidence intervals are similarly transformed. This option affects how results are displayed, not how they are estimated. or may be specified at estimation or when replaying previously estimated results.

nocnsreport; see [R] **estimation options**.

display_options: noomitted, vsquish, noemptycells, baselevels, allbaselevels, cformat(%*fmt*), pformat(%*fmt*), sformat(%*fmt*), and nolstretch; see [R] **estimation options**.

Maximization

maximize_options: difficult, technique(*algorithm_spec*), iterate(#), [no]log, trace, gradient, showstep, hessian, showtolerance, tolerance(#), ltolerance(#), nrtolerance(#), nonrtolerance, and from(*init_specs*); see [R] **maximize**. These options are seldom used.

Setting the optimization type to technique(bhhh) resets the default *vcetype* to vce(opg).

The following option is available with clogit but is not shown in the dialog box:

coeflegend; see [R] **estimation options**.

Remarks

Remarks are presented under the following headings:

> *Introduction*
> *Matched case–control data*
> *Use of weights*
> *Fixed-effects logit*

Introduction

clogit fits maximum likelihood models with a dichotomous dependent variable coded as 0/1 (more precisely, clogit interprets 0 and not 0 to indicate the dichotomy). Conditional logistic analysis differs from regular logistic regression in that the data are grouped and the likelihood is calculated relative to each group; that is, a conditional likelihood is used. See *Methods and formulas* at the end of this entry.

Biostatisticians and epidemiologists fit these models when analyzing matched case–control studies with $1:1$ matching, $1:k_{2i}$ matching, or $k_{1i}:k_{2i}$ matching, where i denotes the ith matched group for $i = 1, 2, \ldots, n$, where n is the total number of groups. clogit fits a model appropriate for all these matching schemes or for any mix of the schemes because the matching $k_{1i}:k_{2i}$ can vary from group to group. clogit always uses the true conditional likelihood, not an approximation. Biostatisticians and epidemiologists sometimes refer to the matched groups as "strata", but we will stick to the more generic term "group".

Economists and other social scientists fitting fixed-effects logit models have data that look exactly like the data biostatisticians and epidemiologists call $k_{1i}:k_{2i}$ matched case–control data. In terms of how the data are arranged, $k_{1i}:k_{2i}$ matching means that in the ith group, the dependent variable is 1 a total of k_{1i} times and 0 a total of k_{2i} times. There are a total of $T_i = k_{1i} + k_{2i}$ observations for the ith group. This data arrangement is what economists and other social scientists call "panel data", "longitudinal data", or "cross-sectional time-series data".

So no matter what terminology you use, the computation and the use of the clogit command is the same. The following example shows how your data should be arranged to use clogit.

▷ Example 1

Suppose that we have grouped data with the variable id containing a unique identifier for each group. Our outcome variable, y, contains 0s and 1s. If we were biostatisticians, $y = 1$ would indicate a case, $y = 0$ would be a control, and id would be an identifier variable that indicates the groups of matched case–control subjects.

If we were economists, $y = 1$ might indicate that a person was unemployed at any time during a year and $y = 0$, that a person was employed all year, and id would be an identifier variable for persons.

If we list the first few observations of this dataset, it looks like

```
. use http://www.stata-press.com/data/r12/clogitid
. list y x1 x2 id in 1/11
```

	y	x1	x2	id
1.	0	0	4	1014
2.	0	1	4	1014
3.	0	1	6	1014
4.	1	1	8	1014
5.	0	0	1	1017
6.	0	0	7	1017
7.	1	1	10	1017
8.	0	0	1	1019
9.	0	1	7	1019
10.	1	1	7	1019
11.	1	1	9	1019

Pretending that we are biostatisticians, we describe our data as follows. The first group ($id = 1014$) consists of four matched persons: 1 case ($y = 1$) and three controls ($y = 0$), that is, $1:3$ matching. The second group has $1:2$ matching, and the third $2:2$.

Pretending that we are economists, we describe our data as follows. The first group consists of 4 observations (one per year) for person 1014. This person had a period of unemployment during 1 year of 4. The second person had a period of unemployment during 1 year of 3, and the third had a period of 2 years of 4.

Our independent variables are x1 and x2. To fit the conditional (fixed-effects) logistic model, we type

```
. clogit y x1 x2, group(id)
note: multiple positive outcomes within groups encountered.
Iteration 0:   log likelihood = -123.42828
Iteration 1:   log likelihood = -123.41386
Iteration 2:   log likelihood = -123.41386
```

```
Conditional (fixed-effects) logistic regression      Number of obs   =        369
                                                     LR chi2(2)      =       9.07
                                                     Prob > chi2     =     0.0107
Log likelihood = -123.41386                          Pseudo R2       =     0.0355
```

y	Coef.	Std. Err.	z	P>|z|	[95% Conf. Interval]	
x1	.653363	.2875215	2.27	0.023	.0898312	1.216895
x2	.0659169	.0449555	1.47	0.143	-.0221943	.1540281

◁

❑ Technical note

The message "note: multiple positive outcomes within groups encountered" at the top of the clogit output for the previous example merely informs us that we have $k_{1i} : k_{2i}$ matching with $k_{1i} > 1$ for at least one group. If your data should be $1 : k_{2i}$ matched, this message tells you that there is an error in the data somewhere.

We can see the distribution of k_{1i} and $T_i = k_{1i} + k_{2i}$ for the data of the previous example by using the following steps:

```
. by id, sort: gen k1 = sum(y)
. by id: replace k1 = . if _n < _N
(303 real changes made, 303 to missing)
. by id: gen T = sum(y < .)
. by id: replace T = . if _n < _N
(303 real changes made, 303 to missing)
. tab k1
```

k1	Freq.	Percent	Cum.
1	48	72.73	72.73
2	12	18.18	90.91
3	4	6.06	96.97
4	2	3.03	100.00
Total	66	100.00	

```
. tab T
```

T	Freq.	Percent	Cum.
2	5	7.58	7.58
3	5	7.58	15.15
4	12	18.18	33.33
5	11	16.67	50.00
6	13	19.70	69.70
7	8	12.12	81.82
8	3	4.55	86.36
9	7	10.61	96.97
10	2	3.03	100.00
Total	66	100.00	

We see that k_{1i} ranges from 1 to 4 and T_i ranges from 2 to 10 for these data.

❑

❑ Technical note

For $k_{1i} : k_{2i}$ matching (and hence in the general case of fixed-effects logit), clogit uses a recursive algorithm to compute the likelihood, which means that there are no limits on the size of T_i. However, computation time is proportional to $\sum T_i \min(k_{1i}, k_{2i})$, so clogit will take roughly 10 times longer to fit a model with $10:10$ matching than one with $1:10$ matching. But clogit is fast, so computation time becomes an issue only when $\min(k_{1i}, k_{2i})$ is around 100 or more. See *Methods and formulas* for details.

❑

Matched case–control data

Here we give a more detailed example of matched case–control data.

▷ Example 2

Hosmer and Lemeshow (2000, 25) present data on matched pairs of infants, each pair having one with low birthweight and another with regular birthweight. The data are matched on age of the mother. Several possible maternal exposures are considered: race (three categories), smoking status, presence of hypertension, presence of uterine irritability, previous preterm delivery, and weight at the last menstrual period.

```
. use http://www.stata-press.com/data/r12/lowbirth2, clear
(Applied Logistic Regression, Hosmer & Lemeshow)
. describe
Contains data from http://www.stata-press.com/data/r12/lowbirth2.dta
  obs:           112                          Applied Logistic Regression,
                                                Hosmer & Lemeshow
 vars:             9                          26 Apr 2011 09:33
 size:         1,456
```

variable name	storage type	display format	value label	variable label
pairid	byte	%8.0g		Case-control pair ID
low	byte	%8.0g		Baby has low birthweight
age	byte	%8.0g		Age of mother
lwt	int	%8.0g		Mother's last menstrual weight
smoke	byte	%8.0g		Mother smoked during pregnancy
ptd	byte	%8.0g		Mother had previous preterm baby
ht	byte	%8.0g		Mother has hypertension
ui	byte	%8.0g		Uterine irritability
race	float	%9.0g		race of mother: 1=white, 2=black, 3=other

```
Sorted by:
```

We list the case–control indicator variable, low; the match identifier variable, pairid; and two of the covariates, lwt and smoke, for the first 10 observations.

```
. list low lwt smoke pairid in 1/10
```

	low	lwt	smoke	pairid
1.	0	135	0	1
2.	1	101	1	1
3.	0	98	0	2
4.	1	115	0	2
5.	0	95	0	3
6.	1	130	0	3
7.	0	103	0	4
8.	1	130	1	4
9.	0	122	1	5
10.	1	110	1	5

We fit a conditional logistic model of low birthweight on mother's weight, race, smoking behavior, and history.

```
. clogit low lwt smoke ptd ht ui i.race, group(pairid) nolog
```

Conditional (fixed-effects) logistic regression

		Number of obs	=	112
		LR chi2(7)	=	26.04
		Prob > chi2	=	0.0005
Log likelihood = -25.794271		Pseudo R2	=	0.3355

low	Coef.	Std. Err.	z	P>|z|	[95% Conf. Interval]	
lwt	-.0183757	.0100806	-1.82	0.068	-.0381333	.0013819
smoke	1.400656	.6278396	2.23	0.026	.1701131	2.631199
ptd	1.808009	.7886502	2.29	0.022	.2622828	3.353735
ht	2.361152	1.086128	2.17	0.030	.2323796	4.489924
ui	1.401929	.6961585	2.01	0.044	.0374836	2.766375
race						
2	.5713643	.689645	0.83	0.407	-.7803149	1.923044
3	-.0253148	.6992044	-0.04	0.971	-1.39573	1.345101

We might prefer to see results presented as odds ratios. We could have specified the or option when we first fit the model, or we can now redisplay results and specify or:

```
. clogit, or
```

Conditional (fixed-effects) logistic regression

		Number of obs	=	112
		LR chi2(7)	=	26.04
		Prob > chi2	=	0.0005
Log likelihood = -25.794271		Pseudo R2	=	0.3355

low	Odds Ratio	Std. Err.	z	P>|z|	[95% Conf. Interval]	
lwt	.9817921	.009897	-1.82	0.068	.9625847	1.001383
smoke	4.057862	2.547686	2.23	0.026	1.185439	13.89042
ptd	6.098293	4.80942	2.29	0.022	1.299894	28.60938
ht	10.60316	11.51639	2.17	0.030	1.261599	89.11467
ui	4.06303	2.828513	2.01	0.044	1.038195	15.90088
race						
2	1.770681	1.221141	0.83	0.407	.4582617	6.84175
3	.975003	.6817263	-0.04	0.971	.2476522	3.838573

Smoking, previous preterm delivery, hypertension, uterine irritability, and possibly the mother's weight all contribute to low birthweight. 2.race (mother black) and 3.race (mother other) are statistically insignificant when compared with the 1.race (mother white) omitted group, although the 2.race effect is large. We can test the joint statistical significance of 2.race and 3.race by using test:

```
. test 2.race 3.race

 ( 1)  [low]2.race = 0
 ( 2)  [low]3.race = 0

           chi2(  2) =    0.88
         Prob > chi2 =    0.6436
```

For a more complete description of test, see [R] **test**. test presents results in coefficients rather than odds ratios. Jointly testing that the coefficients on 2.race and 3.race are 0 is equivalent to jointly testing that the odds ratios are 1.

Here one case was matched to one control, that is, $1:1$ matching. From clogit's point of view, that was not important—k_1 cases could have been matched to k_2 controls ($k_1 : k_2$ matching), and we would have fit the model in the same way. Furthermore, the matching can change from group to group, which we have denoted as $k_{1i} : k_{2i}$ matching, where i denotes the group. clogit does not care. To fit the conditional logistic regression model, we specified the group(*varname*) option, group(pairid). The case and control are stored in separate observations. clogit knew that they were linked (in the same group) because the related observations share the same value of pairid.

◁

❑ Technical note

clogit provides a way to extend McNemar's test to multiple controls per case ($1:k_{2i}$ matching) and to multiple controls matched with multiple cases ($k_{1i}:k_{2i}$ matching).

In Stata, McNemar's test is calculated by the mcc command; see [ST] **epitab**. The mcc command, however, requires that the matched case and control appear in one observation, so the data will need to be manipulated from 1 to 2 observations per stratum before using clogit. Alternatively, if you begin with clogit's 2-observations-per-group organization, you will have to change it to 1 observation per group if you wish to use mcc. In either case, reshape provides an easy way to change the organization of the data. We will demonstrate its use below, but we direct you to [D] **reshape** for a more thorough discussion.

In the example above, we used clogit to analyze the relationship between low birthweight and various characteristics of the mother. Assume that we now want to assess the relationship between low birthweight and smoking, ignoring the mother's other characteristics. Using clogit, we obtain the following results:

```
. clogit low smoke, group(pairid) or
Iteration 0:   log likelihood = -35.425931
Iteration 1:   log likelihood = -35.419283
Iteration 2:   log likelihood = -35.419282
Conditional (fixed-effects) logistic regression   Number of obs   =       112
                                                  LR chi2(1)      =      6.79
                                                  Prob > chi2     =    0.0091
Log likelihood = -35.419282                       Pseudo R2       =    0.0875
```

low	Odds Ratio	Std. Err.	z	P>|z|	[95% Conf. Interval]
smoke	2.75	1.135369	2.45	0.014	1.224347 6.176763

Let's compare our estimated odds ratio and 95% confidence interval with that produced by mcc. We begin by reshaping the data:

```
. keep low smoke pairid
. reshape wide smoke, i(pairid) j(low 0 1)
Data                           long   ->   wide

Number of obs.                  112   ->     56
Number of variables               3   ->      3
j variable (2 values)           low   ->   (dropped)
xij variables:
                              smoke   ->   smoke0 smoke1
```

We now have the variables smoke0 (formed from smoke and low = 0), recording 1 if the control mother smoked and 0 otherwise; and smoke1 (formed from smoke and low = 1), recording 1 if the case mother smoked and 0 otherwise. We can now use mcc:

```
. mcc smoke1 smoke0
```

Cases	Controls Exposed	Unexposed	Total
Exposed	8	22	30
Unexposed	8	18	26
Total	16	40	56

```
McNemar's chi2(1) =       6.53     Prob > chi2 = 0.0106
Exact McNemar significance probability       = 0.0161
```

Proportion with factor			
Cases	.5357143		
Controls	.2857143	[95% Conf. Interval]	
difference	.25	.0519726	.4480274
ratio	1.875	1.148685	3.060565
rel. diff.	.35	.1336258	.5663742
odds ratio	2.75	1.179154	7.143667 (exact)

Both methods estimated the same odds ratio, and the 95% confidence intervals are similar. clogit produced a confidence interval of $[1.22, 6.18]$, whereas mcc produced a confidence interval of $[1.18, 7.14]$.

❑

Use of weights

With clogit, weights apply to groups as a whole, not to individual observations. For example, if there is a group in your dataset with a frequency weight of 3, there are a total of three groups in your sample with the same values of the dependent and independent variables as this one group. Weights must have the same value for all observations belonging to the same group; otherwise, an error message will be displayed.

▷ Example 3

We use the example from the above discussion of the mcc command. Here we have a total of 56 matched case–control groups, each with one case matched to one control. We had 8 matched pairs in which both the case and the control are exposed, 22 pairs in which the case is exposed and the control is unexposed, 8 pairs in which the case is unexposed and the control is exposed, and 18 pairs in which they are both unexposed.

With weights, it is easy to enter these data into Stata and run clogit.

```
. clear
. input id case exposed weight
            id        case      exposed      weight
  1. 1 1 1 8
  2. 1 0 1 8
  3. 2 1 1 22
  4. 2 0 0 22
  5. 3 1 0 8
  6. 3 0 1 8
  7. 4 1 0 18
  8. 4 0 0 18
  9. end
. clogit case exposed [w=weight], group(id) or
(frequency weights assumed)
Iteration 0:   log likelihood = -35.425931
Iteration 1:   log likelihood = -35.419283
Iteration 2:   log likelihood = -35.419282
Conditional (fixed-effects) logistic regression   Number of obs   =        112
                                                   LR chi2(1)      =       6.79
                                                   Prob > chi2     =     0.0091
Log likelihood = -35.419282                        Pseudo R2       =     0.0875
```

case	Odds Ratio	Std. Err.	z	P>|z|	[95% Conf. Interval]	
exposed	2.75	1.135369	2.45	0.014	1.224347	6.176763

◁

Fixed-effects logit

The fixed-effects logit model can be written as

$$\Pr(y_{it} = 1 \mid \mathbf{x}_{it}) = F(\alpha_i + \mathbf{x}_{it}\boldsymbol{\beta})$$

where F is the cumulative logistic distribution

$$F(z) = \frac{\exp(z)}{1 + \exp(z)}$$

$i = 1, 2, \ldots, n$ denotes the independent units (called "groups" by clogit), and $t = 1, 2, \ldots, T_i$ denotes the observations for the ith unit (group).

Fitting this model by using a full maximum-likelihood approach leads to difficulties, however. When T_i is fixed, the maximum likelihood estimates for α_i and β are inconsistent (Andersen 1970; Chamberlain 1980). This difficulty can be circumvented by looking at the probability of $\mathbf{y}_i = (y_{i1}, \ldots, y_{iT_i})$ conditional on $\sum_{t=1}^{T_i} y_{it}$. This conditional probability does not involve the α_i, so they are never estimated when the resulting conditional likelihood is used. See Hamerle and Ronning (1995) for a succinct and lucid development. See *Methods and formulas* for the estimation equation.

▷ Example 4

We are studying unionization of women in the United States by using the union dataset; see [XT] **xt**. We fit the fixed-effects logit model:

```
. use http://www.stata-press.com/data/r12/union, clear
(NLS Women 14-24 in 1968)
. clogit union age grade not_smsa south black, group(idcode)
note: multiple positive outcomes within groups encountered.
note: 2744 groups (14165 obs) dropped because of all positive or
      all negative outcomes.
note: black omitted because of no within-group variance.
Iteration 0:   log likelihood = -4521.3385
Iteration 1:   log likelihood = -4516.1404
Iteration 2:   log likelihood = -4516.1385
Iteration 3:   log likelihood = -4516.1385
Conditional (fixed-effects) logistic regression   Number of obs   =       12035
                                                  LR chi2(4)      =       68.09
                                                  Prob > chi2     =      0.0000
Log likelihood = -4516.1385                       Pseudo R2       =      0.0075
```

union	Coef.	Std. Err.	z	P>|z|	[95% Conf. Interval]	
age	.0170301	.004146	4.11	0.000	.0089042	.0251561
grade	.0853572	.0418781	2.04	0.042	.0032777	.1674368
not_smsa	.0083678	.1127963	0.07	0.941	-.2127088	.2294445
south	-.748023	.1251752	-5.98	0.000	-.9933619	-.5026842
black	0	(omitted)				

We received three messages at the top of the output. The first one, "multiple positive outcomes within groups encountered", we expected. Our data do indeed have multiple positive outcomes (union = 1) in many groups. (Here a group consists of all the observations for a particular individual.)

The second message tells us that 2,744 groups were "dropped" by clogit. When either union = 0 or union = 1 for all observations for an individual, this individual's contribution to the log-likelihood is zero. Although these are perfectly valid observations in every sense, they have no effect on the estimation, so they are not included in the total "Number of obs". Hence, the reported "Number of obs" gives the effective sample size of the estimation. Here it is 12,035 observations—only 46% of the total 26,200.

We can easily check that there are indeed 2,744 groups with union either all 0 or all 1. We will generate a variable that contains the fraction of observations for each individual who has union = 1.

```
. by idcode, sort: generate fraction = sum(union)/sum(union < .)

. by idcode: replace fraction = . if _n < _N
(21766 real changes made, 21766 to missing)

. tabulate fraction
```

fraction	Freq.	Percent	Cum.
0	2,481	55.95	55.95
.0833333	30	0.68	56.63
.0909091	33	0.74	57.37
.1	53	1.20	58.57
(output omitted)			
.9	10	0.23	93.59
.9090909	11	0.25	93.84
.9166667	10	0.23	94.07
1	263	5.93	100.00
Total	4,434	100.00	

Because $2481 + 263 = 2744$, we confirm what clogit did.

The third warning message from clogit said "black omitted because of no within-group variance". Obviously, race stays constant for an individual across time. Any such variables are collinear with the α_i (that is, the fixed effects), and just as the α_i drop out of the conditional likelihood, so do all variables that are unchanging within groups. Thus they cannot be estimated with the conditional fixed-effects model.

There are several other estimators implemented in Stata that we could use with these data:

```
cloglog ... ,   vce(cluster idcode)
logit ... ,   vce(cluster idcode)
probit ... ,   vce(cluster idcode)
scobit ... ,   vce(cluster idcode)
xtcloglog ...
xtgee ... ,   family(binomial) link(logit) corr(exchangeable)
xtlogit ...
xtprobit ...
```

See [R] cloglog, [R] logit, [R] probit, [R] scobit, [XT] xtcloglog, [XT] xtgee, [XT] xtlogit, and [XT] xtprobit for details.

◁

Saved results

clogit saves the following in e():

Scalars

e(N)	number of observations
e(N_drop)	number of observations dropped because of all positive or all negative outcomes
e(N_group_drop)	number of groups dropped because of all positive or all negative outcomes
e(k)	number of parameters
e(k_eq)	number of equations in e(b)
e(k_eq_model)	number of equations in overall model test
e(k_dv)	number of dependent variables
e(df_m)	model degrees of freedom
e(r2_p)	pseudo-R-squared
e(ll)	log likelihood
e(ll_0)	log likelihood, constant-only model
e(N_clust)	number of clusters
e(chi2)	χ^2
e(p)	significance
e(rank)	rank of e(V)
e(ic)	number of iterations
e(rc)	return code
e(converged)	1 if converged, 0 otherwise

Macros

e(cmd)	clogit
e(cmdline)	command as typed
e(depvar)	name of dependent variable
e(group)	name of group() variable
e(multiple)	multiple if multiple positive outcomes within group
e(wtype)	weight type
e(wexp)	weight expression
e(title)	title in estimation output
e(clustvar)	name of cluster variable
e(offset)	linear offset variable
e(chi2type)	Wald or LR; type of model χ^2 test
e(vce)	*vcetype* specified in vce()
e(vcetype)	title used to label Std. Err.
e(opt)	type of optimization
e(which)	max or min; whether optimizer is to perform maximization or minimization
e(ml_method)	type of ml method
e(user)	name of likelihood-evaluator program
e(technique)	maximization technique
e(properties)	b V
e(predict)	program used to implement predict
e(marginsok)	predictions allowed by margins
e(marginsnotok)	predictions disallowed by margins
e(asbalanced)	factor variables fvset as asbalanced
e(asobserved)	factor variables fvset as asobserved

Matrices
  e(b)   coefficient vector
  e(Cns)   constraints matrix
  e(ilog)   iteration log (up to 20 iterations)
  e(V)   variance–covariance matrix of the estimators
  e(V_modelbased) model-based variance
  e(gradient)   gradient vector
Functions
  e(sample)   marks estimation sample

Methods and formulas

`clogit` is implemented as an ado-file.

Breslow and Day (1980, 247–279), Collett (2003, 251–267), and Hosmer and Lemeshow (2000, 223–259) provide a biostatistical point of view on conditional logistic regression. Hamerle and Ronning (1995) give a succinct and lucid review of fixed-effects logit; Chamberlain (1980) is a standard reference for this model. Greene (2012, chap. 17) provides a straightforward textbook description of conditional logistic regression from an economist's point of view, as well as a brief description of choice models.

Let $i = 1, 2, \ldots, n$ denote the groups and let $t = 1, 2, \ldots, T_i$ denote the observations for the ith group. Let y_{it} be the dependent variable taking on values 0 or 1. Let $\mathbf{y}_i = (y_{i1}, \ldots, y_{iT_i})$ be the outcomes for the ith group as a whole. Let $\mathbf{x}_{it}$ be a row vector of covariates. Let

$$k_{1i} = \sum_{t=1}^{T_i} y_{it}$$

be the observed number of ones for the dependent variable in the ith group. Biostatisticians would say that there are k_{1i} cases matched to $k_{2i} = T_i - k_{1i}$ controls in the ith group.

We consider the probability of a possible value of $\mathbf{y}_i$ conditional on $\sum_{t=1}^{T_i} y_{it} = k_{1i}$ (Hamerle and Ronning 1995, eq. 8.33; Hosmer and Lemeshow 2000, eq. 7.4),

$$\Pr\!\left(\mathbf{y}_i \mid \textstyle\sum_{t=1}^{T_i} y_{it} = k_{1i}\right) = \frac{\exp\!\left(\sum_{t=1}^{T_i} y_{it}\mathbf{x}_{it}\boldsymbol{\beta}\right)}{\sum_{\mathbf{d}_i \in S_i} \exp\!\left(\sum_{t=1}^{T_i} d_{it}\mathbf{x}_{it}\boldsymbol{\beta}\right)}$$

where d_{it} is equal to 0 or 1 with $\sum_{t=1}^{T_i} d_{it} = k_{1i}$, and S_i is the set of all possible combinations of k_{1i} ones and k_{2i} zeros. Clearly, there are $\binom{T_i}{k_{1i}}$ such combinations, but we need not count all these combinations to compute the denominator of the above equation. It can be computed recursively.

Denote the denominator by

$$f_i(T_i, k_{1i}) = \sum_{\mathbf{d}_i \in S_i} \exp\!\left(\sum_{t=1}^{T_i} d_{it}\mathbf{x}_{it}\boldsymbol{\beta}\right)$$

Consider, computationally, how f_i changes as we go from a total of 1 observation in the group to 2 observations to 3, etc. Doing this, we derive the recursive formula

$$f_i(T, k) = f_i(T - 1, k) + f_i(T - 1, k - 1)\exp(\mathbf{x}_{iT}\boldsymbol{\beta})$$

where we define $f_i(T, k) = 0$ if $T < k$ and $f_i(T, 0) = 1$.

The conditional log-likelihood is

$$\ln L = \sum_{i=1}^{n} \left\{ \sum_{t=1}^{T_i} y_{it} \mathbf{x}_{it} \boldsymbol{\beta} - \log f_i(T_i, k_{1i}) \right\}$$

The derivatives of the conditional log-likelihood can also be computed recursively by taking derivatives of the recursive formula for f_i.

Computation time is roughly proportional to

$$p^2 \sum_{i=1}^{n} T_i \, \min(k_{1i}, k_{2i})$$

where p is the number of independent variables in the model. If $\min(k_{1i}, k_{2i})$ is small, computation time is not an issue. But if it is large—say, 100 or more—patience may be required.

If T_i is large for all groups, the bias of the unconditional fixed-effects estimator is not a concern, and we can confidently use logit with an indicator variable for each group (provided, of course, that the number of groups does not exceed matsize; see [R] **matsize**).

This command supports the clustered version of the Huber/White/sandwich estimator of the variance using vce(robust) and vce(cluster *clustvar*). See [P] **_robust**, particularly *Maximum likelihood estimators* and *Methods and formulas*. Specifying vce(robust) is equivalent to specifying vce(cluster *groupvar*), where *groupvar* is the variable for the matched groups.

clogit also supports estimation with survey data. For details on VCEs with survey data, see [SVY] **variance estimation**.

References

Andersen, E. B. 1970. Asymptotic properties of conditional maximum likelihood estimators. *Journal of the Royal Statistical Society, Series B* 32: 283–301.

Breslow, N. E., and N. E. Day. 1980. *Statistical Methods in Cancer Research: Vol. 1—The Analysis of Case–Control Studies*. Lyon: IARC.

Chamberlain, G. 1980. Analysis of covariance with qualitative data. *Review of Economic Studies* 47: 225–238.

Collett, D. 2003. *Modelling Binary Data*. 2nd ed. London: Chapman & Hall/CRC.

Greene, W. H. 2012. *Econometric Analysis*. 7th ed. Upper Saddle River, NJ: Prentice Hall.

Hamerle, A., and G. Ronning. 1995. Panel analysis for qualitative variables. In *Handbook of Statistical Modeling for the Social and Behavioral Sciences*, ed. G. Arminger, C. C. Clogg, and M. E. Sobel, 401–451. New York: Plenum.

Hole, A. R. 2007. Fitting mixed logit models by using maximum simulated likelihood. *Stata Journal* 7: 388–401.

Hosmer, D. W., Jr., and S. Lemeshow. 2000. *Applied Logistic Regression*. 2nd ed. New York: Wiley.

Kleinbaum, D. G., and M. Klein. 2010. *Logistic Regression: A Self-Learning Text*. 3rd ed. New York: Springer.

Long, J. S., and J. Freese. 2006. *Regression Models for Categorical Dependent Variables Using Stata*. 2nd ed. College Station, TX: Stata Press.

McFadden, D. L. 1974. Conditional logit analysis of qualitative choice behavior. In *Frontiers in Econometrics*, ed. P. Zarembka, 105–142. New York: Academic Press.

Also see

[R] **clogit postestimation** — Postestimation tools for clogit

[R] **asclogit** — Alternative-specific conditional logit (McFadden's choice) model

[R] **logistic** — Logistic regression, reporting odds ratios

[R] **mlogit** — Multinomial (polytomous) logistic regression

[R] **nlogit** — Nested logit regression

[R] **ologit** — Ordered logistic regression

[R] **scobit** — Skewed logistic regression

[MI] **estimation** — Estimation commands for use with mi estimate

[SVY] **svy estimation** — Estimation commands for survey data

[XT] **xtgee** — Fit population-averaged panel-data models by using GEE

[XT] **xtlogit** — Fixed-effects, random-effects, and population-averaged logit models

[U] **20 Estimation and postestimation commands**

Title

> **clogit postestimation** — Postestimation tools for clogit

Description

The following postestimation commands are available after `clogit`:

Command	Description
contrast	contrasts and ANOVA-style joint tests of estimates
estat	AIC, BIC, VCE, and estimation sample summary
estat (svy)	postestimation statistics for survey data
estimates	cataloging estimation results
hausman	Hausman's specification test
lincom	point estimates, standard errors, testing, and inference for linear combinations of coefficients
linktest	link test for model specification
lrtest[1]	likelihood-ratio test
margins[2]	marginal means, predictive margins, marginal effects, and average marginal effects
marginsplot	graph the results from margins (profile plots, interaction plots, etc.)
nlcom	point estimates, standard errors, testing, and inference for nonlinear combinations of coefficients
predict	predictions, residuals, influence statistics, and other diagnostic measures
predictnl	point estimates, standard errors, testing, and inference for generalized predictions
pwcompare	pairwise comparisons of estimates
suest	seemingly unrelated estimation
test	Wald tests of simple and composite linear hypotheses
testnl	Wald tests of nonlinear hypotheses

[1] `lrtest` is not appropriate with svy estimation results.

[2] The default prediction statistic `pc1` cannot be correctly handled by `margins`; however, `margins` can be used after `clogit` with options `predict(pu0)` and `predict(xb)`.

See the corresponding entries in the *Base Reference Manual* for details, but see [SVY] **estat** for details about `estat` (svy).

Syntax for predict

> predict [*type*] *newvar* [*if*] [*in*] [, *statistic* <u>nooff</u>set]

statistic	Description
Main	
pc1	probability of a positive outcome; the default
pu0	probability of a positive outcome, assuming fixed effect is zero
xb	linear prediction
stdp	standard error of the linear prediction
*<u>dbeta</u>	Delta-β influence statistic
*<u>dx2</u>	Delta-χ^2 lack-of-fit statistic
*<u>gdbeta</u>	Delta-β influence statistic for each group
*<u>gdx2</u>	Delta-χ^2 lack-of-fit statistic for each group
*<u>hat</u>	Hosmer and Lemeshow leverage
*<u>residuals</u>	Pearson residuals
*<u>rstandard</u>	standardized Pearson residuals
<u>score</u>	first derivative of the log likelihood with respect to $\mathbf{x}_j\beta$

Unstarred statistics are available both in and out of sample; type predict ... if e(sample) ... if wanted only for the estimation sample. Starred statistics are calculated only for the estimation sample, even when if e(sample) is not specified.

Starred statistics are available for multiple controls per case-matching design only. They are not available if vce(robust), vce(cluster *clustvar*), or pweights were specified with clogit.

dbeta, dx2, gdbeta, gdx2, hat, and rstandard are not available if constraints() was specified with clogit.

Menu

Statistics > Postestimation > Predictions, residuals, etc.

Options for predict

_____| Main |_____

pc1, the default, calculates the probability of a positive outcome conditional on one positive outcome within group.

pu0 calculates the probability of a positive outcome, assuming that the fixed effect is zero.

xb calculates the linear prediction.

stdp calculates the standard error of the linear prediction.

dbeta calculates the Delta-β influence statistic, a standardized measure of the difference in the coefficient vector that is due to deletion of the observation.

dx2 calculates the Delta-χ^2 influence statistic, reflecting the decrease in the Pearson chi-squared that is due to deletion of the observation.

gdbeta calculates the approximation to the Pregibon stratum-specific Delta-β influence statistic, a standardized measure of the difference in the coefficient vector that is due to deletion of the entire stratum.

gdx2 calculates the approximation to the Pregibon stratum-specific Delta-χ^2 influence statistic, reflecting the decrease in the Pearson chi-squared that is due to deletion of the entire stratum.

hat calculates the Hosmer and Lemeshow leverage or the diagonal element of the hat matrix.

residuals calculates the Pearson residuals.

rstandard calculates the standardized Pearson residuals.

score calculates the equation-level score, $\partial \ln L / \partial(\mathbf{x}_{it}\beta)$.

nooffset is relevant only if you specified offset(*varname*) for clogit. It modifies the calculations made by predict so that they ignore the offset variable; the linear prediction is treated as $\mathbf{x}_j\mathbf{b}$ rather than as $\mathbf{x}_j\mathbf{b} + \text{offset}_j$. This option cannot be specified with dbeta, dx2, gdbeta, gdx2, hat, and rstandard.

Remarks

predict may be used after clogit to obtain predicted values of the index $\mathbf{x}_{it}\beta$. Predicted probabilities for conditional logistic regression must be interpreted carefully. Probabilities are estimated for each group as a whole, not for individual observations. Furthermore, the probabilities are conditional on the number of positive outcomes in the group (that is, the number of cases and the number of controls), or it is assumed that the fixed effect is zero. predict may also be used to obtain influence and lack of fit statistics for an individual observation and for the whole group, to compute Pearson, standardized Pearson residuals, and leverage values.

predict may be used for both within-sample and out-of-sample predictions.

▷ Example 1

Suppose that we have $1 : k_{2i}$ matched data and that we have previously fit the following model:

```
. use http://www.stata-press.com/data/r12/clogitid
. clogit y x1 x2, group(id)
(output omitted)
```

To obtain the predicted values of the index, we could type predict idx, xb to create a new variable called idx. From idx, we could then calculate the predicted probabilities. Easier, however, would be to type

```
. predict phat
(option pc1 assumed; probability of success given one success within group)
```

phat would then contain the predicted probabilities.

As noted previously, the predicted probabilities are really predicted probabilities for the group as a whole (that is, they are the predicted probability of observing $y_{it} = 1$ and $y_{it'} = 0$ for all $t' \neq t$). Thus, if we want to obtain the predicted probabilities for the estimation sample, it is important that, when we make the calculation, predictions be restricted to the same sample on which we estimated the data. We cannot predict the probabilities and then just keep the relevant ones because the entire sample determines each probability. Thus, assuming that we are not attempting to make out-of-sample predictions, we type

```
. predict phat2 if e(sample)
(option pc1 assumed; probability of success given one success within group)
```
◁

Methods and formulas

All postestimation commands listed above are implemented as ado-files.

Recall that $i = 1, \ldots, n$ denote the groups and $t = 1, \ldots, T_i$ denote the observations for the ith group.

predict produces probabilities of a positive outcome within group conditional on there being one positive outcome (pc1),

$$\Pr\left(y_{it} = 1 \,\middle|\, \sum_{t=1}^{T_i} y_{it} = 1\right) = \frac{\exp(\mathbf{x}_{it}\boldsymbol{\beta})}{\sum_{t=1}^{T_i} \exp(\mathbf{x}_{it}\boldsymbol{\beta})}$$

or predict calculates the unconditional pu0:

$$\Pr(y_{it} = 1) = \frac{\exp(\mathbf{x}_{it}\boldsymbol{\beta})}{1 + \exp(\mathbf{x}_{it}\boldsymbol{\beta})}$$

Let $N = \sum_{j=1}^{n} T_j$ denote the total number of observations, p denote the number of covariates, and $\widehat{\theta}_{it}$ denote the conditional predicted probabilities of a positive outcome (pc1).

For the multiple control per case ($1 : k_{2i}$) matching, Hosmer and Lemeshow (2000, 248–251) propose the following diagnostics:

The Pearson residual is

$$r_{it} = \frac{(y_{it} - \widehat{\theta}_{it})}{\sqrt{\widehat{\theta}_{it}}}$$

The leverage (hat) value is defined as

$$h_{it} = \widehat{\theta}_{it}\widetilde{\mathbf{x}}_{it}^{T}(\widetilde{\mathbf{X}}^{T}\mathbf{U}\widetilde{\mathbf{X}})^{-1}\widetilde{\mathbf{x}}_{it}$$

where $\widetilde{\mathbf{x}}_{it} = \mathbf{x}_{it} - \sum_{j=1}^{T_i} \mathbf{x}_{ij}\widehat{\theta}_{ij}$ is the $1 \times p$ row vector of centered by a weighted stratum-specific mean covariate values, $\mathbf{U}_N = \mathrm{diag}\{\widehat{\theta}_{it}\}$, and the rows of $\widetilde{\mathbf{X}}_{N \times p}$ are composed of $\widetilde{\mathbf{x}}_{it}$ values.

The standardized Pearson residual is

$$r_{sit} = \frac{r_{it}}{\sqrt{1 - h_{it}}}$$

The lack of fit and influence diagnostics for an individual observation are (respectively) computed as

$$\Delta\chi_{it}^{2} = r_{sit}^{2}$$

and

$$\Delta\widehat{\beta}_{it} = \Delta\chi_{it}^{2}\frac{h_{it}}{1 - h_{it}}$$

The lack of fit and influence diagnostics for the groups are the group-specific totals of the respective individual diagnostics shown above.

Reference

Hosmer, D. W., Jr., and S. Lemeshow. 2000. *Applied Logistic Regression*. 2nd ed. New York: Wiley.

Also see

Title

> **cloglog** — Complementary log-log regression

Syntax

cloglog *depvar* [*indepvars*] [*if*] [*in*] [*weight*] [, *options*]

options	Description
Model	
<u>nocon</u>stant	suppress constant term
<u>off</u>set(*varname*)	include *varname* in model with coefficient constrained to 1
asis	retain perfect predictor variables
<u>constraints</u>(*constraints*)	apply specified linear constraints
<u>coll</u>inear	keep collinear variables
SE/Robust	
vce(*vcetype*)	*vcetype* may be oim, <u>r</u>obust, <u>cl</u>uster *clustvar*, opg, <u>boot</u>strap, or jackknife
Reporting	
<u>level</u>(#)	set confidence level; default is level(95)
<u>ef</u>orm	report exponentiated coefficients
<u>nocns</u>report	do not display constraints
display_options	control column formats, row spacing, line width, and display of omitted variables and base and empty cells
Maximization	
maximize_options	control the maximization process; seldom used
<u>coefl</u>egend	display legend instead of statistics

indepvars may contain factor variables; see [U] **11.4.3 Factor variables**.
depvar and *indepvars* may contain time-series operators; see [U] **11.4.4 Time-series varlists**.
bootstrap, by, jackknife, mi estimate, nestreg, rolling, statsby, stepwise, and svy are allowed; see
  [U] **11.1.10 Prefix commands**.
vce(bootstrap) and vce(jackknife) are not allowed with the mi estimate prefix; see [MI] **mi estimate**.
Weights are not allowed with the bootstrap prefix; see [R] **bootstrap**.
vce() and weights are not allowed with the svy prefix; see [SVY] **svy**.
fweights, iweights, and pweights are allowed; see [U] **11.1.6 weight**.
coeflegend does not appear in the dialog box.
See [U] **20 Estimation and postestimation commands** for more capabilities of estimation commands.

Menu

Statistics > Binary outcomes > Complementary log-log regression

Description

cloglog fits maximum-likelihood complementary log-log models.

See [R] **logistic** for a list of related estimation commands.

Options

<u>Model</u>

noconstant, offset(*varname*); see [R] **estimation options**.

asis forces retention of perfect predictor variables and their associated perfectly predicted observations and may produce instabilities in maximization; see [R] **probit**.

constraints(*constraints*), collinear; see [R] **estimation options**.

<u>SE/Robust</u>

vce(*vcetype*) specifies the type of standard error reported, which includes types that are derived from asymptotic theory, that are robust to some kinds of misspecification, that allow for intragroup correlation, and that use bootstrap or jackknife methods; see [R] **vce_option**.

<u>Reporting</u>

level(*#*); see [R] **estimation options**.

eform displays the exponentiated coefficients and corresponding standard errors and confidence intervals.

nocnsreport; see [R] **estimation options**.

display_options: <u>noomit</u>ted, vsquish, <u>noempty</u>cells, <u>base</u>levels, <u>allbase</u>levels, cformat(*%fmt*), pformat(*%fmt*), sformat(*%fmt*), and nolstretch; see [R] **estimation options**.

<u>Maximization</u>

maximize_options: <u>diff</u>icult, <u>tech</u>nique(*algorithm_spec*), <u>iter</u>ate(*#*), [<u>no</u>]<u>log</u>, <u>trace</u>, gradient, showstep, <u>hess</u>ian, <u>showtol</u>erance, <u>tol</u>erance(*#*), <u>ltol</u>erance(*#*), <u>nrtol</u>erance(*#*), <u>nonrtol</u>erance, and from(*init_specs*); see [R] **maximize**. These options are seldom used.

Setting the optimization type to technique(bhhh) resets the default *vcetype* to vce(opg).

The following option is available with cloglog but is not shown in the dialog box:

coeflegend; see [R] **estimation options**.

Remarks

Remarks are presented under the following headings:

> *Introduction to complementary log-log regression*
> *Robust standard errors*

Introduction to complementary log-log regression

cloglog fits maximum likelihood models with dichotomous dependent variables coded as 0/1 (or, more precisely, coded as 0 and not 0).

▷ Example 1

We have data on the make, weight, and mileage rating of 22 foreign and 52 domestic automobiles. We wish to fit a model explaining whether a car is foreign based on its weight and mileage. Here is an overview of our data:

```
. use http://www.stata-press.com/data/r12/auto
(1978 Automobile Data)

. keep make mpg weight foreign

. describe
Contains data from http://www.stata-press.com/data/r12/auto.dta
  obs:            74                          1978 Automobile Data
 vars:             4                          13 Apr 2011 17:45
 size:         1,702                          (_dta has notes)
```

variable name	storage type	display format	value label	variable label
make	str18	%-18s		Make and Model
mpg	int	%8.0g		Mileage (mpg)
weight	int	%8.0gc		Weight (lbs.)
foreign	byte	%8.0g	origin	Car type

```
Sorted by:  foreign
     Note:  dataset has changed since last saved

. inspect foreign
foreign:  Car type                          Number of Observations
```

			Total	Integers	Nonintegers
#		Negative	–	–	–
#		Zero	52	52	–
#		Positive	22	22	–
#					
#	#	Total	74	74	–
#	#	Missing	–		

```
0                         1                          74
   (2 unique values)
        foreign is labeled and all values are documented in the label.
```

The variable foreign takes on two unique values, 0 and 1. The value 0 denotes a domestic car, and 1 denotes a foreign car.

The model that we wish to fit is

$$\Pr(\texttt{foreign} = 1) = F(\beta_0 + \beta_1 \texttt{weight} + \beta_2 \texttt{mpg})$$

where $F(z) = 1 - \exp\{-\exp(z)\}$.

To fit this model, we type

```
. cloglog foreign weight mpg
Iteration 0:    log likelihood = -34.054593
Iteration 1:    log likelihood = -27.869915
Iteration 2:    log likelihood = -27.742997
Iteration 3:    log likelihood = -27.742769
Iteration 4:    log likelihood = -27.742769
Complementary log-log regression                Number of obs    =        74
                                                Zero outcomes    =        52
                                                Nonzero outcomes =        22
                                                LR chi2(2)       =     34.58
Log likelihood = -27.742769                     Prob > chi2      =    0.0000
```

foreign	Coef.	Std. Err.	z	P>|z|	[95% Conf. Interval]	
weight	-.0029153	.0006974	-4.18	0.000	-.0042823	-.0015483
mpg	-.1422911	.076387	-1.86	0.062	-.2920069	.0074247
_cons	10.09694	3.351841	3.01	0.003	3.527448	16.66642

We find that heavier cars are less likely to be foreign and that cars yielding better gas mileage are also less likely to be foreign, at least when holding the weight of the car constant.

See [R] **maximize** for an explanation of the output.

◁

□ Technical note

Stata interprets a value of 0 as a negative outcome (failure) and treats all other values (except missing) as positive outcomes (successes). Thus, if your dependent variable takes on the values 0 and 1, 0 is interpreted as failure and 1 as success. If your dependent variable takes on the values 0, 1, and 2, 0 is still interpreted as failure, but both 1 and 2 are treated as successes.

If you prefer a more formal mathematical statement, when you type cloglog y x, Stata fits the model

$$\Pr(y_j \neq 0 \mid \mathbf{x}_j) = 1 - \exp\left\{-\exp(\mathbf{x}_j\boldsymbol{\beta})\right\}$$

□

Robust standard errors

If you specify the vce(robust) option, cloglog reports robust standard errors, as described in [U] **20.20 Obtaining robust variance estimates**. For the model of foreign on weight and mpg, the robust calculation increases the standard error of the coefficient on mpg by 44%:

```
. cloglog foreign weight mpg, vce(robust)

Iteration 0:    log pseudolikelihood = -34.054593
Iteration 1:    log pseudolikelihood = -27.869915
Iteration 2:    log pseudolikelihood = -27.742997
Iteration 3:    log pseudolikelihood = -27.742769
Iteration 4:    log pseudolikelihood = -27.742769
```

Complementary log-log regression

		Number of obs	=	74
		Zero outcomes	=	52
		Nonzero outcomes	=	22
		Wald chi2(2)	=	29.74
Log pseudolikelihood = -27.742769		Prob > chi2	=	0.0000

foreign	Coef.	Robust Std. Err.	z	P>|z|	[95% Conf. Interval]
weight	-.0029153	.0007484	-3.90	0.000	-.0043822 -.0014484
mpg	-.1422911	.1102466	-1.29	0.197	-.3583704 .0737882
_cons	10.09694	4.317305	2.34	0.019	1.635174 18.5587

Without vce(robust), the standard error for the coefficient on mpg was reported to be 0.076, with a resulting confidence interval of $[-0.29, 0.01]$.

The vce(cluster *clustvar*) option can relax the independence assumption required by the complementary log-log estimator to being just independence between clusters. To demonstrate this ability, we will switch to a different dataset.

We are studying unionization of women in the United States by using the union dataset; see [XT] **xt**. We fit the following model, ignoring that women are observed an average of 5.9 times each in this dataset:

```
. use http://www.stata-press.com/data/r12/union, clear
(NLS Women 14-24 in 1968)

. cloglog union age grade not_smsa south##c.year

Iteration 0:    log likelihood = -13606.373
Iteration 1:    log likelihood = -13540.726
Iteration 2:    log likelihood = -13540.607
Iteration 3:    log likelihood = -13540.607
```

Complementary log-log regression

		Number of obs	=	26200
		Zero outcomes	=	20389
		Nonzero outcomes	=	5811
		LR chi2(6)	=	647.24
Log likelihood = -13540.607		Prob > chi2	=	0.0000

union	Coef.	Std. Err.	z	P>|z|	[95% Conf. Interval]
age	.0185346	.0043616	4.25	0.000	.009986 .0270833
grade	.0452772	.0057125	7.93	0.000	.0340809 .0564736
not_smsa	-.1886592	.0317801	-5.94	0.000	-.2509471 -.1263712
1.south	-1.422292	.3949381	-3.60	0.000	-2.196356 -.648227
year	-.0133007	.0049576	-2.68	0.007	-.0230174 -.0035839
south#c.year					
1	.0105659	.0049234	2.15	0.032	.0009161 .0202157
_cons	-1.219801	.2952374	-4.13	0.000	-1.798455 -.6411462

The reported standard errors in this model are probably meaningless. Women are observed repeatedly, and so the observations are not independent. Looking at the coefficients, we find a large southern effect against unionization and a different time trend for the south. The vce(cluster *clustvar*) option provides a way to fit this model and obtains correct standard errors:

```
. cloglog union age grade not_smsa south##c.year, vce(cluster id) nolog
Complementary log-log regression                Number of obs    =      26200
                                                Zero outcomes    =      20389
                                                Nonzero outcomes =       5811
                                                Wald chi2(6)     =     160.76
Log pseudolikelihood = -13540.607               Prob > chi2      =     0.0000
                           (Std. Err. adjusted for 4434 clusters in idcode)
```

union	Coef.	Robust Std. Err.	z	P>|z|	[95% Conf. Interval]	
age	.0185346	.0084873	2.18	0.029	.0018999	.0351694
grade	.0452772	.0125776	3.60	0.000	.0206255	.069929
not_smsa	-.1886592	.0642068	-2.94	0.003	-.3145021	-.0628162
1.south	-1.422292	.506517	-2.81	0.005	-2.415047	-.4295365
year	-.0133007	.0090628	-1.47	0.142	-.0310633	.004462
south#c.year						
1	.0105659	.0063175	1.67	0.094	-.0018162	.022948
_cons	-1.219801	.5175129	-2.36	0.018	-2.234107	-.2054942

These standard errors are larger than those reported by the inappropriate conventional calculation. By comparison, another way we could fit this model is with an equal-correlation population-averaged complementary log-log model:

```
. xtcloglog union age grade not_smsa south##c.year, pa nolog
GEE population-averaged model          Number of obs     =      26200
Group variable:              idcode    Number of groups  =       4434
Link:                       cloglog    Obs per group: min =          1
Family:                    binomial                   avg =        5.9
Correlation:           exchangeable                   max =         12
                                       Wald chi2(6)      =     234.66
Scale parameter:                   1   Prob > chi2       =     0.0000
```

union	Coef.	Std. Err.	z	P>|z|	[95% Conf. Interval]	
age	.0153737	.0081156	1.89	0.058	-.0005326	.03128
grade	.0549518	.0095093	5.78	0.000	.0363139	.0735897
not_smsa	-.1045232	.0431082	-2.42	0.015	-.1890138	-.0200326
1.south	-1.714868	.3384558	-5.07	0.000	-2.378229	-1.051507
year	-.0115881	.0084125	-1.38	0.168	-.0280763	.0049001
south#c.year						
1	.0149796	.0041687	3.59	0.000	.0068091	.0231501
_cons	-1.488278	.4468005	-3.33	0.001	-2.363991	-.6125652

The coefficient estimates are similar, but these standard errors are smaller than those produced by cloglog, vce(cluster *clustvar*). This finding is as we would expect. If the within-panel correlation assumptions are valid, the population-averaged estimator should be more efficient.

In addition to this estimator, we may use the xtgee command to fit a panel estimator (with complementary log-log link) and any number of assumptions on the within-idcode correlation.

cloglog, vce(cluster *clustvar*) is robust to assumptions about within-cluster correlation. That is, it inefficiently sums within cluster for the standard-error calculation rather than attempting to exploit what might be assumed about the within-cluster correlation (as do the xtgee population-averaged models).

Saved results

cloglog saves the following in e():

Scalars
e(N)	number of observations
e(k)	number of parameters
e(k_eq)	number of equations in e(b)
e(k_eq_model)	number of equations in overall model test
e(k_dv)	number of dependent variables
e(N_f)	number of zero outcomes
e(N_s)	number of nonzero outcomes
e(df_m)	model degrees of freedom
e(ll)	log likelihood
e(ll_0)	log likelihood, constant-only model
e(N_clust)	number of clusters
e(chi2)	χ^2
e(p)	significance
e(rank)	rank of e(V)
e(ic)	number of iterations
e(rc)	return code
e(converged)	1 if converged, 0 otherwise

Macros
e(cmd)	cloglog
e(cmdline)	command as typed
e(depvar)	name of dependent variable
e(wtype)	weight type
e(wexp)	weight expression
e(title)	title in estimation output
e(clustvar)	name of cluster variable
e(offset)	linear offset variable
e(chi2type)	Wald or LR; type of model χ^2 test
e(vce)	*vcetype* specified in vce()
e(vcetype)	title used to label Std. Err.
e(opt)	type of optimization
e(which)	max or min; whether optimizer is to perform maximization or minimization
e(ml_method)	type of ml method
e(user)	name of likelihood-evaluator program
e(technique)	maximization technique
e(properties)	b V
e(predict)	program used to implement predict
e(asbalanced)	factor variables fvset as asbalanced
e(asobserved)	factor variables fvset as asobserved

Matrices
e(b)	coefficient vector
e(Cns)	constraints matrix
e(ilog)	iteration log (up to 20 iterations)
e(gradient)	gradient vector
e(V)	variance–covariance matrix of the estimators
e(V_modelbased)	model-based variance

Functions
e(sample)	marks estimation sample

Methods and formulas

cloglog is implemented as an ado-file.

Complementary log-log analysis (related to the gompit model, so named because of its relationship to the Gompertz distribution) is an alternative to logit and probit analysis, but it is unlike these other estimators in that the transformation is not symmetric. Typically, this model is used when the positive (or negative) outcome is rare.

The log-likelihood function for complementary log-log is

$$\ln L = \sum_{j \in S} w_j \ln F(\mathbf{x}_j \mathbf{b}) + \sum_{j \notin S} w_j \ln \left\{ 1 - F(\mathbf{x}_j \mathbf{b}) \right\}$$

where S is the set of all observations j such that $y_j \neq 0$, $F(z) = 1 - \exp\left\{ -\exp(z) \right\}$, and w_j denotes the optional weights. $\ln L$ is maximized as described in [R] **maximize**.

We can fit a gompit model by reversing the success–failure sense of the dependent variable and using cloglog.

This command supports the Huber/White/sandwich estimator of the variance and its clustered version using vce(robust) and vce(cluster *clustvar*), respectively. See [P] **_robust**, particularly *Maximum likelihood estimators* and *Methods and formulas*. The scores are calculated as $\mathbf{u}_j = [\exp(\mathbf{x}_j \mathbf{b}) \exp\left\{ -\exp(\mathbf{x}_j \mathbf{b}) \right\} / F(\mathbf{x}_j \mathbf{b})] \mathbf{x}_j$ for the positive outcomes and $\{ -\exp(\mathbf{x}_j \mathbf{b}) \} \mathbf{x}_j$ for the negative outcomes.

cloglog also supports estimation with survey data. For details on VCEs with survey data, see [SVY] **variance estimation**.

Acknowledgment

We thank Joseph Hilbe of Arizona State University for providing the inspiration for the cloglog command (Hilbe 1996, 1998).

References

Clayton, D. G., and M. Hills. 1993. *Statistical Models in Epidemiology*. Oxford: Oxford University Press.

Hilbe, J. M. 1996. sg53: Maximum-likelihood complementary log-log regression. *Stata Technical Bulletin* 32: 19–20. Reprinted in *Stata Technical Bulletin Reprints*, vol. 6, pp. 129–131. College Station, TX: Stata Press.

———. 1998. sg53.2: Stata-like commands for complementary log-log regression. *Stata Technical Bulletin* 41: 23. Reprinted in *Stata Technical Bulletin Reprints*, vol. 7, pp. 166–167. College Station, TX: Stata Press.

Long, J. S. 1997. *Regression Models for Categorical and Limited Dependent Variables*. Thousand Oaks, CA: Sage.

Long, J. S., and J. Freese. 2006. *Regression Models for Categorical Dependent Variables Using Stata*. 2nd ed. College Station, TX: Stata Press.

Xu, J., and J. S. Long. 2005. Confidence intervals for predicted outcomes in regression models for categorical outcomes. *Stata Journal* 5: 537–559.

Also see

[R] **cloglog postestimation** — Postestimation tools for cloglog

[R] **clogit** — Conditional (fixed-effects) logistic regression

[R] **glm** — Generalized linear models

[R] **logistic** — Logistic regression, reporting odds ratios

[R] **scobit** — Skewed logistic regression

[MI] **estimation** — Estimation commands for use with mi estimate

[SVY] **svy estimation** — Estimation commands for survey data

[XT] **xtcloglog** — Random-effects and population-averaged cloglog models

[U] **20 Estimation and postestimation commands**

Title

> **cloglog postestimation** — Postestimation tools for cloglog

Description

The following postestimation commands are available after `cloglog`:

Command	Description
contrast	contrasts and ANOVA-style joint tests of estimates
estat	AIC, BIC, VCE, and estimation sample summary
estat (svy)	postestimation statistics for survey data
estimates	cataloging estimation results
lincom	point estimates, standard errors, testing, and inference for linear combinations of coefficients
linktest	link test for model specification
lrtest[1]	likelihood-ratio test
margins	marginal means, predictive margins, marginal effects, and average marginal effects
marginsplot	graph the results from margins (profile plots, interaction plots, etc.)
nlcom	point estimates, standard errors, testing, and inference for nonlinear combinations of coefficients
predict	predictions, residuals, influence statistics, and other diagnostic measures
predictnl	point estimates, standard errors, testing, and inference for generalized predictions
pwcompare	pairwise comparisons of estimates
suest	seemingly unrelated estimation
test	Wald tests of simple and composite linear hypotheses
testnl	Wald tests of nonlinear hypotheses

[1] `lrtest` is not appropriate with `svy` estimation results.

See the corresponding entries in the *Base Reference Manual* for details, but see [SVY] **estat** for details about estat (svy).

Syntax for predict

> predict [*type*] *newvar* [*if*] [*in*] [, *statistic* <u>nooff</u>set]

statistic	Description
Main	
<u>pr</u>	probability of a positive outcome; the default
xb	linear prediction
stdp	standard error of the linear prediction
<u>sc</u>ore	first derivative of the log likelihood with respect to $\mathbf{x}_j\beta$

These statistics are available both in and out of sample; type `predict ... if e(sample) ...` if wanted only for the estimation sample.

Menu

Statistics > Postestimation > Predictions, residuals, etc.

Options for predict

    [_Main_]

pr, the default, calculates the probability of a positive outcome.

xb calculates the linear prediction.

stdp calculates the standard error of the linear prediction.

score calculates the equation-level score, $\partial \ln L / \partial (\mathbf{x}_j \boldsymbol{\beta})$.

nooffset is relevant only if you specified offset(*varname*) for cloglog. It modifies the calculations made by predict so that they ignore the offset variable; the linear prediction is treated as $\mathbf{x}_j \mathbf{b}$ rather than as $\mathbf{x}_j \mathbf{b} + \text{offset}_j$.

Remarks

Once you have fit a model, you can obtain the predicted probabilities by using the predict command for both the estimation sample and other samples; see [U] **20 Estimation and postestimation commands** and [R] **predict**. Here we will make only a few comments.

predict without arguments calculates the predicted probability of a positive outcome. With the xb option, it calculates the linear combination $\mathbf{x}_j \mathbf{b}$, where $\mathbf{x}_j$ are the independent variables in the jth observation and $\mathbf{b}$ is the estimated parameter vector.

With the stdp option, predict calculates the standard error of the linear prediction, which is not adjusted for replicated covariate patterns in the data.

▷ Example 1

In example 1 in [R] **cloglog**, we fit the complementary log-log model cloglog foreign weight mpg. To obtain predicted probabilities,

```
. use http://www.stata-press.com/data/r12/auto
(1978 Automobile Data)

. cloglog foreign weight mpg
 (output omitted )
. predict p
(option pr assumed; Pr(foreign))

. summarize foreign p
```

Variable	Obs	Mean	Std. Dev.	Min	Max
foreign	74	.2972973	.4601885	0	1
p	74	.2928348	.29732	.0032726	.9446067

◁

Methods and formulas

All postestimation commands listed above are implemented as ado-files.

Also see

[R] **cloglog** — Complementary log-log regression

[U] **20 Estimation and postestimation commands**

Title

> **cnsreg** — Constrained linear regression

Syntax

> cnsreg *depvar indepvars* [*if*] [*in*] [*weight*] , constraints(*constraints*) [*options*]

options	Description
Model	
* constraints(*constraints*)	apply specified linear constraints
collinear	keep collinear variables
noconstant	suppress constant term
SE/Robust	
vce(*vcetype*)	*vcetype* may be ols, robust, cluster *clustvar*, bootstrap, or jackknife
Reporting	
level(#)	set confidence level; default is level(95)
nocnsreport	do not display constraints
display_options	control column formats, row spacing, line width, and display of omitted variables and base and empty cells
mse1	force MSE to be 1
coeflegend	display legend instead of statistics

* constraints(*constraints*) is required.

indepvars may contain factor variables; see [U] **11.4.3 Factor variables**.

depvar and *indepvars* may contain time-series operators; see [U] **11.4.4 Time-series varlists**.

bootstrap, by, jackknife, mi estimate, rolling, statsby, and svy are allowed; see

  [U] **11.1.10 Prefix commands**.

vce(bootstrap) and vce(jackknife) are not allowed with the mi estimate prefix; see [MI] **mi estimate**.

Weights are not allowed with the bootstrap prefix; see [R] **bootstrap**.

aweights are not allowed with the jackknife prefix; see [R] **jackknife**.

vce(), mse1, and weights are not allowed with the svy prefix; see [SVY] **svy**.

aweights, fweights, pweights, and iweights are allowed; see [U] **11.1.6 weight**.

mse1 and coeflegend do not appear in the dialog.

See [U] **20 Estimation and postestimation commands** for more capabilities of estimation commands.

Menu

Statistics > Linear models and related > Constrained linear regression

Description

cnsreg fits constrained linear regression models.

Options

constraints(*constraints*), collinear, noconstant; see [R] **estimation options**.

vce(*vcetype*) specifies the type of standard error reported, which includes types that are derived from asymptotic theory, that are robust to some kinds of misspecification, that allow for intragroup correlation, and that use bootstrap or jackknife methods; see [R] *vce_option*.

vce(ols), the default, uses the standard variance estimator for ordinary least-squares regression.

level(*#*); see [R] **estimation options**.

nocnsreport; see [R] **estimation options**.

display_options: noomitted, vsquish, noemptycells, baselevels, allbaselevels, cformat(*% fmt*), pformat(*% fmt*), sformat(*% fmt*), and nolstretch; see [R] **estimation options**.

The following options are available with cnsreg but are not shown in the dialog box:

mse1 is used only in programs and ado-files that use cnsreg to fit models other than constrained linear regression. mse1 sets the mean squared error to 1, thus forcing the variance–covariance matrix of the estimators to be $(\mathbf{X'DX})^{-1}$ (see *Methods and formulas* in [R] **regress**) and affecting calculated standard errors. Degrees of freedom for t statistics are calculated as n rather than $n - p + c$, where p is the total number of parameters (prior to restrictions and including the constant) and c is the number of constraints.

mse1 is not allowed with the svy prefix.

coeflegend; see [R] **estimation options**.

Remarks

For a discussion of constrained linear regression, see Greene (2012, 121–122); Hill, Griffiths, and Lim (2011, 231–233); or Davidson and MacKinnon (1993, 17).

▷ Example 1

In principle, we can obtain constrained linear regression estimates by modifying the list of independent variables. For instance, if we wanted to fit the model

$$mpg = \beta_0 + \beta_1 \, price + \beta_2 \, weight + u$$

and constrain $\beta_1 = \beta_2$, we could write

$$mpg = \beta_0 + \beta_1 (price + weight) + u$$

and run a regression of mpg on price + weight. The estimated coefficient on the sum would be the constrained estimate of β_1 and β_2. Using cnsreg, however, is easier:

```
. use http://www.stata-press.com/data/r12/auto
(1978 Automobile Data)

. constraint 1 price = weight

. cnsreg mpg price weight, constraint(1)
```

```
Constrained linear regression                    Number of obs =       74
                                                  F(  1,    72) =    37.59
                                                  Prob > F      =   0.0000
                                                  Root MSE      =    4.722

 ( 1)  price - weight = 0
```

mpg	Coef.	Std. Err.	t	P>|t|	[95% Conf. Interval]	
price	-.0009875	.0001611	-6.13	0.000	-.0013086	-.0006664
weight	-.0009875	.0001611	-6.13	0.000	-.0013086	-.0006664
_cons	30.36718	1.577958	19.24	0.000	27.22158	33.51278

We define constraints by using the constraint command; see [R] **constraint**. We fit the model with cnsreg and specify the constraint number or numbers in the constraints() option.

Just to show that the results above are correct, here is the result of applying the constraint by hand:

```
. generate x = price + weight

. regress mpg x
```

Source	SS	df	MS		Number of obs = 74
Model	838.065767	1	838.065767		F(1, 72) = 37.59
Residual	1605.39369	72	22.2971346		Prob > F = 0.0000
					R-squared = 0.3430
					Adj R-squared = 0.3339
Total	2443.45946	73	33.4720474		Root MSE = 4.722

mpg	Coef.	Std. Err.	t	P>|t|	[95% Conf. Interval]	
x	-.0009875	.0001611	-6.13	0.000	-.0013086	-.0006664
_cons	30.36718	1.577958	19.24	0.000	27.22158	33.51278

◁

▷ Example 2

Models can be fit subject to multiple simultaneous constraints. We simply define the constraints and then include the constraint numbers in the constraints() option. For instance, say that we wish to fit the model

$$\text{mpg} = \beta_0 + \beta_1 \text{price} + \beta_2 \text{weight} + \beta_3 \text{displ} + \beta_4 \text{gear\_ratio} + \beta_5 \text{foreign} +$$
$$\beta_6 \text{length} + u$$

subject to the constraints

$$\beta_1 = \beta_2 = \beta_3 = \beta_6$$
$$\beta_4 = -\beta_5 = \beta_0/20$$

(This model, like the one in example 1, is admittedly senseless.) We fit the model by typing

```
. constraint 1 price=weight

. constraint 2 displ=weight

. constraint 3 length=weight
```

```
. constraint 5 gear_ratio = -foreign
. constraint 6 gear_ratio = _cons/20
. cnsreg mpg price weight displ gear_ratio foreign length, c(1-3,5-6)
```

```
Constrained linear regression          Number of obs =      74
                                        F(  2,    72) =  785.20
                                        Prob > F      =  0.0000
                                        Root MSE      =  4.6823

( 1)  price - weight = 0
( 2)  - weight + displacement = 0
( 3)  - weight + length = 0
( 4)  gear_ratio + foreign = 0
( 5)  gear_ratio - .05 _cons = 0
```

mpg	Coef.	Std. Err.	t	P>|t|	[95% Conf.	Interval]
price	-.000923	.0001534	-6.02	0.000	-.0012288	-.0006172
weight	-.000923	.0001534	-6.02	0.000	-.0012288	-.0006172
displacement	-.000923	.0001534	-6.02	0.000	-.0012288	-.0006172
gear_ratio	1.326114	.0687589	19.29	0.000	1.189046	1.463183
foreign	-1.326114	.0687589	-19.29	0.000	-1.463183	-1.189046
length	-.000923	.0001534	-6.02	0.000	-.0012288	-.0006172
_cons	26.52229	1.375178	19.29	0.000	23.78092	29.26365

There are many ways we could have specified the constraints() option (which we abbreviated c() above). We typed c(1-3,5-6), meaning that we want constraints 1 through 3 and 5 and 6; those numbers correspond to the constraints we defined. The only reason we did not use the number 4 was to emphasize that constraints do not have to be consecutively numbered. We typed c(1-3,5-6), but we could have typed c(1,2,3,5,6) or c(1-3,5,6) or c(1-2,3,5,6) or even c(1-6), which would have worked as long as constraint 4 was not defined. If we had previously defined a constraint 4, then c(1-6) would have included it.

◁

Saved results

cnsreg saves the following in e():

Scalars

e(N)	number of observations
e(df_m)	model degrees of freedom
e(df_r)	residual degrees of freedom
e(F)	F statistic
e(rmse)	root mean squared error
e(ll)	log likelihood
e(N_clust)	number of clusters
e(rank)	rank of e(V)

Macros

e(cmd)	cnsreg
e(cmdline)	command as typed
e(depvar)	name of dependent variable
e(wtype)	weight type
e(wexp)	weight expression
e(title)	title in estimation output
e(clustvar)	name of cluster variable
e(vce)	*vcetype* specified in vce()
e(vcetype)	title used to label Std. Err.
e(properties)	b V
e(predict)	program used to implement predict
e(asbalanced)	factor variables fvset as asbalanced
e(asobserved)	factor variables fvset as asobserved

Matrices

e(b)	coefficient vector
e(Cns)	constraints matrix
e(V)	variance–covariance matrix of the estimators
e(V_modelbased)	model-based variance

Functions

e(sample)	marks estimation sample

Methods and formulas

cnsreg is implemented as an ado-file.

Let n be the number of observations, p be the total number of parameters (prior to restrictions and including the constant), and c be the number of constraints. The coefficients are calculated as $\mathbf{b}' = \mathbf{T}\{(\mathbf{T}'\mathbf{X}'\mathbf{W}\mathbf{X}\mathbf{T})^{-1}(\mathbf{T}'\mathbf{X}'\mathbf{W}\mathbf{y} - \mathbf{T}'\mathbf{X}'\mathbf{W}\mathbf{X}\mathbf{a}')\} + \mathbf{a}'$, where $\mathbf{T}$ and $\mathbf{a}$ are as defined in [P] **makecns**. $\mathbf{W} = \mathbf{I}$ if no weights are specified. If weights are specified, let $\mathbf{v}: 1 \times n$ be the specified weights. If fweight frequency weights are specified, $\mathbf{W} = \text{diag}(\mathbf{v})$. If aweight analytic weights are specified, then $\mathbf{W} = \text{diag}[\mathbf{v}/(\mathbf{1}'\mathbf{v})(\mathbf{1}'\mathbf{1})]$, meaning that the weights are normalized to sum to the number of observations.

The mean squared error is $s^2 = (\mathbf{y}'\mathbf{W}\mathbf{y} - 2\mathbf{b}'\mathbf{X}'\mathbf{W}\mathbf{y} + \mathbf{b}'\mathbf{X}'\mathbf{W}\mathbf{X}\mathbf{b})/(n-p+c)$. The variance–covariance matrix is $s^2\mathbf{T}(\mathbf{T}'\mathbf{X}'\mathbf{W}\mathbf{X}\mathbf{T})^{-1}\mathbf{T}'$.

This command supports the Huber/White/sandwich estimator of the variance and its clustered version using vce(robust) and vce(cluster *clustvar*), respectively. See [P] **_robust**, particularly *Introduction* and *Methods and formulas*.

cnsreg also supports estimation with survey data. For details on VCEs with survey data, see [SVY] **variance estimation**.

References

Davidson, R., and J. G. MacKinnon. 1993. *Estimation and Inference in Econometrics*. New York: Oxford University Press.

Greene, W. H. 2012. *Econometric Analysis*. 7th ed. Upper Saddle River, NJ: Prentice Hall.

Hill, R. C., W. E. Griffiths, and G. C. Lim. 2011. *Principles of Econometrics*. 4th ed. Hoboken, NJ: Wiley.

Also see

[R] **cnsreg postestimation** — Postestimation tools for cnsreg

[R] **regress** — Linear regression

[MI] **estimation** — Estimation commands for use with mi estimate

[SVY] **svy estimation** — Estimation commands for survey data

[U] **20 Estimation and postestimation commands**

Title

> **cnsreg postestimation** — Postestimation tools for cnsreg

Description

The following postestimation commands are available after `cnsreg`:

Command	Description
contrast	contrasts and ANOVA-style joint tests of estimates
estat	AIC, BIC, VCE, and estimation sample summary
estat (svy)	postestimation statistics for survey data
estimates	cataloging estimation results
lincom	point estimates, standard errors, testing, and inference for linear combinations of coefficients
linktest	link test for model specification
lrtest[1]	likelihood-ratio test
margins	marginal means, predictive margins, marginal effects, and average marginal effects
marginsplot	graph the results from margins (profile plots, interaction plots, etc.)
nlcom	point estimates, standard errors, testing, and inference for nonlinear combinations of coefficients
predict	predictions, residuals, influence statistics, and other diagnostic measures
predictnl	point estimates, standard errors, testing, and inference for generalized predictions
pwcompare	pairwise comparisons of estimates
suest	seemingly unrelated estimation
test	Wald tests of simple and composite linear hypotheses
testnl	Wald tests of nonlinear hypotheses

[1] `lrtest` is not appropriate with svy estimation results.

See the corresponding entries in the *Base Reference Manual* for details, but see [SVY] **estat** for details about `estat` (svy).

Syntax for predict

> predict [*type*] *newvar* [*if*] [*in*] [, *statistic*]

statistic	Description
Main	
xb	linear prediction; the default
residuals	residuals
stdp	standard error of the prediction
stdf	standard error of the forecast
pr(*a*,*b*)	$\Pr(a < y_j < b)$
e(*a*,*b*)	$E(y_j \mid a < y_j < b)$
ystar(*a*,*b*)	$E(y_j^*)$, $y_j^* = \max\{a, \min(y_j, b)\}$
score	equivalent to residuals

308

These statistics are available both in and out of sample; type predict ... if e(sample) ... if wanted only for the estimation sample.

stdf is not allowed with svy estimation results.

where a and b may be numbers or variables; a missing ($a \geq .$) means $-\infty$, and b missing ($b \geq .$) means $+\infty$; see [U] **12.2.1 Missing values**.

Menu

Statistics > Postestimation > Predictions, residuals, etc.

Options for predict

⌐ Main ⌐

xb, the default, calculates the linear prediction.

residuals calculates the residuals, that is, $y_j - \mathbf{x}_j \mathbf{b}$.

stdp calculates the standard error of the prediction, which can be thought of as the standard error of the predicted expected value or mean for the observation's covariate pattern. The standard error of the prediction is also referred to as the standard error of the fitted value.

stdf calculates the standard error of the forecast, which is the standard error of the point prediction for 1 observation. It is commonly referred to as the standard error of the future or forecast value. By construction, the standard errors produced by stdf are always larger than those produced by stdp; see *Methods and formulas* in [R] **regress postestimation**.

pr(a,b) calculates $\Pr(a < \mathbf{x}_j \mathbf{b} + u_j < b)$, the probability that $y_j | \mathbf{x}_j$ would be observed in the interval (a, b).

a and b may be specified as numbers or variable names; *lb* and *ub* are variable names;
pr(20,30) calculates $\Pr(20 < \mathbf{x}_j \mathbf{b} + u_j < 30)$;
pr(*lb*,*ub*) calculates $\Pr(lb < \mathbf{x}_j \mathbf{b} + u_j < ub)$; and
pr(20,*ub*) calculates $\Pr(20 < \mathbf{x}_j \mathbf{b} + u_j < ub)$.

a missing ($a \geq .$) means $-\infty$; pr(.,30) calculates $\Pr(-\infty < \mathbf{x}_j \mathbf{b} + u_j < 30)$;
pr(*lb*,30) calculates $\Pr(-\infty < \mathbf{x}_j \mathbf{b} + u_j < 30)$ in observations for which $lb \geq .$
and calculates $\Pr(lb < \mathbf{x}_j \mathbf{b} + u_j < 30)$ elsewhere.

b missing ($b \geq .$) means $+\infty$; pr(20,.) calculates $\Pr(+\infty > \mathbf{x}_j \mathbf{b} + u_j > 20)$;
pr(20,*ub*) calculates $\Pr(+\infty > \mathbf{x}_j \mathbf{b} + u_j > 20)$ in observations for which $ub \geq .$
and calculates $\Pr(20 < \mathbf{x}_j \mathbf{b} + u_j < ub)$ elsewhere.

e(a,b) calculates $E(\mathbf{x}_j \mathbf{b} + u_j \,|\, a < \mathbf{x}_j \mathbf{b} + u_j < b)$, the expected value of $y_j | \mathbf{x}_j$ conditional on $y_j | \mathbf{x}_j$ being in the interval (a, b), meaning that $y_j | \mathbf{x}_j$ is truncated. a and b are specified as they are for pr().

ystar(a,b) calculates $E(y_j^*)$, where $y_j^* = a$ if $\mathbf{x}_j \mathbf{b} + u_j \leq a$, $y_j^* = b$ if $\mathbf{x}_j \mathbf{b} + u_j \geq b$, and $y_j^* = \mathbf{x}_j \mathbf{b} + u_j$ otherwise, meaning that y_j^* is censored. a and b are specified as they are for pr().

score is equivalent to residuals for linear regression models.

Methods and formulas

All postestimation commands listed above are implemented as ado-files.

Also see

[R] **cnsreg** — Constrained linear regression

[U] **20 Estimation and postestimation commands**

Title

constraint — Define and list constraints

Syntax

Define constraints

<u>cons</u>traint [<u>def</u>ine] # [*exp=exp* | *coeflist*]

List constraints

<u>cons</u>traint <u>d</u>ir [*numlist* | _all]

<u>cons</u>traint <u>l</u>ist [*numlist* | _all]

Drop constraints

<u>cons</u>traint drop { *numlist* | _all }

Programmer's commands

<u>cons</u>traint get #

<u>cons</u>traint free

where *coeflist* is as defined in [R] **test** and # is restricted to the range 1–1,999, inclusive.

Menu

Statistics > Other > Manage constraints

Description

constraint defines, lists, and drops linear constraints. Constraints are for use by models that allow constrained estimation.

Constraints are defined by the constraint command. The currently defined constraints can be listed by either constraint list or constraint dir; both do the same thing. Existing constraints can be eliminated by constraint drop.

constraint get and constraint free are programmer's commands. constraint get returns the contents of the specified constraint in macro r(contents) and returns in scalar r(defined) 0 or 1—1 being returned if the constraint was defined. constraint free returns the number of a free (unused) constraint in macro r(free).

Remarks

Using constraints is discussed in [R] **cnsreg**, [R] **mlogit**, and [R] **reg3**; this entry is concerned only with practical aspects of defining and manipulating constraints.

▷ Example 1

Constraints are numbered from 1 to 1,999, and we assign the number when we define the constraint:

```
. use http://www.stata-press.com/data/r12/sysdsn1
(Health insurance data)
. constraint 2 [Indemnity]2.site = 0
```

The currently defined constraints can be listed by `constraint list`:

```
. constraint list
      2:   [indemnity]2.site = 0
```

`constraint drop` drops constraints:

```
. constraint drop 2
. constraint list
```

The empty list after `constraint list` indicates that no constraints are defined. Below we demonstrate the various syntaxes allowed by `constraint`:

```
. constraint 1 [Indemnity]
. constraint 10 [Indemnity]: 1.site 2.site
. constraint 11 [Indemnity]: 3.site
. constraint 21 [Prepaid=Uninsure]: nonwhite
. constraint 30 [Prepaid]
. constraint 31 [Insure]
. constraint list
      1:   [Indemnity]
     10:   [Indemnity]: 1.site 2.site
     11:   [Indemnity]: 3.site
     21:   [Prepaid=Uninsure]: nonwhite
     30:   [Prepaid]
     31:   [Insure]
. constraint drop 21-25, 31
. constraint list
      1:   [Indemnity]
     10:   [Indemnity]: 1.site 2.site
     11:   [Indemnity]: 3.site
     30:   [Prepaid]
. constraint drop _all
. constraint list
```

◁

❏ Technical note

The `constraint` command does not check the syntax of the constraint itself because a constraint can be interpreted only in the context of a model. Thus `constraint` is willing to define constraints that later will not make sense. Any errors in the constraints will be detected and mentioned at the time of estimation.

❏

Reference

Weesie, J. 1999. sg100: Two-stage linear constrained estimation. *Stata Technical Bulletin* 47: 24–30. Reprinted in *Stata Technical Bulletin Reprints*, vol. 8, pp. 217–225. College Station, TX: Stata Press.

Also see

[R] **cnsreg** — Constrained linear regression

Title

> **contrast** — Contrasts and linear hypothesis tests after estimation

Syntax

> contrast *termlist* [, *options*]

where *termlist* is a list of factor variables or interactions that appear in the current estimation results. The variables may be typed with or without contrast operators, and you may use any factor-variable syntax:

> . contrast sex group sex#group

> . contrast r.sex

See the *operators (op.)* table below for the list of contrast operators.

options	Description
Main	
<u>over</u>all	add a joint hypothesis test for all specified contrasts
<u>asobs</u>erved	treat all factor variables as observed
lincom	treat user-defined contrasts as linear combinations
Equations	
equation(*eqspec*)	perform contrasts in *termlist* for equation *eqspec*
atequations	perform contrasts in *termlist* within each equation
Advanced	
emptycells(*empspec*)	treatment of empty cells for balanced factors
noestimcheck	suppress estimability checks
Reporting	
<u>l</u>evel(#)	confidence level; default is level(95)
mcompare(*method*)	adjust for multiple comparisons; default is mcompare(noadjust)
<u>noeff</u>ects	suppress table of individual contrasts
<u>ci</u>effects	show effects table with confidence intervals
<u>pv</u>effects	show effects table with *p*-values
<u>eff</u>ects	show effects table with confidence intervals and *p*-values
nowald	suppress table of Wald tests
<u>noatl</u>evels	report only the overall Wald test for terms that use the within @ or nested | operator
nosvyadjust	compute unadjusted Wald tests for survey results
sort	sort the individual contrast values in each term
post	post contrasts and their VCEs as estimation results
display_options	control column formats, row spacing, and line width
eform_option	report exponentiated contrasts

Term	Description
Main effects	
A	joint test of the main effects of A
r.A	individual contrasts that decompose A using r.
Interaction effects	
A#B	joint test of the two-way interaction effects of A and B
A#B#C	joint test of the three-way interaction effects of A, B, and C
r.A#g.B	individual contrasts for each interaction of A and B defined by r. and g.
Partial interaction effects	
r.A#B	joint tests of interactions of A and B within each contrast defined by r.A
A#r.B	joint tests of interactions of A and B within each contrast defined by r.B
Simple effects	
A@B	joint tests of the effects of A within each level of B
A@B#C	joint tests of the effects of A within each combination of the levels of B and C
r.A@B	individual contrasts of A that decompose A@B using r.
r.A@B#C	individual contrasts of A that decompose A@B#C using r.
Other conditional effects	
A#B@C	joint tests of the interaction effects of A and B within each level of C
A#B@C#D	joint tests of the interaction effects of A and B within each combination of the levels of C and D
r.A#g.B@C	individual contrasts for each interaction of A and B that decompose A#B@C using r. and g.
Nested effects	
A|B	joint tests of the effects of A nested in each level of B
A|B#C	joint tests of the effects of A nested in each combination of the levels of B and C
A#B|C	joint tests of the interaction effects of A and B nested in each level of C
A#B|C#D	joint tests of the interaction effects of A and B nested in each combination of the levels of C and D
r.A|B	individual contrasts of A that decompose A|B using r.
r.A|B#C	individual contrasts of A that decompose A|B#C using r.
r.A#g.B|C	individual contrasts for each interaction of A and B defined by r. and g. nested in each level of C
Slope effects	
A#c.x	joint test of the effects of A on the slopes of x
A#c.x#c.y	joint test of the effects of A on the slopes of the product (interaction) of x and y
A#B#c.x	joint test of the interaction effects of A and B on the slopes of x
A#B#c.x#c.y	joint test of the interaction effects of A and B on the slopes of the product (interaction) of x and y
r.A#c.x	individual contrasts of A's effects on the slopes of x using r.
Denominators	
... / *term2*	use *term2* as the denominator in the F tests of the preceding terms
... /	use the residual as the denominator in the F tests of the preceding terms (the default if no other /s are specified)

A, B, C, and D represent any factor variable in the current estimation results.

x and y represent any continuous variable in the current estimation results.

r. and *g.* represent any contrast operator. See the table below.

c. specifies that a variable be treated as continuous; see [U] **11.4.3 Factor variables**.

Operators are allowed on any factor variable that does not appear to the right of @ or |. Operators decompose the effects of the associated factor variable into one-degree-of-freedom effects (contrasts).

Higher-level interactions are allowed anywhere an interaction operator (#) appears in the table.

Time-series operators are allowed if they were used in the estimation.

_eqns designates the equations in manova, mlogit, mprobit, and mvreg and can be specified anywhere a factor variable appears.

/ is allowed only after anova, cnsreg, manova, mvreg, or regress.

operators (op.)	Description
r.	differences from a reference (base) level; the default
a.	differences from the next level (adjacent contrasts)
ar.	differences from the previous level (reverse adjacent contrasts)
As-balanced operators	
g.	differences from the balanced grand mean
h.	differences from the balanced mean of subsequent levels (Helmert contrasts)
j.	differences from the balanced mean of previous levels (reverse Helmert contrasts)
p.	orthogonal polynomial in the level values
q.	orthogonal polynomial in the level sequence
As-observed operators	
gw.	differences from the observation-weighted grand mean
hw.	differences from the observation-weighted mean of subsequent levels
jw.	differences from the observation-weighted mean of previous levels
pw.	observation-weighted orthogonal polynomial in the level values
qw.	observation-weighted orthogonal polynomial in the level sequence

One or more individual contrasts may be selected by using the *op#.* or *op*(*numlist*). syntax. For example, a3.A selects the adjacent contrast for level 3 of A, and p(1/2).B selects the linear and quadratic effects of B. Also see *Orthogonal polynomial contrasts* and *Beyond linear models*.

Custom contrasts	Description
{A *numlist*}	user-defined contrast on the levels of factor A
{A#B *numlist*}	user-defined contrast on the levels of the interaction between A and B

Custom contrasts may be part of a term, such as {A *numlist*}#B, {A *numlist*}@B, {A *numlist*}|B, {A#B *numlist*}, and {A *numlist*}#{B *numlist*}. The same is true of higher-order custom contrasts, such as {A#B *numlist*}@C, {A#B *numlist*}#*r*.C, and {A#B *numlist*}#c.x.

Higher-order interactions with at most eight factor variables are allowed with custom contrasts.

method	Description
<u>no</u>adjust	do not adjust for multiple comparisons; the default
<u>bonf</u>erroni [adjustall]	Bonferroni's method; adjust across all terms
<u>sid</u>ak [adjustall]	Šidák's method; adjust across all terms
<u>sch</u>effe	Scheffé's method

Menu

Statistics > Postestimation > Contrasts

Description

contrast tests linear hypotheses and forms contrasts involving factor variables and their interactions from the most recently fit model. The tests include ANOVA-style tests of main effects, simple effects, interactions, and nested effects. contrast can use named contrasts to decompose these effects into comparisons against reference categories, comparisons of adjacent levels, comparisons against the grand mean, orthogonal polynomials, and such. Custom contrasts may also be specified.

contrast can be used with svy estimation results; see [SVY] **svy postestimation**.

Contrasts can also be computed for margins of linear and nonlinear responses; see [R] **margins, contrast**.

Options

    ⌐ Main ⌐

overall specifies that a joint hypothesis test over all terms be performed.

asobserved specifies that factor covariates be evaluated using the cell frequencies observed in the estimation sample. The default is to treat all factor covariates as though there were an equal number of observations in each level.

lincom specifies that user-defined contrasts be treated as linear combinations. The default is to require that all user-defined contrasts sum to zero. (Summing to zero is part of the definition of a contrast.)

    ⌐ Equations ⌐

equation(*eqspec*) specifies the equation from which contrasts are to be computed. The default is to compute contrasts from the first equation.

atequations specifies that the contrasts be computed within each equation.

    ⌐ Advanced ⌐

emptycells(*empspec*) specifies how empty cells are handled in interactions involving factor variables that are being treated as balanced.

    emptycells(strict) is the default; it specifies that contrasts involving empty cells be treated as not estimable.

    emptycells(reweight) specifies that the effects of the observed cells be increased to accommodate any missing cells. This makes the contrast estimable but changes its interpretation.

noestimcheck specifies that contrast not check for estimability. By default, the requested contrasts are checked and those found not estimable are reported as such. Nonestimability is usually caused by empty cells. If noestimcheck is specified, estimates are computed in the usual way and reported even though the resulting estimates are manipulable, which is to say they can differ across equivalent models having different parameterizations.

⎤ Reporting ⎣

level(#) specifies the confidence level, as a percentage, for confidence intervals. The default is level(95) or as set by set level; see [U] **20.7 Specifying the width of confidence intervals**.

mcompare(*method*) specifies the method for computing p-values and confidence intervals that account for multiple comparisons within a factor-variable term.

Most methods adjust the comparisonwise error rate, α_c, to achieve a prespecified experimentwise error rate, α_e.

mcompare(noadjust) is the default; it specifies no adjustment.

$$\alpha_c = \alpha_e$$

mcompare(bonferroni) adjusts the comparisonwise error rate based on the upper limit of the Bonferroni inequality

$$\alpha_e \leq m\alpha_c$$

where m is the number of comparisons within the term.

The adjusted comparisonwise error rate is

$$\alpha_c = \alpha_e/m$$

mcompare(sidak) adjusts the comparisonwise error rate based on the upper limit of the probability inequality

$$\alpha_e \leq 1 - (1 - \alpha_c)^m$$

where m is the number of comparisons within the term.

The adjusted comparisonwise error rate is

$$\alpha_c = 1 - (1 - \alpha_e)^{1/m}$$

This adjustment is exact when the m comparisons are independent.

mcompare(scheffe) controls the experimentwise error rate using the F or χ^2 distribution with degrees of freedom equal to the rank of the term.

mcompare(*method* adjustall) specifies that the multiple-comparison adjustments count all comparisons across all terms rather than performing multiple comparisons term by term. This leads to more conservative adjustments when multiple variables or terms are specified in *marginslist*. This option is compatible only with the bonferroni and sidak methods.

noeffects suppresses the table of individual contrasts with confidence intervals. This table is produced by default when the mcompare() option is specified or when a term in *termlist* implies all individual contrasts.

cieffects specifies that a table containing a confidence interval for each individual contrast be reported.

pveffects specifies that a table containing a p-value for each individual contrast be reported.

effects specifies that a single table containing a confidence interval and p-value for each individual contrast be reported.

`nowald` suppresses the table of Wald tests.

`noatlevels` indicates that only the overall Wald test be reported for each term containing within or nested (@ or |) operators.

`nosvyadjust` is for use with `svy` estimation commands. It specifies that the Wald test be carried out without the default adjustment for the design degrees of freedom. That is to say the test is carried out as $W/k \sim F(k, d)$ rather than as $(d - k + 1)W/(kd) \sim F(k, d - k + 1)$, where k is the dimension of the test and d is the total number of sampled PSUs minus the total number of strata.

`sort` specifies that the table of individual contrasts be sorted by the contrast values within each term.

`post` causes `contrast` to behave like a Stata estimation (e-class) command. `contrast` posts the vector of estimated contrasts along with the estimated variance–covariance matrix to `e()`, so you can treat the estimated contrasts just as you would results from any other estimation command. For example, you could use `test` to perform simultaneous tests of hypotheses on the contrasts, or you could use `lincom` to create linear combinations.

display_options: `vsquish`, `cformat(%fmt)`, `pformat(%fmt)`, `sformat(%fmt)`, and `nolstretch`.

`vsquish` specifies that the blank space separating factor-variable terms or time-series–operated variables from other variables in the model be suppressed.

`cformat(%fmt)` specifies how to format contrasts, standard errors, and confidence limits in the table of estimated contrasts.

`pformat(%fmt)` specifies how to format p-values in the table of estimated contrasts.

`sformat(%fmt)` specifies how to format test statistics in the table of estimated contrasts.

`nolstretch` specifies that the width of the table of estimated contrasts not be automatically widened to accommodate longer variable names. The default, `lstretch`, is to automatically widen the table of estimated contrasts up to the width of the Results window. To change the default, use `set lstretch off`. `nolstretch` is not shown in the dialog box.

eform_option specifies that the contrasts table be displayed in exponentiated form. e^{contrast} is displayed rather than contrast. Standard errors and confidence intervals are also transformed. See [R] **eform_option** for the list of available options.

Remarks

Remarks are presented under the following headings:

Introduction

contrast performs ANOVA-style tests of main effects, interactions, simple effects, and nested effects. It can easily decompose these tests into constituent contrasts using either named contrasts (codings) or user-specified contrasts. Comparing levels of factor variables—whether as main effects, interactions, or simple effects—is as easy as adding a contrast operator to the variable. The operators can compare each level with the previous level, each level with a reference level, each level with the mean of previous levels, and more.

contrast tests and estimates contrasts. A contrast of the parameters $\mu_1, \mu_2, \ldots, \mu_p$ is a linear combination $\sum_i c_i \mu_i$ whose c_i sum to zero. A difference of population means that $\mu_1 - \mu_2$ is a contrast, as are most other comparisons of population or model quantities (Coster 2005). Some contrasts may be estimated with lincom, but contrast is much more powerful. contrast can handle multiple contrasts simultaneously, and the command's contrast operators make it easy to specify complicated linear combinations.

Both the contrast operation and the creation of the margins for comparison can be performed as though the data were balanced (typical for experimental designs) or using the observed frequencies in the estimation sample (typical for observational studies). contrast can perform these analyses on the results of almost all of Stata's estimators, not just the linear-models estimators.

Most of `contrast`'s computations can be considered comparisons of estimated cell means from a model fit. Tests of interactions are tests of whether the cell means for the interaction are all equal. Tests of main effects are tests of whether the marginal cell means for the factor are all equal. More focused comparisons of cell means (for example, is level 2 equal to level 1) are specified using contrast operators. More formally, all of `contrast`'s computations are comparisons of conditional expectations; cell means are one type of conditional expectation.

All contrasts can also easily be graphed; see [R] **marginsplot**.

For a discussion of contrasts and testing for linear models, see Searle (1971) and Searle (1997). For discussions specifically related to experimental design, see Kuehl (2000), Winer, Brown, and Michels (1991), and Milliken and Johnson (2009). Rosenthal, Rosnow, and Rubin (2000) focus on contrasts with applications in behavioral sciences.

`contrast` is a flexible tool for understanding the effects of categorical covariates. If your model contains categorical covariates, and especially if it contains interactions, you will want to use `contrast`.

One-way models

Suppose we have collected data on cholesterol levels for individuals from five age groups. To study the effect of age group on cholesterol, we can begin by fitting a one-way model using `regress`:

```
. use http://www.stata-press.com/data/r12/cholesterol
(Artificial cholesterol data)

. label list ages
ages:
           1 10-19
           2 20-29
           3 30-39
           4 40-59
           5 60-79

. regress chol i.agegrp
```

Source	SS	df	MS		Number of obs =	75
					F(4, 70) =	35.02
Model	14943.3997	4	3735.84993		Prob > F =	0.0000
Residual	7468.21971	70	106.688853		R-squared =	0.6668
					Adj R-squared =	0.6477
Total	22411.6194	74	302.859722		Root MSE =	10.329

chol	Coef.	Std. Err.	t	P>|t|	[95% Conf. Interval]	
agegrp						
2	8.203575	3.771628	2.18	0.033	.6812991	15.72585
3	21.54105	3.771628	5.71	0.000	14.01878	29.06333
4	30.15067	3.771628	7.99	0.000	22.6284	37.67295
5	38.76221	3.771628	10.28	0.000	31.23993	46.28448
_cons	180.5198	2.666944	67.69	0.000	175.2007	185.8388

Estimated cell means

margins will show us the estimated cell means for each age group based on our fitted model:

```
. margins agegrp

Adjusted predictions                            Number of obs    =         75
Model VCE      : OLS

Expression     : Linear prediction, predict()
```

	Margin	Delta-method Std. Err.	z	P>|z|	[95% Conf. Interval]	
agegrp						
1	180.5198	2.666944	67.69	0.000	175.2926	185.7469
2	188.7233	2.666944	70.76	0.000	183.4962	193.9504
3	202.0608	2.666944	75.76	0.000	196.8337	207.2879
4	210.6704	2.666944	78.99	0.000	205.4433	215.8975
5	219.282	2.666944	82.22	0.000	214.0548	224.5091

We can graph those means with marginsplot:

```
. marginsplot
    Variables that uniquely identify margins: agegrp
```

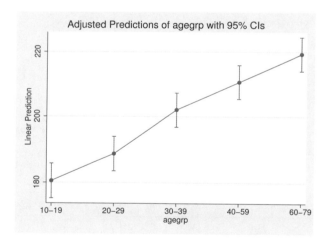

Testing equality of cell means

Are all the means equal? That is to say is there an effect of age group on cholesterol level? We can answer that by asking `contrast` to test whether the means of the age groups are identical.

```
. contrast agegrp
Contrasts of marginal linear predictions
Margins        : asbalanced
```

	df	F	P>F
agegrp	4	35.02	0.0000
Residual	70		

The means are clearly different. We could have obtained this same test directly had we fit our model using `anova` rather than `regress`.

```
. anova chol agegrp
```

	Number of obs = 75		R-squared = 0.6668		
	Root MSE = 10.329		Adj R-squared = 0.6477		
Source	Partial SS	df	MS	F	Prob > F
Model	14943.3997	4	3735.84993	35.02	0.0000
agegrp	14943.3997	4	3735.84993	35.02	0.0000
Residual	7468.21971	70	106.688853		
Total	22411.6194	74	302.859722		

Achieving a more direct test result is why we recommend using `anova` instead of `regress` for models where our focus is on the categorical covariates. The models fit by `anova` and `regress` are identical; they merely parameterize the effects differently. The results of `contrast` will be identical regardless of which command is used to fit the model. If, however, we were fitting models whose responses are nonlinear functions of the covariates, such as logistic regression, then there would be no analogue to `anova`, and we would appreciate `contrast`'s ability to quickly test main effects and interactions.

Reference category contrasts

Now that we know that the overall effect of age group is statistically significant, we can explore the effects of each age group. One way to do that is to use the reference category operator, r.:

```
. contrast r.agegrp
Contrasts of marginal linear predictions
Margins        : asbalanced
```

	df	F	P>F
agegrp			
(2 vs 1)	1	4.73	0.0330
(3 vs 1)	1	32.62	0.0000
(4 vs 1)	1	63.91	0.0000
(5 vs 1)	1	105.62	0.0000
Joint	4	35.02	0.0000
Residual	70		

	Contrast	Std. Err.	[95% Conf.	Interval]
agegrp				
(2 vs 1)	8.203575	3.771628	.6812991	15.72585
(3 vs 1)	21.54105	3.771628	14.01878	29.06333
(4 vs 1)	30.15067	3.771628	22.6284	37.67295
(5 vs 1)	38.76221	3.771628	31.23993	46.28448

The cell mean of each age group is compared against the base age group (group 1, ages 10–19). The first table shows that each difference is significant. The second table gives an estimate and confidence interval for each contrast. These are the comparisons that linear regression gives with a factor covariate and no interactions. The contrasts are identical to the coefficients from our linear regression.

Reverse adjacent contrasts

We have far more flexibility with contrast. Age group is ordinal, so it is interesting to compare each age group with the preceding age group (rather than against one reference group). We specify that analysis by using the reverse adjacent operator, ar.:

```
. contrast ar.agegrp
Contrasts of marginal linear predictions
Margins        : asbalanced
```

	df	F	P>F
agegrp			
(2 vs 1)	1	4.73	0.0330
(3 vs 2)	1	12.51	0.0007
(4 vs 3)	1	5.21	0.0255
(5 vs 4)	1	5.21	0.0255
Joint	4	35.02	0.0000
Residual	70		

	Contrast	Std. Err.	[95% Conf. Interval]	
agegrp				
(2 vs 1)	8.203575	3.771628	.6812991	15.72585
(3 vs 2)	13.33748	3.771628	5.815204	20.85976
(4 vs 3)	8.60962	3.771628	1.087345	16.1319
(5 vs 4)	8.611533	3.771628	1.089257	16.13381

Age group 2's cholesterol level is 8.2 points higher than age group 1's; age group 3's is 13.3 points higher than age group 2's; and so on. Each age group is statistically different from the preceding age group at the 5% level.

Orthogonal polynomial contrasts

The relationship between age group and cholesterol level looked almost linear in our graph. We can examine that relationship further by using the orthogonal polynomial operator, p.:

```
. contrast p.agegrp, noeffects
Contrasts of marginal linear predictions
Margins        : asbalanced
```

	df	F	P>F
agegrp			
(linear)	1	139.11	0.0000
(quadratic)	1	0.15	0.6962
(cubic)	1	0.37	0.5448
(quartic)	1	0.43	0.5153
Joint	4	35.02	0.0000
Residual	70		

Only the linear effect is statistically significant.

We can even perform the joint test that all effects beyond linear are zero. We do that by selecting all polynomial contrasts above linear—that is, polynomial contrasts 2, 3, and 4.

```
. contrast p(2 3 4).agegrp, noeffects
Contrasts of marginal linear predictions
Margins        : asbalanced
```

	df	F	P>F
agegrp			
(quadratic)	1	0.15	0.6962
(cubic)	1	0.37	0.5448
(quartic)	1	0.43	0.5153
Joint	3	0.32	0.8129
Residual	70		

The joint test has three degrees of freedom and is clearly insignificant. A linear effect of age group seems adequate for this model.

Two-way models

Suppose we are investigating the effects of different dosages of a blood pressure medication and believe that the effects may be different for men and women. We can fit the following ANOVA model for bpchange, the change in diastolic blood pressure. Change is defined as the after measurement minus the before measurement, so that negative values of bpchange correspond to decreases in blood pressure.

```
. use http://www.stata-press.com/data/r12/bpchange
(Artificial blood pressure data)

. label list gender
gender:
           1 male
           2 female

. anova bpchange dose##gender
```

	Number of obs = 30	R-squared = 0.9647
	Root MSE = 1.4677	Adj R-squared = 0.9573

Source	Partial SS	df	MS	F	Prob > F
Model	1411.9087	5	282.381741	131.09	0.0000
dose	963.481795	2	481.740897	223.64	0.0000
gender	355.118817	1	355.118817	164.85	0.0000
dose#gender	93.3080926	2	46.6540463	21.66	0.0000
Residual	51.699253	24	2.15413554		
Total	1463.60796	29	50.4692399		

Estimated interaction cell means

Everything is significant, including the interaction. So increasing dosage is effective and differs by gender. Let's explore the effects. First, let's look at the estimated cell mean of blood pressure change for each combination of gender and dosage.

```
. margins dose#gender
Adjusted predictions                              Number of obs   =         30
Expression   : Linear prediction, predict()
```

	Margin	Delta-method Std. Err.	z	P>|z|	[95% Conf.	Interval]
dose#gender						
250 1	-7.35384	.6563742	-11.20	0.000	-8.64031	-6.06737
250 2	3.706567	.6563742	5.65	0.000	2.420097	4.993037
500 1	-13.73386	.6563742	-20.92	0.000	-15.02033	-12.44739
500 2	-6.584167	.6563742	-10.03	0.000	-7.870637	-5.297697
750 1	-16.82108	.6563742	-25.63	0.000	-18.10754	-15.53461
750 2	-14.38795	.6563742	-21.92	0.000	-15.67442	-13.10148

Our data are balanced, so these results will not be affected by the many different ways that margins can compute cell means. Moreover, because our model consists of only dose and gender, these are also the point estimates for each combination.

We can graph the results:

```
. marginsplot
Variables that uniquely identify margins: dose gender
```

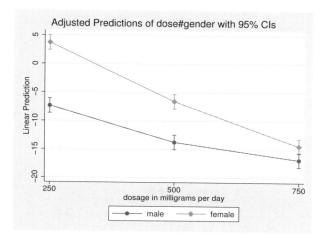

The lines are not parallel, which we expected because the interaction term is significant. Males experience a greater decline in blood pressure at every dosage level, but the effect of increasing dosage is greater for females. In fact, it is not clear if we can tell the difference between male and female response at the maximum dosage.

Simple effects

We can contrast the male and female responses within dosage to see the simple effects of gender. Because there are only two levels in gender, the choice of contrast operator is largely irrelevant. Aside from orthogonal polynomials, all operators produce the same estimates, although the effects can change signs.

```
. contrast r.gender@dose
Contrasts of marginal linear predictions
Margins        : asbalanced
```

	df	F	P>F
gender@dose			
(2 vs 1) 250	1	141.97	0.0000
(2 vs 1) 500	1	59.33	0.0000
(2 vs 1) 750	1	6.87	0.0150
Joint	3	69.39	0.0000
Residual	24		

	Contrast	Std. Err.	[95% Conf. Interval]	
gender@dose				
(2 vs 1) 250	11.06041	.9282533	9.144586	12.97623
(2 vs 1) 500	7.149691	.9282533	5.23387	9.065512
(2 vs 1) 750	2.433124	.9282533	.5173031	4.348944

The effect for males is about 11 points higher than for females at a dosage of 250, and that shrinks to 2.4 points higher at the maximum dosage of 750.

We can form the simple effects the other way by contrasting the effect of dose at each level of gender:

```
. contrast ar.dose@gender

Contrasts of marginal linear predictions

Margins       : asbalanced
```

	df	F	P>F
dose@gender			
(500 vs 250) 1	1	47.24	0.0000
(500 vs 250) 2	1	122.90	0.0000
(750 vs 500) 1	1	11.06	0.0028
(750 vs 500) 2	1	70.68	0.0000
Joint	4	122.65	0.0000
Residual	24		

	Contrast	Std. Err.	[95% Conf. Interval]
dose@gender			
(500 vs 250) 1	-6.380018	.9282533	-8.295839 -4.464198
(500 vs 250) 2	-10.29073	.9282533	-12.20655 -8.374914
(750 vs 500) 1	-3.087217	.9282533	-5.003038 -1.171396
(750 vs 500) 2	-7.803784	.9282533	-9.719605 -5.887963

Here we use the `ar.` reverse adjacent contrast operator so that first we are comparing a dosage of 500 with a dosage of 250, and then we are comparing 750 with 500. We see that increasing the dosage has a larger effect on females—10.3 points when going from 250 to 500 compared with 6.4 points for males, and 7.8 points when going from 500 to 750 versus 3.1 points for males.

Interaction effects

By specifying contrast operators on both factors, we can decompose the interaction effect into separate interaction contrasts.

```
. contrast ar.dose#r.gender

Contrasts of marginal linear predictions

Margins       : asbalanced
```

	df	F	P>F
dose#gender			
(500 vs 250) (2 vs 1)	1	8.87	0.0065
(750 vs 500) (2 vs 1)	1	12.91	0.0015
Joint	2	21.66	0.0000
Residual	24		

	Contrast	Std. Err.	[95% Conf. Interval]	
dose#gender				
(500 vs 250) (2 vs 1)	-3.910716	1.312748	-6.620095	-1.201336
(750 vs 500) (2 vs 1)	-4.716567	1.312748	-7.425947	-2.007187

Look for departures from zero to indicate an interaction effect between dose and gender. Both contrasts are significantly different from zero. Of course, we already knew the overall interaction was significant from our ANOVA results. The effect of increasing dose from 250 to 500 is 3.9 points greater in females than in males, and the effect of increasing dose from 500 to 750 is 4.7 points greater in females than in males. The confidence intervals for both estimates easily exclude zero, meaning that there is an interaction effect.

The joint test of these two interaction effects reproduces the test of interaction effects in the anova output. We can see that the F statistic of 21.66 matches the statistic from our original ANOVA results.

Main effects

We can perform tests of the main effects by listing each variable individually in contrast.

```
. contrast dose gender
Contrasts of marginal linear predictions
Margins       : asbalanced
```

	df	F	P>F
dose	2	223.64	0.0000
gender	1	164.85	0.0000
Residual	24		

The F tests are equivalent to the tests of main effects in the anova output. This is true only for linear models. contrast provides an easy way to obtain main effects and other ANOVA-style tests for models whose responses are not linear in the parameters—logistic, probit, glm, etc.

If we include contrast operators on the variables, we can also decompose the main effects into individual contrasts:

```
. contrast ar.dose r.gender
Contrasts of marginal linear predictions
Margins       : asbalanced
```

	df	F	P>F
dose			
(500 vs 250)	1	161.27	0.0000
(750 vs 500)	1	68.83	0.0000
Joint	2	223.64	0.0000
gender	1	164.85	0.0000
Residual	24		

	Contrast	Std. Err.	[95% Conf.	Interval]
dose				
(500 vs 250)	-8.335376	.6563742	-9.690066	-6.980687
(750 vs 500)	-5.4455	.6563742	-6.80019	-4.090811
gender				
(2 vs 1)	6.881074	.5359273	5.774974	7.987173

By specifying the `ar.` operator on `dose`, we decompose the main effect for dose into two one-degree-of-freedom contrasts, comparing the marginal mean of blood pressure change for each dosage level with that of the previous level. Because `gender` has only two levels, we cannot decompose this main effect any further. However, specifying a contrast operator on `gender` allowed us to calculate the difference in the marginal means for women and men.

Partial interaction effects

At this point, we have looked at the total interaction effects and at the main effects of each variable. The partial interaction effects are a midpoint between these two types of effects where we collect the individual interaction effects along the levels of one of the variables and perform a joint test of those interactions. If we think of the interaction effects as forming a table, with the levels of one factor variable forming the rows and the levels of the other forming the columns, partial interaction effects are joint tests of the interactions in a row or a column. To perform these tests, we specify a contrast operator on only one of the variables in our interaction. For this particular model, these are not very interesting because our variables have only two and three levels. Therefore, the tests of the partial interaction effects reproduce the tests that we obtained for the total interaction effects. We specify a contrast operator only on `dose` to decompose the overall test for interaction effects into joint tests for each `ar.dose` contrast:

```
. contrast ar.dose#gender
Contrasts of marginal linear predictions
Margins      : asbalanced
```

	df	F	P>F
dose#gender			
(500 vs 250) (joint)	1	8.87	0.0065
(750 vs 500) (joint)	1	12.91	0.0015
Joint	2	21.66	0.0000
Residual	24		

The first row is a joint test of all the interaction effects involving the (500 vs 250) comparison of dosages. The second row is a joint test of all the interaction effects involving the (750 vs 500) comparison. If we look back at our output in *Interaction effects*, we can see that there was only one of each of these interaction effects. Therefore, each test labeled (`joint`) has only one degree-of-freedom.

We could have instead included a contrast operator on `gender` to compute the partial interaction effects along the other dimension:

```
. contrast dose#r.gender
Contrasts of marginal linear predictions
Margins      : asbalanced
```

	df	F	P>F
dose#gender	2	21.66	0.0000
Residual	24		

Here we obtain a joint test of all the interaction effects involving the (2 vs 1) comparison for gender. Because gender has only two levels, the (2 vs 1) contrast is the only reference category contrast possible. Therefore, we obtain a single joint test of all the interaction effects.

Clearly, the partial interaction effects are not interesting for this particular model. However, if our factors had more levels, the partial interaction effects would produce tests that are not available in the total interaction effects. For example, if our model included factors for four dosage levels and three races, then typing

```
. contrast ar.dose#race
```

would produce three joint tests, one for each of the reverse adjacent contrasts for dosage. Each of these tests would be a two-degree-of-freedom test because race has three levels.

Three-way and higher-order models

All the contrasts and tests that we reviewed above for two-way models can be used with models that have more terms. For instance, we could fit a three-way full factorial model by using the anova command:

```
. use http://www.stata-press.com/data/r12/cont3way
. anova y race##sex##group
```

We could then test the simple effects of race within each level of the interaction between sex and group:

```
. contrast race@sex#group
```

To see the reference category contrasts that decompose these simple effects, type

```
. contrast r.race@sex#group
```

We could test the three-way interaction effects by typing

```
. contrast race#sex#group
```

or the interaction effects for the interaction of race and sex by typing

```
. contrast race#sex
```

To see the individual reference category contrasts that decompose this interaction effect, type

```
. contrast r.race#r.sex
```

We could even obtain joint tests for the interaction of race and sex within each level of group by typing

    . contrast race#sex@group

For tests of the main effects of each factor, we can type

    . contrast race sex group

We can calculate the individual reference category contrasts that decompose these main effects:

    . contrast r.race r.sex r.group

For the partial interaction effects, we could type

    . contrast r.race#group

to obtain a joint test of the two-way interaction effects of race and group for each of the individual r.race contrasts.

We could type

    . contrast r.race#sex#group

to obtain a joint test of all the three-way interaction terms for each of the individual r.race contrasts.

Contrast operators

contrast recognizes a set of contrast operators that are used to specify commonly used contrasts. When these operators are used, contrast will report a test for each individual contrast in addition to the joint test for the term. We have already seen a few of these, like r. and ar., in the previous examples. Here we will take a closer look at each of the unweighted operators.

Here we use the cholesterol dataset and the one-way ANOVA model from the example in *One-way models*:

    . use http://www.stata-press.com/data/r12/cholesterol
    (Artificial cholesterol data)
    . anova chol agegrp
    (output omitted)

The margins command reports the estimated cell means, $\widehat{\mu}_1, \ldots, \widehat{\mu}_5$, for each of the five age groups.

    . margins agegrp
    Adjusted predictions   Number of obs   =   75
    Expression   : Linear prediction, predict()

		Delta-method				
	Margin	Std. Err.	z	P>|z|	[95% Conf. Interval]	
agegrp						
1	180.5198	2.666944	67.69	0.000	175.2926	185.7469
2	188.7233	2.666944	70.76	0.000	183.4962	193.9504
3	202.0608	2.666944	75.76	0.000	196.8337	207.2879
4	210.6704	2.666944	78.99	0.000	205.4433	215.8975
5	219.282	2.666944	82.22	0.000	214.0548	224.5091

Contrast operators provide an easy way to make certain types of comparisons of these cell means. We use the ordinal factor agegrp to demonstrate these operators because some types of contrasts are only meaningful when the levels of the factor have a natural ordering. We demonstrate these contrast operators using a one-way model; however, they are equally applicable to main effects, simple effects, and interactions for more complicated models.

Differences from a reference level (r.)

The r. operator specifies that each level of the attached factor variable be compared with a reference level. These are referred to as reference-level or reference-category contrasts (or effects), and r. is the reference-level operator.

In the following, we use the r. operator to test the effect of each category of age group when that category is compared with a reference category.

```
. contrast r.agegrp
Contrasts of marginal linear predictions
Margins        : asbalanced
```

	df	F	P>F
agegrp			
(2 vs 1)	1	4.73	0.0330
(3 vs 1)	1	32.62	0.0000
(4 vs 1)	1	63.91	0.0000
(5 vs 1)	1	105.62	0.0000
Joint	4	35.02	0.0000
Residual	70		

	Contrast	Std. Err.	[95% Conf.	Interval]
agegrp				
(2 vs 1)	8.203575	3.771628	.6812991	15.72585
(3 vs 1)	21.54105	3.771628	14.01878	29.06333
(4 vs 1)	30.15067	3.771628	22.6284	37.67295
(5 vs 1)	38.76221	3.771628	31.23993	46.28448

In the first table, the row labeled (2 vs 1) is a test of $\mu_2 = \mu_1$, a test that the mean cholesterol levels for the 10–19 age group (agegrp = 1) and the 20–29 age group (agegrp = 2) are equal. The tests in the next three rows are defined similarly. The row labeled Joint provides the joint test for these four hypotheses, which is just the test of the main effects of age group.

The second table provides the contrasts of each category with the reference category along with confidence intervals. The contrast in the row labeled (2 vs 1) is the difference in the cell means of the second age group and the first age group, $\widehat{\mu}_2 - \widehat{\mu}_1$.

The first level of a factor is the default reference level, but we can specify a different reference level by using the b. operator; see [U] **11.4.3.2 Base levels**. Here we use the last age group, agegrp = 5, instead of the first as the reference category. We also include the nowald option so that only the table of contrasts and their confidence intervals is produced.

```
. contrast rb5.agegrp, nowald
Contrasts of marginal linear predictions
Margins       : asbalanced
```

	Contrast	Std. Err.	[95% Conf. Interval]	
agegrp				
(1 vs 5)	-38.76221	3.771628	-46.28448	-31.23993
(2 vs 5)	-30.55863	3.771628	-38.08091	-23.03636
(3 vs 5)	-17.22115	3.771628	-24.74343	-9.698877
(4 vs 5)	-8.611533	3.771628	-16.13381	-1.089257

Now the first row is labeled (1 vs 5) and is the difference in the cell means of the first and fifth age groups.

Differences from the next level (a.)

The a. operator specifies that each level of the attached factor variable be compared with the next level. These are referred to as adjacent contrasts (or effects), and a. is the adjacent operator. This operator is only meaningful with factor variables that have a natural ordering in the levels.

We can use the a. operator to perform tests that each level of age group differs from the next adjacent level.

```
. contrast a.agegrp
Contrasts of marginal linear predictions
Margins       : asbalanced
```

	df	F	P>F
agegrp			
(1 vs 2)	1	4.73	0.0330
(2 vs 3)	1	12.51	0.0007
(3 vs 4)	1	5.21	0.0255
(4 vs 5)	1	5.21	0.0255
Joint	4	35.02	0.0000
Residual	70		

	Contrast	Std. Err.	[95% Conf. Interval]	
agegrp				
(1 vs 2)	-8.203575	3.771628	-15.72585	-.6812991
(2 vs 3)	-13.33748	3.771628	-20.85976	-5.815204
(3 vs 4)	-8.60962	3.771628	-16.1319	-1.087345
(4 vs 5)	-8.611533	3.771628	-16.13381	-1.089257

In the first table, the row labeled (1 vs 2) tests the effect of belonging to the 10–19 age group instead of the 20–29 age group. Likewise, the rows labeled (2 vs 3), (3 vs 4), and (4 vs 5) are tests for the effects of being in the younger of the two age groups instead of the older one.

In the second table, the contrast in the row labeled (1 vs 2) is the difference in the cell means of the first and second age groups, $\hat{\mu}_1 - \hat{\mu}_2$. The contrasts in the other rows are defined similarly.

Differences from the previous level (ar.)

The ar. operator specifies that each level of the attached factor variable be compared with the previous level. These are referred to as reverse adjacent contrasts (or effects), and ar. is the reverse adjacent operator. As with the a. operator, this operator is only meaningful with factor variables that have a natural ordering in the levels.

In the following, we use the ar. operator to report tests for the individual reverse adjacent effects of agegrp.

```
. contrast ar.agegrp
Contrasts of marginal linear predictions
Margins       : asbalanced
```

	df	F	P>F
agegrp			
(2 vs 1)	1	4.73	0.0330
(3 vs 2)	1	12.51	0.0007
(4 vs 3)	1	5.21	0.0255
(5 vs 4)	1	5.21	0.0255
Joint	4	35.02	0.0000
Residual	70		

	Contrast	Std. Err.	[95% Conf. Interval]	
agegrp				
(2 vs 1)	8.203575	3.771628	.6812991	15.72585
(3 vs 2)	13.33748	3.771628	5.815204	20.85976
(4 vs 3)	8.60962	3.771628	1.087345	16.1319
(5 vs 4)	8.611533	3.771628	1.089257	16.13381

Here the Wald tests in the first table for the individual reverse adjacent effects are equivalent to the tests for the adjacent effects in the previous example. However, if we compare values of the contrasts in the bottom tables, we see the difference between the r. and the ar. operators. This time, the contrast in the first row is labeled (2 vs 1) and is the difference in the cell means of the second and first age groups, $\widehat{\mu}_2 - \widehat{\mu}_1$. This is the estimated effect of belonging to the 20–29 age group instead of the 10–19 age group. The remaining rows make similar comparisons to the previous level.

Differences from the grand mean (g.)

The g. operator specifies that each level of a factor variable be compared with the grand mean of all levels. For this operator, the grand mean is computed using a simple average of the cell means.

Here are the grand mean effects of `agegrp`:

```
. contrast g.agegrp
Contrasts of marginal linear predictions
Margins        : asbalanced
```

	df	F	P>F
agegrp			
(1 vs mean)	1	68.42	0.0000
(2 vs mean)	1	23.36	0.0000
(3 vs mean)	1	0.58	0.4506
(4 vs mean)	1	19.08	0.0000
(5 vs mean)	1	63.65	0.0000
Joint	4	35.02	0.0000
Residual	70		

	Contrast	Std. Err.	[95% Conf.	Interval]
agegrp				
(1 vs mean)	-19.7315	2.385387	-24.48901	-14.974
(2 vs mean)	-11.52793	2.385387	-16.28543	-6.770423
(3 vs mean)	1.809552	2.385387	-2.947953	6.567057
(4 vs mean)	10.41917	2.385387	5.661668	15.17668
(5 vs mean)	19.0307	2.385387	14.2732	23.78821

There are five age groups in our estimation sample. Thus the row labeled (1 vs mean) tests $\mu_1 = (\mu_1+\mu_2+\mu_3+\mu_4+\mu_5)/5$. The row labeled (2 vs mean) tests $\mu_2 = (\mu_1+\mu_2+\mu_3+\mu_4+\mu_5)/5$. The remaining rows perform similar tests for the third, fourth, and fifth age groups. In our example, the means for all age groups except group 3 (30–39 age group) are statistically different from the grand mean.

Differences from the mean of subsequent levels (h.)

The `h.` operator specifies that each level of the attached factor variable be compared with the mean of subsequent levels. These are referred to as Helmert contrasts (or effects), and `h.` is the Helmert operator. For this operator, the mean is computed using a simple average of the cell means. This operator is only meaningful with factor variables that have a natural ordering in the levels.

Here are the Helmert contrasts for `agegrp`:

```
. contrast h.agegrp
Contrasts of marginal linear predictions
Margins        : asbalanced
```

	df	F	P>F
agegrp			
(1 vs >1)	1	68.42	0.0000
(2 vs >2)	1	50.79	0.0000
(3 vs >3)	1	15.63	0.0002
(4 vs 5)	1	5.21	0.0255
Joint	4	35.02	0.0000
Residual	70		

	Contrast	Std. Err.	[95% Conf. Interval]	
agegrp				
(1 vs >1)	-24.66438	2.981734	-30.61126	-18.7175
(2 vs >2)	-21.94774	3.079522	-28.08965	-15.80583
(3 vs >3)	-12.91539	3.266326	-19.42987	-6.400905
(4 vs 5)	-8.611533	3.771628	-16.13381	-1.089257

The row labeled (1 vs >1) tests $\mu_1 = (\mu_2 + \mu_3 + \mu_4 + \mu_5)/4$, that is, that the cell mean for the youngest age group is equal to the average of the cell means for the older age groups. The row labeled (2 vs >2) tests $\mu_2 = (\mu_3 + \mu_4 + \mu_5)/3$. The tests in the other rows are defined similarly.

Differences from the mean of previous levels (j.)

The j. operator specifies that each level of the attached factor variable be compared with the mean of the previous levels. These are referred to as reverse Helmert contrasts (or effects), and j. is the reverse Helmert operator. For this operator, the mean is computed using a simple average of the cell means. This operator is only meaningful with factor variables that have a natural ordering in the levels.

Here are the reverse Helmert contrasts of agegrp:

```
. contrast j.agegrp
Contrasts of marginal linear predictions
Margins      : asbalanced
```

	df	F	P>F
agegrp			
(2 vs 1)	1	4.73	0.0330
(3 vs <3)	1	28.51	0.0000
(4 vs <4)	1	43.18	0.0000
(5 vs <5)	1	63.65	0.0000
Joint	4	35.02	0.0000
Residual	70		

	Contrast	Std. Err.	[95% Conf. Interval]	
agegrp				
(2 vs 1)	8.203575	3.771628	.6812991	15.72585
(3 vs <3)	17.43927	3.266326	10.92479	23.95375
(4 vs <4)	20.2358	3.079522	14.09389	26.37771
(5 vs <5)	23.78838	2.981734	17.8415	29.73526

The row labeled (2 vs 1) tests $\mu_2 = \mu_1$, that is, that the cell means for the 20–29 and the 10–19 age groups are equal. The row labeled (3 vs <3) tests $\mu_3 = (\mu_1 + \mu_2)/2$, that is, that the cell mean for the 30–39 age group is equal to the average of the cell means for the 10–19 and 20–29 age groups. The tests in the remaining rows are defined similarly.

Orthogonal polynomials (p. and q.)

The p. and q. operators specify that orthogonal polynomials be applied to the attached factor variable. Orthogonal polynomial contrasts allow us to partition the effects of a factor variable into linear, quadratic, cubic, and higher-order polynomial components. The p. operator applies orthogonal polynomials using the values of the factor variable. The q. operator applies orthogonal polynomials using the level indices. If the level values of the factor variable are equally spaced, as with our agegrp variable, then the p. and q. operators yield the same result. These operators are only meaningful with factor variables that have a natural ordering in the levels.

Because agegrp has five levels, contrast can test the linear, quadratic, cubic, and quartic effects of agegrp.

```
. contrast p.agegrp, noeffects
Contrasts of marginal linear predictions
Margins        : asbalanced
```

	df	F	P>F
agegrp			
(linear)	1	139.11	0.0000
(quadratic)	1	0.15	0.6962
(cubic)	1	0.37	0.5448
(quartic)	1	0.43	0.5153
Joint	4	35.02	0.0000
Residual	70		

The row labeled (linear) tests the linear effect of agegrp, the only effect that appears to be significant in this case.

The labels for our agegrp variable show the age ranges that correspond to each level.

```
. label list ages
ages:
           1 10-19
           2 20-29
           3 30-39
           4 40-59
           5 60-79
```

Notice that these groups do not have equal widths. Now let's refit our model using the agemidpt variable. The values of agemidpt indicate the midpoint of each age group that was defined by the agegrp variable and are, therefore, not equally spaced.

```
. anova chol agemidpt
```

| | Number of obs = | 75 | R-squared | = 0.6668 |
| | Root MSE   = | 10.329 | Adj R-squared | = 0.6477 |

Source	Partial SS	df	MS	F	Prob > F
Model	14943.3997	4	3735.84993	35.02	0.0000
agemidpt	14943.3997	4	3735.84993	35.02	0.0000
Residual	7468.21971	70	106.688853		
Total	22411.6194	74	302.859722		

Now if we use the q. operator, we will obtain the same results as above because the level indices of agemidpt are equivalent to the values of agegrp.

. contrast q.agemidpt, noeffects

Contrasts of marginal linear predictions

Margins   : asbalanced

	df	F	P>F
agemidpt			
(linear)	1	139.11	0.0000
(quadratic)	1	0.15	0.6962
(cubic)	1	0.37	0.5448
(quartic)	1	0.43	0.5153
Joint	4	35.02	0.0000
Residual	70		

However, if we use the p. operator, we will instead fit an orthogonal polynomial to the midpoint values.

. contrast p.agemidpt, noeffects

Contrasts of marginal linear predictions

Margins   : asbalanced

	df	F	P>F
agemidpt			
(linear)	1	133.45	0.0000
(quadratic)	1	5.40	0.0230
(cubic)	1	0.05	0.8198
(quartic)	1	1.16	0.2850
Joint	4	35.02	0.0000
Residual	70		

Using the values of the midpoints, the quadratic effect is also significant at the 5% level.

❑ Technical note

We used the noeffects option when working with orthogonal polynomial contrasts. Apart from perhaps the sign of the contrast, the values of the individual contrasts are not meaningful for orthogonal polynomial contrasts. In addition, many textbooks provide tables with contrast coefficients that can be used to compute orthogonal polynomial contrasts where the levels of a factor are equally spaced. If we use these coefficients and calculate the contrasts manually with user-defined contrasts, as described below, the Wald tests for the polynomial terms will be equivalent, but the values of the individual contrasts will not necessarily match those that we obtain when using the polynomial contrast operator. When we use one of these contrast operators, an algorithm is used to calculate the coefficients of the polynomial contrast that will allow for unequal spacing in the levels of the factor as well as in the weights for the cell frequencies (when using pw. or qw.), as described in *Methods and formulas*.

❑

User-defined contrasts

In the previous examples, we performed tests using contrast operators. When there is not a contrast operator available to calculate the contrast in which we are interested, we can specify custom contrasts.

Here we fit a one-way model for cholesterol on the factor `race`, which has three levels:

```
. label list race
race:
           1 black
           2 white
           3 other

. anova chol race
```

		Number of obs =	75	R-squared	= 0.0299
		Root MSE	= 17.3775	Adj R-squared	= 0.0029

Source	Partial SS	df	MS	F	Prob > F
Model	669.278235	2	334.639117	1.11	0.3357
race	669.278235	2	334.639117	1.11	0.3357
Residual	21742.3412	72	301.976961		
Total	22411.6194	74	302.859722		

`margins` calculates the estimated cell mean cholesterol level for each race:

```
. margins race
```

Adjusted predictions		Number of obs	=	75

Expression   : Linear prediction, predict()

		Delta-method				
	Margin	Std. Err.	z	P>|z|	[95% Conf. Interval]	
race						
1	204.4279	3.475497	58.82	0.000	197.6161	211.2398
2	197.6132	3.475497	56.86	0.000	190.8014	204.425
3	198.7127	3.475497	57.18	0.000	191.9008	205.5245

Suppose we want to test the following linear combination:

$$\sum_{i=1}^{3} c_i \mu_i$$

where μ_i is the cell mean of `chol` when `race` is equal to its ith level (the means estimated using `margins` above). Assuming the c_i elements sum to zero, this linear combination is a contrast. We can specify this type of custom contrast by using the following syntax:

$$\{\text{race } c_1 \ c_2 \ c_3\}$$

The null hypothesis for the test of the main effects of `race` is

$$H_{0_{\text{race}}}: \mu_1 = \mu_2 = \mu_3$$

Although $H_{0_{race}}$ can be tested using any of several different contrasts on the cell means, we will test it by comparing the second and third cell means with the first. To test that the cell means for blacks and whites are equal, $\mu_1 = \mu_2$, we can specify the contrast

$$\{race\ -1\ 1\ 0\}$$

To test that the cell means for blacks and other races are equal, $\mu_1 = \mu_3$, we can specify the contrast

$$\{race\ -1\ 0\ 1\}$$

We can use both in a single call to contrast.

```
. contrast {race -1 1 0} {race -1 0 1}
Contrasts of marginal linear predictions
Margins       : asbalanced
```

	df	F	P>F
race			
(1)	1	1.92	0.1699
(2)	1	1.35	0.2488
Joint	2	1.11	0.3357
Residual	72		

	Contrast	Std. Err.	[95% Conf. Interval]	
race				
(1)	-6.814717	4.915095	-16.61278	2.983345
(2)	-5.715261	4.915095	-15.51332	4.082801

The row labeled (1) is the test for $\mu_1 = \mu_2$, the first specified contrast. The row labeled (2) is the test for $\mu_1 = \mu_3$, the second specified contrast. The row labeled Joint is the overall test for the main effects of race.

Now let's fit a model with two factors, race and age group:

```
. anova chol race##agegrp
```

| | | | Number of obs = | 75 | R-squared   = 0.7524 |
| | | | Root MSE   = 9.61785 | Adj R-squared = 0.6946 |

Source	Partial SS	df	MS	F	Prob > F
Model	16861.438	14	1204.38843	13.02	0.0000
race	669.278235	2	334.639117	3.62	0.0329
agegrp	14943.3997	4	3735.84993	40.39	0.0000
race#agegrp	1248.76005	8	156.095006	1.69	0.1201
Residual	5550.18143	60	92.5030238		
Total	22411.6194	74	302.859722		

The null hypothesis for the test of the main effects of race is now

$$H_{0_{race}}: \mu_{1\cdot} = \mu_{2\cdot} = \mu_{3\cdot}.$$

where $\mu_{i\cdot}$ is the marginal mean of chol when race is equal to its ith level.

We can use the same syntax as above to perform this test by specifying contrasts on the marginal means of `race`:

```
. contrast {race -1 1 0} {race -1 0 1}
Contrasts of marginal linear predictions
Margins        : asbalanced
```

	df	F	P>F
race			
(1)	1	6.28	0.0150
(2)	1	4.41	0.0399
Joint	2	3.62	0.0329
Residual	60		

	Contrast	Std. Err.	[95% Conf. Interval]	
race				
(1)	-6.814717	2.720339	-12.2562	-1.37323
(2)	-5.715261	2.720339	-11.15675	-.2737739

Custom contrasts may be specified on the cell means of interactions, too. Here we use `margins` to calculate the mean of `chol` for each cell in the interaction of `race` and `agegrp`:

```
. margins race#agegrp
Adjusted predictions                          Number of obs   =        75
Expression     : Linear prediction, predict()
```

	Margin	Delta-method Std. Err.	z	P>|z|	[95% Conf. Interval]	
race#agegrp						
1 1	179.2309	4.301233	41.67	0.000	170.8006	187.6611
1 2	196.4777	4.301233	45.68	0.000	188.0474	204.908
1 3	210.6694	4.301233	48.98	0.000	202.2391	219.0996
1 4	214.097	4.301233	49.78	0.000	205.6668	222.5273
1 5	221.6646	4.301233	51.54	0.000	213.2344	230.0949
2 1	186.0727	4.301233	43.26	0.000	177.6425	194.503
2 2	184.6714	4.301233	42.93	0.000	176.2411	193.1017
2 3	196.2633	4.301233	45.63	0.000	187.833	204.6936
2 4	209.9953	4.301233	48.82	0.000	201.5651	218.4256
2 5	211.0633	4.301233	49.07	0.000	202.633	219.4935
3 1	176.2556	4.301233	40.98	0.000	167.8254	184.6859
3 2	185.0209	4.301233	43.02	0.000	176.5906	193.4512
3 3	199.2498	4.301233	46.32	0.000	190.8195	207.68
3 4	207.9189	4.301233	48.34	0.000	199.4887	216.3492
3 5	225.118	4.301233	52.34	0.000	216.6877	233.5483

Now we are interested in testing the following linear combination of these cell means:

$$\sum_{i=1}^{3}\sum_{j=1}^{5} c_{ij}\mu_{ij}$$

We can specify this type of custom contrast using the following syntax:

$$\{\text{race\#agegrp } c_{11} \ c_{12} \ \cdots \ c_{15} \ c_{21} \ c_{22} \ \cdots \ c_{25} \ c_{31} \ c_{32} \ \cdots \ c_{35}\}$$

Because the marginal means of chol for each level of race are linear combinations of the cell means, we can compose the test for the main effects of race in terms of the cell means directly. The constraint that the marginal means for blacks and whites are equal, $\mu_{1.} = \mu_{2.}$, translates to the following constraint on the cell means:

$$\frac{1}{5}(\mu_{11} + \mu_{12} + \mu_{13} + \mu_{14} + \mu_{15}) = \frac{1}{5}(\mu_{21} + \mu_{22} + \mu_{23} + \mu_{24} + \mu_{25})$$

Ignoring the common factor, we can specify this contrast as

$$\{\text{race\#agegrp } -1 \ -1 \ -1 \ -1 \ -1 \ 1 \ 1 \ 1 \ 1 \ 1 \ 0 \ 0 \ 0 \ 0 \ 0\}$$

contrast will fill in the trailing zeros for us if we neglect to specify them, so

$$\{\text{race\#agegrp } -1 \ -1 \ -1 \ -1 \ -1 \ 1 \ 1 \ 1 \ 1 \ 1\}$$

is also allowed. The other constraint, $\mu_{1.} = \mu_{3.}$, translates to

$$\frac{1}{5}(\mu_{11} + \mu_{12} + \mu_{13} + \mu_{14} + \mu_{15}) = \frac{1}{5}(\mu_{31} + \mu_{32} + \mu_{33} + \mu_{34} + \mu_{35})$$

This can be specified to contrast as

$$\{\text{race\#agegrp } -1 \ -1 \ -1 \ -1 \ -1 \ 0 \ 0 \ 0 \ 0 \ 0 \ 1 \ 1 \ 1 \ 1 \ 1\}$$

The following call to contrast yields the same test results as above.

```
. contrast {race#agegrp -1 -1 -1 -1 -1 1 1 1 1 1}
>           {race#agegrp -1 -1 -1 -1 -1 0 0 0 0 0 1 1 1 1 1}, noeffects
Contrasts of marginal linear predictions
Margins        : asbalanced
```

	df	F	P>F
race#agegrp			
(1) (1)	1	6.28	0.0150
(2) (2)	1	4.41	0.0399
Joint	2	3.62	0.0329
Residual	60		

The row labeled (1) (1) is the test for

$$\mu_{11} + \mu_{12} + \mu_{13} + \mu_{14} + \mu_{15} = \mu_{21} + \mu_{22} + \mu_{23} + \mu_{24} + \mu_{25}$$

It was the first specified contrast. The row labeled (2) (2) is the test for

$$\mu_{11} + \mu_{12} + \mu_{13} + \mu_{14} + \mu_{15} = \mu_{31} + \mu_{32} + \mu_{33} + \mu_{34} + \mu_{35}$$

It was the second specified contrast. The row labeled Joint tests (1) (1) and (2) (2) simultaneously.

We used the `noeffects` option above to suppress the table of contrasts. We can omit the $1/5$ from the equations for $\mu_1. = \mu_2.$ and $\mu_1. = \mu_3.$ and still obtain the appropriate tests. However, if we want to calculate the differences in the marginal means, we must include the $1/5 = 0.2$ on each of the contrast coefficients as follows:

```
. contrast {race#agegrp -0.2 -0.2 -0.2 -0.2 -0.2
                         0.2  0.2  0.2  0.2  0.2}
           {race#agegrp -0.2 -0.2 -0.2 -0.2 -0.2
                          0    0    0    0    0
                         0.2  0.2  0.2  0.2  0.2}
```

So far, we have reproduced the reference category contrasts by specifying user-defined contrasts on the marginal means and then on the cell means. For this test, it would have been easier to use the `r.` contrast operator:

```
. contrast r.race, noeffects
```

Contrasts of marginal linear predictions

Margins   : asbalanced

	df	F	P>F
race			
(2 vs 1)	1	6.28	0.0150
(3 vs 1)	1	4.41	0.0399
Joint	2	3.62	0.0329
Residual	60		

In most cases, we can use contrast operators to perform tests. However, if we want to compare, for instance, the second and third age groups with the fourth and fifth age groups with the test

$$\frac{1}{2}(\mu_{.2} + \mu_{.3}) = \frac{1}{2}(\mu_{.4} + \mu_{.5})$$

there is not a contrast operator that corresponds to this particular contrast. A custom contrast is necessary.

```
. contrast {agegrp 0 -0.5 -0.5 0.5 0.5}
```

Contrasts of marginal linear predictions

Margins   : asbalanced

	df	F	P>F
agegrp	1	62.19	0.0000
Residual	60		

	Contrast	Std. Err.	[95% Conf. Interval]	
agegrp				
(1)	19.58413	2.483318	14.61675	24.5515

Empty cells

An empty cell is a combination of the levels of factor variables that is not observed in the estimation sample. In the previous examples, we have seen data with three levels of race, five levels of agegrp, and all level combinations of race and agegrp present. Suppose there are no observations for white individuals in the second age group (ages 20–29).

```
. use http://www.stata-press.com/data/r12/cholesterol2
(Artificial cholesterol data, empty cells)
. label list
race:
           1 black
           2 white
           3 other
ages:
           1 10-19
           2 20-29
           3 30-39
           4 40-59
           5 60-79
. regress chol race##agegrp
note: 2.race#2.agegrp identifies no observations in the sample
```

Source	SS	df	MS		Number of obs =	70
					F(13, 56) =	13.51
Model	15751.6113	13	1211.66241		Prob > F =	0.0000
Residual	5022.71559	56	89.6913498		R-squared =	0.7582
					Adj R-squared =	0.7021
Total	20774.3269	69	301.077201		Root MSE =	9.4706

chol	Coef.	Std. Err.	t	P>|t|	[95% Conf. Interval]	
race						
2	12.84185	5.989703	2.14	0.036	.8430383	24.84067
3	-.167627	5.989703	-0.03	0.978	-12.16644	11.83119
agegrp						
2	17.24681	5.989703	2.88	0.006	5.247991	29.24562
3	31.43847	5.989703	5.25	0.000	19.43966	43.43729
4	34.86613	5.989703	5.82	0.000	22.86732	46.86495
5	44.43374	5.989703	7.42	0.000	32.43492	56.43256
race#agegrp						
2 2	0	(empty)				
2 3	-22.83983	8.470719	-2.70	0.009	-39.80872	-5.870939
2 4	-14.67558	8.470719	-1.73	0.089	-31.64447	2.293306
2 5	-10.51115	8.470719	-1.24	0.220	-27.48004	6.457735
3 2	-6.054425	8.470719	-0.71	0.478	-23.02331	10.91446
3 3	-11.48083	8.470719	-1.36	0.181	-28.44971	5.488063
3 4	-.6796112	8.470719	-0.08	0.936	-17.6485	16.28928
3 5	-1.578052	8.470719	-0.19	0.853	-18.54694	15.39084
_cons	175.2309	4.235359	41.37	0.000	166.7464	183.7153

Now let's use contrast to test the main effects of race:

```
. contrast race
Contrasts of marginal linear predictions
Margins      : asbalanced
```

	df	F	P>F
race	(not testable)		
Residual	56		

By "not testable", contrast means that it cannot form a test for the main effects of race based on estimable functions of the model coefficients. agegrp has five levels, so contrast constructs an estimate of the ith margin for race as

$$\widehat{\mu}_{i\cdot} = \frac{1}{5} \sum_{j=1}^{5} \widehat{\mu}_{ij} = \widehat{\mu}_0 + \widehat{\alpha}_i + \frac{1}{5} \sum_{j=1}^{5} \left\{ \widehat{\beta}_j + \widehat{(\alpha\beta)}_{ij} \right\}$$

but $\widehat{(\alpha\beta)}_{22}$ was constrained to zero because of the empty cell, so $\widehat{\mu}_{2\cdot}$ is not an estimable function of the model coefficients.

See *Estimable functions* in *Methods and formulas* of [R] **margins** for a technical description of estimable functions. The emptycells(reweight) option causes contrast to estimate $\mu_{2\cdot}$ by

$$\widehat{\mu}_{2\cdot} = \frac{\widehat{\mu}_{21} + \widehat{\mu}_{23} + \widehat{\mu}_{24} + \widehat{\mu}_{25}}{4}$$

which is an estimable function of the model coefficients.

```
. contrast race, emptycells(reweight)
Contrasts of marginal linear predictions
Margins      : asbalanced
Empty cells  : reweight
```

	df	F	P>F
race	2	3.17	0.0498
Residual	56		

We can reconstruct the effect of the `emptycells(reweight)` option by using custom contrasts.

```
. contrast {race#agegrp -4 -4 -4 -4 -4 5 0 5 5 5}
>          {race#agegrp -1 -1 -1 -1 -1 0 0 0 0 0 1 1 1 1 1}, noeffects
Contrasts of marginal linear predictions
Margins      : asbalanced
```

	df	F	P>F
race#agegrp			
(1) (1)	1	1.06	0.3080
(2) (2)	1	2.37	0.1291
Joint	2	3.17	0.0498
Residual	56		

The row labeled (1) (1) is the test for

$$\frac{1}{5}(\mu_{11} + \mu_{12} + \mu_{13} + \mu_{14} + \mu_{15}) = \frac{1}{4}(\mu_{21} + \mu_{23} + \mu_{24} + \mu_{25})$$

It was the first specified contrast. The row labeled (2) (2) is the test for

$$\mu_{11} + \mu_{12} + \mu_{13} + \mu_{14} + \mu_{15} = \mu_{31} + \mu_{32} + \mu_{33} + \mu_{34} + \mu_{35}$$

It was the second specified contrast. The row labeled Joint is the overall test of the main effects of race.

Empty cells, ANOVA style

Let's refit the linear model from the previous example with anova to compare with contrast's test for the main effects of race.

```
. anova chol race##agegrp
```

		Number of obs =	70	R-squared	=	0.7582
		Root MSE	= 9.47055	Adj R-squared =		0.7021
Source	Partial SS	df	MS	F	Prob > F	
Model	15751.6113	13	1211.66241	13.51	0.0000	
race	305.49046	2	152.74523	1.70	0.1914	
agegrp	14387.8559	4	3596.96397	40.10	0.0000	
race#agegrp	795.807574	7	113.686796	1.27	0.2831	
Residual	5022.71559	56	89.6913498			
Total	20774.3269	69	301.077201			

contrast and anova handled the empty cell differently; the F statistic reported by contrast was 3.17, but anova reported 1.70. To see how they differ, consider the following table of the cell means and margins for our situation.

				agegrp			
		1	2	3	4	5	
	1	μ_{11}	μ_{12}	μ_{13}	μ_{14}	μ_{15}	$\mu_{1\cdot}$
race	2	μ_{21}		μ_{23}	μ_{24}	μ_{25}	
	3	μ_{31}	μ_{32}	μ_{33}	μ_{34}	μ_{35}	$\mu_{3\cdot}$
		$\mu_{\cdot 1}$		$\mu_{\cdot 3}$	$\mu_{\cdot 4}$	$\mu_{\cdot 5}$	

For testing the main effects of `race`, we know that we will be testing the equality of the marginal means for rows 1 and 3, that is, $\mu_{1\cdot} = \mu_{3\cdot}$. This translates into the following constraint:

$$\mu_{11} + \mu_{12} + \mu_{13} + \mu_{14} + \mu_{15} = \mu_{31} + \mu_{32} + \mu_{33} + \mu_{34} + \mu_{35}$$

Because row 2 contains an empty cell in column 2, `anova` dropped column 2 and tested the equality of the marginal mean for row 2 with the average of the marginal means from rows 1 and 3, using only the remaining cell means. This translates into the following constraint:

$$2(\mu_{21} + \mu_{23} + \mu_{24} + \mu_{25}) = \mu_{11} + \mu_{13} + \mu_{14} + \mu_{15} + \mu_{31} + \mu_{33} + \mu_{34} + \mu_{35} \qquad (1)$$

Now that we know the constraints that `anova` used to test for the main effects of `race`, we can use custom contrasts to reproduce the `anova` test result.

```
. contrast {race#agegrp -1 -1 -1 -1 -1  0 0  0  0  0 1 1 1 1 1}
>          {race#agegrp  1  0  1  1  1 -2 0 -2 -2 -2 1 0 1 1 1}, noeffects
Contrasts of marginal linear predictions
Margins        : asbalanced
```

	df	F	P>F
race#agegrp			
(1) (1)	1	2.37	0.1291
(2) (2)	1	1.03	0.3138
Joint	2	1.70	0.1914
Residual	56		

The row labeled (1) (1) is the test for $\mu_{1\cdot} = \mu_{3\cdot}$; it was the first specified contrast. The row labeled (2) (2) is the test for the constraint in (1); it was the second specified contrast. The row labeled `Joint` is an overall test for the main effects of `race`.

Nested effects

`contrast` has the | operator for computing simple effects when the levels of one factor are nested within the levels of another. Here is a fictional example where we are interested in the effect of five methods of teaching algebra on students' scores for the math portion of the SAT. Suppose three algebra classes are randomly sampled from classes using each of the five methods so that `class` is nested in `method` as demonstrated in the following tabulation.

```
. use http://www.stata-press.com/data/r12/SAT
(Artificial SAT data)
. tabulate class method
```

class	method 1	2	3	4	5	Total
1	5	0	0	0	0	5
2	5	0	0	0	0	5
3	5	0	0	0	0	5
4	0	5	0	0	0	5
5	0	5	0	0	0	5
6	0	5	0	0	0	5
7	0	0	5	0	0	5
8	0	0	5	0	0	5
9	0	0	5	0	0	5
10	0	0	0	5	0	5
11	0	0	0	5	0	5
12	0	0	0	5	0	5
13	0	0	0	0	5	5
14	0	0	0	0	5	5
15	0	0	0	0	5	5
Total	15	15	15	15	15	75

We will consider method as fixed and class nested in method as random. To use class nested in method as the error term for method, we can specify the following anova model:

```
. anova score method / class|method /
```

	Number of obs = 75		R-squared	= 0.7599
	Root MSE = 71.8517		Adj R-squared =	0.7039

Source	Partial SS	df	MS	F	Prob > F
Model	980312	14	70022.2857	13.56	0.0000
method	905872	4	226468	30.42	0.0000
class|method	74440	10	7444		
class|method	74440	10	7444	1.44	0.1845
Residual	309760	60	5162.66667		
Total	1290072	74	17433.4054		

Like anova, contrast allows the | operator, which specifies that one variable is nested in the levels of another. We can use contrast to test the main effects of method and the simple effects of class within method.

```
. contrast method class|method
```

Contrasts of marginal linear predictions

Margins   : asbalanced

	df	F	P>F
method	(not testable)		
class|method			
1	2	2.80	0.0687
2	2	0.91	0.4089
3	2	1.10	0.3390
4	2	0.22	0.8025
5	2	2.18	0.1221
Joint	10	1.44	0.1845
Residual	60		

Although contrast was able to perform the individual tests for the simple effects of class within method, empty cells in the interaction between method and class prevented contrast from testing for a main effect of method. Here we add the emptycells(reweight) option so that contrast can take the empty cells into account when computing the marginal means for method.

```
. contrast method class|method, emptycells(reweight)
```

Contrasts of marginal linear predictions

Margins   : asbalanced
Empty cells   : reweight

	df	F	P>F
method	4	43.87	0.0000
class|method			
1	2	2.80	0.0687
2	2	0.91	0.4089
3	2	1.10	0.3390
4	2	0.22	0.8025
5	2	2.18	0.1221
Joint	10	1.44	0.1845
Residual	60		

Now contrast does report a test for the main effects of method. However, if we compare this with the anova results, we will see that the results are different. They are different because contrast uses the residual error term to compute the F test by default. Using notation similar to anova, we can use the / operator to specify a different error term for the test. Therefore, we can reproduce the test of main effects from our anova command by typing

```
. contrast method / class|method /, emptycells(reweight)
```
Contrasts of marginal linear predictions

Margins   : asbalanced
Empty cells : reweight

	df	F	P>F
method	4	30.42	0.0000
class|method	10	(denominator)	
class|method			
1	2	2.80	0.0687
2	2	0.91	0.4089
3	2	1.10	0.3390
4	2	0.22	0.8025
5	2	2.18	0.1221
Joint	10	1.44	0.1845
Residual	60		

Multiple comparisons

We have seen that contrast can report the individual linear combinations that make up the requested effects. Depending upon the specified option, contrast will report confidence intervals, p-values, or both in the effects table. By default, the reported confidence intervals and p-values are not adjusted for multiple comparisons. Use the mcompare() option to adjust the confidence intervals and p-values for multiple comparisons of the individual effects.

Let's compute the grand mean effects of race using the g. operator. We also specify the mcompare(bonferroni) option to compute p-values and confidence intervals using Bonferroni's adjustment.

```
. use http://www.stata-press.com/data/r12/cholesterol
(Artificial cholesterol data)
. anova chol race##agegrp
  (output omitted)
. contrast g.race, mcompare(bonferroni)
```
Contrasts of marginal linear predictions

Margins   : asbalanced

				Bonferroni
	df	F	P>F	P>F
race				
(1 vs mean)	1	7.07	0.0100	0.0301
(2 vs mean)	1	2.82	0.0982	0.2947
(3 vs mean)	1	0.96	0.3312	0.9936
Joint	2	3.62	0.0329	
Residual	60			

Note: Bonferroni-adjusted p-values are reported for tests
      on individual contrasts only.

	Number of Comparisons
race	3

			Bonferroni	
	Contrast	Std. Err.	[95% Conf. Interval]	
race				
(1 vs mean)	4.17666	1.570588	.3083743	8.044945
(2 vs mean)	-2.638058	1.570588	-6.506343	1.230227
(3 vs mean)	-1.538602	1.570588	-5.406887	2.329684

The last table reports a Bonferroni-adjusted confidence interval for each individual contrast. (Use the effects option to add p-values to the last table.) The first table includes a Bonferroni-adjusted p-value for each test that is not a joint test.

Joint tests are never adjusted for multiple comparisons. For example,

```
. contrast race@agegrp, mcompare(bonferroni)
Contrasts of marginal linear predictions
Margins        : asbalanced
```

	df	F	P>F
race@agegrp			
1	2	1.37	0.2620
2	2	2.44	0.0958
3	2	3.12	0.0512
4	2	0.53	0.5889
5	2	2.90	0.0628
Joint	10	2.07	0.0409
Residual	60		

Note: Bonferroni-adjusted p-values are reported
      for tests on individual contrasts only.

	Number of Comparisons
race@agegrp	10

	Contrast	Std. Err.	Bonferroni [95% Conf. Interval]	
race@agegrp				
(2 vs base) 1	6.841855	6.082862	-10.88697	24.57068
(2 vs base) 2	-11.80631	6.082862	-29.53513	5.922513
(2 vs base) 3	-14.40607	6.082862	-32.13489	3.322751
(2 vs base) 4	-4.101691	6.082862	-21.83051	13.62713
(2 vs base) 5	-10.60137	6.082862	-28.33019	7.127448
(3 vs base) 1	-2.975244	6.082862	-20.70407	14.75358
(3 vs base) 2	-11.45679	6.082862	-29.18561	6.272031
(3 vs base) 3	-11.41958	6.082862	-29.1484	6.309244
(3 vs base) 4	-6.17807	6.082862	-23.90689	11.55075
(3 vs base) 5	3.453375	6.082862	-14.27545	21.1822

Here we have five tests of simple effects with two degrees of freedom each. No Bonferroni-adjusted p-values are available for these tests, but the confidence intervals for the individual contrasts are adjusted.

Unbalanced data

By default, contrast treats all factors as balanced when computing marginal means. By balanced, we mean that contrast assumes an equal number of observations in each level of each factor and an equal number of observations in each cell of each interaction. If our data are balanced, there is no issue. If, however, our data are not balanced, we might prefer that contrast use the actual cell frequencies from our data in computing marginal means. We instruct contrast to use observed frequencies by adding the asobserved option.

Even if our data are unbalanced, we might still want contrast to compute balanced marginal means. It depends on what we want to test and what our data represent. If we have data from a designed experiment that started with an equal number of males and females but the data became unbalanced because the data from a few males were unusable, we might still want our margins computed as though the data were balanced. If, however, we have a representative sample of individuals from Los Angeles with 40% of European descent, 34% African-American, 25% Hispanic, and 1% Australian, we probably want our margins computed using these representative frequencies. We do not want Australians receiving the same weight as Europeans.

The following examples will use an unbalanced version of our dataset.

```
. use http://www.stata-press.com/data/r12/cholesterol3
(Artificial cholesterol data, unbalanced)
. tab race agegrp
```

race	agegrp 10-19	20-29	30-39	40-59	60-79	Total
black	1	5	5	4	3	18
white	4	5	7	4	4	24
other	3	7	6	5	4	25
Total	8	17	18	13	11	67

The row labeled Total gives observed cell frequencies for age group. These can be obtained by summing frequencies from the cells in the corresponding column. In this respect, we can also refer to them as marginal frequencies. We use the terms marginal frequencies and cell frequencies interchangeably below.

We begin by fitting the two-factor model with an interaction.

```
. anova chol race##agegrp
```

| | Number of obs = 67 | R-squared = 0.8179 |
| | Root MSE = 8.37496 | Adj R-squared = 0.7689 |

Source	Partial SS	df	MS	F	Prob > F
Model	16379.9926	14	1169.99947	16.68	0.0000
race	230.754396	2	115.377198	1.64	0.2029
agegrp	13857.9877	4	3464.49693	49.39	0.0000
race#agegrp	857.815209	8	107.226901	1.53	0.1701
Residual	3647.2774	52	70.13995		
Total	20027.27	66	303.443485		

Using observed cell frequencies

Recall that the marginal means are computed from the cell means. Treating the factors as balanced yields the following marginal means for race:

$$\eta_{1\cdot} = \frac{1}{5}(\mu_{11} + \mu_{12} + \mu_{13} + \mu_{14} + \mu_{15})$$

$$\eta_{2\cdot} = \frac{1}{5}(\mu_{21} + \mu_{22} + \mu_{23} + \mu_{24} + \mu_{25})$$

$$\eta_{3\cdot} = \frac{1}{5}(\mu_{31} + \mu_{32} + \mu_{33} + \mu_{34} + \mu_{35})$$

If we have a fixed population and unbalanced cells, then the $\eta_{i\cdot}$ do not represent population means. If, however, our data are representative of the population, we can use the frequencies from our estimation sample to estimate the population marginal means, denoted $\mu_{i\cdot}$.

Here are the results of testing for a main effect of race, treating all the factors as balanced.

```
. contrast r.race
```
Contrasts of marginal linear predictions
Margins   : asbalanced

	df	F	P>F
race			
(2 vs 1)	1	3.28	0.0757
(3 vs 1)	1	1.50	0.2263
Joint	2	1.64	0.2029
Residual	52		

	Contrast	Std. Err.	[95% Conf. Interval]	
race				
(2 vs 1)	-5.324254	2.93778	-11.21934	.5708338
(3 vs 1)	-3.596867	2.93778	-9.491955	2.298221

The row labeled (2 vs 1) is the test for $\eta_{2\cdot} = \eta_{1\cdot}$. The row labeled (3 vs 1) is the test for $\eta_{3\cdot} = \eta_{1\cdot}$.

If the observed marginal frequencies are representative of the distribution of the levels of `agegrp`, we can use them to form the marginal means of `chol` for each of the levels of `race` from the cell means.

$$\mu_{1.} = \frac{1}{67}(8\mu_{11} + 17\mu_{12} + 18\mu_{13} + 13\mu_{14} + 11\mu_{15})$$

$$\mu_{2.} = \frac{1}{67}(8\mu_{21} + 17\mu_{22} + 18\mu_{23} + 13\mu_{24} + 11\mu_{25})$$

$$\mu_{3.} = \frac{1}{67}(8\mu_{31} + 17\mu_{32} + 18\mu_{33} + 13\mu_{34} + 11\mu_{35})$$

Here are the results of testing for the main effects of `race`, using the observed marginal frequencies:

```
. contrast r.race, asobserved
Contrasts of marginal linear predictions
Margins        : asobserved
```

	df	F	P>F
race			
(2 vs 1)	1	7.25	0.0095
(3 vs 1)	1	3.89	0.0538
Joint	2	3.74	0.0304
Residual	52		

	Contrast	Std. Err.	[95% Conf. Interval]	
race				
(2 vs 1)	-7.232433	2.686089	-12.62246	-1.842402
(3 vs 1)	-5.231198	2.651203	-10.55123	.0888295

The row labeled (2 vs 1) is the test for $\mu_{2.} = \mu_{1.}$. The row labeled (3 vs 1) is the test for $\mu_{3.} = \mu_{1.}$. Both tests were insignificant when we tested the cell means resulting from balanced frequencies; however, when we tested the cell means from observed frequencies, the first test is significant beyond the 5% level (and the second test is nearly so).

Here we reproduce the results of the `asobserved` option with custom contrasts. Because we are modifying the way that the marginal means are constructed from the cell means, we will specify the contrasts on the predicted cell means. We use macro expansion, `=exp`, to evaluate the fractions instead of approximating them with decimals. Macro expansion guarantees that the contrast coefficients sum to zero. For more information, see *Macro expansion operators and function* in [P] **macro**.

```
. contrast {race#agegrp -'=8/67' -'=17/67' -'=18/67' -'=13/67' -'=11/67'
>                        '=8/67'  '=17/67'  '=18/67'  '=13/67'  '=11/67'}
>           {race#agegrp -'=8/67' -'=17/67' -'=18/67' -'=13/67' -'=11/67'
>                           0         0         0         0         0
>                        '=8/67'  '=17/67'  '=18/67'  '=13/67'  '=11/67'}
Contrasts of marginal linear predictions

Margins        : asbalanced
```

	df	F	P>F
race#agegrp			
(1) (1)	1	7.25	0.0095
(2) (2)	1	3.89	0.0538
Joint	2	3.74	0.0304
Residual	52		

	Contrast	Std. Err.	[95% Conf. Interval]	
race#agegrp				
(1) (1)	-7.232433	2.686089	-12.62246	-1.842402
(2) (2)	-5.231198	2.651203	-10.55123	.0888295

Weighted contrast operators

contrast provides observation-weighted versions of five of the contrast operators—gw., hw., jw., pw., and qw.. The first three of these operators perform comparisons of means across cells, and like the marginal means just discussed, these means can be computed in two ways: 1) as though the cell frequencies were equal or 2) using the observed cell frequencies from the estimation sample. The weighted operators provide versions of the standard (as balanced) operators that weight these means by their cell frequencies. The two orthogonal polynomial operators involve similar adjustments for weighting.

Let's examine what this means by using the gw. operator. The gw. operator is a weighted version of the g. operator. The gw. operator computes the grand mean using the cell frequencies for race obtained from the model fit.

Here we test the effects of race, comparing each level with the weighted grand mean but otherwise treating the factors as balanced in the marginal mean calculations.

```
. contrast gw.race
Contrasts of marginal linear predictions

Margins        : asbalanced
```

	df	F	P>F
race			
(1 vs mean)	1	2.78	0.1014
(2 vs mean)	1	2.06	0.1573
(3 vs mean)	1	0.06	0.8068
Joint	2	1.64	0.2029
Residual	52		

	Contrast	Std. Err.	[95% Conf. Interval]	
race				
(1 vs mean)	3.24931	1.948468	-.6605779	7.159198
(2 vs mean)	-2.074944	1.44618	-4.976915	.8270276
(3 vs mean)	-.347557	1.414182	-3.18532	2.490206

The observed marginal frequencies of race are 18, 24, and 25. Thus the row labeled (1 vs mean) tests $\eta_{1.} = (18\eta_{1.} + 24\eta_{2.} + 25\eta_{3.})/67$; the row labeled (2 vs mean) tests $\eta_{2.} = (18\eta_{1.} + 24\eta_{2.} + 25\eta_{3.})/67$; and the row labeled (3 vs mean) tests $\eta_{3.} = (18\eta_{1.} + 24\eta_{2.} + 25\eta_{3.})/67$.

Now we reproduce the above results using custom contrasts. We are weighting the calculation of the grand mean from the marginal means for each of the races, but we are not weighting the calculation of the marginal means themselves. Therefore, we can specify the custom contrast on the marginal means for race instead of on the cell means.

```
. contrast {race  '=49/67' -'=24/67' -'=25/67'}
>          {race -'=18/67'  '=43/67' -'=25/67'}
>          {race -'=18/67' -'=24/67'  '=42/67'}
Contrasts of marginal linear predictions
Margins        : asbalanced
```

	df	F	P>F
race			
(1)	1	2.78	0.1014
(2)	1	2.06	0.1573
(3)	1	0.06	0.8068
Joint	2	1.64	0.2029
Residual	52		

	Contrast	Std. Err.	[95% Conf. Interval]	
race				
(1)	3.24931	1.948468	-.6605779	7.159198
(2)	-2.074944	1.44618	-4.976915	.8270276
(3)	-.347557	1.414182	-3.18532	2.490206

Now we will test for each race the difference between the marginal mean and the weighted grand mean, treating the factors as observed in the marginal mean calculations.

```
. contrast gw.race, asobserved wald ci
```
Contrasts of marginal linear predictions

Margins   : asobserved

	df	F	P>F
race			
(1 vs mean)	1	6.81	0.0118
(2 vs mean)	1	3.74	0.0587
(3 vs mean)	1	0.26	0.6099
Joint	2	3.74	0.0304
Residual	52		

	Contrast	Std. Err.	[95% Conf. Interval]	
race				
(1 vs mean)	4.542662	1.740331	1.050432	8.034891
(2 vs mean)	-2.689771	1.39142	-5.481859	.1023172
(3 vs mean)	-.6885363	1.341261	-3.379973	2.002901

The row labeled (1 vs mean) tests $\mu_{1.} = (18\mu_{1.} + 24\mu_{2.} + 25\mu_{3.})/67$; the row labeled (2 vs mean) tests $\mu_{2.} = (18\mu_{1.} + 24\mu_{2.} + 25\mu_{3.})/67$; and the row labeled (3 vs mean) tests $\mu_{3.} = (18\mu_{1.} + 24\mu_{2.} + 25\mu_{3.})/67$.

Here we use a custom contrast to reproduce the above result testing $\mu_{1.} = (18\mu_{1.} + 24\mu_{2.} + 25\mu_{3.})/67$. Because both the calculation of the marginal means and the calculation of the grand mean are adjusted, we specify the custom contrast on the cell means.

```
. contrast {race#agegrp  '=49/67*8/67'   '=49/67*17/67'   '=49/67*18/67'
>                        '=49/67*13/67'  '=49/67*11/67'
>                        -'=24/67*8/67'  -'=24/67*17/67'  -'=24/67*18/67'
>                        -'=24/67*13/67' -'=24/67*11/67'
>                        -'=25/67*8/67'  -'=25/67*17/67'  -'=25/67*18/67'
>                        -'=25/67*13/67' -'=25/67*11/67'}, nowald
```
Contrasts of marginal linear predictions

Margins   : asbalanced

	Contrast	Std. Err.	[95% Conf. Interval]	
race#agegrp				
(1) (1)	4.542662	1.740331	1.050432	8.034891

The Helmert and reverse Helmert contrasts also involve calculating averages of the marginal means; therefore, weighted versions of these parameters are available as well. The hw. operator is a weighted version of the h. operator that computes the mean of the subsequent levels using the cell frequencies obtained from the model fit. The jw. operator is a weighted version of the j. operator that computes the mean of the previous levels using the cell frequencies obtained from the model fit.

For orthogonal polynomials, we can use the pw. and qw. operators, which are the weighted versions of the p. and q. operators. In this case, the cell frequencies from the model fit are used in the calculation of the orthogonal polynomial contrast coefficients.

Testing factor effects on slopes

For linear models where the independent variables are all factor variables, the linear prediction at fixed levels of the factor variables turns out to be a cell mean. With these models, contrast computes and tests the effects of the factor variables on the expected mean of the dependent variable. When factor variables are interacted with continuous variables, contrast distinguishes factor effects on the intercept from factor effects on the slope.

Here we have 1980 census data including information on the birth rate (brate), the median age (medage), and the region of the country (region) for each of the 50 states. We can fit an ANCOVA model for brate using main effects of the factor variable region and the continuous variable medage.

```
. use http://www.stata-press.com/data/r12/census3
(1980 Census data by state)
. label list cenreg
cenreg:
           1 NE
           2 N Cntrl
           3 South
           4 West
. anova brate i.region c.medage
```

| | | Number of obs = | 50 | R-squared | = 0.8264 |
| | | Root MSE | = 12.7575 | Adj R-squared = | 0.8110 |

Source	Partial SS	df	MS	F	Prob > F
Model	34872.8589	4	8718.21473	53.57	0.0000
region	2197.75453	3	732.584844	4.50	0.0076
medage	15327.423	1	15327.423	94.18	0.0000
Residual	7323.96108	45	162.754691		
Total	42196.82	49	861.159592		

For those more comfortable with linear regression, this is equivalent to the regression model

```
. regress brate i.region c.medage
```

You may use either.

We can use contrast to compute reference category effects for region. These contrasts compare the adjusted means of regions 2, 3, and 4 with the adjusted mean of region 1.

```
. contrast r.region
Contrasts of marginal linear predictions
Margins        : asbalanced
```

	df	F	P>F
region			
(2 vs 1)	1	2.24	0.1417
(3 vs 1)	1	0.78	0.3805
(4 vs 1)	1	10.33	0.0024
Joint	3	4.50	0.0076
Residual	45		

	Contrast	Std. Err.	[95% Conf. Interval]	
region				
(2 vs 1)	9.061063	6.057484	-3.139337	21.26146
(3 vs 1)	5.06991	5.72396	-6.458738	16.59856
(4 vs 1)	21.71328	6.755616	8.106774	35.31979

Let's add the interaction between region and medage to the model.

```
. anova brate region##c.medage
```

	Number of obs =	50	R-squared	=	0.9000
	Root MSE	= 10.0244	Adj R-squared =		0.8833

Source	Partial SS	df	MS	F	Prob > F
Model	37976.3149	7	5425.18784	53.99	0.0000
region	3405.07044	3	1135.02348	11.30	0.0000
medage	5279.71448	1	5279.71448	52.54	0.0000
region#medage	3103.45597	3	1034.48532	10.29	0.0000
Residual	4220.5051	42	100.488217		
Total	42196.82	49	861.159592		

The parameterization for the expected value of brate as a function of region and medage is given by

$$E(\text{brate}|\text{region} = i, \text{medage}) = \alpha_0 + \alpha_i + \beta_0\text{medage} + \beta_i\text{medage}$$

where α_0 is the intercept and β_0 is the slope of medage. We are modeling the effects of region in two different ways. The α_i parameters measure the effect of region on the intercept, and the β_i parameters measure the effect of region on the slope of medage.

contrast computes and tests effects on slopes separately from effects on intercepts. First, we will compute the reference category effects of region on the intercept:

```
. contrast r.region
```
Contrasts of marginal linear predictions

Margins   : asbalanced

	df	F	P>F
region			
(2 vs 1)	1	0.09	0.7691
(3 vs 1)	1	0.01	0.9389
(4 vs 1)	1	8.50	0.0057
Joint	3	11.30	0.0000
Residual	42		

	Contrast	Std. Err.	[95% Conf. Interval]	
region				
(2 vs 1)	-49.38396	167.1281	-386.6622	287.8942
(3 vs 1)	-9.058983	117.424	-246.0302	227.9123
(4 vs 1)	343.0024	117.6547	105.5656	580.4393

Now we will compute the reference category effects of region on the slope of medage:

```
. contrast r.region#c.medage
Contrasts of marginal linear predictions
Margins      : asbalanced
```

	df	F	P>F
region#c.medage			
(2 vs 1)	1	0.16	0.6917
(3 vs 1)	1	0.03	0.8558
(4 vs 1)	1	8.18	0.0066
Joint	3	10.29	0.0000
Residual	42		

	Contrast	Std. Err.	[95% Conf. Interval]	
region#c.medage				
(2 vs 1)	2.208539	5.530981	-8.953432	13.37051
(3 vs 1)	.6928008	3.788735	-6.953175	8.338777
(4 vs 1)	-10.94649	3.827357	-18.67041	-3.22257

At the 5% level, the slope of medage for the fourth region differs from that of the first region, but at that level of significance, we cannot say that the slope for the second or third region differs from that of the first.

This model is simple enough that the reference category contrasts reproduce the coefficients for region and for the interactions in an equivalent model fit by regress.

```
. regress brate region##c.medage
```

Source	SS	df	MS		
Model	37976.3149	7	5425.18784		
Residual	4220.5051	42	100.488217		
Total	42196.82	49	861.159592		

Number of obs = 50
F(7, 42) = 53.99
Prob > F = 0.0000
R-squared = 0.9000
Adj R-squared = 0.8833
Root MSE = 10.024

brate	Coef.	Std. Err.	t	P>|t|	[95% Conf. Interval]	
region						
2	-49.38396	167.1281	-0.30	0.769	-386.6622	287.8942
3	-9.058983	117.424	-0.08	0.939	-246.0302	227.9123
4	343.0024	117.6547	2.92	0.006	105.5656	580.4393
medage	-8.802707	3.462865	-2.54	0.015	-15.79105	-1.814362
region#						
c.medage						
2	2.208539	5.530981	0.40	0.692	-8.953432	13.37051
3	.6928008	3.788735	0.18	0.856	-6.953175	8.338777
4	-10.94649	3.827357	-2.86	0.007	-18.67041	-3.22257
_cons	411.8268	108.2084	3.81	0.000	193.4533	630.2002

This will not be the case for models that are more complicated.

Chow tests

Now let's suppose we are fitting a model for birth rates on median age and marriage rate. We are also interested in whether the regression coefficients differ for states in the east versus states in the west. We use census divisions to create a new variable, west, that indicates which states are in the western half of the United States.

```
. generate west = inlist(division, 4, 7, 8, 9)
```

We fit a model that includes a separate intercept for west as well as an interaction between west and each of the other variables in our model.

```
. regress brate i.west##c.medage i.west##c.mrgrate
```

Source	SS	df	MS
Model	38516.2172	5	7703.24344
Residual	3680.60281	44	83.6500639
Total	42196.82	49	861.159592

```
Number of obs =      50
F(  5,    44) =   92.09
Prob > F      =  0.0000
R-squared     =  0.9128
Adj R-squared =  0.9029
Root MSE      =   9.146
```

brate	Coef.	Std. Err.	t	P>|t|	[95% Conf. Interval]
1.west	327.8733	58.71793	5.58	0.000	209.5351 446.2115
medage	-7.532304	1.387624	-5.43	0.000	-10.32888 -4.735731
west# c.medage 1	-10.11443	1.849103	-5.47	0.000	-13.84105 -6.387808
mrgrate	828.6813	643.3443	1.29	0.204	-467.8939 2125.257
west# c.mrgrate 1	-800.8036	645.488	-1.24	0.221	-2101.699 500.092
_cons	366.5325	47.08904	7.78	0.000	271.6308 461.4343

We can test the effects of west on the intercept and on the slopes of medage and mrgrate. We will specify all these effects in a single contrast command and include the overall option to obtain a joint test of effects, that is, a test that the coefficients for eastern states and for western states are equal.

```
. contrast west west#c.medage west#c.mrgrate, overall
Contrasts of marginal linear predictions
Margins       : asbalanced
```

	df	F	P>F
west	1	31.18	0.0000
west#c.medage	1	29.92	0.0000
west#c.mrgrate	1	1.54	0.2213
Overall	3	22.82	0.0000
Residual	44		

This overall test is referred to as a Chow test in econometrics (Chow 1960).

Beyond linear models

contrast may be used after almost any estimation command, with the added benefit that contrast provides direct support for testing main and interaction effects that is not available in most estimation commands. To illustrate, we will use contrast with results from a logistic regression. Stata's logit command fits logistic regression models, reporting the fitted regression coefficients. The logistic command fits the same models but reports odds ratios. Although contrast can report odds ratios for the computed effects, the tests are all computed from linear combinations of the model coefficients regardless of which estimation command we used.

Suppose we have data on patient satisfaction for three hospitals in a city. Let's begin by fitting a model for satisfied, whether the patient was satisfied with his or her treatment, using the main effects of hospital:

```
. use http://www.stata-press.com/data/r12/hospital, clear
(Artificial hospital satisfaction data)

. logit satisfied i.hospital

Iteration 0:   log likelihood = -393.72216
Iteration 1:   log likelihood = -387.55736
Iteration 2:   log likelihood =  -387.4768
Iteration 3:   log likelihood = -387.47679

Logistic regression                               Number of obs    =       802
                                                  LR chi2(2)       =     12.49
                                                  Prob > chi2      =    0.0019
Log likelihood = -387.47679                       Pseudo R2        =    0.0159
```

satisfied	Coef.	Std. Err.	z	P>|z|	[95% Conf. Interval]	
hospital						
2	.5348129	.2136021	2.50	0.012	.1161604	.9534654
3	.7354519	.2221929	3.31	0.001	.2999618	1.170942
_cons	1.034708	.1391469	7.44	0.000	.7619855	1.307431

Because there are no other independent variables in this model, the reference category effects of hospital computed by contrast will match the fitted model coefficients, assuming a common reference level.

```
. contrast r.hospital

Contrasts of marginal linear predictions

Margins      : asbalanced
```

	df	chi2	P>chi2
hospital			
(2 vs 1)	1	6.27	0.0123
(3 vs 1)	1	10.96	0.0009
Joint	2	12.55	0.0019

	Contrast	Std. Err.	[95% Conf. Interval]	
hospital				
(2 vs 1)	.5348129	.2136021	.1161604	.9534654
(3 vs 1)	.7354519	.2221929	.2999618	1.170942

We see that the reference category effects are equal to the fitted coefficients. They also have the same interpretation, the difference in log odds from the reference category. The top table also provides a joint test of these effects, a test of the main effects of hospital.

We also have information on the condition for which each patient is being treated in the variable illness. Here we fit a logistic regression using a two-way crossed model of hospital and illness.

```
. label list illness
illness:
           1 heart attack
           2 stroke
           3 pneumonia
           4 lung disease
           5 kidney failure
. logistic satisfied hospital##illness
```

Logistic regression

				Number of obs	=	802
				LR chi2(14)	=	38.51
				Prob > chi2	=	0.0004
Log likelihood = -374.46865				Pseudo R2	=	0.0489

satisfied	Odds Ratio	Std. Err.	z	P>|z|	[95% Conf. Interval]	
hospital						
2	1.226496	.5492177	0.46	0.648	.509921	2.950049
3	1.711111	.8061016	1.14	0.254	.6796395	4.308021
illness						
2	1.328704	.6044214	0.62	0.532	.544779	3.240678
3	.7993827	.3408305	-0.53	0.599	.3466015	1.843653
4	1.231481	.5627958	0.46	0.649	.5028318	3.016012
5	1.25	.5489438	0.51	0.611	.5285676	2.956102
hospital# illness						
2 2	2.434061	1.768427	1.22	0.221	.5860099	10.11016
2 3	4.045805	2.868559	1.97	0.049	1.008058	16.23769
2 4	.54713	.3469342	-0.95	0.342	.1578866	1.89599
2 5	1.594425	1.081104	0.69	0.491	.4221288	6.022312
3 2	.5416535	.3590089	-0.93	0.355	.1477555	1.985635
3 3	1.579502	1.042504	0.69	0.489	.4332209	5.758783
3 4	3.137388	2.595748	1.38	0.167	.6198955	15.87881
3 5	1.672727	1.226149	0.70	0.483	.3976256	7.036812
_cons	2.571429	.8099239	3.00	0.003	1.386983	4.767358

Using contrast, we can obtain an ANOVA-style table of tests for the main effects and interaction effects of hospital and illness.

```
. contrast hospital##illness
Contrasts of marginal linear predictions
Margins      : asbalanced
```

	df	chi2	P>chi2
hospital	2	14.92	0.0006
illness	4	4.09	0.3937
hospital#illness	8	20.45	0.0088

Our interaction effect is significant, so we decide to evaluate the simple reference category effects of hospital within illness. We are particularly interested in patient satisfaction when being treated for a heart attack or stroke, so we will use the i. operator to limit our output to simple effects within the first two illnesses.

```
. contrast r.hospital@i(1 2).illness, nowald
Contrasts of marginal linear predictions
Margins      : asbalanced
```

	Contrast	Std. Err.	[95% Conf. Interval]	
hospital@illness				
(2 vs 1) 1	.2041611	.4477942	-.6734995	1.081822
(2 vs 1) 2	1.093722	.5721288	-.0276296	2.215074
(3 vs 1) 1	.5371429	.4710983	-.3861928	1.460479
(3 vs 1) 2	-.0759859	.4662325	-.9897847	.8378129

The row labeled (2 vs 1) 1 estimates simple effects on the log odds when comparing hospital 2 with hospital 1 for patients having heart attacks. These effects are differences in the cell means of the linear predictions.

We can add the or option to report an odds ratio for each of these simple effects:

```
. contrast r.hospital@i(1 2).illness, nowald or
Contrasts of marginal linear predictions
Margins      : asbalanced
```

	Odds Ratio	Std. Err.	[95% Conf. Interval]	
hospital@illness				
(2 vs 1) 1	1.226496	.5492177	.509921	2.950049
(2 vs 1) 2	2.985366	1.708014	.9727486	9.162089
(3 vs 1) 1	1.711111	.8061016	.6796395	4.308021
(3 vs 1) 2	.9268293	.4321179	.3716567	2.311306

These odds ratios are just the exponentiated version of the contrasts in the previous table.

For contrasts of the margins of nonlinear predictions, such as predicted probabilities, see [R] **margins, contrast**.

Multiple equations

contrast works with models containing multiple equations. Commands such as intreg and gnbreg allow their ancillary parameters to be modeled as functions of independent variables, and contrast can compute and test effects within these equations. In addition, contrast allows a special pseudofactor for equation—called _eqns—when working with results from manova, mvreg, mlogit, and mprobit.

In example 4 of [MV] **manova**, we fit a two-way MANOVA model using data from Woodard (1931). Here we will fit this model using mvreg. The data represent patients with jaw fractures. y1 is the patient's age, y2 is blood lymphocytes, and y3 is blood polymorphonuclears. Two factor variables, gender and fracture, are used as independent variables.

```
. use http://www.stata-press.com/data/r12/jaw
(Table 4.6 Two-Way Unbalanced Data for Fractures of the Jaw -- Rencher (1998))
. mvreg y1 y2 y3 = gender##fracture, vsquish
```

Equation	Obs	Parms	RMSE	"R-sq"	F	P
y1	27	6	10.21777	0.4086	2.902124	0.0382
y2	27	6	5.268768	0.4743	3.78967	0.0133
y3	27	6	4.993647	0.4518	3.460938	0.0195

	Coef.	Std. Err.	t	P>|t|	[95% Conf. Interval]	
y1						
2.gender	-17.5	11.03645	-1.59	0.128	-40.45156	5.451555
fracture						
2	-12.625	5.518225	-2.29	0.033	-24.10078	-1.149222
3	5.666667	5.899231	0.96	0.348	-6.601456	17.93479
gender#						
fracture						
2 2	21.375	12.68678	1.68	0.107	-5.008595	47.75859
2 3	8.833333	13.83492	0.64	0.530	-19.93796	37.60463
_cons	39.5	4.171386	9.47	0.000	30.82513	48.17487
y2						
2.gender	20.5	5.69092	3.60	0.002	8.665083	32.33492
fracture						
2	-3.125	2.84546	-1.10	0.285	-9.042458	2.792458
3	.6666667	3.041925	0.22	0.829	-5.659362	6.992696
gender#						
fracture						
2 2	-19.625	6.541907	-3.00	0.007	-33.22964	-6.02036
2 3	-23.66667	7.133946	-3.32	0.003	-38.50252	-8.830813
_cons	35.5	2.150966	16.50	0.000	31.02682	39.97318
y3						
2.gender	-18.16667	5.393755	-3.37	0.003	-29.38359	-6.949739
fracture						
2	1.083333	2.696877	0.40	0.692	-4.52513	6.691797
3	-3	2.883083	-1.04	0.310	-8.9957	2.9957
gender#						
fracture						
2 2	19.91667	6.200305	3.21	0.004	7.022426	32.81091
2 3	23.5	6.76143	3.48	0.002	9.438837	37.56116
_cons	61.16667	2.038648	30.00	0.000	56.92707	65.40627

contrast computes Wald tests using the coefficients from the first equation by default.

```
. contrast gender##fracture
```
Contrasts of marginal linear predictions
Margins   : asbalanced

	df	F	P>F
y1			
gender	1	2.16	0.1569
fracture	2	2.74	0.0880
gender#fracture	2	1.69	0.2085
Residual	21		

Here we use the equation() option to compute the Wald tests in the y2 equation:

```
. contrast gender##fracture, equation(y2)
```
Contrasts of marginal linear predictions
Margins   : asbalanced

	df	F	P>F
y2			
gender	1	5.41	0.0301
fracture	2	7.97	0.0027
gender#fracture	2	5.97	0.0088
Residual	21		

Here we use the equation index to compute the Wald tests in the third equation:

```
. contrast gender##fracture, equation(#3)
```
Contrasts of marginal linear predictions
Margins   : asbalanced

	df	F	P>F
y3			
gender	1	2.23	0.1502
fracture	2	6.36	0.0069
gender#fracture	2	6.66	0.0058
Residual	21		

Here we use the atequations option to compute Wald tests for each equation in the model. We also use the vsquish option to suppress the extra blank lines between terms.

```
. contrast gender##fracture, atequations vsquish
```
Contrasts of marginal linear predictions
Margins   : asbalanced

	df	F	P>F
y1			
gender	1	2.16	0.1569
fracture	2	2.74	0.0880
gender#fracture	2	1.69	0.2085
y2			
gender	1	5.41	0.0301
fracture	2	7.97	0.0027
gender#fracture	2	5.97	0.0088
y3			
gender	1	2.23	0.1502
fracture	2	6.36	0.0069
gender#fracture	2	6.66	0.0058
Residual	21		

Because we are investigating the results from mvreg, we can use the special _eqns factor to test for a marginal effect on the means among the dependent variables:

```
. contrast _eqns
```
Contrasts of marginal linear predictions
Margins   : asbalanced

	df	F	P>F
_eqns	2	49.19	0.0000
Residual	21		

Here we test whether the main effects of gender differ among the dependent variables:

```
. contrast gender#_eqns
```
Contrasts of marginal linear predictions
Margins   : asbalanced

	df	F	P>F
gender#_eqns	2	3.61	0.0448
Residual	21		

Although it is not terribly interesting in this case, we can even calculate contrasts across equations:

```
. contrast gender#r._eqns
Contrasts of marginal linear predictions
Margins      : asbalanced
```

	df	F	P>F
gender#_eqns			
(joint) (2 vs 1)	1	5.82	0.0251
(joint) (3 vs 1)	1	0.40	0.5352
Joint	2	3.61	0.0448
Residual	21		

Saved results

contrast saves the following in r():

Scalars
  r(df_r)   variance degrees of freedom, from original estimation results
  r(k_terms)   number of terms in *termlist*
  r(level)   confidence level of confidence intervals

Macros
  r(cmd)   contrast
  r(cmdline)   command as typed
  r(est_cmd)   e(cmd) from original estimation results
  r(est_cmdline)   e(cmdline) from original estimation results
  r(title)   title in output
  r(overall)   overall or empty
  r(emptycells)   *empspec* from emptycells()
  r(mcmethod)   *method* from mcompare()
  r(mctitle)   title for *method* from mcompare()
  r(mcadjustall)   adjustall or empty
  r(margin_method)   asbalanced or asobserved

Matrices
  r(b)   contrast estimates
  r(V)   variance–covariance matrix of the contrast estimates
  r(error)   contrast estimability codes;
                        0 means estimable,
                        8 means not estimable
  r(L)   matrix of contrasts applied to the model coefficients
  r(table)   matrix containing the contrasts with their standard errors,
                        test statistics, *p*-values, and confidence intervals
  r(F)   vector of *F* statistics; r(df_r) present
  r(chi2)   vector of χ^2 statistics; r(df_r) not present
  r(p)   vector of *p*-values corresponding to r(F) or r(chi2)
  r(df)   vector of degrees of freedom corresponding to r(p)
  r(df2)   vector of denominator degrees of freedom corresponding to r(F)

contrast with the post option saves the following in e():

Scalars
e(df_r)	variance degrees of freedom, from original estimation results
e(k_terms)	number of terms in *termlist*

Macros
e(cmd)	contrast
e(cmdline)	command as typed
e(est_cmd)	e(cmd) from original estimation results
e(est_cmdline)	e(cmdline) from original estimation results
e(title)	title in output
e(overall)	overall or empty
e(emptycells)	*empspec* from emptycells()
e(margin_method)	asbalanced or asobserved
e(properties)	b V

Matrices
e(b)	contrast estimates
e(V)	variance–covariance matrix of the contrast estimates
e(error)	contrast estimability codes;
	0 means estimable,
	8 means not estimable
e(L)	matrix of contrasts applied to the model coefficients
e(F)	vector of F statistics; e(df_r) present
e(chi2)	vector of χ^2 statistics; e(df_r) not present
e(p)	vector of p-values corresponding to e(F) or e(chi2)
e(df)	vector of degrees of freedom corresponding to e(p)
e(df2)	vector of denominator degrees of freedom corresponding to e(F)

Methods and formulas

contrast is implemented as an ado-file.

Methods and formulas are presented under the following headings:

> *Marginal linear predictions*
> *Contrast operators*
>   *Reference level contrasts*
>   *Adjacent contrasts*
>   *Grand mean contrasts*
>   *Helmert contrasts*
>   *Reverse Helmert contrasts*
>   *Orthogonal polynomial contrasts*
> *Contrasts within interactions*
> *Multiple comparisons*

Marginal linear predictions

contrast treats intercept effects separately from slope effects. To illustrate, consider the following parameterization for a quadratic regression of y on x that also models the effects of two factor variables A and B, where the levels of A are indexed by $i = 1, \ldots, k_a$ and the levels of B are indexed by $j = 1, \ldots, k_b$.

$$E(y|A = i, B = j, x) = \eta_{0ij} + \eta_{1ij}x + \eta_{2ij}x^2$$

$$\eta_{0ij} = \eta_0 + \alpha_{0i} + \beta_{0j} + (\alpha\beta)_{0ij}$$

$$\eta_{1ij} = \eta_1 + \alpha_{1i} + \beta_{1j} + (\alpha\beta)_{1ij}$$

$$\eta_{2ij} = \eta_2 + \alpha_{2i} + \beta_{2j} + (\alpha\beta)_{2ij}$$

We have partitioned the coefficients into three groups of parameters: η_{0ij} is a cell prediction for the intercept, η_{1ij} is a cell prediction for the slope on x, and η_{2ij} is a cell prediction for the slope on x^2. For the intercept parameters, η_0 is the intercept, α_{0i} represents a main effect for factor A at its ith level, β_{0j} represents a main effect for factor B at its jth level, and $(\alpha\beta)_{0ij}$ represents an effect for the interaction of A and B at the ijth level. The individual coefficients in η_{1ij} and η_{2ij} have similar interpretations, but the effects are on the slopes of x and x^2, respectively.

The marginal intercepts for A are given by

$$\eta_{0i.} = \sum_{j=1}^{k_b} f_{ij}\eta_{0ij}$$

where f_{ij} is a marginal relative frequency of the jth level of B and is controlled by the asobserved and emptycells(reweight) options according to

$$f_{ij} = \begin{cases} 1/k_b, & \text{default} \\ w_{.j}/w_{..}, & \text{asobserved} \\ 1/(k_b - e_{i.}), & \text{emptycells(reweight)} \\ w_{ij}/w_{i.}, & \text{emptycells(reweight) and asobserved} \end{cases}$$

Above, w_{ij} is the number of individuals with A at its ith level and B at its jth,

$$w_{i.} = \sum_{j=1}^{k_b} w_{ij}$$

$$w_{.j} = \sum_{i=1}^{k_a} w_{ij}$$

$$w_{..} = \sum_{i=1}^{k_a}\sum_{j=1}^{k_b} w_{ij}$$

and $e_{i.}$ is the number of empty cells where A is at its ith level. The marginal intercepts for B and marginal slopes on x and x^2 are similarly defined.

Estimates for the cell intercepts and slopes are computed using the corresponding linear combination of the coefficients from the fitted model. For example, the estimated cell intercepts are computed using

$$\widehat{\eta}_{0ij} = \widehat{\eta}_0 + \widehat{\alpha}_{0i} + \widehat{\beta}_{0j} + \widehat{(\alpha\beta)}_{0ij}$$

and the estimated marginal intercepts for A are computed as

$$\widehat{\eta}_{0i\cdot} = \sum_{j=1}^{k_b} f_{ij}\widehat{\eta}_{0ij}$$

Contrast operators

`contrast` performs Wald tests using linear combinations of marginal linear predictions. For example, the following linear combination can be used to test for a specific effect of factor A on the marginal intercepts.

$$\sum_{i=1}^{k_a} c_i \eta_{0i\cdot}$$

If the c_i elements sum to zero, the linear combination is called a contrast. If the factor A is represented by a variable named A, then we specify this contrast using the following syntax:

$$\{A\ c_1\ c_2\ \ldots\ c_{k_a}\}$$

Similarly, the following linear combination can be used to test for a specific interaction effect of factors A and B on the marginal slope of x.

$$\sum_{i=1}^{k_a}\sum_{j=1}^{k_b} c_{ij}\eta_{1ij}$$

If the factor B is represented by a variable named B, then we specify this contrast using the following syntax:

$$\{A\#B\ c_{11}\ c_{12}\ \ldots\ c_{1k_b}\ c_{21}\ \ldots\ c_{k_a k_b}\}$$

`contrast` has variable operators for several commonly used contrasts. Each contrast operator specifies a matrix of linear combinations that yield the requested set of contrasts to be applied to the marginal linear predictions associated with the attached factor variable.

Reference level contrasts

The `r.` operator compares each level with a reference level. Let $\mathbf{R}$ be the corresponding contrast matrix for factor A, and then $\mathbf{R}$ is a $(k_a - 1) \times k_a$ matrix with elements

$$\mathbf{R}_{ij} = \begin{cases} -1, & \text{if } j \text{ is the reference level} \\ 1, & \text{if } i = j \text{ and } j \text{ is less than the reference level} \\ 1, & \text{if } i + 1 = j \text{ and } j \text{ is greater than the reference level} \\ 0, & \text{otherwise} \end{cases}$$

If $k_a = 5$ and the reference level is the third level of A (specified as rb(#3).A), then

$$\mathbf{R} = \begin{pmatrix} 1 & 0 & -1 & 0 & 0 \\ 0 & 1 & -1 & 0 & 0 \\ 0 & 0 & -1 & 1 & 0 \\ 0 & 0 & -1 & 0 & 1 \end{pmatrix}$$

Adjacent contrasts

The a. operator compares each level with the next level. Let $\mathbf{A}$ be the corresponding contrast matrix for factor A, and then $\mathbf{A}$ is a $(k_a - 1) \times k_a$ matrix with elements

$$\mathbf{A}_{ij} = \begin{cases} 1, & \text{if } i = j \\ -1, & \text{if } i + 1 = j \\ 0, & \text{otherwise} \end{cases}$$

If $k_a = 5$, then

$$\mathbf{A} = \begin{pmatrix} 1 & -1 & 0 & 0 & 0 \\ 0 & 1 & -1 & 0 & 0 \\ 0 & 0 & 1 & -1 & 0 \\ 0 & 0 & 0 & 1 & -1 \end{pmatrix}$$

The ar. operator compares each level with the previous level. If $\mathbf{A}$ is the contrast matrix for the a. operator, then $-\mathbf{A}$ is the corresponding contrast matrix for the ar. operator.

Grand mean contrasts

The g. operator compares each level with the mean of all the levels. Let $\mathbf{G}$ be the corresponding contrast matrix for factor A, and then $\mathbf{G}$ is a $k_a \times k_a$ matrix with elements

$$\mathbf{G}_{ij} = \begin{cases} 1 - 1/k_a, & \text{if } i = j \\ -1/k_a, & \text{if } i \neq j \end{cases}$$

If $k_a = 5$, then

$$\mathbf{G} = \begin{pmatrix} 4/5 & -1/5 & -1/5 & -1/5 & -1/5 \\ -1/5 & 4/5 & -1/5 & -1/5 & -1/5 \\ -1/5 & -1/5 & 4/5 & -1/5 & -1/5 \\ -1/5 & -1/5 & -1/5 & 4/5 & -1/5 \\ -1/5 & -1/5 & -1/5 & -1/5 & 4/5 \end{pmatrix}$$

The gw. operator compares each level with the weighted mean of all the levels. The weights are taken from the observed weighted cell frequencies in the estimation sample of the fitted model. Let $\mathbf{G}_w$ be the corresponding contrast matrix for factor A, and then $\mathbf{G}_w$ is a $k_a \times k_a$ matrix with elements

$$\mathbf{G}_{ij} = \begin{cases} 1 - w_i/w., & \text{if } i = j \\ -w_j/w., & \text{if } i \neq j \end{cases}$$

where w_i is a marginal weight representing the number of individuals with A at its ith level and $w. = \sum_i w_i$.

Helmert contrasts

The h. operator compares each level with the mean of the subsequent levels. Let $\mathbf{H}$ be the corresponding contrast matrix for factor A, and then $\mathbf{H}$ is a $(k_a - 1) \times k_a$ matrix with elements

$$\mathbf{H}_{ij} = \begin{cases} 1, & \text{if } i = j \\ -1/(k_a - i), & \text{if } i < j \\ 0, & \text{otherwise} \end{cases}$$

If $k_a = 5$, then

$$\mathbf{H} = \begin{pmatrix} 1 & -1/4 & -1/4 & -1/4 & -1/4 \\ 0 & 1 & -1/3 & -1/3 & -1/3 \\ 0 & 0 & 1 & -1/2 & -1/2 \\ 0 & 0 & 0 & 1 & -1 \end{pmatrix}$$

The hw. operator compares each level with the weighted mean of the subsequent levels. Let $\mathbf{H}_w$ be the corresponding contrast matrix for factor A, and then $\mathbf{H}_w$ is a $(k_a - 1) \times k_a$ matrix with elements

$$\mathbf{H}_{wij} = \begin{cases} 1, & \text{if } i = j \\ -w_j/\sum_{l=j}^{k_a} w_l, & \text{if } i < j \\ 0, & \text{otherwise} \end{cases}$$

Reverse Helmert contrasts

The j. operator compares each level with the mean of the previous levels. Let $\mathbf{J}$ be the corresponding contrast matrix for factor A, and then $\mathbf{J}$ is a $(k_a - 1) \times k_a$ matrix with elements

$$\mathbf{J}_{ij} = \begin{cases} 1, & \text{if } i + 1 = j \\ -1/i, & \text{if } j \leq i \\ 0, & \text{otherwise} \end{cases}$$

If $k_a = 5$, then

$$\mathbf{H} = \begin{pmatrix} -1 & 1 & 0 & 0 & 0 \\ -1/2 & -1/2 & 1 & 0 & 0 \\ -1/3 & -1/3 & -1/3 & 1 & 0 \\ -1/4 & -1/4 & -1/4 & -1/4 & 1 \end{pmatrix}$$

The jw. operator compares each level with the weighted mean of the previous levels. Let $\mathbf{J}_w$ be the corresponding contrast matrix for factor A, and then $\mathbf{J}_w$ is a $(k_a - 1) \times k_a$ matrix with elements

$$
\mathbf{J}_{wij} = \begin{cases} 1, & \text{if } i+1 = j \\ -w_j / \sum_{l=1}^{i} w_l, & \text{if } i \leq j \\ 0, & \text{otherwise} \end{cases}
$$

Orthogonal polynomial contrasts

The p. operator applies orthogonal polynomial contrasts using the level values of the attached factor variable. The q. operator applies orthogonal polynomial contrasts using the level indices of the attached factor variable. These two operators are equivalent when the level values of the attached factor are equally spaced. The pw. and qw. operators are weighted versions of p. and q., where the weights are taken from the observed weighted cell frequencies in the estimation sample of the fitted model. contrast uses the Christoffel–Darboux recurrence formula for computing orthogonal polynomial contrasts (Abramowitz and Stegun 1972). The elements of the contrasts are normalized such that

$$
\mathbf{Q}'\mathbf{W}\mathbf{Q} = \frac{1}{w.}\mathbf{I}
$$

where $\mathbf{W}$ is a diagonal matrix of the marginal cell weights $w_1, w_2, \ldots, w_k$ of the attached factor variable (all 1 for p. and q.), and $w.$ is the sum of the weights (the number of levels k for p. and q.).

Contrasts within interactions

Contrast operators are allowed to be specified on factor variables participating in interactions. In such cases, contrast applies the proper matrix product of the contrast matrices to the cell margins of the interacted factor variables.

For example, consider the contrasts implied by specifying r.A#h.B. Let $\mathbf{M}$ be the matrix of estimated cell margins for the levels of A and B, where the rows of $\mathbf{M}$ are indexed by the levels of A and the columns are indexed by the levels of B. contrast puts the estimated cell margins in the following vector form:

$$
\mathbf{v} = \mathrm{vec}(\mathbf{M}') = \begin{pmatrix} \mathbf{M}_{11} \\ \mathbf{M}_{12} \\ \vdots \\ \mathbf{M}_{1k_b} \\ \mathbf{M}_{21} \\ \mathbf{M}_{22} \\ \vdots \\ \mathbf{M}_{2k_b} \\ \vdots \\ \mathbf{M}_{k_a k_b} \end{pmatrix}
$$

The individual contrasts are then given by the elements of

$$(\mathbf{R} \otimes \mathbf{H})\mathbf{v}$$

where $\otimes$ denotes the Kronecker direct product.

Multiple comparisons

See [R] **pwcompare** for details on the methods and formulas used to adjust p-values and confidence intervals for multiple comparisons. The formulas for Bonferroni's method and Šidák's method are presented with $m = k(k-1)/2$, the number of pairwise comparisons for a factor term with k levels. For contrasts, m is instead the number of contrasts being performed on the factor term; often, $m = k - 1$ for a term with k levels.

References

Abramowitz, M., and I. A. Stegun, ed. 1972. *Handbook of Mathematical Functions with Formulas, Graphs, and Mathematical Tables.* 10th ed. Washington, DC: National Bureau of Standards.

Chow, G. C. 1960. Tests of equality between sets of coefficients in two linear regressions. *Econometrica* 28: 591–605.

Coster, D. 2005. Contrasts. In Vol. 2 of *Encyclopedia of Biostatistics*, ed. P. Armitage and T. Colton, 1153–1157. Chichester, UK: Wiley.

Kuehl, R. O. 2000. *Design of Experiments: Statistical Principles of Research Design and Analysis.* 2nd ed. Belmont, CA: Duxbury.

Milliken, G. A., and D. E. Johnson. 2009. *Analysis of Messy Data, Volume 1: Designed Experiments.* 2nd ed. Boca Raton, FL: CRC Press.

Rosenthal, R., R. L. Rosnow, and D. B. Rubin. 2000. *Contrasts and Effect Sizes in Behavioral Research: A Correlational Approach.* Cambridge: Cambridge University Press.

Searle, S. R. 1971. *Linear Models.* New York: Wiley.

———. 1997. *Linear Models for Unbalanced Data.* New York: Wiley.

Winer, B. J., D. R. Brown, and K. M. Michels. 1991. *Statistical Principles in Experimental Design.* 3rd ed. New York: McGraw–Hill.

Woodard, D. E. 1931. Healing time of fractures of the jaw in relation to delay before reduction, infection, syphilis and blood calcium and phosphorus content. *Journal of the American Dental Association* 18: 419–442.

Also see

[R] **contrast postestimation** — Postestimation tools for contrast

[R] **lincom** — Linear combinations of estimators

[R] **margins** — Marginal means, predictive margins, and marginal effects

[R] **margins, contrast** — Contrasts of margins

[R] **pwcompare** — Pairwise comparisons

[R] **test** — Test linear hypotheses after estimation

[U] **20 Estimation and postestimation commands**

Title

> **contrast postestimation** — Postestimation tools for contrast

Description

The following postestimation commands are available after `contrast, post`:

Command	Description
estat	VCE; `estat vce` only
estat (svy)	postestimation statistics for survey data
estimates	cataloging estimation results
lincom	point estimates, standard errors, testing, and inference for linear combinations of coefficients
nlcom	point estimates, standard errors, testing, and inference for nonlinear combinations of coefficients
test	Wald tests of simple and composite linear hypotheses
testnl	Wald tests of nonlinear hypotheses

See the corresponding entries in the *Base Reference Manual* for details, but see [SVY] **estat** for details about `estat` (svy).

Remarks

In *Orthogonal polynomial contrasts* in [R] **contrast**, we used the `p.` operator to test the orthogonal polynomial effects of age group.

```
. contrast p.agegrp, noeffects
```

We then used a second `contrast` command,

```
. contrast p(2 3 4).agegrp, noeffects
```

selecting levels to test whether the quadratic, cubic, and quartic contrasts were jointly significant.

We can perform the same joint test by using the `test` command after specifying the `post` option with our first `contrast` command.

```
. use http://www.stata-press.com/data/r12/cholesterol
(Artificial cholesterol data)

. anova chol agegrp
 (output omitted )

. contrast p.agegrp, noeffects post
Contrasts of marginal linear predictions
Margins       : asbalanced
```

	df	F	P>F
agegrp			
(linear)	1	139.11	0.0000
(quadratic)	1	0.15	0.6962
(cubic)	1	0.37	0.5448
(quartic)	1	0.43	0.5153
Joint	4	35.02	0.0000
Residual	70		

```
. test p2.agegrp p3.agegrp p4.agegrp

 ( 1)  p2.agegrp = 0
 ( 2)  p3.agegrp = 0
 ( 3)  p4.agegrp = 0

       F(  3,    70) =     0.32
            Prob > F =     0.8129
```

Also see

[R] **contrast** — Contrasts and linear hypothesis tests after estimation

[U] **20 Estimation and postestimation commands**

Title

> **copyright** — Display copyright information

Syntax

```
copyright
```

Description

`copyright` presents copyright notifications concerning tools, libraries, etc., used in the construction of Stata.

Remarks

The correct form for a copyright notice is

Copyright *dates* by *author/owner*

The word "Copyright" is spelled out. You can use the © symbol, but "(C)" has never been given legal recognition. The phrase "All Rights Reserved" was historically required but is no longer needed.

Currently, most works are copyrighted from the moment they are written, and no copyright notice is required. Copyright concerns the protection of the expression and structure of facts and ideas, not the facts and ideas themselves. Copyright concerns the ownership of the expression and not the name given to the expression, which is covered under trademark law.

Copyright law as it exists today began in England in 1710 with the Statute of Anne, *An Act for the Encouragement of Learning, by Vesting the Copies of Printed Books in the Authors or Purchases of Such Copies, during the Times therein mentioned*. In 1672, Massachusetts introduced the first copyright law in what was to become the United States. After the Revolutionary War, copyright was introduced into the U.S. Constitution in 1787 and went into effect on May 31, 1790. On June 9, 1790, the first copyright in the United States was registered for *The Philadelphia Spelling Book* by John Barry.

There are significant differences in the understanding of copyright in the English- and non–English-speaking world. The Napoleonic or Civil Code, the dominant legal system in the non–English-speaking world, splits the rights into two classes: the author's economic rights and the author's moral rights. Moral rights are available only to "natural persons". Legal persons (corporations) have economic rights but not moral rights.

Also see

Copyright page of this book

Title

Description

Stata uses portions of Boost, a library used by JagPDF, which helps create PDF files, with the express permission of the authors pursuant to the following notice:

Also see

[R] **copyright** — Display copyright information

Title

Description

Stata uses portions of FreeType, a library used by JagPDF, which helps create PDF files, with the express permission of the authors.

StataCorp thanks and acknowledges the authors of FreeType for producing FreeType and allowing its use in Stata and other software.

For more information about FreeType, visit http://www.freetype.org/.

The full FreeType copyright notice is

Legal Terms

0. Definitions

Throughout this license, the terms 'package', 'FreeType Project', and 'FreeType archive' refer to the set of files originally distributed by the authors (David Turner, Robert Wilhelm, and Werner Lemberg) as the 'FreeType Project', be they named as alpha, beta or final release.

'You' refers to the licensee, or person using the project, where 'using' is a generic term including compiling the project's source code as well as linking it to form a 'program' or 'executable'. This program is referred to as 'a program using the FreeType engine'.

This license applies to all files distributed in the original FreeType Project, including all source code, binaries and documentation, unless otherwise stated in the file in its original, unmodified form as distributed in the original archive. If you are unsure whether or not a particular file is covered by this license, you must contact us to verify this.

This license applies to all files distributed in the original FreeType Project, including all source code, binaries and documentation, unless otherwise stated in the file in its original, unmodified form as distributed in the original archive. If you are unsure whether or not a particular file is covered by this license, you must contact us to verify this.

The FreeType Project is copyright © 1996–2000 by David Turner, Robert Wilhelm, and Werner Lemberg. All rights reserved except as specified below.

1. No Warranty

THE FREETYPE PROJECT IS PROVIDED 'AS IS' WITHOUT WARRANTY OF ANY KIND, EITHER EXPRESS OR IMPLIED, INCLUDING, BUT NOT LIMITED TO, WARRANTIES OF MERCHANTABILITY AND FITNESS FOR A PARTICULAR PURPOSE. IN NO EVENT WILL ANY OF THE AUTHORS OR COPYRIGHT HOLDERS BE LIABLE FOR ANY DAMAGES CAUSED BY THE USE OR THE INABILITY TO USE, OF THE FREETYPE PROJECT.

2. Redistribution

This license grants a worldwide, royalty-free, perpetual and irrevocable right and license to use, execute, perform, compile, display, copy, create derivative works of, distribute and sublicense the FreeType Project (in both source and object code forms) and derivative works thereof for any purpose; and to authorize others to exercise some or all of the rights granted herein, subject to the following conditions:

- Redistribution of source code must retain this license file ('FTL.TXT') unaltered; any additions, deletions or changes to the original files must be clearly indicated in accompanying documentation. The copyright notices of the unaltered, original files must be preserved in all copies of source files.

- Redistribution in binary form must provide a disclaimer that states that the software is based in part of the work of the FreeType Team, in the distribution documentation. We also encourage you to put an URL to the FreeType web page in your documentation, though this isn't mandatory.

These conditions apply to any software derived from or based on the FreeType Project, not just the unmodified files. If you use our work, you must acknowledge us. However, no fee need be paid to us.

3. Advertising

Neither the FreeType authors and contributors nor you shall use the name of the other for commercial, advertising, or promotional purposes without specific prior written permission.

We suggest, but do not require, that you use one or more of the following phrases to refer to this software in your documentation or advertising materials: 'FreeType Project', 'FreeType Engine', 'FreeType library', or 'FreeType Distribution'.

As you have not signed this license, you are not required to accept it. However, as the FreeType Project is copyrighted material, only this license, or another one contracted with the authors, grants you the right to use, distribute, and modify it. Therefore, by using, distributing, or modifying the FreeType Project, you indicate that you understand and accept all the terms of this license.

4. Contacts

There are two mailing lists related to FreeType:

- freetype@nongnu.org

  Discusses general use and applications of FreeType, as well as future and wanted additions to the library and distribution. If you are looking for support, start in this list if you haven't found anything to help you in the documentation.

- freetype-devel@nongnu.org

  Discusses bugs, as well as engine internals, design issues, specific licenses, porting, etc.

Our home page can be found at

http://www.freetype.org

Also see

[R] **copyright** — Display copyright information

Title

Description

Stata uses portions of ICU, a library used by JagPDF, which helps create PDF files, with the express permission of the authors pursuant to the following notice:

Also see

[R] **copyright** — Display copyright information

Title

> **copyright jagpdf** — JagPDF copyright notification

Description

Stata uses portions of JagPDF, a library for creating PDF files, with the express permission of the author pursuant to the following notice:

Also see

[R] **copyright** — Display copyright information

Title

Description

Stata uses portions of LAPACK, a linear algebra package, with the express permission of the authors pursuant to the following notice:

Also see

[R] **copyright** — Display copyright information

Title

Description

Stata uses portions of libpng, a library used by JagPDF, which helps create PDF files, with the express permission of the authors.

For the purposes of this acknowledgement, "Contributing Authors" is as defined by the copyright notice below.

StataCorp thanks and acknowledges the Contributing Authors of libpng and Group 42, Inc. for producing libpng and allowing its use in Stata and other software.

For more information about libpng, visit http://www.libpng.org/.

The full libpng copyright notice is

libpng versions 0.89, June 1996, through 0.96, May 1997, are Copyright © 1996, 1997 Andreas Dilger Distributed according to the same disclaimer and license as libpng-0.88, with the following individuals added to the list of Contributing Authors:

John Bowler

Kevin Bracey

Sam Bushell

Magnus Holmgren

Greg Roelofs

Tom Tanner

libpng versions 0.5, May 1995, through 0.88, January 1996, are Copyright © 1995, 1996 Guy Eric Schalnat, Group 42, Inc.

For the purposes of this copyright and license, "Contributing Authors" is defined as the following set of individuals:

Andreas Dilger

Dave Martindale

Guy Eric Schalnat

Paul Schmidt

Tim Wegner

The PNG Reference Library is supplied "AS IS". The Contributing Authors and Group 42, Inc. disclaim all warranties, expressed or implied, including, without limitation, the warranties of merchantability and of fitness for any purpose. The Contributing Authors and Group 42, Inc. assume no liability for direct, indirect, incidental, special, exemplary, or consequential damages, which may result from the use of the PNG Reference Library, even if advised of the possibility of such damage.

Permission is hereby granted to use, copy, modify, and distribute this source code, or portions hereof, for any purpose, without fee, subject to the following restrictions:

1. The origin of this source code must not be misrepresented.

2. Altered versions must be plainly marked as such and must not be misrepresented as being the original source.

3. This Copyright notice may not be removed or altered from any source or altered source distribution.

The Contributing Authors and Group 42, Inc. specifically permit, without fee, and encourage the use of this source code as a component to supporting the PNG file format in commercial products. If you use this source code in a product, acknowledgment is not required but would be appreciated.

Also see

[R] **copyright** — Display copyright information

Title

copyright scintilla — Scintilla copyright notification

Description

Stata uses portions of Scintilla with the express permission of the author, pursuant to the following notice:

Also see

[R] **copyright** — Display copyright information

Title

copyright ttf2pt1 — ttf2pt1 copyright notification

Description

Stata uses portions of ttf2pt1 to convert TrueType fonts to PostScript fonts, with express permission of the authors, pursuant to the following notice:

Copyright © 1997–2003 by the AUTHORS:

Andrew Weeks <ccsaw@bath.ac.uk>

Frank M. Siegert <fms@this.net>

Mark Heath <mheath@netspace.net.au>

Thomas Henlich <thenlich@rcs.urz.tu-dresden.de>

Sergey Babkin <babkin@users.sourceforge.net>, <sab123@hotmail.com>

Turgut Uyar <uyar@cs.itu.edu.tr>

Rihardas Hepas <rch@WriteMe.Com>

Szalay Tamas <tomek@elender.hu>

Johan Vromans <jvromans@squirrel.nl>

Petr Titera <P.Titera@sh.cvut.cz>

Lei Wang <lwang@amath8.amt.ac.cn>

Chen Xiangyang <chenxy@sun.ihep.ac.cn>

Zvezdan Petkovic <z.petkovic@computer.org>

Rigel <rigel863@yahoo.com>

All rights reserved.

Redistribution and use in source and binary forms, with or without modification, are permitted provided that the following conditions are met:

1. Redistributions of source code must retain the above copyright notice, this list of conditions and the following disclaimer.

2. Redistributions in binary form must reproduce the above copyright notice, this list of conditions and the following disclaimer in the documentation and/or other materials provided with the distribution.

3. All advertising materials mentioning features or use of this software must display the following acknowledgment: This product includes software developed by the TTF2PT1 Project and its contributors.

THIS SOFTWARE IS PROVIDED BY THE AUTHORS AND CONTRIBUTORS "AS IS" AND ANY EXPRESS OR IMPLIED WARRANTIES, INCLUDING, BUT NOT LIMITED TO, THE IMPLIED WARRANTIES OF MERCHANTABILITY AND FITNESS FOR A PARTICULAR PURPOSE ARE DISCLAIMED. IN NO EVENT SHALL THE AUTHORS OR CONTRIBUTORS BE LIABLE FOR ANY DIRECT, INDIRECT, INCIDENTAL, SPECIAL, EXEMPLARY, OR CONSEQUENTIAL

DAMAGES (INCLUDING, BUT NOT LIMITED TO, PROCUREMENT OF SUBSTITUTE GOODS OR SERVICES; LOSS OF USE, DATA, OR PROFITS; OR BUSINESS INTERRUPTION) HOWEVER CAUSED AND ON ANY THEORY OF LIABILITY, WHETHER IN CONTRACT, STRICT LIABILITY, OR TORT (INCLUDING NEGLIGENCE OR OTHERWISE) ARISING IN ANY WAY OUT OF THE USE OF THIS SOFTWARE, EVEN IF ADVISED OF THE POSSIBILITY OF SUCH DAMAGE.

Also see

[R] **copyright** — Display copyright information

Title

Description

Stata uses portions of zlib, a library used by JagPDF, which helps create PDF files, with the express permission of the authors.

StataCorp thanks and acknowledges the authors of zlib, Jean-loup Gailly and Mark Adler, for producing zlib and allowing its use in Stata and other software.

For more information about zlib, visit http://www.zlib.net/.

The full zlib copyright notice is

Also see

[R] **copyright** — Display copyright information

Title

> **correlate** — Correlations (covariances) of variables or coefficients

Syntax

Display correlation matrix or covariance matrix

> <u>cor</u>relate [*varlist*] [*if*] [*in*] [*weight*] [, *correlate_options*]

Display all pairwise correlation coefficients

> pwcorr [*varlist*] [*if*] [*in*] [*weight*] [, *pwcorr_options*]

correlate_options	Description
Options	
<u>means</u>	display means, standard deviations, minimums, and maximums with matrix
<u>nof</u>ormat	ignore display format associated with variables
<u>c</u>ovariance	display covariances
<u>wrap</u>	allow wide matrices to wrap

pwcorr_options	Description
Main	
<u>o</u>bs	print number of observations for each entry
sig	print significance level for each entry
<u>list</u>wise	use listwise deletion to handle missing values
<u>case</u>wise	synonym for listwise
<u>print</u>(#)	significance level for displaying coefficients
<u>star</u>(#)	significance level for displaying with a star
<u>bon</u>ferroni	use Bonferroni-adjusted significance level
<u>sid</u>ak	use Šidák-adjusted significance level

varlist may contain time-series operators; see [U] **11.4.4 Time-series varlists**.
by is allowed with correlate and pwcorr; see [D] **by**.
aweights and fweights are allowed; see [U] **11.1.6 weight**.

Menu

correlate

Statistics > Summaries, tables, and tests > Summary and descriptive statistics > Correlations and covariances

pwcorr

Statistics > Summaries, tables, and tests > Summary and descriptive statistics > Pairwise correlations

Description

The `correlate` command displays the correlation matrix or covariance matrix for a group of variables. If *varlist* is not specified, the matrix is displayed for all variables in the dataset. Also see the `estat vce` command in [R] **estat**.

`pwcorr` displays all the pairwise correlation coefficients between the variables in *varlist* or, if *varlist* is not specified, all the variables in the dataset.

Options for correlate

`means` displays summary statistics (means, standard deviations, minimums, and maximums) with the matrix.

`noformat` displays the summary statistics requested by the `means` option in g format, regardless of the display formats associated with the variables.

`covariance` displays the covariances rather than the correlation coefficients.

`wrap` requests that no action be taken on wide correlation matrices to make them readable. It prevents Stata from breaking wide matrices into pieces to enhance readability. You might want to specify this option if you are displaying results in a window wider than 80 characters. Then you may need to `set linesize` to however many characters you can display across a line; see [R] **log**.

Options for pwcorr

`obs` adds a line to each row of the matrix reporting the number of observations used to calculate the correlation coefficient.

`sig` adds a line to each row of the matrix reporting the significance level of each correlation coefficient.

`listwise` handles missing values through listwise deletion, meaning that the entire observation is omitted from the estimation sample if any of the variables in *varlist* is missing for that observation. By default, `pwcorr` handles missing values by pairwise deletion; all available observations are used to calculate each pairwise correlation without regard to whether variables outside that pair are missing.

    `correlate` uses listwise deletion. Thus `listwise` allows users of `pwcorr` to mimic `correlate`'s treatment of missing values while retaining access to `pwcorr`'s features.

`casewise` is a synonym for `listwise`.

`print(#)` specifies the significance level of correlation coefficients to be printed. Correlation coefficients with larger significance levels are left blank in the matrix. Typing `pwcorr, print(.10)` would list only correlation coefficients significant at the 10% level or better.

`star(#)` specifies the significance level of correlation coefficients to be starred. Typing `pwcorr, star(.05)` would star all correlation coefficients significant at the 5% level or better.

`bonferroni` makes the Bonferroni adjustment to calculated significance levels. This option affects printed significance levels and the `print()` and `star()` options. Thus `pwcorr, print(.05) bonferroni` prints coefficients with Bonferroni-adjusted significance levels of 0.05 or less.

sidak makes the Šidák adjustment to calculated significance levels. This option affects printed significance levels and the print() and star() options. Thus pwcorr, print(.05) sidak prints coefficients with Šidák-adjusted significance levels of 0.05 or less.

Remarks

Remarks are presented under the following headings:

> *correlate*
> *pwcorr*

correlate

Typing correlate by itself produces a correlation matrix for all variables in the dataset. If you specify the *varlist*, a correlation matrix for just those variables is displayed.

▷ Example 1

We have state data on demographic characteristics of the population. To obtain a correlation matrix, we type

```
. use http://www.stata-press.com/data/r12/census13
(1980 Census data by state)

. correlate
(obs=50)
```

	state	brate	pop	medage	division	region	mrgrate
state	1.0000						
brate	0.0208	1.0000					
pop	-0.0540	-0.2830	1.0000				
medage	-0.0624	-0.8800	0.3294	1.0000			
division	-0.1345	0.6356	-0.1081	-0.5207	1.0000		
region	-0.1339	0.6086	-0.1515	-0.5292	0.9688	1.0000	
mrgrate	0.0509	0.0677	-0.1502	-0.0177	0.2280	0.2490	1.0000
dvcrate	-0.0655	0.3508	-0.2064	-0.2229	0.5522	0.5682	0.7700
medagesq	-0.0621	-0.8609	0.3324	0.9984	-0.5162	-0.5239	-0.0202

	dvcrate	medagesq
dvcrate	1.0000	
medagesq	-0.2192	1.0000

Because we did not specify the wrap option, Stata did its best to make the result readable by breaking the table into two parts.

To obtain the correlations between `mrgrate`, `dvcrate`, and `medage`, we type

```
. correlate mrgrate dvcrate medage
(obs=50)
```

	mrgrate	dvcrate	medage
mrgrate	1.0000		
dvcrate	0.7700	1.0000	
medage	-0.0177	-0.2229	1.0000

◁

▷ Example 2

The `pop` variable in our previous example represents the total population of the state. Thus, to obtain population-weighted correlations among `mrgrate`, `dvcrate`, and `medage`, we type

```
. correlate mrgrate dvcrate medage [w=pop]
(analytic weights assumed)
(sum of wgt is    2.2591e+08)
(obs=50)
```

	mrgrate	dvcrate	medage
mrgrate	1.0000		
dvcrate	0.5854	1.0000	
medage	-0.1316	-0.2833	1.0000

◁

With the `covariance` option, `correlate` can be used to obtain covariance matrices, as well as correlation matrices, for both weighted and unweighted data.

▷ Example 3

To obtain the matrix of covariances between `mrgrate`, `dvcrate`, and `medage`, we type `correlate mrgrate dvcrate medage, covariance`:

```
. correlate mrgrate dvcrate medage, covariance
(obs=50)
```

	mrgrate	dvcrate	medage
mrgrate	.000662		
dvcrate	.000063	1.0e-05	
medage	-.000769	-.001191	2.86775

We could have obtained the pop-weighted covariance matrix by typing `correlate mrgrate dvcrate medage [w=pop], covariance`.

◁

pwcorr

`correlate` calculates correlation coefficients by using casewise deletion; when you request correlations of variables $x_1, x_2, \ldots, x_k$, any observation for which any of $x_1, x_2, \ldots, x_k$ is missing is not used. Thus if x_3 and x_4 have no missing values, but x_2 is missing for half the data, the correlation between x_3 and x_4 is calculated using only the half of the data for which x_2 is not missing. Of course, you can obtain the correlation between x_3 and x_4 by using all the data by typing `correlate` x_3 x_4.

`pwcorr` makes obtaining such pairwise correlation coefficients easier.

▷ Example 4

Using auto.dta, we investigate the correlation between several of the variables.

```
. use http://www.stata-press.com/data/r12/auto1
(Automobile Models)
. pwcorr mpg price rep78 foreign, obs sig
```

	mpg	price	rep78	foreign
mpg	1.0000			
	74			
price	-0.4594	1.0000		
	0.0000			
	74	74		
rep78	0.3739	0.0066	1.0000	
	0.0016	0.9574		
	69	69	69	
foreign	0.3613	0.0487	0.5922	1.0000
	0.0016	0.6802	0.0000	
	74	74	69	74

```
. pwcorr mpg price headroom rear_seat trunk rep78 foreign, print(.05) star(.01)
```

	mpg	price	headroom	rear_s~t	trunk	rep78	foreign
mpg	1.0000						
price	-0.4594*	1.0000					
headroom	-0.4220*		1.0000				
rear_seat	-0.5213*	0.4194*	0.5238*	1.0000			
trunk	-0.5703*	0.3143*	0.6620*	0.6480*	1.0000		
rep78	0.3739*					1.0000	
foreign	0.3613*		-0.2939	-0.2409	-0.3594*	0.5922*	1.0000

```
. pwcorr mpg price headroom rear_seat trunk rep78 foreign, print(.05) bon
```

	mpg	price	headroom	rear_s~t	trunk	rep78	foreign
mpg	1.0000						
price	-0.4594	1.0000					
headroom	-0.4220		1.0000				
rear_seat	-0.5213	0.4194	0.5238	1.0000			
trunk	-0.5703		0.6620	0.6480	1.0000		
rep78	0.3739					1.0000	
foreign	0.3613				-0.3594	0.5922	1.0000

◁

❑ Technical note

The correlate command will report the correlation matrix of the data, but there are occasions when you need the matrix stored as a Stata matrix so that you can further manipulate it. You can obtain the matrix by typing

```
. matrix accum R = varlist, nocons dev
. matrix R = corr(R)
```

The first line places the cross-product matrix of the data in matrix R. The second line converts that to a correlation matrix. Also see [P] **matrix define** and [P] **matrix accum**.

❑

Saved results

correlate saves the following in r():

Scalars

r(N)	number of observations
r(rho)	ρ (first and second variables)
r(cov_12)	covariance (covariance only)
r(Var_1)	variance of first variable (covariance only)
r(Var_2)	variance of second variable (covariance only)

Matrices

r(C)	correlation or covariance matrix

pwcorr will leave in its wake only the results of the last call that it makes internally to correlate for the correlation between the last variable and itself. Only rarely is this feature useful.

Methods and formulas

pwcorr is implemented as an ado-file.

For a discussion of correlation, see, for instance, Snedecor and Cochran (1989, 177–195); for an introductory explanation using Stata examples, see Acock (2010, 186–192).

According to Snedecor and Cochran (1989, 180), the term "co-relation" was first proposed by Galton (1888). The product-moment correlation coefficient is often called the Pearson product-moment correlation coefficient because Pearson (1896) and Pearson and Filon (1898) were partially responsible for popularizing its use. See Stigler (1986) for information on the history of correlation.

The estimate of the product-moment correlation coefficient, ρ, is

$$\widehat{\rho} = \frac{\sum_{i=1}^{n} w_i(x_i - \overline{x})(y_i - \overline{y})}{\sqrt{\sum_{i=1}^{n} w_i(x_i - \overline{x})^2}\sqrt{\sum_{i=1}^{n} w_i(y_i - \overline{y})^2}}$$

where w_i are the weights, if specified, or $w_i = 1$ if weights are not specified. $\overline{x} = (\sum w_i x_i)/(\sum w_i)$ is the mean of x, and $\overline{y}$ is similarly defined.

The unadjusted significance level is calculated by pwcorr as

$$p = 2 * \texttt{ttail}(n - 2, |\widehat{\rho}|\sqrt{n - 2}/\sqrt{1 - \widehat{\rho}^2})$$

Let v be the number of variables specified so that $k = v(v-1)/2$ correlation coefficients are to be estimated. If bonferroni is specified, the adjusted significance level is $p' = \min(1, kp)$. If sidak is specified, $p' = \min\left\{1, 1 - (1 - p)^k\right\}$. In both cases, see *Methods and formulas* in [R] **oneway** for a more complete description of the logic behind these adjustments.

Carlo Emilio Bonferroni (1892–1960) studied in Turin and taught there and in Bari and Florence. He published on actuarial mathematics, probability, statistics, analysis, geometry, and mechanics. His work on probability inequalities has been applied to simultaneous statistical inference, although the method known as Bonferroni adjustment usually relies only on an inequality established earlier by Boole.

Karl Pearson (1857–1936) studied mathematics at Cambridge. He was professor of applied mathematics (1884–1911) and eugenics (1911–1933) at University College London. His publications include literary, historical, philosophical, and religious topics. Statistics became his main interest in the early 1890s after he learned about its application to biological problems. His work centered on distribution theory, the method of moments, correlation, and regression. Pearson introduced the chi-squared test and the terms coefficient of variation, contingency table, heteroskedastic, histogram, homoskedastic, kurtosis, mode, random sampling, random walk, skewness, standard deviation, and truncation. Despite many strong qualities, he also fell into prolonged disagreements with others, most notably, William Bateson and R. A. Fisher.

Zbyněk Šidák (1933–1999) was a notable Czech statistician and probabilist. He worked on Markov chains, rank tests, multivariate distribution theory and multiple-comparison methods, and he served as the chief editor of *Applications of Mathematics*.

References

Acock, A. C. 2010. *A Gentle Introduction to Stata*. 3rd ed. College Station, TX: Stata Press.

Dewey, M. E., and E. Seneta. 2001. Carlo Emilio Bonferroni. In *Statisticians of the Centuries*, ed. C. C. Heyde and E. Seneta, 411–414. New York: Springer.

Eisenhart, C. 1974. Pearson, Karl. In Vol. 10 of *Dictionary of Scientific Biography*, ed. C. C. Gillispie, 447–473. New York: Charles Scribner's Sons.

Galton, F. 1888. Co-relations and their measurement, chiefly from anthropometric data. *Proceedings of the Royal Society of London* 45: 135–145.

Gleason, J. R. 1996. sg51: Inference about correlations using the Fisher z-transform. *Stata Technical Bulletin* 32: 13–18. Reprinted in *Stata Technical Bulletin Reprints*, vol. 6, pp. 121–128. College Station, TX: Stata Press.

Goldstein, R. 1996. sg52: Testing dependent correlation coefficients. *Stata Technical Bulletin* 32: 18. Reprinted in *Stata Technical Bulletin Reprints*, vol. 6, pp. 128–129. College Station, TX: Stata Press.

Pearson, K. 1896. Mathematical contributions to the theory of evolution—III. Regression, heredity, and panmixia. *Philosophical Transactions of the Royal Society of London, Series A* 187: 253–318.

Pearson, K., and L. N. G. Filon. 1898. Mathematical contributions to the theory of evolution. IV. On the probable errors of frequency constants and on the influence of random selection on variation and correlation. *Philosophical Transactions of the Royal Society of London, Series A* 191: 229–311.

Porter, T. M. 2004. *Karl Pearson: The Scientific Life in a Statistical Age*. Princeton, NJ: Princeton University Press.

Rodgers, J. L., and W. A. Nicewander. 1988. Thirteen ways to look at the correlation coefficient. *American Statistician* 42: 59–66.

Rovine, M. J., and A. von Eye. 1997. A 14th way to look at the correlation coefficient: Correlation as the proportion of matches. *American Statistician* 51: 42–46.

Seed, P. T. 2001. sg159: Confidence intervals for correlations. *Stata Technical Bulletin* 59: 27–28. Reprinted in *Stata Technical Bulletin Reprints*, vol. 10, pp. 267–269. College Station, TX: Stata Press.

Seidler, J., J. Vondráček, and I. Saxl. 2000. The life and work of Zbyněk Šidák (1933–1999). *Applications of Mathematics* 45: 321–336.

Snedecor, G. W., and W. G. Cochran. 1989. *Statistical Methods*. 8th ed. Ames, IA: Iowa State University Press.

Stigler, S. M. 1986. *The History of Statistics: The Measurement of Uncertainty before 1900*. Cambridge, MA: Belknap Press.

Verardi, V., and C. Dehon. 2010. Multivariate outlier detection in Stata. *Stata Journal* 10: 259–266.

Weber, S. 2010. bacon: An effective way to detect outliers in multivariate data using Stata (and Mata). *Stata Journal* 10: 331–338.

Wolfe, F. 1997. sg64: pwcorrs: Enhanced correlation display. *Stata Technical Bulletin* 35: 22–25. Reprinted in *Stata Technical Bulletin Reprints*, vol. 6, pp. 163–167. College Station, TX: Stata Press.

——. 1999. sg64.1: Update to pwcorrs. *Stata Technical Bulletin* 49: 17. Reprinted in *Stata Technical Bulletin Reprints*, vol. 9, p. 159. College Station, TX: Stata Press.

Also see

[R] **pcorr** — Partial and semipartial correlation coefficients

[R] **spearman** — Spearman's and Kendall's correlations

[R] **summarize** — Summary statistics

[R] **tetrachoric** — Tetrachoric correlations for binary variables

Title

cumul — Cumulative distribution

Syntax

cumul *varname* [*if*] [*in*] [*weight*] , generate(*newvar*) [*options*]

options	Description
Main	
*generate(*newvar*)	create variable *newvar*
freq	use frequency units for cumulative
equal	generate equal cumulatives for tied values

*generate(*newvar*) is required.

by is allowed; see [D] **by**.

fweights and aweights are allowed; see [U] **11.1.6 weight**.

Menu

Statistics > Summaries, tables, and tests > Distributional plots and tests > Generate cumulative distribution

Description

cumul creates *newvar*, defined as the empirical cumulative distribution function of *varname*.

Options

```
                Main
```

generate(*newvar*) is required. It specifies the name of the new variable to be created.

freq specifies that the cumulative be in frequency units; otherwise, it is normalized so that *newvar* is 1 for the largest value of *varname*.

equal requests that observations with equal values in *varname* get the same cumulative value in *newvar*.

Jean Baptiste Joseph Fourier (1768–1830) was born in Auxerre in France. As a young man, Fourier became entangled in the complications of the French Revolution. As a result, he was arrested and put into prison, where he feared he might meet his end at the guillotine. When he was not in prison, he was studying, researching, and teaching mathematics. Later, he served Napolean's army in Egypt as a scientific adviser. Upon his return to France in 1801, he was appointed Prefect of the Department of Isère. While prefect, Fourier worked on the mathematical basis of the theory of heat, which is based on what are now called Fourier series. This work was published in 1822, despite the skepticism of Lagrange, Laplace, Legendre, and others—who found the work lacking in generality and even rigor—and disagreements of both priority and substance with Biot and Poisson.

Remarks

▷ Example 1

cumul is most often used with graph to graph the empirical cumulative distribution. For instance, we have data on the median family income of 957 U.S. cities:

```
. use http://www.stata-press.com/data/r12/hsng
(1980 Census housing data)

. cumul faminc, gen(cum)

. sort cum

. line cum faminc, ylab(, grid) ytitle("") xlab(, grid)
> title("Cumulative of median family income")
> subtitle("1980 Census, 957 U.S. Cities")
```

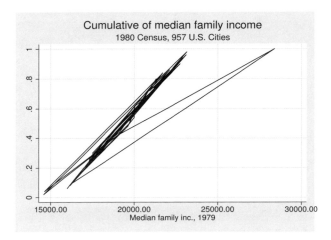

It would have been enough to type line cum faminc, but we wanted to make the graph look better; see [G-2] **graph twoway line**.

If we had wanted a weighted cumulative, we would have typed cumul faminc [w=pop] at the first step.

◁

▷ Example 2

To graph two (or more) cumulatives on the same graph, use cumul and stack; see [D] **stack**. For instance, we have data on the average January and July temperatures of 956 U.S. cities:

```
. use http://www.stata-press.com/data/r12/citytemp, clear
(City Temperature Data)

. cumul tempjan, gen(cjan)

. cumul tempjuly, gen(cjuly)

. stack  cjan tempjan  cjuly tempjuly, into(c temp) wide clear

. line cjan cjuly temp, sort ylab(, grid) ytitle("") xlab(, grid)
> xtitle("Temperature (F)")
> title("Cumulatives:" "Average January and July Temperatures")
> subtitle("956 U.S. Cities") clstyle(. dot)
```

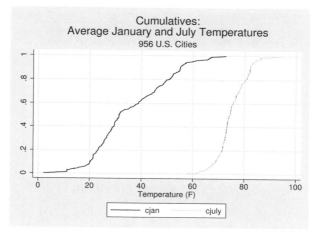

As before, it would have been enough to type `line cjan cjuly temp, sort`. See [D] **stack** for an explanation of how the `stack` command works.

◁

❑ Technical note

According to Beniger and Robyn (1978), Fourier (1821) published the first graph of a cumulative frequency distribution, which was later given the name "ogive" by Galton (1875).

❑

Methods and formulas

`cumul` is implemented as an ado-file.

Acknowledgment

The `equal` option was added by Nicholas J. Cox, Durham University, Durham, UK.

References

Beniger, J. R., and D. L. Robyn. 1978. Quantitative graphics in statistics: A brief history. *American Statistician* 32: 1–11.

Clayton, D. G., and M. Hills. 1999. gr37: Cumulative distribution function plots. *Stata Technical Bulletin* 49: 10–12. Reprinted in *Stata Technical Bulletin Reprints*, vol. 9, pp. 96–98. College Station, TX: Stata Press.

Cox, N. J. 1999. gr41: Distribution function plots. *Stata Technical Bulletin* 51: 12–16. Reprinted in *Stata Technical Bulletin Reprints*, vol. 9, pp. 108–112. College Station, TX: Stata Press.

Fourier, J. B. J. 1821. Notions générales, sur la population. *Recherches Statistiques sur la Ville de Paris et le Département de la Seine* 1: 1–70.

Galton, F. 1875. Statistics by intercomparison, with remarks on the law of frequency of error. *Philosophical Magazine* 49: 33–46.

Wilk, M. B., and R. Gnanadesikan. 1968. Probability plotting methods for the analysis of data. *Biometrika* 55: 1–17.

Also see

[D] **stack** — Stack data

[R] **diagnostic plots** — Distributional diagnostic plots

[R] **kdensity** — Univariate kernel density estimation

Title

cusum — Graph cumulative spectral distribution

Syntax

cusum *yvar* *xvar* [*if*] [*in*] [, *options*]

options	Description
Main	
generate(*newvar*)	save cumulative sum in *newvar*
yfit(*fitvar*)	calculate cumulative sum against *fitvar*
nograph	suppress the plot
nocalc	suppress cusum test statistics
Cusum plot	
connect_options	affect the rendition of the plotted line
Add plots	
addplot(*plot*)	add plots to the generated graph
Y axis, X axis, Titles, Legend, Overall	
twoway_options	any options other than by() documented in [G-3] ***twoway_options***

Menu

Statistics > Other > Quality control > Cusum plots and tests for binary variables

Description

cusum graphs the cumulative sum (cusum) of a binary (0/1) variable, *yvar*, against a (usually) continuous variable, *xvar*.

Options

Main

generate(*newvar*) saves the cusum in *newvar*.

yfit(*fitvar*) calculates a cusum against *fitvar*, that is, the running sums of the "residuals" *fitvar* minus *yvar*. Typically, *fitvar* is the predicted probability of a positive outcome obtained from a logistic regression analysis.

nograph suppresses the plot.

nocalc suppresses calculation of the cusum test statistics.

Cusum plot

connect_options affect the rendition of the plotted line; see [G-3] ***connect_options***.

405

⌐ Add plots ⌐

addplot(*plot*) provides a way to add other plots to the generated graph. See [G-3] ***addplot_option***.

⌐ Y axis, X axis, Titles, Legend, Overall ⌐

twoway_options are any of the options documented in [G-3] ***twoway_options***, excluding by(). These include options for titling the graph (see [G-3] ***title_options***) and for saving the graph to disk (see [G-3] ***saving_option***).

Remarks

The cusum is the running sum of the proportion of ones in the sample, a constant number, minus *yvar*,

$$c_j = \sum_{k=1}^{j} f - yvar_{(k)}, \qquad 1 \le j \le N$$

where $f = (\sum yvar)/N$ and $yvar_{(k)}$ refers to the corresponding value of *yvar* when *xvar* is placed in ascending order: $xvar_{(k+1)} \ge xvar_{(k)}$. Tied values of *xvar* are broken at random. If you want them broken the same way in two runs, you must set the random-number seed to the same value before giving the cusum command; see [R] **set seed**.

A U-shaped or inverted U-shaped cusum indicates, respectively, a negative or a positive trend of *yvar* with *xvar*. A sinusoidal shape is evidence of a nonmonotonic (for example, quadratic) trend. cusum displays the maximum absolute cusum for monotonic and nonmonotonic trends of *yvar* on *xvar*. These are nonparametric tests of departure from randomness of *yvar* with respect to *xvar*. Approximate values for the tests are given.

▷ Example 1

For the automobile dataset, auto.dta, we wish to investigate the relationship between foreign (0 = domestic, 1 = foreign) and car weight as follows:

```
. use http://www.stata-press.com/data/r12/auto
(1978 Automobile Data)

. cusum foreign weight
```

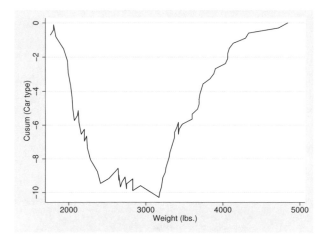

Variable	Obs	Pr(1)	CusumL	zL	Pr>zL	CusumQ	zQ	Pr>zQ
foreign	74	0.2973	10.30	3.963	0.000	3.32	0.469	0.320

The resulting plot, which is U-shaped, suggests a negative monotonic relationship. The trend is confirmed by a highly significant linear cusum statistic, labeled CusumL in the output above.

Some 29.73% of the cars are foreign (coded 1). The proportion of foreign cars diminishes with increasing weight. The domestic cars are crudely heavier than the foreign ones. We could have discovered that by typing table foreign, stats(mean weight), but such an approach does not give the full picture of the relationship. The quadratic cusum (CusumQ) is not significant, so we do not suspect any tendency for the very heavy cars to be foreign rather than domestic. A slightly enhanced version of the plot shows the preponderance of domestic (coded 0) cars at the heavy end of the weight axis:

```
. label values foreign
. cusum foreign weight, s(none) recast(scatter) mlabel(foreign) mlabp(0)
```

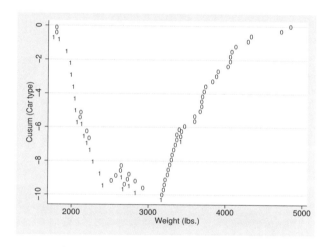

Variable	Obs	Pr(1)	CusumL	zL	Pr>zL	CusumQ	zQ	Pr>zQ
foreign	74	0.2973	10.30	3.963	0.000	2.92	0.064	0.475

The example is, of course, artificial, because we would not really try to model the probability of a car being foreign given its weight.

◁

Saved results

cusum saves the following in r():

Scalars

r(N)	number of observations	r(P_zl)	*p*-value for test (linear)
r(prop1)	proportion of positive outcomes	r(cusumq)	quadratic cusum
r(cusuml)	cusum	r(zq)	test (quadratic)
r(zl)	test (linear)	r(P_zq)	*p*-value for test (quadratic)

Methods and formulas

cusum is implemented as an ado-file.

Acknowledgment

cusum was written by Patrick Royston, MRC Clinical Trials Unit, London.

References

Royston, P. 1992. The use of cusums and other techniques in modelling continuous covariates in logistic regression. *Statistics in Medicine* 11: 1115–1129.

——. 1993. sqv7: Cusum plots and tests for binary variables. *Stata Technical Bulletin* 12: 16–17. Reprinted in *Stata Technical Bulletin Reprints*, vol. 2, pp. 175–177. College Station, TX: Stata Press.

Also see

[R] **logistic** — Logistic regression, reporting odds ratios

[R] **logit** — Logistic regression, reporting coefficients

[R] **probit** — Probit regression

Title

> **db** — Launch dialog

Syntax

Syntax for db

> db *commandname*

For programmers

> db *commandname* $\left[\right.$, message(*string*) debug dryrun $\left.\right]$

Set system parameter

> set maxdb # $\left[\right.$, <u>permanently</u> $\left.\right]$

where # must be between 5 and 1,000.

Description

db is the command-line way to launch a dialog for a Stata command.

The second syntax (which is the same but includes options) is for use by programmers.

If you wish to allow the launching of dialogs from a help file, see [P] **smcl** for information on the dialog SMCL directive.

set maxdb sets the maximum number of dialog boxes whose contents are remembered from one invocation to the next during a session. The default value of maxdb is 50.

Options

message(*string*) specifies that *string* be passed to the dialog box, where it can be referred to from the __MESSAGE STRING property.

debug specifies that the underlying dialog box be loaded with debug messaging turned on.

dryrun specifies that, rather than launching the dialog, db show the commands it would issue to launch the dialog.

permanently specifies that, in addition to making the change right now, the maxdb setting be remembered and become the default setting when you invoke Stata.

Remarks

The usual way to launch a dialog is to open the **Data**, **Graphics**, or **Statistics** menu and to make your selection from there. When you know the name of the command that you want to run, however, db provides a way to invoke the dialog from the command line.

db follows the same abbreviation rules that Stata's command-line interface follows. So, to launch the dialog for regress, you can type

```
. db regress
```

or

```
. db reg
```

Say that you use the dialog box for `regress`, either by selecting

Statistics > Linear models and related > Linear regression

or by typing

```
. db regress
```

You fit a regression.

Much later during the session, you return to the `regress` dialog box. It will have the contents as you left them if 1) you have not typed `clear all` between the first and second invocations; 2) you have not typed `discard` between the two invocations; and 3) you have not used more than 50 different dialog boxes—regardless of how many times you have used each—between the first and second invocations of `regress`. If you use 51 or more, the contents of the `regress` dialog box will be forgotten.

`set maxdb` determines how many different dialog boxes are remembered. A dialog box takes, on average, about 20 KB of memory, so the 50 default corresponds to allowing dialog boxes to consume about 1 MB of memory.

Methods and formulas

db is implemented as an ado-file.

Also see

[R] **query** — Display system parameters

Title

diagnostic plots — Distributional diagnostic plots

Syntax

Symmetry plot

    symplot *varname* $\big[$*if*$\big]$ $\big[$*in*$\big]$ $\big[$, *options*$_1$ $\big]$

Ordered values of varname against quantiles of uniform distribution

    quantile *varname* $\big[$*if*$\big]$ $\big[$*in*$\big]$ $\big[$, *options*$_1$ $\big]$

Quantiles of varname$_1$ against quantiles of varname$_2$

    qqplot *varname*$_1$ *varname*$_2$ $\big[$*if*$\big]$ $\big[$*in*$\big]$ $\big[$, *options*$_1$ $\big]$

Quantiles of varname against quantiles of normal distribution

    qnorm *varname* $\big[$*if*$\big]$ $\big[$*in*$\big]$ $\big[$, *options*$_2$ $\big]$

Standardized normal probability plot

    pnorm *varname* $\big[$*if*$\big]$ $\big[$*in*$\big]$ $\big[$, *options*$_2$ $\big]$

Quantiles of varname against quantiles of χ^2 distribution

    qchi *varname* $\big[$*if*$\big]$ $\big[$*in*$\big]$ $\big[$, *options*$_3$ $\big]$

χ^2 probability plot

    pchi *varname* $\big[$*if*$\big]$ $\big[$*in*$\big]$ $\big[$, *options*$_3$ $\big]$

options$_1$	Description
Plot	
marker_options	change look of markers (color, size, etc.)
marker_label_options	add marker labels; change look or position
Reference line	
rlopts(*cline_options*)	affect rendition of the reference line
Add plots	
addplot(*plot*)	add other plots to the generated graph
Y axis, X axis, Titles, Legend, Overall	
twoway_options	any options other than by() documented in [G-3] **twoway_options**

411

options₂	Description
Main	
grid	add grid lines
Plot	
marker_options	change look of markers (color, size, etc.)
marker_label_options	add marker labels; change look or position
Reference line	
rlopts(*cline_options*)	affect rendition of the reference line
Add plots	
addplot(*plot*)	add other plots to the generated graph
Y axis, X axis, Titles, Legend, Overall	
twoway_options	any options other than by() documented in [G-3] *twoway_options*

options₃	Description
Main	
grid	add grid lines
df(*#*)	degrees of freedom of χ^2 distribution; default is df(1)
Plot	
marker_options	change look of markers (color, size, etc.)
marker_label_options	add marker labels; change look or position
Reference line	
rlopts(*cline_options*)	affect rendition of the reference line
Add plots	
addplot(*plot*)	add other plots to the generated graph
Y axis, X axis, Titles, Legend, Overall	
twoway_options	any options other than by() documented in [G-3] *twoway_options*

Menu

symplot

Statistics > Summaries, tables, and tests > Distributional plots and tests > Symmetry plot

quantile

Statistics > Summaries, tables, and tests > Distributional plots and tests > Quantiles plot

qqplot

Statistics > Summaries, tables, and tests > Distributional plots and tests > Quantile-quantile plot

qnorm

Statistics > Summaries, tables, and tests > Distributional plots and tests > Normal quantile plot

pnorm

Statistics > Summaries, tables, and tests > Distributional plots and tests > Normal probability plot, standardized

qchi

Statistics > Summaries, tables, and tests > Distributional plots and tests > Chi-squared quantile plot

pchi

Statistics > Summaries, tables, and tests > Distributional plots and tests > Chi-squared probability plot

Description

symplot graphs a symmetry plot of *varname*.

quantile plots the ordered values of *varname* against the quantiles of a uniform distribution.

qqplot plots the quantiles of *varname*$_1$ against the quantiles of *varname*$_2$ (Q–Q plot).

qnorm plots the quantiles of *varname* against the quantiles of the normal distribution (Q–Q plot).

pnorm graphs a standardized normal probability plot (P–P plot).

qchi plots the quantiles of *varname* against the quantiles of a χ^2 distribution (Q–Q plot).

pchi graphs a χ^2 probability plot (P–P plot).

See [R] **regress postestimation** for regression diagnostic plots and [R] **logistic postestimation** for logistic regression diagnostic plots.

Options for symplot, quantile, and qqplot

⌐ Plot ⌐

marker_options affect the rendition of markers drawn at the plotted points, including their shape, size, color, and outline; see [G-3] **marker_options**.

marker_label_options specify if and how the markers are to be labeled; see [G-3] **marker_label_options**.

⌐ Reference line ⌐

rlopts(*cline_options*) affect the rendition of the reference line; see [G-3] **cline_options**.

⌐ Add plots ⌐

addplot(*plot*) provides a way to add other plots to the generated graph; see [G-3] **addplot_option**.

⌐ Y axis, X axis, Titles, Legend, Overall ⌐

twoway_options are any of the options documented in [G-3] **twoway_options**, excluding by(). These include options for titling the graph (see [G-3] **title_options**) and for saving the graph to disk (see [G-3] **saving_option**).

Options for qnorm and pnorm

⌐ Main ⌐

grid adds grid lines at the 0.05, 0.10, 0.25, 0.50, 0.75, 0.90, and 0.95 quantiles when specified with qnorm. With pnorm, grid is equivalent to yline(.25,.5,.75) xline(.25,.5,.75).

⌐ Plot ⌐

marker_options affect the rendition of markers drawn at the plotted points, including their shape, size, color, and outline; see [G-3] *marker_options*.

marker_label_options specify if and how the markers are to be labeled; see [G-3] *marker_label_options*.

⌐ Reference line ⌐

rlopts(*cline_options*) affect the rendition of the reference line; see [G-3] *cline_options*.

⌐ Add plots ⌐

addplot(*plot*) provides a way to add other plots to the generated graph; see [G-3] *addplot_option*.

⌐ Y axis, X axis, Titles, Legend, Overall ⌐

twoway_options are any of the options documented in [G-3] *twoway_options*, excluding by(). These include options for titling the graph (see [G-3] *title_options*) and for saving the graph to disk (see [G-3] *saving_option*).

Options for qchi and pchi

⌐ Main ⌐

grid adds grid lines at the 0.05, 0.10, 0.25, 0.50, 0.75, 0.90, and .95 quantiles when specified with qchi. With pchi, grid is equivalent to yline(.25,.5,.75) xline(.25,.5,.75).

df(*#*) specifies the degrees of freedom of the χ^2 distribution. The default is df(1).

⌐ Plot ⌐

marker_options affect the rendition of markers drawn at the plotted points, including their shape, size, color, and outline; see [G-3] *marker_options*.

marker_label_options specify if and how the markers are to be labeled; see [G-3] *marker_label_options*.

⌐ Reference line ⌐

rlopts(*cline_options*) affect the rendition of the reference line; see [G-3] *cline_options*.

⌐ Add plots ⌐

addplot(*plot*) provides a way to add other plots to the generated graph; see [G-3] *addplot_option*.

Y axis, X axis, Titles, Legend, Overall

twoway_options are any of the options documented in [G-3] ***twoway_options***, excluding by(). These include options for titling the graph (see [G-3] ***title_options***) and for saving the graph to disk (see [G-3] ***saving_option***).

Remarks

Remarks are presented under the following headings:

> *symplot*
> *quantile*
> *qqplot*
> *qnorm*
> *pnorm*
> *qchi*
> *pchi*

symplot

▷ Example 1

We have data on 74 automobiles. To make a symmetry plot of the variable price, we type

```
. use http://www.stata-press.com/data/r12/auto
(1978 Automobile Data)

. symplot price
```

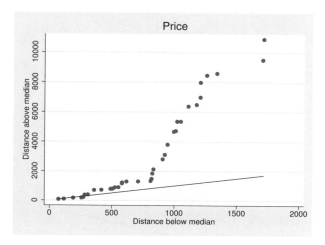

All points would lie along the reference line (defined as $y = x$) if car prices were symmetrically distributed. The points in this plot lie above the reference line, indicating that the distribution of car prices is skewed to the right—the most expensive cars are far more expensive than the least expensive cars are inexpensive.

The logic works as follows: a variable, z, is distributed symmetrically if

$$\text{median} - z_{(i)} = z_{(N+1-i)} - \text{median}$$

where $z_{(i)}$ indicates the ith-order statistic of z. symplot graphs $y_i = \text{median} - z_{(i)}$ versus $x_i = z_{(N+1-i)} - \text{median}$.

For instance, consider the largest and smallest values of price in the example above. The most expensive car costs \$15,906 and the least expensive, \$3,291. Let's compare these two cars with the typical car in the data and see how much more it costs to buy the most expensive car, and compare that with how much less it costs to buy the least expensive car. If the automobile price distribution is symmetric, the price differences would be the same.

Before we can make this comparison, we must agree on a definition for the word "typical". Let's agree that "typical" means median. The price of the median car is \$5,006.50, so the most expensive car costs \$10,899.50 more than the median car, and the least expensive car costs \$1,715.50 less than the median car. We now have one piece of evidence that the car price distribution is not symmetric. We can repeat the experiment for the second-most-expensive car and the second-least-expensive car. We find that the second-most-expensive car costs \$9,494.50 more than the median car, and the second-least-expensive car costs \$1,707.50 less than the median car. We now have more evidence. We can continue doing this with the third most expensive and the third least expensive, and so on.

Once we have all these numbers, we want to compare each pair and ask how similar, on average, they are. The easiest way to do that is to plot all the pairs.

◁

quantile

▷ Example 2

We have data on the prices of 74 automobiles. To make a quantile plot of price, we type

```
. use http://www.stata-press.com/data/r12/auto, clear
(1978 Automobile Data)
. quantile price, rlopts(clpattern(dash))
```

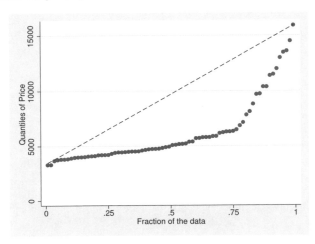

We changed the pattern of the reference line by specifying rlopts(clpattern(dash)).

In a quantile plot, each value of the variable is plotted against the fraction of the data that have values less than that fraction. The diagonal line is a reference line. If automobile prices were rectangularly distributed, all the data would be plotted along the line. Because all the points are below the reference line, we know that the price distribution is skewed right.

◁

qqplot

▷ Example 3

We have data on the weight and country of manufacture of 74 automobiles. We wish to compare the distributions of weights for domestic and foreign automobiles:

```
. use http://www.stata-press.com/data/r12/auto
(1978 Automobile Data)
. generate weightd=weight if !foreign
(22 missing values generated)
. generate weightf=weight if foreign
(52 missing values generated)
. qqplot weightd weightf
```

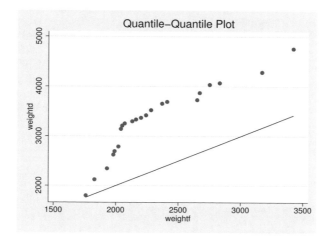

◁

qnorm

▷ Example 4

Continuing with our price data on 74 automobiles, we now wish to compare the distribution of price with the normal distribution:

```
. qnorm price, grid ylabel(, angle(horizontal) axis(1))
> ylabel(, angle(horizontal) axis(2))
```

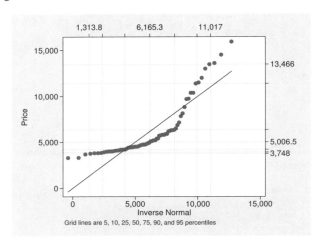

The result shows that the distributions are different.

◁

❏ Technical note

The idea behind qnorm is recommended strongly by Miller (1997): he calls it probit plotting. His recommendations from much practical experience should interest many users. "My recommendation for detecting nonnormality is *probit plotting*" (Miller 1997, 10). "If a deviation from normality cannot be spotted by eye on probit paper, it is not worth worrying about. I never use the Kolmogorov–Smirnov test (or one of its cousins) or the χ^2 test as a preliminary test of normality. They do not tell you how the sample is differing from normality, and I have a feeling they are more likely to detect irregularities in the middle of the distribution than in the tails" (Miller 1997, 13–14).

❏

pnorm

▷ Example 5

Quantile–normal plots emphasize the tails of the distribution. Normal probability plots put the focus on the center of the distribution:

```
. pnorm price, grid
```

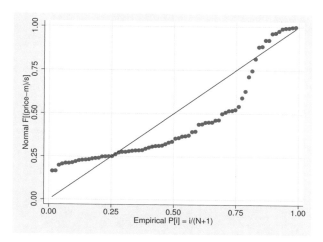

qchi

▷ Example 6

Suppose that we want to examine the distribution of the sum of squares of `price` and `mpg`, standardized for their variances.

```
. egen c1 = std(price)
. egen c2 = std(mpg)
. generate ch = c1^2 + c2^2
. qchi ch, df(2) grid ylabel(, alt axis(2)) xlabel(, alt axis(2))
```

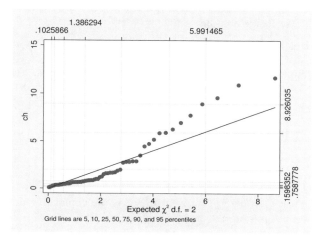

The quadratic form is clearly not χ^2 with 2 degrees of freedom.

pchi

▷ Example 7

We can focus on the center of the distribution by doing a probability plot:

```
. pchi ch, df(2) grid
```

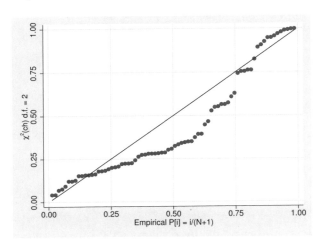

Methods and formulas

symplot, quantile, qqplot, qnorm, pnorm, qchi, and pchi are implemented as ado-files. Let $x_{(1)}, x_{(2)}, \ldots, x_{(N)}$ be the data sorted in ascending order.

If a continuous variable, x, has a cumulative distribution function $F(x) = P(X \leq x) = p$, the quantiles x_{p_i} are such that $F(x_{p_i}) = p_i$. For example, if $p_i = 0.5$, then $x_{0.5}$ is the median. When we plot data, the probabilities, p_i, are often referred to as plotting positions. There are many different conventions for choice of plotting positions, given $x_{(1)} \leq \cdots \leq x_{(N)}$. Most belong to the family $(i - a)/(N - 2a + 1)$. $a = 0.5$ (suggested by Hazen) and $a = 0$ (suggested by Weibull) are popular choices.

For a wider discussion of the calculation of plotting positions, see Cox (2002).

symplot plots median $- x_{(i)}$ versus $x_{(N+1-i)} -$ median.

quantile plots $x_{(i)}$ versus $(i - 0.5)/N$ (the Hazen position).

qnorm plots $x_{(i)}$ against q_i, where $q_i = \Phi^{-1}(p_i)$, Φ is the cumulative normal distribution, and $p_i = i/(N + 1)$ (the Weibull position).

pnorm plots $\Phi\{(x_i - \widehat{\mu})/\widehat{\sigma}\}$ versus $p_i = i/(N + 1)$, where $\widehat{\mu}$ is the mean of the data and $\widehat{\sigma}$ is the standard deviation.

qchi and pchi are similar to qnorm and pnorm; the cumulative χ^2 distribution is used in place of the cumulative normal distribution.

`qqplot` is just a two-way scatterplot of one variable against the other after both variables have been sorted into ascending order, and both variables have the same number of nonmissing observations. If the variables have unequal numbers of nonmissing observations, interpolated values of the variable with more data are plotted against the variable with fewer data.

> Ramanathan Gnanadesikan (1932–) was born in Madras. He obtained degrees from the Universities of Madras and North Carolina. He worked in industry at Procter & Gamble, Bell Labs, and Bellcore, as well as in universities, retiring from Rutgers in 1998. Among many contributions to statistics he is especially well known for work on probability plotting, robustness, outlier detection, clustering, classification, and pattern recognition.
>
> Martin Bradbury Wilk (1922–) was born in Montreal. He obtained degrees in chemical engineering and statistics from McGill and Iowa State Universities. After holding several statistics-related posts in industry and at universities (including periods at Princeton, Bell Labs, and Rutgers), Wilk was appointed Chief Statistician at Statistics Canada (1980–1986). He is especially well known for his work with Gnanadesikan on probability plotting and with Shapiro on tests for normality.

Acknowledgments

We thank Peter A. Lachenbruch of the Department of Public Health, Oregon State University, for writing the original version of `qchi` and `pchi`. Patrick Royston of the MRC Clinical Trials Unit, London, also published a similar command in the *Stata Technical Bulletin* (Royston 1996).

References

Chambers, J. M., W. S. Cleveland, B. Kleiner, and P. A. Tukey. 1983. *Graphical Methods for Data Analysis*. Belmont, CA: Wadsworth.

Cox, N. J. 1999. gr42: Quantile plots, generalized. *Stata Technical Bulletin* 51: 16–18. Reprinted in *Stata Technical Bulletin Reprints*, vol. 9, pp. 113–116. College Station, TX: Stata Press.

——. 2001. gr42.1: Quantile plots, generalized: Update to Stata 7. *Stata Technical Bulletin* 61: 10. Reprinted in *Stata Technical Bulletin Reprints*, vol. 10, pp. 55–56. College Station, TX: Stata Press.

——. 2002. Speaking Stata: On getting functions to do the work. *Stata Journal* 2: 411–427.

——. 2004a. Speaking Stata: Graphing distributions. *Stata Journal* 4: 66–88.

——. 2004b. gr42_2: Software update: Quantile plots, generalized. *Stata Journal* 4: 97.

——. 2005a. Speaking Stata: Density probability plots. *Stata Journal* 5: 259–273.

——. 2005b. Speaking Stata: The protean quantile plot. *Stata Journal* 5: 442–460.

——. 2005c. Speaking Stata: Smoothing in various directions. *Stata Journal* 5: 574–593.

——. 2007. Stata tip 47: Quantile–quantile plots without programming. *Stata Journal* 7: 275–279.

Daniel, C., and F. S. Wood. 1980. *Fitting Equations to Data: Computer Analysis of Multifactor Data*. 2nd ed. New York: Wiley.

Gan, F. F., K. J. Koehler, and J. C. Thompson. 1991. Probability plots and distribution curves for assessing the fit of probability models. *American Statistician* 45: 14–21.

Hamilton, L. C. 1992. *Regression with Graphics: A Second Course in Applied Statistics*. Belmont, CA: Duxbury.

——. 2009. *Statistics with Stata (Updated for Version 10)*. Belmont, CA: Brooks/Cole.

Hoaglin, D. C. 1985. Using quantiles to study shape. In *Exploring Data Tables, Trends, and Shapes*, ed. D. C. Hoaglin, F. Mosteller, and J. W. Tukey, 417–460. New York: Wiley.

Kettenring, J. R. 2001. A conversation with Ramanathan Gnanadesikan. *Statistical Science* 16: 295–309.

Miller, R. G., Jr. 1997. *Beyond ANOVA: Basics of Applied Statistics.* London: Chapman & Hall.

Nolan, D., and T. Speed. 2000. *Stat Labs: Mathematical Statistics Through Applications.* New York: Springer.

Royston, P. 1996. sg47: A plot and a test for the χ^2 distribution. *Stata Technical Bulletin* 29: 26–27. Reprinted in *Stata Technical Bulletin Reprints*, vol. 5, pp. 142–144. College Station, TX: Stata Press.

Scotto, M. G. 2000. sg140: The Gumbel quantile plot and a test for choice of extreme models. *Stata Technical Bulletin* 55: 23–25. Reprinted in *Stata Technical Bulletin Reprints*, vol. 10, pp. 156–159. College Station, TX: Stata Press.

Wilk, M. B., and R. Gnanadesikan. 1968. Probability plotting methods for the analysis of data. *Biometrika* 55: 1–17.

Also see

[R] **cumul** — Cumulative distribution

[R] **kdensity** — Univariate kernel density estimation

[R] **logistic postestimation** — Postestimation tools for logistic

[R] **lv** — Letter-value displays

[R] **regress postestimation** — Postestimation tools for regress

Title

> **display** — Substitute for a hand calculator

Syntax

<u>di</u>splay *exp*

Description

display displays strings and values of scalar expressions.

display really has many more features and a more complex syntax diagram, but the diagram shown above is adequate for interactive use. For a full discussion of display's capabilities, see [P] **display**.

Remarks

display can be used as a substitute for a hand calculator.

▷ Example 1

display 2+2 produces the output 4. Stata variables may also appear in the expression, such as in display myvar/2. Because display works only with scalars, the resulting calculation is performed only for the first observation. You could type display myvar[10]/2 to display the calculation for the 10th observation. Here are more examples:

```
. display sqrt(2)/2
.70710678
. display normal(-1.1)
.13566606
. di (57.2-3)/(12-2)
5.42
. display myvar/10
7
. display myvar[10]/2
3.5
```

◁

Also see

[P] **display** — Display strings and values of scalar expressions

[U] **13 Functions and expressions**

Title

> **do** — Execute commands from a file

Syntax

$\{\text{do} \,|\, \underline{\text{run}}\}$ *filename* $\big[$ *arguments* $\big]$ $\big[$, nostop $\big]$

Menu

File > Do...

Description

do and run cause Stata to execute the commands stored in *filename* just as if they were entered from the keyboard. do echoes the commands as it executes them, whereas run is silent. If *filename* is specified without an extension, .do is assumed.

Option

nostop allows the do-file to continue executing even if an error occurs. Normally, Stata stops executing the do-file when it detects an error (nonzero return code).

Remarks

You can create *filename* (called a *do-file*) using Stata's Do-file Editor; see [R] **doedit**. This file will be a standard ASCII (text) file. A complete discussion of do-files can be found in [U] **16 Do-files**.

You can also create *filename* by using a non-Stata text editor; see [D] **shell** for a way to invoke your favorite editor from inside Stata. Make sure that you save the file in ASCII format.

If the path or *filename* contains spaces, it should be enclosed in double quotes.

Reference

Jenkins, S. P. 2006. Stata tip 32: Do not stop. *Stata Journal* 6: 281.

Also see

[R] **doedit** — Edit do-files and other text files

[P] **include** — Include commands from file

[GSM] **13 Using the Do-file Editor—automating Stata**

[GSU] **13 Using the Do-file Editor—automating Stata**

[GSW] **13 Using the Do-file Editor—automating Stata**

[U] **15 Saving and printing output—log files**

[U] **16 Do-files**

Title

doedit — Edit do-files and other text files

Syntax

<u>doedit</u> [*filename*]

Menu

Window > Do-file Editor

Description

doedit opens a text editor that lets you edit do-files and other text files.

The Do-file Editor lets you submit several commands to Stata at once.

Remarks

Clicking on the **Do-file Editor** button is equivalent to typing **doedit**.

doedit, typed by itself, invokes the Editor with an empty document. If you specify *filename*, that file is displayed in the Editor.

You may have more than one Do-file Editor open at once. Each time you submit the **doedit** command, a new window will be opened.

A tutorial discussion of **doedit** can be found in the *Getting Started with Stata* manual. Read [U] **16 Do-files** for an explanation of do-files, and then read [GSW] **13 Using the Do-file Editor—automating Stata** to learn how to use the Do-file Editor to create and execute do-files.

Also see

[GSM] **13 Using the Do-file Editor—automating Stata**

[GSU] **13 Using the Do-file Editor—automating Stata**

[GSW] **13 Using the Do-file Editor—automating Stata**

[U] **16 Do-files**

Title

> **dotplot** — Comparative scatterplots

Syntax

Dotplot of varname, with one column per value of groupvar

dotplot *varname* [*if*] [*in*] [, *options*]

Dotplot for each variable in varlist, with one column per variable

dotplot *varlist* [*if*] [*in*] [, *options*]

options	Description
Options	
over(*groupvar*)	display one columnar dotplot for each value of *groupvar*
nx(#)	horizontal dot density; default is nx(0)
ny(#)	vertical dot density; default is ny(35)
incr(#)	label every # group; default is incr(1)
mean | median	plot a horizontal line of pluses at the mean or median
bounded	use minimum and maximum as boundaries
bar	plot horizontal dashed lines at shoulders of each group
nogroup	use the actual values of *yvar*
center	center the dot for each column
Plot	
marker_options	change look of markers (color, size, etc.)
marker_label_options	add marker labels; change look or position
Y axis, X axis, Titles, Legend, Overall	
twoway_options	any options other than by() documented in [G-3] ***twoway_options***

Menu

Graphics > Distributional graphs > Distribution dotplot

Description

A dotplot is a scatterplot with values grouped together vertically ("binning", as in a histogram) and with plotted points separated horizontally. The aim is to display all the data for several variables or groups in one compact graphic.

In the first syntax, dotplot produces a columnar dotplot of *varname*, with one column per value of *groupvar*. In the second syntax, dotplot produces a columnar dotplot for each variable in *varlist*, with one column per variable; over(*groupvar*) is not allowed. In each case, the "dots" are plotted as small circles to increase readability.

Options

```
Options
```

over(*groupvar*) identifies the variable for which dotplot will display one columnar dotplot for each value of *groupvar*.

nx(*#*) sets the horizontal dot density. A larger value of *#* will increase the dot density, reducing the horizontal separation between dots. This option will increase the separation between columns if two or more groups or variables are used.

ny(*#*) sets the vertical dot density (number of "bins" on the y axis). A larger value of *#* will result in more bins and a plot that is less spread out horizontally. *#* should be determined in conjunction with nx() to give the most pleasing appearance.

incr(*#*) specifies how the x axis is to be labeled. incr(1), the default, labels all groups. incr(2) labels every second group.

[mean | median] plots a horizontal line of pluses at the mean or median of each group.

bounded forces the minimum and maximum of the variable to be used as boundaries of the smallest and largest bins. It should be used with one variable whose support is not the whole of the real line and whose density does not tend to zero at the ends of its support, for example, a uniform random variable or an exponential random variable.

bar plots horizontal dashed lines at the "shoulders" of each group. The shoulders are taken to be the upper and lower quartiles unless mean has been specified; here they will be the mean plus or minus the standard deviation.

nogroup uses the actual values of *yvar* rather than grouping them (the default). This option may be useful if *yvar* takes on only a few values.

center centers the dots for each column on a hidden vertical line.

```
Plot
```

marker_options affect the rendition of markers drawn at the plotted points, including their shape, size, color, and outline; see [G-3] **marker_options**.

marker_label_options specify if and how the markers are to be labeled; see [G-3] **marker_label_options**.

```
Y axis, X axis, Titles, Legend, Overall
```

twoway_options are any of the options documented in [G-3] **twoway_options**, excluding by(). These include options for titling the graph (see [G-3] **title_options**) and for saving the graph to disk (see [G-3] **saving_option**).

Remarks

dotplot produces a figure that has elements of a boxplot, a histogram, and a scatterplot. Like a boxplot, it is most useful for comparing the distributions of several variables or the distribution of 1 variable in several groups. Like a histogram, the figure provides a crude estimate of the density, and, as with a scatterplot, each symbol (dot) represents 1 observation.

▷ Example 1

dotplot may be used as an alternative to Stata's histogram graph for displaying the distribution of one variable.

```
. set seed 123456789
. set obs 1000
. generate norm = rnormal()
. dotplot norm, title("Normal distribution, sample size 1000")
```

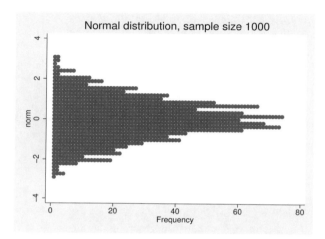

◁

▷ Example 2

The over() option lets us use dotplot to compare the distribution of one variable within different levels of a grouping variable. The center, median, and bar options create a graph that may be compared with Stata's boxplot; see [G-2] **graph box**. The next graph illustrates this option with Stata's automobile dataset.

```
. use http://www.stata-press.com/data/r12/auto, clear
(1978 Automobile Data)

. dotplot mpg, over(foreign) nx(25) ny(10) center median bar
```

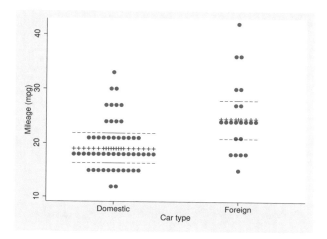

◁

▷ Example 3

The second version of dotplot lets us compare the distribution of several variables. In the next graph, all 10 variables contain measurements on tumor volume.

```
. use http://www.stata-press.com/data/r12/dotgr

. dotplot g1r1-g1r10, ytitle("Tumor volume, cu mm")
```

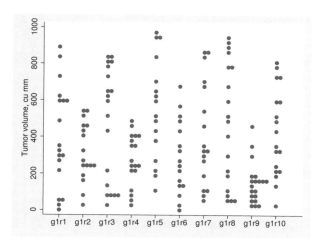

◁

▷ Example 4

When using the first form with the over() option, we can encode a third dimension in a dotplot by using a different plotting symbol for different groups. The third dimension cannot be encoded with a varlist. The example is of a hypothetical matched case–control study. The next graph shows the exposure of each individual in each matched stratum. Cases are marked by the letter 'x', and controls are marked by the letter 'o'.

```
. use http://www.stata-press.com/data/r12/dotdose
. label define symbol 0 "o" 1 "x"
. label values case symbol
. dotplot dose, over(strata) m(none) mlabel(case) mlabp(0) center
```

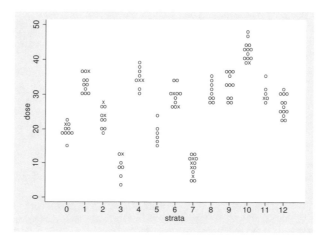

                                                                                                          ◁

▷ Example 5

dotplot can also be used with two virtually continuous variables as an alternative to jittering the data to distinguish ties. We must use the xlabel() option, because otherwise dotplot will attempt to label too many points on the x axis. It is often useful in such instances to use a value of nx that is smaller than the default. That was not necessary in this example, partly because of our choice of symbols.

```
. use http://www.stata-press.com/data/r12/auto
(1978 Automobile Data)
. generate byte hi_price = (price>10000) if price < .
. label define symbol 0 "|" 1 "o"
. label values hi_price symbol
```

```
. dotplot weight, over(gear_ratio) m(none) mlabel(hi_price) mlabp(0) center
> xlabel(#5)
```

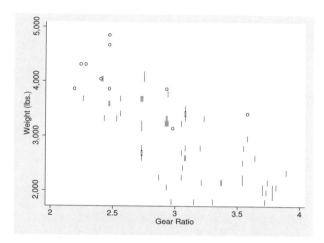

◁

▷ Example 6

The following figure is included mostly for aesthetic reasons. It also demonstrates dotplot's ability to cope with even very large datasets. The sample size for each variable is 10,000, so it may take a long time to print.

```
. clear all
. set seed 123456789
. set obs 10000
. gen norm0 = rnormal()
. gen norm1 = rnormal() + 1
. gen norm2 = rnormal() + 2
. label variable norm0 "N(0,1)"
. label variable norm1 "N(1,1)"
. label variable norm2 "N(2,1)"
. dotplot norm0 norm1 norm2
```

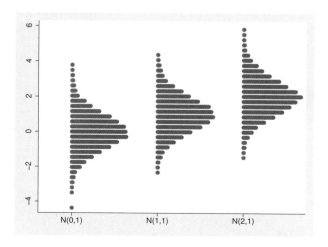

◁

Saved results

dotplot saves the following in r():

Scalars
r(nx)	horizontal dot density
r(ny)	vertical dot density

Methods and formulas

dotplot is implemented as an ado-file.

Acknowledgments

dotplot was written by Peter Sasieni of the Wolfson Institute of Preventive Medicine, London, and Patrick Royston of the MRC Clinical Trials Unit, London.

References

Sasieni, P., and P. Royston. 1994. gr14: dotplot: Comparative scatterplots. *Stata Technical Bulletin* 19: 8–10. Reprinted in *Stata Technical Bulletin Reprints*, vol. 4, pp. 50–54. College Station, TX: Stata Press.

——. 1996. Dotplots. *Applied Statistics* 45: 219–234.

Title

> **dstdize** — Direct and indirect standardization

Syntax

Direct standardization

> dstdize *charvar popvar stratavars* [*if*] [*in*] , by(*groupvars*) [*dstdize_options*]

Indirect standardization

> istdize *casevar$_s$ popvar$_s$ stratavars* [*if*] [*in*] using *filename*,
>
> { <u>pop</u>vars(*casevar$_p$ popvar$_p$*) | rate(*ratevar$_p$* { # | *crudevar$_p$* }) }
>
> [*istdize_options*]

dstdize_options	Description
Main	
* by(*groupvars*)	study populations
<u>us</u>ing(*filename*)	use standard population from Stata dataset
<u>base</u>(#|*string*)	use standard population from a value of grouping variable
<u>level</u>(#)	set confidence level; default is level(95)
Options	
<u>sav</u>ing(*filename*)	save computed standard population distribution as a Stata dataset
<u>f</u>ormat(%*fmt*)	final summary table display format; default is %10.0g
<u>print</u>	include table summary of standard population in output
<u>nores</u>	suppress saving results in r()

* by(*groupvars*) is required.

istdize_options	Description
Main	
* <u>pop</u>vars(*casevar$_p$ popvar$_p$*)	for standard population, *casevar$_p$* is number of cases and *popvar$_p$* is number of individuals
* rate(*ratevar$_p$* { #|*crudevar$_p$* })	*ratevar$_p$* is stratum-specific rates and # or *crudevar$_p$* is the crude case rate value or variable
<u>level</u>(#)	set confidence level; default is level(95)
Options	
by(*groupvars*)	variables identifying study populations
<u>f</u>ormat(%*fmt*)	final summary table display format; default is %10.0g
<u>print</u>	include table summary of standard population in output

* Either popvars(*casevar$_p$ popvar$_p$*) or rate(*ratevar$_p$* { #\|*crudevar$_p$* }) must be specified.

433

Menu

dstdize

Statistics > Epidemiology and related > Other > Direct standardization

istdize

Statistics > Epidemiology and related > Other > Indirect standardization

Description

dstdize produces standardized rates for *charvar*, which are defined as a weighted average of the stratum-specific rates. These rates can be used to compare the characteristic *charvar* across different populations identified by *groupvars*. Weights used in the standardization are given by *popvar*; the strata across which the weights are to be averaged are defined by *stratavars*.

istdize produces indirectly standardized rates for a study population based on a standard population. This standardization method is appropriate when the stratum-specific rates for the population being studied are either unavailable or based on small samples and thus are unreliable. The standardization uses the stratum-specific rates of a standard population to calculate the expected number of cases in the study population(s), sums them, and then compares them with the actual number of cases observed. The standard population is in another Stata data file specified by using *filename*, and it must contain *popvar* and *stratavars*.

In addition to calculating rates, the indirect standardization command produces point estimates and exact confidence intervals of the study population's standardized mortality ratio (SMR), if death is the event of interest, or the standardized incidence ratio (SIR) for studies of incidence. Here we refer to both ratios as SMR.

casevar$_s$ is the variable name for the study population's number of cases (usually deaths). It must contain integers, and for each group, defined by *groupvar*, each subpopulation identified by *stratavars* must have the same values or missing.

popvar$_s$ identifies the number of subjects represented by each observation in the study population.

stratavars define the strata.

Options for dstdize

> Main

by(*groupvars*) is required for the dstdize command; it specifies the variables identifying the study populations. If base() is also specified, there must be only one variable in the by() group. If you do not have a variable for this option, you can generate one by using something like gen newvar=1 and then use newvar as the argument to this option.

using(*filename*) or base(*# | string*) may be used to specify the standard population. You may not specify both options. using(*filename*) supplies the name of a .dta file containing the standard population. The standard population must contain the *popvar* and the *stratavars*. If using() is not specified, the standard population distribution will be obtained from the data. base(*# | string*) lets you specify one of the values of *groupvar*—either a numeric value or a string—to be used as the standard population. If neither base() nor using() is specified, the entire dataset is used to determine an estimate of the standard population.

level(*#*) specifies the confidence level, as a percentage, for a confidence interval of the adjusted rate. The default is level(95) or as set by set level; see [U] **20.7 Specifying the width of confidence intervals**.

saving(*filename*) saves the computed standard population distribution as a Stata dataset that can be used in further analyses.

format(%*fmt*) specifies the format in which to display the final summary table. The default is %10.0g.

print includes a table summary of the standard population before displaying the study population results.

nores suppresses saving results in r(). This option is seldom specified. Some saved results are stored in matrices. If there are more groups than matsize, dstdize will report "matsize too small". Then you can either increase matsize or specify nores. The nores option does not change how results are calculated but specifies that results need not be left behind for use by other programs.

Options for istdize

popvars(*casevar$_p$ popvar$_p$*) or rate(*ratevar$_p$ #* | *ratevar$_p$ crudevar$_p$*) must be specified with istdize. Only one of these two options is allowed. These options are used to describe the standard population's data.

With popvars(*casevar$_p$ popvar$_p$*), *casevar$_p$* records the number of cases (deaths) for each stratum in the standard population, and *popvar$_p$* records the total number of individuals in each stratum (individuals at risk).

With rate(*ratevar$_p$* {*#* | *crudevar$_p$*}), *ratevar$_p$* contains the stratum-specific rates. *#* | *crudevar$_p$* specifies the crude case rate either by a variable name or by the crude case rate value. If a crude rate variable is used, it must be the same for all observations, although it could be missing for some.

level(*#*) specifies the confidence level, as a percentage, for a confidence interval of the adjusted rate. The default is level(95) or as set by set level; see [U] **20.7 Specifying the width of confidence intervals**.

by(*groupvars*) specifies variables identifying study populations when more than one exists in the data. If this option is not specified, the entire study population is treated as one group.

format(%*fmt*) specifies the format in which to display the final summary table. The default is %10.0g.

print outputs a table summary of the standard population before displaying the study population results.

Remarks

Remarks are presented under the following headings:

>*Direct standardization*
>*Indirect standardization*

In epidemiology and other fields, you will often need to compare rates for some characteristic across different populations. These populations often differ on factors associated with the characteristic under study; thus directly comparing overall rates may be misleading.

See van Belle et al. (2004, 642–684), Fleiss, Levin, and Paik (2003, chap. 19), or Kirkwood and Sterne (2003, chap. 25) for a discussion of direct and indirect standardization.

Direct standardization

The direct method of adjusting for differences among populations involves computing the overall rates that would result if, instead of having different distributions, all populations had the same standard distribution. The standardized rate is defined as a weighted average of the stratum-specific rates, with the weights taken from the standard distribution. Direct standardization may be applied only when the specific rates for a given population are available.

dstdize generates adjusted summary measures of occurrence, which can be used to compare prevalence, incidence, or mortality rates between populations that may differ on certain characteristics (for example, age, gender, race). These underlying differences may affect the crude prevalence, mortality, or incidence rates.

▷ Example 1

We have data (Rothman 1986, 42) on mortality rates for Sweden and Panama for 1962, and we wish to compare mortality in these two countries:

```
. use http://www.stata-press.com/data/r12/mortality
(1962 Mortality, Sweden & Panama)

. describe

Contains data from http://www.stata-press.com/data/r12/mortality.dta
  obs:             6                          1962 Mortality, Sweden & Panama
 vars:             4                          14 Apr 2011 16:18
 size:            90

              storage   display    value
variable name   type    format     label      variable label

nation          str6    %9s                   Nation
age_category    byte    %9.0g      age_lbl     Age Category
population       float   %10.0gc               Population in Age Category
deaths          float   %9.0gc                Deaths in Age Category

Sorted by:
```

```
. list, sepby(nation) abbrev(12) divider
```

	nation	age_category	population	deaths
1.	Sweden	0 - 29	3145000	3,523
2.	Sweden	30 - 59	3057000	10,928
3.	Sweden	60+	1294000	59,104
4.	Panama	0 - 29	741,000	3,904
5.	Panama	30 - 59	275,000	1,421
6.	Panama	60+	59,000	2,456

We divide the total number of cases in the population by the population to obtain the *crude rate*:

```
. collapse (sum) pop deaths, by(nation)
. list, abbrev(10) divider
```

	nation	population	deaths
1.	Panama	1075000	7,781
2.	Sweden	7496000	73,555

```
. generate crude = deaths/pop
. list, abbrev(10) divider
```

	nation	population	deaths	crude
1.	Panama	1075000	7,781	.0072381
2.	Sweden	7496000	73,555	.0098126

If we examine the total number of deaths in the two nations, the total crude mortality rate in Sweden is higher than that in Panama. From the original data, we see one possible explanation: Swedes are older than Panamanians, making direct comparison of the mortality rates difficult.

Direct standardization lets us remove the distortion caused by the different age distributions. The adjusted rate is defined as the weighted sum of the crude rates, where the weights are given by the standard distribution. Suppose that we wish to standardize these mortality rates to the following age distribution:

```
. use http://www.stata-press.com/data/r12/1962, clear
(Standard Population Distribution)
. list, abbrev(12) divider
```

	age_category	population
1.	0 - 29	.35
2.	30 - 59	.35
3.	60+	.3

```
. sort age_cat
. save 1962, replace
file 1962.dta saved
```

If we multiply the above weights for the age strata by the crude rate for the corresponding age category, the sum gives us the standardized rate.

```
. use http://www.stata-press.com/data/r12/mortality
(1962 Mortality, Sweden & Panama)
. generate crude=deaths/pop
. drop pop
. sort age_cat
. merge m:1 age_cat using 1962
age_category was byte now float
```

Result	# of obs.
not matched	0
matched	6 (_merge==3)

```
. list, sepby(age_category) abbrev(12)
```

	nation	age_category	deaths	crude	population	_merge
1.	Sweden	0 - 29	3,523	.0011202	.35	matched (3)
2.	Panama	0 - 29	3,904	.0052686	.35	matched (3)
3.	Sweden	30 - 59	10,928	.0035747	.35	matched (3)
4.	Panama	30 - 59	1,421	.0051673	.35	matched (3)
5.	Panama	60+	2,456	.0416271	.3	matched (3)
6.	Sweden	60+	59,104	.0456754	.3	matched (3)

```
. generate product = crude*pop
. by nation, sort: egen adj_rate = sum(product)
. drop _merge
. list, sepby(nation)
```

	nation	age_ca~y	deaths	crude	popula~n	product	adj_rate
1.	Panama	30 - 59	1,421	.0051673	.35	.0018085	.0161407
2.	Panama	0 - 29	3,904	.0052686	.35	.001844	.0161407
3.	Panama	60+	2,456	.0416271	.3	.0124881	.0161407
4.	Sweden	0 - 29	3,523	.0011202	.35	.0003921	.0153459
5.	Sweden	60+	59,104	.0456754	.3	.0137026	.0153459
6.	Sweden	30 - 59	10,928	.0035747	.35	.0012512	.0153459

Comparing the standardized rates indicates that the Swedes have a slightly lower mortality rate.

To perform the above analysis with dstdize, type

```
. use http://www.stata-press.com/data/r12/mortality, clear
(1962 Mortality, Sweden & Panama)

. dstdize deaths pop age_cat, by(nation) using(1962)
```

```
-> nation= Panama

                        ————Unadjusted————   Std.
                          Pop.   Stratum    Pop.
    Stratum     Pop.    Cases  Dist. Rate[s]  Dst[P]  s*P

     0 - 29    741000    3904  0.689 0.0053   0.350 0.0018
    30 - 59    275000    1421  0.256 0.0052   0.350 0.0018
       60+      59000    2456  0.055 0.0416   0.300 0.0125

Totals:      1075000    7781      Adjusted Cases:   17351.2
                                     Crude Rate:     0.0072
                                  Adjusted Rate:     0.0161
                        95% Conf. Interval: [0.0156, 0.0166]
```

```
-> nation= Sweden

                        ————Unadjusted————   Std.
                          Pop.   Stratum    Pop.
    Stratum     Pop.    Cases  Dist. Rate[s]  Dst[P]  s*P

     0 - 29   3145000    3523  0.420 0.0011   0.350 0.0004
    30 - 59   3057000   10928  0.408 0.0036   0.350 0.0013
       60+    1294000   59104  0.173 0.0457   0.300 0.0137

Totals:      7496000   73555      Adjusted Cases: 115032.5
                                     Crude Rate:     0.0098
                                  Adjusted Rate:     0.0153
                        95% Conf. Interval: [0.0152, 0.0155]
```

```
Summary of Study Populations:
   nation         N      Crude    Adj_Rate      Confidence Interval

   Panama     1075000  0.007238   0.016141   [  0.015645,   0.016637]
   Sweden     7496000  0.009813   0.015346   [  0.015235,   0.015457]
```

The summary table above lets us make a quick inspection of the results within the study populations, and the detail tables give the behavior among the strata within the study populations.

◁

▷ Example 2

We have individual-level data on persons in four cities over several years. Included in the data is a variable indicating whether the person has high blood pressure, together with information on the person's age, sex, and race. We wish to obtain standardized high blood pressure rates for each city for 1990 and 1992, using, as the standard, the age, sex, and race distribution of the four cities and two years combined.

Our dataset contains

```
. use http://www.stata-press.com/data/r12/hbp
. describe
Contains data from http://www.stata-press.com/data/r12/hbp.dta
  obs:         1,130
  vars:            7                          21 Feb 2011 06:42
  size:       19,210
```

variable name	storage type	display format	value label	variable label
id	str10	%10s		Record identification number
city	byte	%8.0g		
year	int	%8.0g		
sex	byte	%8.0g	sexfmt	
age_group	byte	%8.0g	agefmt	
race	byte	%8.0g	racefmt	
hbp	byte	%8.0g	yn	high blood pressure

```
Sorted by:
```

The dstdize command is designed to work with aggregate data but will work with individual-level data only if we create a variable recording the population represented by each observation. For individual-level data, this is one:

```
. gen pop = 1
```

On the next page, we specify print to obtain a listing of the standard population and level(90) to request 90% rather than 95% confidence intervals. Typing if year==1990 | year==1992 restricts the data to the two years for both summary tables and the standard population.

```
. dstdize hbp pop age race sex if year==1990 | year==1992, by(city year) print
> level(90)
```

```
——————————Standard Population——————————
                        Stratum        Pop.     Dist.

15 - 19      Black    Female           35      0.077
15 - 19      Black     Male            44      0.097
15 - 19   Hispanic   Female            5       0.011
15 - 19   Hispanic    Male            10       0.022
15 - 19      White    Female           7       0.015
15 - 19      White     Male            5       0.011
20 - 24      Black    Female          43       0.095
20 - 24      Black     Male           67       0.147
20 - 24   Hispanic   Female           14       0.031
20 - 24   Hispanic    Male           13       0.029
20 - 24      White    Female           4       0.009
20 - 24      White     Male           21       0.046
25 - 29      Black    Female          17       0.037
25 - 29      Black     Male           44       0.097
25 - 29   Hispanic   Female            7       0.015
25 - 29   Hispanic    Male           13       0.029
25 - 29      White    Female           9       0.020
25 - 29      White     Male           16       0.035
30 - 34      Black    Female          16       0.035
30 - 34      Black     Male           32       0.070
30 - 34   Hispanic   Female            2       0.004
30 - 34   Hispanic    Male            3       0.007
30 - 34      White    Female           5       0.011
30 - 34      White     Male           23       0.051
```

```
Total:                             455
(6 observations excluded because of missing values)
```

```
-> city year= 1 1990
```

	Stratum	Pop.	Cases	Unadjusted Pop. Dist.	Stratum Rate[s]	Std. Pop. Dst[P]	s*P
15 - 19 Black	Female	6	2	0.128	0.3333	0.077	0.0256
15 - 19 Black	Male	6	0	0.128	0.0000	0.097	0.0000
15 - 19 Hispanic	Male	1	0	0.021	0.0000	0.022	0.0000
20 - 24 Black	Female	3	0	0.064	0.0000	0.095	0.0000
20 - 24 Black	Male	11	0	0.234	0.0000	0.147	0.0000
25 - 29 Black	Female	4	0	0.085	0.0000	0.037	0.0000
25 - 29 Black	Male	6	1	0.128	0.1667	0.097	0.0161
25 - 29 Hispanic	Female	2	0	0.043	0.0000	0.015	0.0000
25 - 29 White	Female	1	0	0.021	0.0000	0.020	0.0000
30 - 34 Black	Female	1	0	0.021	0.0000	0.035	0.0000
30 - 34 Black	Male	6	0	0.128	0.0000	0.070	0.0000

```
Totals:                            47       3       Adjusted Cases:        2.0
                                                       Crude Rate:      0.0638
                                                    Adjusted Rate:      0.0418
                                        90% Conf. Interval: [0.0074, 0.0761]
```

-> city year= 1 1992

		Stratum	Pop.	Cases	Unadjusted Pop. Dist.	Stratum Rate[s]	Std. Pop. Dst[P]	s*P
15 - 19	Black	Female	3	0	0.054	0.0000	0.077	0.0000
15 - 19	Black	Male	9	0	0.161	0.0000	0.097	0.0000
15 - 19	Hispanic	Male	1	0	0.018	0.0000	0.022	0.0000
20 - 24	Black	Female	7	0	0.125	0.0000	0.095	0.0000
20 - 24	Black	Male	9	0	0.161	0.0000	0.147	0.0000
20 - 24	Hispanic	Female	1	0	0.018	0.0000	0.031	0.0000
25 - 29	Black	Female	2	0	0.036	0.0000	0.037	0.0000
25 - 29	Black	Male	11	1	0.196	0.0909	0.097	0.0088
25 - 29	Hispanic	Male	1	0	0.018	0.0000	0.029	0.0000
30 - 34	Black	Female	7	0	0.125	0.0000	0.035	0.0000
30 - 34	Black	Male	4	0	0.071	0.0000	0.070	0.0000
30 - 34	White	Female	1	0	0.018	0.0000	0.011	0.0000

Totals: 56   1   Adjusted Cases: 0.5
                                                Crude Rate: 0.0179
                                             Adjusted Rate: 0.0088
                           90% Conf. Interval: [0.0000, 0.0226]

-> city year= 2 1990

		Stratum	Pop.	Cases	Unadjusted Pop. Dist.	Stratum Rate[s]	Std. Pop. Dst[P]	s*P
15 - 19	Black	Female	5	0	0.078	0.0000	0.077	0.0000
15 - 19	Black	Male	7	1	0.109	0.1429	0.097	0.0138
15 - 19	Hispanic	Male	1	0	0.016	0.0000	0.022	0.0000
20 - 24	Black	Female	7	1	0.109	0.1429	0.095	0.0135
20 - 24	Black	Male	8	0	0.125	0.0000	0.147	0.0000
20 - 24	Hispanic	Female	5	0	0.078	0.0000	0.031	0.0000
20 - 24	Hispanic	Male	2	0	0.031	0.0000	0.029	0.0000
20 - 24	White	Male	2	0	0.031	0.0000	0.046	0.0000
25 - 29	Black	Female	3	0	0.047	0.0000	0.037	0.0000
25 - 29	Black	Male	9	0	0.141	0.0000	0.097	0.0000
25 - 29	Hispanic	Female	2	0	0.031	0.0000	0.015	0.0000
25 - 29	White	Female	1	0	0.016	0.0000	0.020	0.0000
25 - 29	White	Male	2	1	0.031	0.5000	0.035	0.0176
30 - 34	Black	Female	1	0	0.016	0.0000	0.035	0.0000
30 - 34	Black	Male	5	0	0.078	0.0000	0.070	0.0000
30 - 34	Hispanic	Female	2	0	0.031	0.0000	0.004	0.0000
30 - 34	White	Female	1	0	0.016	0.0000	0.011	0.0000
30 - 34	White	Male	1	0	0.016	0.0000	0.051	0.0000

Totals: 64   3   Adjusted Cases: 2.9
                                                Crude Rate: 0.0469
                                             Adjusted Rate: 0.0449
                           90% Conf. Interval: [0.0091, 0.0807]

-> city year= 2 1992

	Stratum	Pop.	Cases	Unadjusted Pop. Dist.	Stratum Rate[s]	Std. Pop. Dst[P]	s*P
15 - 19 Black	Female	1	0	0.015	0.0000	0.077	0.0000
15 - 19 Black	Male	5	0	0.075	0.0000	0.097	0.0000
15 - 19 Hispanic	Female	3	0	0.045	0.0000	0.011	0.0000
15 - 19 Hispanic	Male	1	0	0.015	0.0000	0.022	0.0000
15 - 19 White	Male	1	0	0.015	0.0000	0.011	0.0000
20 - 24 Black	Female	8	0	0.119	0.0000	0.095	0.0000
20 - 24 Black	Male	11	0	0.164	0.0000	0.147	0.0000
20 - 24 Hispanic	Female	6	0	0.090	0.0000	0.031	0.0000
20 - 24 Hispanic	Male	4	2	0.060	0.5000	0.029	0.0143
20 - 24 White	Female	1	0	0.015	0.0000	0.009	0.0000
20 - 24 White	Male	2	0	0.030	0.0000	0.046	0.0000
25 - 29 Black	Female	2	0	0.030	0.0000	0.037	0.0000
25 - 29 Black	Male	3	0	0.045	0.0000	0.097	0.0000
25 - 29 Hispanic	Female	2	0	0.030	0.0000	0.015	0.0000
25 - 29 Hispanic	Male	4	0	0.060	0.0000	0.029	0.0000
25 - 29 White	Female	4	0	0.060	0.0000	0.020	0.0000
25 - 29 White	Male	2	0	0.030	0.0000	0.035	0.0000
30 - 34 Black	Female	1	0	0.015	0.0000	0.035	0.0000
30 - 34 Black	Male	2	0	0.030	0.0000	0.070	0.0000
30 - 34 Hispanic	Male	1	0	0.015	0.0000	0.007	0.0000
30 - 34 White	Female	2	0	0.030	0.0000	0.011	0.0000
30 - 34 White	Male	1	0	0.015	0.0000	0.051	0.0000

Totals: 67 2

Adjusted Cases: 1.0
Crude Rate: 0.0299
Adjusted Rate: 0.0143
90% Conf. Interval: [0.0025, 0.0260]

-> city year= 3 1990

	Stratum	Pop.	Cases	Unadjusted Pop. Dist.	Stratum Rate[s]	Std. Pop. Dst[P]	s*P
15 - 19 Black	Female	3	0	0.043	0.0000	0.077	0.0000
15 - 19 Black	Male	1	0	0.014	0.0000	0.097	0.0000
15 - 19 Hispanic	Female	1	0	0.014	0.0000	0.011	0.0000
15 - 19 White	Female	3	0	0.043	0.0000	0.015	0.0000
15 - 19 White	Male	1	0	0.014	0.0000	0.011	0.0000
20 - 24 Black	Female	1	0	0.014	0.0000	0.095	0.0000
20 - 24 Black	Male	9	0	0.130	0.0000	0.147	0.0000
20 - 24 Hispanic	Male	3	0	0.043	0.0000	0.029	0.0000
20 - 24 White	Female	2	0	0.029	0.0000	0.009	0.0000
20 - 24 White	Male	8	1	0.116	0.1250	0.046	0.0058
25 - 29 Black	Female	1	0	0.014	0.0000	0.037	0.0000
25 - 29 Black	Male	8	3	0.116	0.3750	0.097	0.0363
25 - 29 Hispanic	Male	4	0	0.058	0.0000	0.029	0.0000
25 - 29 White	Female	1	0	0.014	0.0000	0.020	0.0000
25 - 29 White	Male	6	0	0.087	0.0000	0.035	0.0000
30 - 34 Black	Male	6	2	0.087	0.3333	0.070	0.0234
30 - 34 White	Male	11	5	0.159	0.4545	0.051	0.0230

Totals: 69 11

Adjusted Cases: 6.1
Crude Rate: 0.1594
Adjusted Rate: 0.0885
90% Conf. Interval: [0.0501, 0.1268]

-> city year= 3 1992

| | | | | | ———Unadjusted——— | | Std. | |
| | | | | | Pop. | Stratum | Pop. | |
		Stratum	Pop.	Cases	Dist.	Rate[s]	Dst[P]	s*P
15 - 19	Black	Female	2	0	0.054	0.0000	0.077	0.0000
15 - 19	Hispanic	Male	3	0	0.081	0.0000	0.022	0.0000
15 - 19	White	Female	2	0	0.054	0.0000	0.015	0.0000
15 - 19	White	Male	1	0	0.027	0.0000	0.011	0.0000
20 - 24	Black	Male	3	0	0.081	0.0000	0.147	0.0000
20 - 24	Hispanic	Female	1	0	0.027	0.0000	0.031	0.0000
20 - 24	Hispanic	Male	3	0	0.081	0.0000	0.029	0.0000
20 - 24	White	Female	1	0	0.027	0.0000	0.009	0.0000
20 - 24	White	Male	6	1	0.162	0.1667	0.046	0.0077
25 - 29	Hispanic	Male	1	0	0.027	0.0000	0.029	0.0000
25 - 29	White	Male	5	1	0.135	0.2000	0.035	0.0070
30 - 34	Black	Male	1	0	0.027	0.0000	0.070	0.0000
30 - 34	White	Male	8	5	0.216	0.6250	0.051	0.0316

Totals:		37	7	Adjusted Cases: 1.7
				Crude Rate: 0.1892
				Adjusted Rate: 0.0463
			90% Conf. Interval:	[0.0253, 0.0674]

-> city year= 5 1990

| | | | | | ———Unadjusted——— | | Std. | |
| | | | | | Pop. | Stratum | Pop. | |
		Stratum	Pop.	Cases	Dist.	Rate[s]	Dst[P]	s*P
15 - 19	Black	Female	9	0	0.196	0.0000	0.077	0.0000
15 - 19	Black	Male	7	0	0.152	0.0000	0.097	0.0000
15 - 19	Hispanic	Male	1	0	0.022	0.0000	0.022	0.0000
15 - 19	White	Male	1	0	0.022	0.0000	0.011	0.0000
20 - 24	Black	Female	4	0	0.087	0.0000	0.095	0.0000
20 - 24	Black	Male	6	0	0.130	0.0000	0.147	0.0000
20 - 24	Hispanic	Female	1	0	0.022	0.0000	0.031	0.0000
25 - 29	Black	Female	3	1	0.065	0.3333	0.037	0.0125
25 - 29	Black	Male	5	0	0.109	0.0000	0.097	0.0000
25 - 29	Hispanic	Female	1	0	0.022	0.0000	0.015	0.0000
25 - 29	White	Female	2	1	0.043	0.5000	0.020	0.0099
30 - 34	Black	Female	2	0	0.043	0.0000	0.035	0.0000
30 - 34	Black	Male	3	0	0.065	0.0000	0.070	0.0000
30 - 34	White	Male	1	0	0.022	0.0000	0.051	0.0000

Totals:		46	2	Adjusted Cases: 1.0
				Crude Rate: 0.0435
				Adjusted Rate: 0.0223
			90% Conf. Interval:	[0.0020, 0.0426]

```
-> city year= 5 1992
```

	Stratum		Pop.	Cases	Unadjusted Pop. Dist.	Stratum Rate[s]	Std. Pop. Dst[P]	s*P
15 - 19	Black	Female	6	0	0.087	0.0000	0.077	0.0000
15 - 19	Black	Male	9	0	0.130	0.0000	0.097	0.0000
15 - 19	Hispanic	Female	1	0	0.014	0.0000	0.011	0.0000
15 - 19	Hispanic	Male	2	0	0.029	0.0000	0.022	0.0000
15 - 19	White	Female	2	0	0.029	0.0000	0.015	0.0000
15 - 19	White	Male	1	0	0.014	0.0000	0.011	0.0000
20 - 24	Black	Female	13	0	0.188	0.0000	0.095	0.0000
20 - 24	Black	Male	10	0	0.145	0.0000	0.147	0.0000
20 - 24	Hispanic	Male	1	0	0.014	0.0000	0.029	0.0000
20 - 24	White	Male	3	0	0.043	0.0000	0.046	0.0000
25 - 29	Black	Female	2	0	0.029	0.0000	0.037	0.0000
25 - 29	Black	Male	2	0	0.029	0.0000	0.097	0.0000
25 - 29	Hispanic	Male	3	0	0.043	0.0000	0.029	0.0000
25 - 29	White	Male	1	0	0.014	0.0000	0.035	0.0000
30 - 34	Black	Female	4	0	0.058	0.0000	0.035	0.0000
30 - 34	Black	Male	5	0	0.072	0.0000	0.070	0.0000
30 - 34	Hispanic	Male	2	0	0.029	0.0000	0.007	0.0000
30 - 34	White	Female	1	0	0.014	0.0000	0.011	0.0000
30 - 34	White	Male	1	1	0.014	1.0000	0.051	0.0505

```
Totals:                       69        1      Adjusted Cases:        3.5
                                                  Crude Rate:      0.0145
                                               Adjusted Rate:      0.0505
                              90% Conf. Interval: [0.0505, 0.0505]
```

Summary of Study Populations:

city year	N	Crude	Adj_Rate	Confidence Interval	
1 1990	47	0.063830	0.041758	[0.007427,	0.076089]
1 1992	56	0.017857	0.008791	[0.000000,	0.022579]
2 1990	64	0.046875	0.044898	[0.009072,	0.080724]
2 1992	67	0.029851	0.014286	[0.002537,	0.026035]
3 1990	69	0.159420	0.088453	[0.050093,	0.126813]
3 1992	37	0.189189	0.046319	[0.025271,	0.067366]
5 1990	46	0.043478	0.022344	[0.002044,	0.042644]
5 1992	69	0.014493	0.050549	[0.050549,	0.050549]

Indirect standardization

Standardization of rates can be performed via the indirect method whenever the stratum-specific rates are either unknown or unreliable. If the stratum-specific rates are known, the direct standardization method is preferred.

To apply the indirect method, you must have the following information:

- The observed number of cases in each population to be standardized, O. For example, if death rates in two states are being standardized using the U.S. death rate for the same period, you must know the total number of deaths in each state.

- The distribution across the various strata for the population being studied, $n_1, \ldots, n_k$. If you are standardizing the death rate in the two states, adjusting for age, you must know the number of individuals in each of the k age groups.

- The stratum-specific rates for the standard population, $p_1, \ldots, p_k$. For example, you must have the U.S. death rate for each stratum (age group).

- The crude rate of the standard population, C. For example, you must have the U.S. mortality rate for the year.

The indirect adjusted rate is then

$$R_{\text{indirect}} = C \frac{O}{E}$$

where E is the expected number of cases (deaths) in each population. See *Methods and formulas* for a more detailed description of calculations.

▷ Example 3

This example is borrowed from Kahn and Sempos (1989, 95–105). We want to compare 1970 mortality rates in California and Maine, adjusting for age. Although we have age-specific population counts for the two states, we lack age-specific death rates. Direct standardization is not feasible here. We can use the U.S. population census data for the same year to produce indirectly standardized rates for these two states.

From the U.S. census, the standard population for this example was entered into Stata and saved in popkahn.dta.

```
. use http://www.stata-press.com/data/r12/popkahn, clear
. list age pop deaths rate, sep(4)
```

	age	population	deaths	rate
1.	<15	57,900,000	103,062	.00178
2.	15-24	35,441,000	45,261	.00128
3.	25-34	24,907,000	39,193	.00157
4.	35-44	23,088,000	72,617	.00315
5.	45-54	23,220,000	169,517	.0073
6.	55-64	18,590,000	308,373	.01659
7.	65-74	12,436,000	445,531	.03583
8.	75+	7,630,000	736,758	.09656

The standard population contains for each age stratum the total number of individuals (pop) and both the age-specific mortality rate (rate) and the number of deaths. The standard population need not contain all three. If we have only the age-specific mortality rate, we can use the $\text{rate}(\textit{ratevar}_p\ \textit{crudevar}_p)$ or $\text{rate}(\textit{ratevar}_p\ \#)$ option, where $\textit{crudevar}_p$ refers to the variable containing the total population's crude death rate or # is the total population's crude death rate.

Now let's look at the states' data (study population):

```
. use http://www.stata-press.com/data/r12/kahn
. list, sep(4)
```

	state	age	populat~n	death	st	death_~e
1.	California	<15	5,524,000	166,285	1	.0016
2.	California	15-24	3,558,000	166,285	1	.0013
3.	California	25-34	2,677,000	166,285	1	.0015
4.	California	35-44	2,359,000	166,285	1	.0028
5.	California	45-54	2,330,000	166,285	1	.0067
6.	California	55-64	1,704,000	166,285	1	.0154
7.	California	65-74	1,105,000	166,285	1	.0328
8.	California	75+	696,000	166,285	1	.0917
9.	Maine	<15	286,000	11,051	2	.0019
10.	Maine	15-24	168,000	.	2	.0011
11.	Maine	25-34	110,000	.	2	.0014
12.	Maine	35-44	109,000	.	2	.0029
13.	Maine	45-54	110,000	.	2	.0069
14.	Maine	55-64	94,000	.	2	.0173
15.	Maine	65-74	69,000	.	2	.039
16.	Maine	75+	46,000	.	2	.1041

For each state, the number of individuals in each stratum (age group) is contained in the pop variable. The death variable is the total number of deaths observed in the state during the year. It must have the same value for all observations in the group, as for California, or it could be missing in all but one observation per group, as for Maine.

To match these two datasets, the strata variables must have the same name in both datasets and ideally the same levels. If a level is missing from either dataset, that level will not be included in the standardization.

With kahn.dta in memory, we now execute the command. We will use the print option to obtain the standard population's summary table, and because we have both the standard population's age-specific count and deaths, we will specify the popvars($\textit{casevar}_p\ \textit{popvar}_p$) option. Or, we could specify the rate(rate 0.00945) option because we know that 0.00945 is the U.S. crude death rate for 1970.

```
. istdize death pop age using http://www.stata-press.com/data/r12/popkahn,
> by(state) pop(deaths pop) print
```

```
————Standard Population————
```

Stratum	Rate
<15	0.00178
15-24	0.00128
25-34	0.00157
35-44	0.00315
45-54	0.00730
55-64	0.01659
65-74	0.03583
75+	0.09656

Standard population's crude rate: 0.00945

-> state= California

Indirect Standardization

Stratum	Standard Population Rate	Observed Population	Cases Expected
<15	0.0018	5524000	9832.72
15-24	0.0013	3558000	4543.85
25-34	0.0016	2677000	4212.46
35-44	0.0031	2359000	7419.59
45-54	0.0073	2330000	17010.10
55-64	0.0166	1704000	28266.14
65-74	0.0358	1105000	39587.63
75+	0.0966	696000	67206.23

| Totals: | | 19953000 | 178078.73 |

Observed Cases: 166285
SMR (Obs/Exp):   0.93
SMR exact 95% Conf. Interval: [0.9293, 0.9383]
Crude Rate: 0.0083
Adjusted Rate: 0.0088
95% Conf. Interval: [0.0088, 0.0089]

-> state= Maine

Indirect Standardization

Stratum	Standard Population Rate	Observed Population	Cases Expected
<15	0.0018	286000	509.08
15-24	0.0013	168000	214.55
25-34	0.0016	110000	173.09
35-44	0.0031	109000	342.83
45-54	0.0073	110000	803.05
55-64	0.0166	94000	1559.28
65-74	0.0358	69000	2471.99
75+	0.0966	46000	4441.79

| Totals: | | 992000 | 10515.67 |

Observed Cases: 11051
SMR (Obs/Exp):   1.05
SMR exact 95% Conf. Interval: [1.0314, 1.0707]
Crude Rate: 0.0111
Adjusted Rate: 0.0099
95% Conf. Interval: [0.0097, 0.0101]

```
Summary of Study Populations (Rates):
                   Cases
      state      Observed     Crude   Adj_Rate  Confidence Interval
 ──────────────────────────────────────────────────────────────────
 California        166285   0.008334  0.008824  [0.008782, 0.008866]
      Maine         11051   0.011140  0.009931  [0.009747, 0.010118]
Summary of Study Populations (SMR):
                   Cases       Cases                      Exact
      state      Observed    Expected     SMR      Confidence Interval
 ──────────────────────────────────────────────────────────────────
 California        166285   178078.73    0.934   [0.929290, 0.938271]
      Maine         11051    10515.67    1.051   [1.031405, 1.070688]
```

◁

Saved results

dstdize saves the following in r():

Scalars
    r(k)   number of populations

Macros
    r(by)   variable names specified in by()
    r(c#)   values of r(by) for #th group

Matrices
    r(se)   standard errors of adjusted rates
    r(ub)   upper bounds of confidence intervals for adjusted rates
    r(lb)   lower bounds of confidence intervals for adjusted rates
    r(Nobs) $1 \times k$ vector of number of observations
    r(crude) $1 \times k$ vector of crude rates (*)
    r(adj) $1 \times k$ vector of adjusted rates (*)

(*) If, in a group, the number of observations is 0, then 9 is stored for the corresponding crude and adjusted rates.

Methods and formulas

dstdize and istdize are implemented as ado-files.

The directly standardized rate, S_{R}, is defined by

$$S_{\mathrm{R}} = \frac{\displaystyle\sum_{i=1}^{k} w_i R_i}{\displaystyle\sum_{i=1}^{k} w_i}$$

(Rothman 1986, 44), where R_i is the stratum-specific rate in stratum i and w_i is the weight for stratum i derived from the standard population.

If n_i is the population of stratum i, the standard error, $\mathrm{se}(S_{\mathrm{R}})$, in stratified sampling for proportions (ignoring the finite population correction) is

$$\mathrm{se}(S_{\mathrm{R}}) = \frac{1}{\sum w_i} \sqrt{\sum_{i=1}^{k} \frac{w_i{}^2 R_i (1 - R_i)}{n_i}}$$

(Cochran 1977, 108), from which the confidence intervals are calculated.

For indirect standardization, define O as the observed number of cases in each population to be standardized; $n_1, \ldots, n_k$ as the distribution across the various strata for the population being studied; $R_1, \ldots, R_k$ as the stratum-specific rates for the standard population; and C as the crude rate of the standard population. The expected number of cases (deaths), E, in each population is obtained by applying the standard population stratum-specific rates, $R_1, \ldots, R_k$, to the study populations:

$$E = \sum_{i=1}^{k} n_i R_i$$

The indirectly adjusted rate is then

$$R_{\text{indirect}} = C \frac{O}{E}$$

and O/E is the study population's SMR if death is the event of interest or the SIR for studies of disease (or other) incidence.

The exact confidence interval is calculated for each estimated SMR by assuming a Poisson process as described in Breslow and Day (1987, 69–71). These intervals are obtained by first calculating the upper and lower bounds for the confidence interval of the Poisson-distributed observed events, O—say, L and U, respectively—and then computing $\text{SMR}_L = L/E$ and $\text{SMR}_U = U/E$.

Acknowledgments

We gratefully acknowledge the collaboration of Dr. Joel A. Harrison, consultant; Dr. José Maria Pacheco from the Departamento de Epidemiologia, Faculdade de Saúde Pública/USP, Sao Paulo, Brazil; and Dr John L. Moran from The Queen Elizabeth Hospital, Woodville, Australia.

References

Breslow, N. E., and N. E. Day. 1987. *Statistical Methods in Cancer Research: Vol. 2—The Design and Analysis of Cohort Studies.* Lyon: IARC.

Cleves, M. A. 1998. sg80: Indirect standardization. *Stata Technical Bulletin* 42: 43–47. Reprinted in *Stata Technical Bulletin Reprints*, vol. 7, pp. 224–228. College Station, TX: Stata Press.

Cochran, W. G. 1977. *Sampling Techniques.* 3rd ed. New York: Wiley.

Fleiss, J. L., B. Levin, and M. C. Paik. 2003. *Statistical Methods for Rates and Proportions.* 3rd ed. New York: Wiley.

Forthofer, R. N., and E. S. Lee. 1995. *Introduction to Biostatistics: A Guide to Design, Analysis, and Discovery.* New York: Academic Press.

Juul, S., and M. Frydenberg. 2010. *An Introduction to Stata for Health Researchers.* 3rd ed. College Station, TX: Stata Press.

Kahn, H. A., and C. T. Sempos. 1989. *Statistical Methods in Epidemiology.* New York: Oxford University Press.

Kirkwood, B. R., and J. A. C. Sterne. 2003. *Essential Medical Statistics.* 2nd ed. Malden, MA: Blackwell.

McGuire, T. J., and J. A. Harrison. 1994. sbe11: Direct standardization. *Stata Technical Bulletin* 21: 5–9. Reprinted in *Stata Technical Bulletin Reprints*, vol. 4, pp. 88–94. College Station, TX: Stata Press.

Pagano, M., and K. Gauvreau. 2000. *Principles of Biostatistics.* 2nd ed. Belmont, CA: Duxbury.

Rothman, K. J. 1986. *Modern Epidemiology.* Boston: Little, Brown.

van Belle, G., L. D. Fisher, P. J. Heagerty, and T. S. Lumley. 2004. *Biostatistics: A Methodology for the Health Sciences.* 2nd ed. New York: Wiley.

Wang, D. 2000. sbe40: Modeling mortality data using the Lee–Carter model. *Stata Technical Bulletin* 57: 15–17. Reprinted in *Stata Technical Bulletin Reprints*, vol. 10, pp. 118–121. College Station, TX: Stata Press.

Also see

[ST] **epitab** — Tables for epidemiologists

[SVY] **direct standardization** — Direct standardization of means, proportions, and ratios

Title

> **dydx** — Calculate numeric derivatives and integrals

Syntax

Derivatives of numeric functions

> dydx *yvar* *xvar* [*if*] [*in*] , generate(*newvar*) [*dydx_options*]

Integrals of numeric functions

> integ *yvar* *xvar* [*if*] [*in*] [, *integ_options*]

dydx_options	Description
Main	
*<u>g</u>enerate(*newvar*)	create variable named *newvar*
<u>r</u>eplace	overwrite the existing variable

*generate(*newvar*) is required.

integ_options	Description
Main	
<u>g</u>enerate(*newvar*)	create variable named *newvar*
<u>t</u>rapezoid	use trapezoidal rule to compute integrals; default is cubic splines
<u>i</u>nitial(#)	initial value of integral; default is initial(0)
<u>r</u>eplace	overwrite the existing variable

by is allowed with dydx and integ; see [D] **by**.

Menu

dydx

Data > Create or change data > Other variable-creation commands > Calculate numerical derivatives

integ

Data > Create or change data > Other variable-creation commands > Calculate numeric integrals

Description

dydx and integ calculate derivatives and integrals of numeric "functions".

Options

Main

generate(*newvar*) specifies the name of the new variable to be created. It must be specified with
dydx.

trapezoid requests that the trapezoidal rule [the sum of $(x_i - x_{i-1})(y_i + y_{i-1})/2$] be used to
compute integrals. The default is cubic splines, which give superior results for most smooth
functions; for irregular functions, trapezoid may give better results.

initial(*#*) specifies the initial condition for calculating definite integrals; see *Methods and formulas*
below. The default is initial(0).

replace specifies that if an existing variable is specified for generate(), it should be overwritten.

Remarks

dydx and integ lets you extend Stata's graphics capabilities beyond data analysis and into
mathematics. (See Gould [1993] for another command that draws functions.)

▷ Example 1

We graph $y = e^{-x/6}\sin(x)$ over the interval $[0, 12.56]$:

```
. range x 0 12.56 100
obs was 0, now 100
. generate y = exp(-x/6)*sin(x)
. label variable y "exp(-x/6)*sin(x)"
. twoway connected y x, connect(i) yline(0)
```

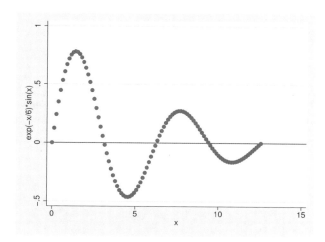

We estimate the derivative by using dydx and compute the relative difference between this estimate
and the true derivative.

```
. dydx y x, gen(dy)
. generate dytrue = exp(-x/6)*(cos(x) - sin(x)/6)
. generate error = abs(dy - dytrue)/dytrue
```

The error is greatest at the endpoints, as we would expect. The error is approximately 0.5% at each endpoint, but the error quickly falls to less than 0.01%.

```
. label variable error "Error in derivative estimate"
. twoway line error x, ylabel(0(.002).006)
```

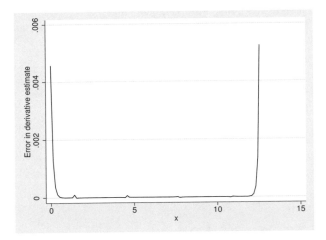

We now estimate the integral by using `integ`:

```
. integ y x, gen(iy)
number of points = 100
integral        = .85316396
. generate iytrue = (36/37)*(1 - exp(-x/6)*(cos(x) + sin(x)/6))
. display iytrue[_N]
.85315901
. display abs(r(integral) - iytrue[_N])/iytrue[_N]
5.799e-06
. generate diff = iy - iytrue
```

The relative difference between the estimate [stored in `r(integral)`] and the true value of the integral is about 6×10^{-6}. A graph of the absolute difference (`diff`) is shown below. Here error is cumulative. Again most of the error is due to a relatively poorer fit near the endpoints.

```
. label variable diff "Error in integral estimate"
. twoway line diff x, ylabel(0(5.00e-06).00001)
```

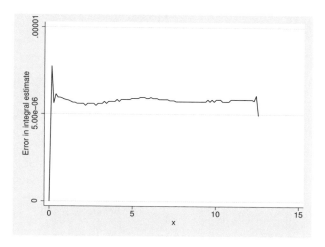

◁

Saved results

dydx saves the following in `r()`:

Macros
    `r(y)`   name of *yvar*

`integ` saves the following in `r()`:

Scalars
    `r(N_points)`   number of unique x points
    `r(integral)`   estimate of the integral

Methods and formulas

`dydx` and `integ` are implemented as ado-files.

Consider a set of data points, $(x_1, y_1), \ldots, (x_n, y_n)$, generated by a function $y = f(x)$. `dydx` and `integ` first fit these points with a cubic spline, which is then analytically differentiated (integrated) to give an approximation for the derivative (integral) of f.

The cubic spline (see, for example, Press et al. [2007]) consists of $n-1$ cubic polynomials $P_i(x)$, with the ith one defined on the interval $[x_i, x_{i+1}]$,

$$P_i(x) = y_i a_i(x) + y_{i+1} b_i(x) + y_i'' c_i(x) + y_{i+1}'' d_i(x)$$

where

$$a_i(x) = \frac{x_{i+1} - x}{x_{i+1} - x_i} \qquad\qquad b_i(x) = \frac{x - x_i}{x_{i+1} - x_i}$$

$$c_i(x) = \frac{1}{6}(x_{i+1} - x_i)^2 a_i(x)[\{a_i(x)\}^2 - 1] \qquad d_i(x) = \frac{1}{6}(x_{i+1} - x_i)^2 b_i(x)[\{b_i(x)\}^2 - 1]$$

and y_i'' and y_{i+1}'' are constants whose values will be determined as described below. The notation for these constants is justified because $P_i''(x_i) = y_i''$ and $P_i''(x_{i+1}) = y_{i+1}''$.

Because $a_i(x_i) = 1$, $a_i(x_{i+1}) = 0$, $b_i(x_i) = 0$, and $b_i(x_{i+1}) = 1$. Therefore, $P_i(x_i) = y_i$, and $P_i(x_{i+1}) = y_{i+1}$. Thus the P_i jointly define a function that is continuous at the interval boundaries. The first derivative should be continuous at the interval boundaries; that is,

$$P_i'(x_{i+1}) = P_{i+1}'(x_{i+1})$$

The above $n - 2$ equations (one equation for each point except the two endpoints) and the values of the first derivative at the endpoints, $P_1'(x_1)$ and $P_{n-1}'(x_n)$, determine the n constants y_i''.

The value of the first derivative at an endpoint is set to the value of the derivative obtained by fitting a quadratic to the endpoint and the two adjacent points; namely, we use

$$P_1'(x_1) = \frac{y_1 - y_2}{x_1 - x_2} + \frac{y_1 - y_3}{x_1 - x_3} - \frac{y_2 - y_3}{x_2 - x_3}$$

and a similar formula for the upper endpoint.

dydx approximates $f'(x_i)$ by using $P_i'(x_i)$.

integ approximates $F(x_i) = F(x_1) + \int_{x_1}^{x_i} f(x)\,dx$ by using

$$I_0 + \sum_{k=1}^{i-1} \int_{x_k}^{x_{k+1}} P_k(x)\,dx$$

where I_0 (an estimate of $F(x_1)$) is the value specified by the initial(#) option. If the trapezoid option is specified, integ approximates the integral by using the trapezoidal rule:

$$I_0 + \sum_{k=1}^{i-1} \frac{1}{2}(x_{k+1} - x_k)(y_{k+1} + y_k)$$

If there are ties among the x_i, the mean of y_i is computed at each set of ties and the cubic spline is fit to these values.

Acknowledgment

The present versions of dydx and integ were inspired by the dydx2 command written by Patrick Royston of the MRC Clinical Trials Unit, London.

References

Gould, W. W. 1993. ssi5.1: Graphing functions. *Stata Technical Bulletin* 16: 23–26. Reprinted in *Stata Technical Bulletin Reprints*, vol. 3, pp. 188–193. College Station, TX: Stata Press.

———. 1997. crc46: Better numerical derivatives and integrals. *Stata Technical Bulletin* 35: 3–5. Reprinted in *Stata Technical Bulletin Reprints*, vol. 6, pp. 8–12. College Station, TX: Stata Press.

Press, W. H., S. A. Teukolsky, W. T. Vetterling, and B. P. Flannery. 2007. *Numerical Recipes in C: The Art of Scientific Computing*. 3rd ed. Cambridge: Cambridge University Press.

Also see

[D] **obs** — Increase the number of observations in a dataset

[D] **range** — Generate numerical range

Title

eform_option — Displaying exponentiated coefficients

Description

An *eform_option* causes the coefficient table to be displayed in exponentiated form: for each coefficient, e^b rather than b is displayed. Standard errors and confidence intervals (CIs) are also transformed.

An *eform_option* is one of the following:

eform_option	Description
eform(*string*)	use *string* for the column title
eform	exponentiated coefficient, *string* is exp(b)
hr	hazard ratio, *string* is Haz. Ratio
shr	subhazard ratio, *string* is SHR
irr	incidence-rate ratio, *string* is IRR
or	odds ratio, *string* is Odds Ratio
rrr	relative-risk ratio, *string* is RRR

Remarks

▷ Example 1

Here is a simple example of the or option with svy: logit. The CI for the odds ratio is computed by transforming (by exponentiating) the endpoints of the CI for the corresponding coefficient.

```
. use http://www.stata-press.com/data/r12/nhanes2d
. svy, or: logit highbp female black
(running logit on estimation sample)
(output omitted)
```

highbp	Odds Ratio	Linearized Std. Err.	t	P>|t|	[95% Conf. Interval]	
female	.693628	.048676	-5.21	0.000	.6011298	.8003593
black	1.509156	.2089571	2.97	0.006	1.137873	2.001588
_cons	.1350642	.0107783	-25.09	0.000	.1147774	.1589367

We also could have specified the following command and received the same results as above:

```
. svy: logit highbp female black, or
```

◁

Also see

[R] **ml** — Maximum likelihood estimation

Title

eivreg — Errors-in-variables regression

Syntax

eivreg *depvar* [*indepvars*] [*if*] [*in*] [*weight*] [, *options*]

options	Description
Model	
<u>reliab</u>(*indepvar #* [*indepvar #* [...]])	
	specify measurement reliability for each *indepvar* measured with error
Reporting	
<u>level</u>(*#*)	set confidence level; default is level(95)
display_options	control column formats, row spacing, line width, and display of omitted variables and base and empty cells
<u>coefl</u>egend	display legend instead of statistics

indepvars may contain factor variables; see [U] **11.4.3 Factor variables**.

bootstrap, by, jackknife, rolling, and statsby are allowed; see [U] **11.1.10 Prefix commands**.

Weights are not allowed with the bootstrap prefix; see [R] **bootstrap**.

aweights are not allowed with the jackknife prefix; see [R] **jackknife**.

aweights and fweights are allowed; see [U] **11.1.6 weight**.

coeflegend does not appear in the dialog box.

See [U] **20 Estimation and postestimation commands** for more capabilities of estimation commands.

Menu

Statistics > Linear models and related > Errors-in-variables regression

Description

eivreg fits errors-in-variables regression models.

Options

> Model

reliab(*indepvar #* [*indepvar #* [...]]) specifies the measurement reliability for each independent variable measured with error. Reliabilities are specified as pairs consisting of an independent variable name (a name that appears in *indepvars*) and the corresponding reliability r, $0 < r \leq 1$. Independent variables for which no reliability is specified are assumed to have reliability 1. If the option is not specified, all variables are assumed to have reliability 1, and the result is thus the same as that produced by regress (the ordinary least-squares results).

459

⌐ Reporting └
level(#); see [R] **estimation options**.

display_options: <u>noomit</u>ted, vsquish, <u>noempty</u>cells, <u>base</u>levels, <u>allbase</u>levels,
    cformat(% *fmt*), pformat(% *fmt*), sformat(% *fmt*), and nolstretch; see [R] **estimation options**.

The following option is available with eivreg but is not shown in the dialog box:

coeflegend; see [R] **estimation options**.

Remarks

For an introduction to errors-in-variables regression, see Draper and Smith (1998, 89–91) or Kmenta (1997, 352–357). Treiman (2009, 258–261) compares the results of errors-in-variables regression with conventional regression.

Errors-in-variables regression models are useful when one or more of the independent variables are measured with additive noise. Standard regression (as performed by regress) would underestimate the effect of the variable, and the other coefficients in the model can be biased to the extent that they are correlated with the poorly measured variable. You can adjust for the biases if you know the reliability:

$$r = 1 - \frac{\text{noise variance}}{\text{total variance}}$$

That is, given the model $\mathbf{y} = \mathbf{X}\boldsymbol{\beta} + \mathbf{u}$, for some variable $\mathbf{x}_i$ in $\mathbf{X}$, the $\mathbf{x}_i$ is observed with error, $\mathbf{x}_i = \mathbf{x}_i^* + \mathbf{e}$, and the noise variance is the variance of $\mathbf{e}$. The total variance is the variance of $\mathbf{x}_i$.

▷ Example 1

Say that in our automobile data, the weight of cars was measured with error, and the reliability of our measured weight is 0.85. The result of this would be to underestimate the effect of weight in a regression of, say, price on weight and foreign, and it would also bias the estimate of the coefficient on foreign (because being of foreign manufacture is correlated with the weight of cars). We would ignore all of this if we fit the model with regress:

```
. use http://www.stata-press.com/data/r12/auto
(1978 Automobile Data)

. regress price weight foreign
```

Source	SS	df	MS		Number of obs =	74
					F(2, 71) =	35.35
Model	316859273	2	158429637		Prob > F =	0.0000
Residual	318206123	71	4481776.38		R-squared =	0.4989
					Adj R-squared =	0.4848
Total	635065396	73	8699525.97		Root MSE =	2117

price	Coef.	Std. Err.	t	P>|t|	[95% Conf. Interval]	
weight	3.320737	.3958784	8.39	0.000	2.531378	4.110096
foreign	3637.001	668.583	5.44	0.000	2303.885	4970.118
_cons	-4942.844	1345.591	-3.67	0.000	-7625.876	-2259.812

With `eivreg`, we can account for our measurement error:

```
. eivreg price weight foreign, r(weight .85)
```

variable	assumed reliability
weight	0.8500
*	1.0000

```
Errors-in-variables regression

Number of obs =      74
F(  2,     71) =   50.37
Prob > F       =  0.0000
R-squared      =  0.6483
Root MSE       = 1773.54
```

price	Coef.	Std. Err.	t	P>|t|	[95% Conf. Interval]	
weight	4.31985	.431431	10.01	0.000	3.459601	5.180099
foreign	4637.32	624.5362	7.43	0.000	3392.03	5882.609
_cons	-8257.017	1452.086	-5.69	0.000	-11152.39	-5361.639

The effect of weight is increased—as we knew it would be—and here the effect of foreign manufacture is also increased. A priori, we knew only that the estimate of `foreign` might be biased; we did not know the direction.

◁

❏ Technical note

Swept under the rug in our example is how we would determine the reliability, r. We can easily see that a variable is measured with error, but we may not know the reliability because the ingredients for calculating r depend on the unobserved noise.

For our example, we made up a value for r, and in fact we do not believe that weight is measured with error at all, so the reported `eivreg` results have no validity. The `regress` results were the statistically correct results here.

But let's say that we do suspect that weight is measured with error and that we do not know r. We could then experiment with various values of r to describe the sensitivity of our estimates to possible error levels. We may not know r, but r does have a simple interpretation, and we could probably produce a sensible range for r by thinking about how the data were collected.

If the reliability, r, is less than the R^2 from a regression of the poorly measured variable on all the other variables, including the dependent variable, the information might as well not have been collected; no adjustment to the final results is possible. For our automobile data, running a regression of `weight` on `foreign` and `price` would result in an R^2 of 0.6743. Thus the reliability must be at least 0.6743 here. If we specify a reliability that is too small, `eivreg` will inform us and refuse to fit the model:

```
. eivreg price weight foreign, r(weight .6742)
reliability r() too small
r(399);
```

Returning to our problem of how to estimate r, too small or not, if the measurements are summaries of scaled items, the reliability may be estimated using the `alpha` command; see [R] **alpha**. If the score is computed from factor analysis and the data are scored using `predict`'s default options (see [MV] **factor postestimation**), the square of the standard deviation of the score is an estimate of the reliability.

❏

❏ Technical note

Consider a model with more than one variable measured with error. For instance, say that our model is that price is a function of weight, foreign, and mpg and that both weight and mpg are measured with error.

```
. eivreg price weight foreign mpg, r(weight .85 mpg .9)
```

variable	assumed reliability
weight	0.8500
mpg	0.9000
*	1.0000

Errors-in-variables regression

Number of obs	=	74
F(3, 70)	=	429.14
Prob > F	=	0.0000
R-squared	=	0.9728
Root MSE	=	496.41

price	Coef.	Std. Err.	t	P>|t|	[95% Conf.	Interval]
weight	12.88302	.6820532	18.89	0.000	11.52271	14.24333
foreign	8268.951	352.8719	23.43	0.000	7565.17	8972.732
mpg	999.2043	73.60037	13.58	0.000	852.413	1145.996
_cons	-56473.19	3710.015	-15.22	0.000	-63872.58	-49073.8

❏

Saved results

eivreg saves the following in e():

Scalars

e(N)	number of observations
e(df_m)	model degrees of freedom
e(df_r)	residual degrees of freedom
e(r2)	R-squared
e(F)	F statistic
e(rmse)	root mean squared error
e(rank)	rank of e(V)

Macros

e(cmd)	eivreg
e(cmdline)	command as typed
e(depvar)	name of dependent variable
e(rellist)	*indepvars* and associated reliabilities
e(wtype)	weight type
e(wexp)	weight expression
e(properties)	b V
e(predict)	program used to implement predict
e(asbalanced)	factor variables fvset as asbalanced
e(asobserved)	factor variables fvset as asobserved

Matrices

e(b)	coefficient vector
e(Cns)	constraints matrix
e(V)	variance–covariance matrix of the estimators

Functions

e(sample)	marks estimation sample

Methods and formulas

`eivreg` is implemented as an ado-file.

Let the model to be fit be

$$\mathbf{y} = \mathbf{X}^* \beta + \mathbf{e}$$

$$\mathbf{X} = \mathbf{X}^* + \mathbf{U}$$

where $\mathbf{X}^*$ are the true values and $\mathbf{X}$ are the observed values. Let $\mathbf{W}$ be the user-specified weights. If no weights are specified, $\mathbf{W} = \mathbf{I}$. If weights are specified, let $\mathbf{v}$ be the specified weights. If `fweight` frequency weights are specified, then $\mathbf{W} = \text{diag}(\mathbf{v})$. If `aweight` analytic weights are specified, then $\mathbf{W} = \text{diag}\{\mathbf{v}/(\mathbf{1}'\mathbf{v})(\mathbf{1}'\mathbf{1})\}$, meaning that the weights are normalized to sum to the number of observations.

The estimates $\mathbf{b}$ of β are obtained as $\mathbf{A}^{-1}\mathbf{X}'\mathbf{W}\mathbf{y}$, where $\mathbf{A} = \mathbf{X}'\mathbf{W}\mathbf{X} - \mathbf{S}$. $\mathbf{S}$ is a diagonal matrix with elements $N(1 - r_i)s_i^2$. N is the number of observations, r_i is the user-specified reliability coefficient for the ith explanatory variable or 1 if not specified, and s_i^2 is the (appropriately weighted) variance of the variable.

The variance–covariance matrix of the estimators is obtained as $s^2\mathbf{A}^{-1}\mathbf{X}'\mathbf{W}\mathbf{X}\mathbf{A}^{-1}$, where the root mean squared error $s^2 = (\mathbf{y}'\mathbf{W}\mathbf{y} - \mathbf{b}\mathbf{A}\mathbf{b}')/(N - p)$, where p is the number of estimated parameters.

References

Draper, N., and H. Smith. 1998. *Applied Regression Analysis*. 3rd ed. New York: Wiley.

Kmenta, J. 1997. *Elements of Econometrics*. 2nd ed. Ann Arbor: University of Michigan Press.

Treiman, D. J. 2009. *Quantitative Data Analysis: Doing Social Research to Test Ideas*. San Francisco, CA: Jossey-Bass.

Also see

[R] **eivreg postestimation** — Postestimation tools for eivreg

[R] **regress** — Linear regression

Stata Structural Equation Modeling Reference Manual

[U] **20 Estimation and postestimation commands**

Title

eivreg postestimation — Postestimation tools for eivreg

Description

The following postestimation commands are available after `eivreg`:

Command	Description
contrast	contrasts and ANOVA-style joint tests of estimates
estat	VCE and estimation sample summary
estimates	cataloging estimation results
lincom	point estimates, standard errors, testing, and inference for linear combinations of coefficients
linktest	link test for model specification
margins	marginal means, predictive margins, marginal effects, and average marginal effects
marginsplot	graph the results from margins (profile plots, interaction plots, etc.)
nlcom	point estimates, standard errors, testing, and inference for nonlinear combinations of coefficients
predict	predictions, residuals, influence statistics, and other diagnostic measures
predictnl	point estimates, standard errors, testing, and inference for generalized predictions
pwcompare	pairwise comparisons of estimates
test	Wald tests of simple and composite linear hypotheses
testnl	Wald tests of nonlinear hypotheses

See the corresponding entries in the *Base Reference Manual* for details.

Syntax for predict

predict [*type*] *newvar* [*if*] [*in*] [, *statistic*]

statistic	Description
Main	
xb	linear prediction; the default
<u>r</u>esiduals	residuals
stdp	standard error of the prediction
stdf	standard error of the forecast
<u>pr</u>(*a*,*b*)	$\Pr(a < y_j < b)$
e(*a*,*b*)	$E(y_j \mid a < y_j < b)$
<u>ystar</u>(*a*,*b*)	$E(y_j^*)$, $y_j^* = \max\{a, \min(y_j, b)\}$

These statistics are available both in and out of sample; type `predict ... if e(sample) ...` if wanted only for the estimation sample.

where *a* and *b* may be numbers or variables; *a* missing ($a \geq .$) means $-\infty$, and *b* missing ($b \geq .$) means $+\infty$; see [U] **12.2.1 Missing values**.

464

Menu

Statistics > Postestimation > Predictions, residuals, etc.

Options for predict

Main

xb, the default, calculates the linear prediction.

residuals calculates the residuals, that is, $y_j - \mathbf{x}_j\mathbf{b}$.

stdp calculates the standard error of the prediction, which can be thought of as the standard error of the predicted expected value or mean for the observation's covariate pattern. The standard error of the prediction is also referred to as the standard error of the fitted value.

stdf calculates the standard error of the forecast, which is the standard error of the point prediction for 1 observation and is commonly referred to as the standard error of the future or forecast value. By construction, the standard errors produced by stdf are always larger than those produced by stdp; see *Methods and formulas* in [R] **regress postestimation**.

pr(a,b) calculates $\Pr(a < \mathbf{x}_j\mathbf{b} + u_j < b)$, the probability that $y_j|\mathbf{x}_j$ would be observed in the interval (a, b).

a and b may be specified as numbers or variable names; *lb* and *ub* are variable names;
pr(20,30) calculates $\Pr(20 < \mathbf{x}_j\mathbf{b} + u_j < 30)$;
pr(*lb,ub*) calculates $\Pr(lb < \mathbf{x}_j\mathbf{b} + u_j < ub)$; and
pr(20,*ub*) calculates $\Pr(20 < \mathbf{x}_j\mathbf{b} + u_j < ub)$.

a missing ($a \geq .$) means $-\infty$; pr(.,30) calculates $\Pr(-\infty < \mathbf{x}_j\mathbf{b} + u_j < 30)$;
pr(*lb*,30) calculates $\Pr(-\infty < \mathbf{x}_j\mathbf{b} + u_j < 30)$ in observations for which *lb* $\geq .$
and calculates $\Pr(lb < \mathbf{x}_j\mathbf{b} + u_j < 30)$ elsewhere.

b missing ($b \geq .$) means $+\infty$; pr(20,.) calculates $\Pr(+\infty > \mathbf{x}_j\mathbf{b} + u_j > 20)$;
pr(20,*ub*) calculates $\Pr(+\infty > \mathbf{x}_j\mathbf{b} + u_j > 20)$ in observations for which *ub* $\geq .$
and calculates $\Pr(20 < \mathbf{x}_j\mathbf{b} + u_j < ub)$ elsewhere.

e(a,b) calculates $E(\mathbf{x}_j\mathbf{b} + u_j \mid a < \mathbf{x}_j\mathbf{b} + u_j < b)$, the expected value of $y_j|\mathbf{x}_j$ conditional on $y_j|\mathbf{x}_j$ being in the interval (a, b), meaning that $y_j|\mathbf{x}_j$ is truncated. a and b are specified as they are for pr().

ystar(a,b) calculates $E(y_j^*)$, where $y_j^* = a$ if $\mathbf{x}_j\mathbf{b} + u_j \leq a$, $y_j^* = b$ if $\mathbf{x}_j\mathbf{b} + u_j \geq b$, and $y_j^* = \mathbf{x}_j\mathbf{b} + u_j$ otherwise, meaning that y_j^* is censored. a and b are specified as they are for pr().

Methods and formulas

All postestimation commands listed above are implemented as ado-files.

Also see

[R] **eivreg** — Errors-in-variables regression

[U] **20 Estimation and postestimation commands**

Title

error messages — Error messages and return codes

Description

Whenever Stata detects that something is wrong—that what you typed is uninterpretable, that you are trying to do something you should not be trying to do, or that you requested the impossible—Stata responds by typing a message describing the problem, together with a *return code*. For instance,

```
. lsit
unrecognized command:  lsit
r(199);
. list myvar
variable myvar not found
r(111);
. test a=b
last estimates not found
r(301);
```

In each case, the message is probably sufficient to guide you to a solution. When we typed lsit, Stata responded with "unrecognized command". We meant to type list. When we typed list myvar, Stata responded with "variable myvar not found". There is no variable named myvar in our data. When we typed test a=b, Stata responded with "last estimates not found". test tests hypotheses about previously fit models, and we have not yet fit a model.

The numbers in parentheses in the r(199), r(111), and r(301) messages are called the *return codes*. To find out more about these messages, type search rc #, where # is the number returned in the parentheses.

▷ Example 1

```
. search rc 301
[P]     error messages . . . . . . . . . . . . . . . . . . . . . . . Return code 301
        last estimates not found;
        You typed an estimation command such as regress without arguments
        or attempted to perform a test or typed predict, but there were no
        previous estimation results.
```

◁

Programmers should see [P] error for details on programming error messages.

Also see

[R] search — Search Stata documentation

466

Title

> **estat** — Postestimation statistics

Syntax

Common subcommands

  Display information criteria

    estat ic $\big[$, n(#) $\big]$

  Summarize estimation sample

    estat <u>summ</u>arize $\big[$ *eqlist* $\big]$ $\big[$, *estat_summ_options* $\big]$

  Display covariance matrix estimates

    estat vce $\big[$, *estat_vce_options* $\big]$

Command-specific subcommands

  estat *subcommand*$_1$ $\big[$, *options*$_1$ $\big]$

  estat *subcommand*$_2$ $\big[$, *options*$_2$ $\big]$

    ...

estat_summ_options	Description
<u>equation</u>	display summary by equation
<u>group</u>	display summary by group; only after sem
<u>labels</u>	display variable labels
<u>nohea</u>der	suppress the header
<u>noweights</u>	ignore weights
display_options	control spacing and display of omitted variables and base and empty cells

eqlist is rarely used and specifies the variables, with optional equation name, to be summarized. *eqlist* may be *varlist* or (*eqname*$_1$: *varlist*) (*eqname*$_2$: *varlist*) .... *varlist* may contain time-series operators; see [U] **11.4.4 Time-series varlists**.

estat_vce_options	Description
<u>cov</u>ariance	display as covariance matrix; the default
<u>corr</u>elation	display as correlation matrix
<u>eq</u>uation(*spec*)	display only specified equations
<u>b</u>lock	display submatrices by equation
<u>di</u>ag	display submatrices by equation; diagonal blocks only
<u>f</u>ormat(%*fmt*)	display format for covariances and correlations
<u>nol</u>ines	suppress lines between equations
display_options	control display of omitted variables and base and empty cells

Menu

Statistics > Postestimation > Reports and statistics

Description

estat displays scalar- and matrix-valued statistics after estimation; it complements predict, which calculates variables after estimation. Exactly what statistics estat can calculate depends on the previous estimation command.

Three sets of statistics are so commonly used that they are available after all estimation commands that store the model log likelihood. estat ic displays Akaike's and Schwarz's Bayesian information criteria. estat summarize summarizes the variables used by the command and automatically restricts the sample to e(sample); it also summarizes the weight variable and cluster structure, if specified. estat vce displays the covariance or correlation matrix of the parameter estimates of the previous model.

Option for estat ic

n(#) specifies the N to be used in calculating BIC; see [R] **BIC note**.

Options for estat summarize

equation requests that the dependent variables and the independent variables in the equations be displayed in the equation-style format of estimation commands, repeating the summary information about variables entered in more than one equation.

group displays summary information separately for each group. group is only allowed after sem with a group() variable specified.

labels displays variable labels.

noheader suppresses the header.

noweights ignores the weights, if any, from the previous estimation command. The default when weights are present is to perform a weighted summarize on all variables except the weight variable itself. An unweighted summarize is performed on the weight variable.

display_options: <u>noomit</u>ted, vsquish, <u>noempty</u>cells, <u>base</u>levels, <u>allbase</u>levels; see [R] **estimation options**.

Options for estat vce

covariance displays the matrix as a variance–covariance matrix; this is the default.

correlation displays the matrix as a correlation matrix rather than a variance–covariance matrix. rho is a synonym.

equation(*spec*) selects part of the VCE to be displayed. If *spec* is *eqlist*, the VCE for the listed equations is displayed. If *spec* is *eqlist1* \ *eqlist2*, the part of the VCE associated with the equations in *eqlist1* (rowwise) and *eqlist2* (columnwise) is displayed. If *spec* is *, all equations are displayed. equation() implies block if diag is not specified.

block displays the submatrices pertaining to distinct equations separately.

diag displays the diagonal submatrices pertaining to distinct equations separately.

format(%*fmt*) specifies the number format for displaying the elements of the matrix. The default is format(%10.0g) for covariances and format(%8.4f) for correlations. See [U] **12.5 Formats: Controlling how data are displayed** for more information.

nolines suppresses lines between equations.

display_options: <u>noomit</u>ted, <u>noemptyc</u>ells, <u>basel</u>evels, <u>allbasel</u>evels; see [R] **estimation options**.

Remarks

estat displays a variety of scalar- and matrix-valued statistics after you have estimated the parameters of a model. Exactly what statistics estat can calculate depends on the estimation command used, and command-specific statistics are detailed in that command's postestimation manual entry. The rest of this entry discusses three sets of statistics that are available after all estimation commands.

Remarks are presented under the following headings:

> *estat ic*
> *estat summarize*
> *estat vce*

estat ic

estat ic calculates two information criteria used to compare models. Unlike likelihood-ratio, Wald, and similar testing procedures, the models need not be nested to compare the information criteria. Because they are based on the log-likelihood function, information criteria are available only after commands that report the log likelihood.

In general, "smaller is better": given two models, the one with the smaller AIC fits the data better than the one with the larger AIC. As with the AIC, a smaller BIC indicates a better-fitting model. For AIC and BIC formulas, see *Methods and formulas*.

▷ Example 1

In [R] **mlogit**, we fit a model explaining the type of insurance a person has on the basis of age, gender, race, and site of study. Here we refit the model with and without the site dummies and compare the models.

```
. use http://www.stata-press.com/data/r12/sysdsn1
(Health insurance data)

. mlogit insure age male nonwhite
(output omitted)

. estat ic
```

Model	Obs	ll(null)	ll(model)	df	AIC	BIC
.	615	-555.8545	-545.5833	8	1107.167	1142.54

Note: N=Obs used in calculating BIC; see **[R] BIC note**

```
. mlogit insure age male nonwhite i.site
(output omitted)
```

```
. estat ic
```

Model	Obs	ll(null)	ll(model)	df	AIC	BIC
.	615	−555.8545	−534.3616	12	1092.723	1145.783

Note: N=Obs used in calculating BIC; see **[R] BIC note**

The AIC indicates that the model including the site dummies fits the data better, whereas the BIC indicates the opposite. As is often the case, different model-selection criteria have led to conflicting conclusions.

◁

❏ Technical note

glm and binreg, ml report a slightly different version of AIC and BIC; see [R] **glm** for the formulas used. That version is commonly used within the GLM literature; see, for example, Hardin and Hilbe (2007). The literature on information criteria is vast; see, among others, Akaike (1973), Sawa (1978), and Raftery (1995). Judge et al. (1985) contains a discussion of using information criteria in econometrics. Royston and Sauerbrei (2008, chap. 2) examine the use of information criteria as an alternative to stepwise procedures for selecting model variables.

❏

estat summarize

Often when fitting a model, you will also be interested in obtaining summary statistics, such as the sample means and standard deviations of the variables in the model. estat summarize makes this process simple. The output displayed is similar to that obtained by typing

```
. summarize varlist if e(sample)
```

without the need to type the *varlist* containing the dependent and independent variables.

▷ Example 2

Continuing with the previous multinomial logit model, here we summarize the variables by using estat summarize.

```
. estat summarize, noomitted
Estimation sample mlogit               Number of obs =     615
```

Variable	Mean	Std. Dev.	Min	Max
insure	1.596748	.6225846	1	3
age	44.46832	14.18523	18.1109	86.0725
male	.2504065	.4335998	0	1
nonwhite	.196748	.3978638	0	1
site				
2	.3707317	.4833939	0	1
3	.3138211	.4644224	0	1

◁

The output in the previous example contains all the variables in one table, though mlogit presents its results in a multiple-equation format. For models in which the same variables appear in all equations, that is fine; but for other multiple-equation models, we may prefer to have the variables separated by the equation in which they appear. The equation option makes this possible.

▷ Example 3

Systems of simultaneous equations typically have different variables in each equation, and the equation option of estat summarize is helpful in such situations. In example 2 of [R] **reg3**, we have a model of supply and demand. We first refit the model and then call estat summarize.

```
. use http://www.stata-press.com/data/r12/supDem
. reg3 (Demand:quantity price pcompete income) (Supply:quantity price praw),
> endog(price)
  (output omitted )
. estat summarize, equation
```

Estimation sample reg3   Number of obs =   49

Variable	Mean	Std. Dev.	Min	Max
depvar				
quantity	12.61818	2.774952	7.71069	20.0477
quantity	12.61818	2.774952	7.71069	20.0477
Demand				
price	32.70944	2.882684	26.3819	38.4769
pcompete	5.929975	3.508264	.207647	11.5549
income	7.811735	4.18859	.570417	14.0077
Supply				
price	32.70944	2.882684	26.3819	38.4769
praw	4.740891	2.962565	.151028	9.79881

The first block of the table contains statistics on the dependent (or, more accurately, left-hand-side) variables, and because we specified quantity as the left-hand-side variable in both equations, it is listed twice. The second block refers to the variables in the first equation we specified, which we labeled "Demand" in our call to reg3; and the final block refers to the supply equation.

◁

estat vce

estat vce allows you to display the VCE of the parameters of the previously fit model, as either a covariance matrix or a correlation matrix.

▷ Example 4

Returning to the mlogit example, we type

```
. use http://www.stata-press.com/data/r12/sysdsn1
(Health insurance data)
. mlogit insure age male nonwhite, nolog
```

Multinomial logistic regression

Number of obs =	615
LR chi2(6) =	20.54
Prob > chi2 =	0.0022
Pseudo R2 =	0.0185

Log likelihood = -545.58328

insure	Coef.	Std. Err.	z	P>|z|	[95% Conf. Interval]	
Indemnity	(base outcome)					
Prepaid						
age	-.0111915	.0060915	-1.84	0.066	-.0231305	.0007475
male	.5739825	.2005221	2.86	0.004	.1809665	.9669985
nonwhite	.7312659	.218978	3.34	0.001	.302077	1.160455
_cons	.1567003	.2828509	0.55	0.580	-.3976773	.7110778
Uninsure						
age	-.0058414	.0114114	-0.51	0.609	-.0282073	.0165245
male	.5102237	.3639793	1.40	0.161	-.2031626	1.22361
nonwhite	.4333141	.4106255	1.06	0.291	-.371497	1.238125
_cons	-1.811165	.5348606	-3.39	0.001	-2.859473	-.7628578

```
. estat vce, block
```

Covariance matrix of coefficients of mlogit model

covariances of equation Indemnity

	age	male	nonwhite	_cons
age	0			
male	0	0		
nonwhite	0	0	0	
_cons	0	0	0	0

covariances of equation Prepaid (row) by equation Indemnity (column)

	age	male	nonwhite	_cons
age	0			
male	0	0		
nonwhite	0	0	0	
_cons	0	0	0	0

covariances of equation Prepaid

	age	male	nonwhite	_cons
age	.00003711			
male	-.00015303	.0402091		
nonwhite	-.00008948	.00470608	.04795135	
_cons	-.00159095	-.00398961	-.00628886	.08000462

covariances of equation Uninsure (row) by equation Indemnity (column)

	age	male	nonwhite	_cons
age	0			
male	0	0		
nonwhite	0	0	0	
_cons	0	0	0	0

```
        covariances of equation Uninsure (row) by equation Prepaid (column)
                           age         male      nonwhite         _cons

             age  |    .00001753  -.00007926   -.00004564   -.00076886
            male  |   -.00007544    .02188398     .0023186   -.00145923
        nonwhite  |   -.00004577    .00250588    .02813553   -.00263872
           _cons  |   -.00077045   -.00130535   -.00257593    .03888032
        covariances of equation Uninsure
                           age         male      nonwhite         _cons

             age  |    .00013022
            male  |   -.00050406    .13248095
        nonwhite  |   -.00026145    .01505449    .16861327
           _cons  |   -.00562159   -.01686629   -.02474852    .28607591
```

The block option is particularly useful for multiple-equation estimators. The first block of output here corresponds to the VCE of the estimated parameters for the first equation—the square roots of the diagonal elements of this matrix are equal to the standard errors of the first equation's parameters. Similarly, the final block corresponds to the VCE of the parameters for the second equation. The middle block shows the covariances between the estimated parameters of the first and second equations. ◁

Saved results

estat ic saves the following in r():

Matrices
    r(S) 1×6 matrix of results:
                               1. sample size
                               2. log likelihood of null model
                               3. log likelihood of full model
                               4. degrees of freedom
                               5. AIC
                               6. BIC

estat summarize saves the following in r():

Scalars
    r(N_groups)   number of groups (group only)
Matrices
    r(stats) $k \times 4$ matrix of means, standard deviations, minimums, and maximums
    r(stats[_#]) $k \times 4$ matrix of means, standard deviations, minimums, and maximums for group # (group only)

estat vce saves the following in r():

Matrices
    r(V)   VCE or correlation matrix

Methods and formulas

estat is implemented as an ado-file.

Akaike's (1974) information criterion is defined as

$$\text{AIC} = -2 \ln L + 2k$$

where $\ln L$ is the maximized log-likelihood of the model and k is the number of parameters estimated. Some authors define the AIC as the expression above divided by the sample size.

Schwarz's (1978) Bayesian information criterion is another measure of fit defined as

$$BIC = -2\ln L + k\ln N$$

where N is the sample size. See [R] **BIC note** for additional information on calculating and interpreting BIC.

Hirotugu Akaike (1927–2009) was born in Fujinomiya City, Shizuoka Prefecture, Japan. He was the son of a silkworm farmer. He gained BA and DSc degrees from the University of Tokyo. Akaike's career from 1952 at the Institute of Statistical Mathematics in Japan culminated in service as Director General; after 1994, he was Professor Emeritus. His best known work in a prolific career is on what is now known as the Akaike information criterion (AIC), which was formulated to help selection of the most appropriate model from a number of candidates.

Gideon E. Schwarz (1933–2007) was a professor of Statistics at the Hebrew University, Jerusalem. He was born in Salzburg, Austria, and obtained an MSc in 1956 from the Hebrew University and a PhD in 1961 from Columbia University. His interests included stochastic processes, sequential analysis, probability, and geometry. He is best known for the Bayesian information criterion (BIC).

References

Akaike, H. 1973. Information theory and an extension of the maximum likelihood principle. In *Second International Symposium on Information Theory*, ed. B. N. Petrov and F. Csaki, 267–281. Budapest: Akailseoniai–Kiudo.

——. 1974. A new look at the statistical model identification. *IEEE Transactions on Automatic Control* 19: 716–723.

Belsley, D. A., E. Kuh, and R. E. Welsch. 1980. *Regression Diagnostics: Identifying Influential Data and Sources of Collinearity*. New York: Wiley.

Findley, D. F., and E. Parzen. 1995. A conversation with Hirotugu Akaike. *Statistical Science* 10: 104–117.

Hardin, J. W., and J. M. Hilbe. 2007. *Generalized Linear Models and Extensions*. 2nd ed. College Station, TX: Stata Press.

Judge, G. G., W. E. Griffiths, R. C. Hill, H. Lütkepohl, and T.-C. Lee. 1985. *The Theory and Practice of Econometrics*. 2nd ed. New York: Wiley.

Raftery, A. 1995. Bayesian model selection in social research. In Vol. 25 of *Sociological Methodology*, ed. P. V. Marsden, 111–163. Oxford: Blackwell.

Royston, P., and W. Sauerbrei. 2008. *Multivariable Model-building: A Pragmatic Approach to Regression Analysis Based on Fractional Polynomials for Modelling Continuous Variables*. Chichester, UK: Wiley.

Sawa, T. 1978. Information criteria for discriminating among alternative regression models. *Econometrica* 46: 1273–1291.

Schwarz, G. 1978. Estimating the dimension of a model. *Annals of Statistics* 6: 461–464.

Tong, H. 2010. Professor Hirotugu Akaike, 1927–2009. *Journal of the Royal Statistical Society, Series A* 173: 451–454.

Also see

[R] **estimates** — Save and manipulate estimation results

[R] **summarize** — Summary statistics

[P] **estat programming** — Controlling estat after user-written commands

[U] **20 Estimation and postestimation commands**

Title

> **estimates** — Save and manipulate estimation results

Syntax

Command	Reference
Save and use results from disk	
estimates save *filename*	[R] **estimates save**
estimates use *filename*	[R] **estimates save**
estimates describe using *filename*	[R] **estimates describe**
estimates esample: ...	[R] **estimates save**
Store and restore estimates in memory	
estimates store *name*	[R] **estimates store**
estimates restore *name*	[R] **estimates store**
estimates query	[R] **estimates store**
estimates dir	[R] **estimates store**
estimates drop *namelist*	[R] **estimates store**
estimates clear	[R] **estimates store**
Set titles and notes	
estimates title: *text*	[R] **estimates title**
estimates title	[R] **estimates title**
estimates notes: *text*	[R] **estimates notes**
estimates notes	[R] **estimates notes**
estimates notes list ...	[R] **estimates notes**
estimates notes drop ...	[R] **estimates notes**
Report	
estimates describe [*name*]	[R] **estimates describe**
estimates replay [*namelist*]	[R] **estimates replay**
Tables and statistics	
estimates table [*namelist*]	[R] **estimates table**
estimates stats [*namelist*]	[R] **estimates stats**
estimates for *namelist*: ...	[R] **estimates for**

Description

estimates allows you to store and manipulate estimation results:

- You can save estimation results in a file for use in later sessions.
- You can store estimation results in memory so that you can
    a. switch among separate estimation results and
    b. form tables combining separate estimation results.

Remarks

estimates is for use after you have fit a model, be it with regress, logistic, etc. You can use estimates after any estimation command, whether it be an official estimation command of Stata or a user-written one.

estimates has three separate but related capabilities:

1. You can save estimation results in a file on disk so that you can use them later, even in a different Stata session.

2. You can store up to 300 estimation results in memory so that they are at your fingertips.

3. You can make tables comparing any results you have stored in memory.

Remarks are presented under the following headings:

> *Saving and using estimation results*
> *Storing and restoring estimation results*
> *Comparing estimation results*
> *Jargon*

Saving and using estimation results

After you have fit a model, say, with regress, type

    . use http://www.stata-press.com/data/r12/auto
    (1978 Automobile Data)
    . regress mpg weight displ foreign
      (output omitted)

You can save the results in a file:

    . estimates save basemodel
    (file basemodel.ster saved)

Later, say, in a different session, you can reload those results:

    . estimates use basemodel

The situation is now nearly identical to what it was immediately after you fit the model. You can replay estimation results:

    . regress
      (output omitted)

You can perform tests:

    . test foreign==0
      (output omitted)

And you can use any postestimation command or postestimation capability of Stata. The only difference is that Stata no longer knows what the estimation sample, e(sample) in Stata jargon, was. When you reload the estimation results, you might not even have the original data in memory. That is okay. Stata will know to refuse to calculate anything that can be calculated only on the original estimation sample.

If it is important that you use a postestimation command that can be used only on the original estimation sample, there is a way you can do that. You use the original data and then use estimates esample to tell Stata what the original sample was.

See [R] **estimates save** for details.

Storing and restoring estimation results

Storing and restoring estimation results in memory is much like saving them to disk. You type

```
. estimates store base
```

to save the current estimation results under the name base, and you type

```
. estimates restore base
```

to get them back later. You can find out what you have stored by typing

```
. estimates dir
```

Saving estimation results to disk is more permanent than storing them in memory, so why would you want merely to store them? The answer is that, once they are stored, you can use other estimates commands to produce tables and reports from them.

See [R] **estimates store** for details about the estimates store and restore commands.

Comparing estimation results

Let's say that you have done the following:

```
. use http://www.stata-press.com/data/r12/auto
(1978 Automobile Data)
. regress mpg weight displ
(output omitted )
. estimates store base
. regress mpg weight displ foreign
(output omitted )
. estimates store alt
```

You can now get a table comparing the coefficients:

```
. estimates table base alt
```

Variable	base	alt
weight	-.00656711	-.00677449
displacement	.00528078	.00192865
foreign		-1.6006312
_cons	40.084522	41.847949

estimates table can do much more; see [R] **estimates table**. Also see [R] **estimates stats**. estimates stats works similarly to estimates table but produces model comparisons in terms of BIC and AIC.

Jargon

You know that if you fit a model, say, by typing

. regress mpg weight displacement

then you can later replay the results by typing

. regress

and you can do tests and calculate other postestimation statistics by typing

. test displacement==0

. estat vif

. predict mpghat

As a result, we often refer to the *estimation results* or the *current estimation results* or the *most recent estimation results* or the *last estimation results* or the *estimation results in memory*.

With estimates store and estimates restore, you can have many estimation results in memory. One set of those, the set most recently estimated, or the set most recently restored, are the *current* or *active* estimation results, which you can replay, which you can test, or from which you can calculate postestimation statistics.

Current and *active* are the two words we will use interchangeably from now on.

Also see

[P] _estimates — Manage estimation results

Title

> **estimates describe** — Describe estimation results

Syntax

estimates describe

estimates describe *name*

estimates describe using *filename* $\left[\, , \text{number}(\#) \,\right]$

Menu

Statistics > Postestimation > Manage estimation results > Describe results

Description

estimates describe describes the current (active) estimates. Reported are the command line that produced the estimates, any title that was set by estimates title (see [R] **estimates title**), and any notes that were added by estimates notes (see [R] **estimates notes**).

estimates describe *name* does the same but reports results for estimates stored by estimates store (see [R] **estimates store**).

estimates describe using *filename* does the same but reports results for estimates saved by estimates save (see [R] **estimates save**). If *filename* contains multiple sets of estimates (saved in it by estimates save, append), the number of sets of estimates is also reported. If *filename* is specified without an extension, .ster is assumed.

Option

number(#) specifies that the #th set of estimation results from *filename* be described. This assumes that multiple sets of estimation results have been saved in *filename* by estimates save, append. The default is number(1).

Remarks

estimates describe can be used to describe the estimation results currently in memory,

```
. estimates describe
  Estimation results produced by
    . regress mpg weight displ if foreign
```

or to describe results saved by estimates save in a .ster file:

```
. estimates describe using final
  Estimation results "Final results" saved on 12apr2011 14:20, produced by
    . logistic myopic age sex drug1 drug2 if complete==1
  Notes:
    1.  Used file patient.dta
    2.  "datasignature myopic age sex drug1 drug2 if complete==1"
        reports 148:5(58763):2252897466:3722318443
    3.  must be reviewed by rgg
```

479

▷ Example 1

```
. use http://www.stata-press.com/data/r12/auto
(1978 Automobile Data)
. regress mpg weight displ if foreign
  (output omitted )
. estimates notes: file `c(filename)'
. datasignature
  74:12(71728):3831085005:1395876116
. estimates notes: datasignature report `r(datasignature)'
. estimates save foreign
file foreign.ster saved
. regress mpg weight displ if !foreign
  (output omitted )
. estimates describe using foreign
  Estimation results saved on 02may2011 10:33, produced by

      . regress mpg weight displ if foreign

  Notes:
    1.  file http://www.stata-press.com/data/r12/auto.dta
    2.  datasignature report 74:12(71728):3831085005:1395876116
```

◁

Saved results

estimates describe and estimates describe *name* save the following in r():

Macros
        r(title)   title
        r(cmdline)   original command line

estimates describe using *filename* saves the above and the following in r():

Scalars
        r(datetime)   %tc value of date/time file saved
        r(nestresults) number of sets of estimation results in file

Methods and formulas

estimates describe is implemented as an ado-file.

Also see

[R] **estimates** — Save and manipulate estimation results

Title

> **estimates for** — Repeat postestimation command across models

Syntax

estimates for *namelist* [, *options*] : *postestimation_command*

where *namelist* is a name, a list of names, _all, or *. A name may be ., meaning the current (active) estimates. _all and * mean the same thing.

options	Description
noheader	do not display title
nostop	do not stop if command fails

Description

estimates for performs *postestimation_command* on each estimation result specified.

Options

noheader suppresses the display of the header as *postestimation_command* is executed each time.

nostop specifies that execution of *postestimation_command* is to be performed on the remaining models even if it fails on some.

Remarks

In the example that follows, we fit a model two different ways, store the results, and then use estimates for to perform the same test on both of them:

▷ Example 1

```
. use http://www.stata-press.com/data/r12/auto
(1978 Automobile Data)
. gen fwgt = foreign*weight
. gen dwgt = !foreign*weight
. gen gpm = 1/mpg
. regress gpm fwgt dwgt displ foreign
(output omitted )
. estimates store reg
. qreg gpm fwgt dwgt displ foreign
(output omitted )
. estimates store qreg
```

```
. estimates for reg qreg: test fwgt==dwgt
```

Model **reg**

```
( 1)  fwgt - dwgt = 0
       F(  1,    69) =     4.87
            Prob > F =   0.0307
```

Model **qreg**

```
( 1)  fwgt - dwgt = 0
       F(  1,    69) =     0.07
            Prob > F =   0.7937
```

◁

Methods and formulas

estimates for is implemented as an ado-file.

Also see

[R] **estimates** — Save and manipulate estimation results

Title

> **estimates notes** — Add notes to estimation results

Syntax

estimates notes: *text*

estimates notes

estimates notes list [in *noterange*]

estimates notes drop in *noterange*

where *noterange* is *#* or *#/#* and where *#* may be a number, the letter f (meaning first), or the letter l (meaning last).

Description

estimates notes: *text* adds a note to the current (active) estimation results.

estimates notes and estimates notes list list the current notes.

estimates notes drop in *noterange* eliminates the specified notes.

Remarks

After adding or removing notes, if estimates have been stored, do not forget to store them again. If estimates have been saved, do not forget to save them again.

Notes are most useful when you intend to save estimation results in a file; see [R] **estimates save**. For instance, after fitting a model, you might type

```
. estimates note:  I think these are final
. estimates save lock2
```

and then, later when going through your files, you could type

```
. estimates use lock2
. estimates notes
  1.  I think these are final
```

Up to 9,999 notes can be attached to estimation results. If estimation results are important, we recommend that you add a note identifying the .dta dataset you used. The best way to do that is to type

```
. estimates notes:  file 'c(filename)'
```

because 'c(filename)' will expand to include not just the name of the file but also its full path; see [P] **creturn**.

If estimation results took a long time to estimate—say, they were produced by asmprobit or gllamm (see [R] **asmprobit** and http://www.gllamm.org)—it is also a good idea to add a data signature. A data signature takes less time to compute than reestimation when you need proof that you really have the right dataset. The easy way to do that is to type

```
. datasignature
74:12(71728):3831085005:1395876116
. estimates notes: datasignature reports 'r(datasignature)'
```

Now when you ask to see the notes, you will see

```
. estimates notes
1.  I think these are final
2.  file C:\project\one\pat4.dta
3.  datasignature reports 74:12(71728):3831085005:1395876116
```

See [D] **datasignature**.

Notes need not be positive. You might set a note to be, "I need to check that age is defined correctly."

▷ Example 1

```
. use http://www.stata-press.com/data/r12/auto
(1978 Automobile Data)
. regress mpg weight displ if foreign
(output omitted)
. estimates notes: file 'c(filename)'
. datasignature
74:12(71728):3831085005:1395876116
. estimates notes: datasignature report 'r(datasignature)'
. estimates save foreign
file foreign.ster saved
. estimates notes list in 1/2
1.  file http://www.stata-press.com/data/r12/auto.dta
2.  datasignature report 74:12(71728):3831085005:1395876116
. estimates notes drop in 2
(1 note dropped)
. estimates notes
1.  file http://www.stata-press.com/data/r12/auto.dta
```

◁

Methods and formulas

`estimates notes` is implemented as an ado-file.

Also see

[R] **estimates** — Save and manipulate estimation results

Title

> **estimates replay** — Redisplay estimation results

Syntax

estimates replay

estimates replay *namelist*

where *namelist* is a name, a list of names, _all, or *. A name may be ., meaning the current (active) estimates. _all and * mean the same thing.

Menu

Statistics > Postestimation > Manage estimation results > Redisplay estimation output

Description

estimates replay redisplays the current (active) estimation results, just as typing the name of the estimation command would do.

estimates replay *namelist* redisplays each specified estimation result. The active estimation results are left unchanged.

Remarks

In the example that follows, we fit a model two different ways, store the results, use estimates for to perform the same test on both of them, and then replay the results:

▷ Example 1

```
. use http://www.stata-press.com/data/r12/auto
(1978 Automobile Data)
. gen fwgt = foreign*weight
. gen dwgt = !foreign*weight
. gen gpm = 1/mpg
. regress gpm fwgt dwgt displ foreign
(output omitted )
. estimates store reg
. qreg gpm fwgt dwgt displ foreign
(output omitted )
. estimates store qreg
. estimates for reg qreg: test fwgt==dwgt
```

Model **reg**
_____

```
( 1)  fwgt - dwgt = 0
      F(  1,    69) =      4.87
            Prob > F =    0.0307
```

```
Model qreg

( 1)  fwgt - dwgt = 0
       F(  1,    69) =    0.07
            Prob > F =    0.7937
. estimates replay
```

```
Model qreg
```

```
Median regression                              Number of obs =        74
  Raw sum of deviations .7555689 (about .05)
  Min sum of deviations .3201479                Pseudo R2     =     0.5763
```

gpm	Coef.	Std. Err.	t	P>|t|	[95% Conf. Interval]	
fwgt	.0000155	2.87e-06	5.40	0.000	9.76e-06	.0000212
dwgt	.0000147	1.88e-06	7.81	0.000	.0000109	.0000184
displacement	.0000179	.0000147	1.22	0.226	-.0000113	.0000471
foreign	.0065352	.0078098	0.84	0.406	-.009045	.0221153
_cons	.0003134	.0042851	0.07	0.942	-.0082351	.0088618

```
. estimates replay reg
```

```
Model reg
```

Source	SS	df	MS		Number of obs =	74
Model	.009342436	4	.002335609		F(4, 69) =	61.62
Residual	.002615192	69	.000037901		Prob > F =	0.0000
					R-squared =	0.7813
					Adj R-squared =	0.7686
Total	.011957628	73	.000163803		Root MSE =	.00616

gpm	Coef.	Std. Err.	t	P>|t|	[95% Conf. Interval]	
fwgt	.00002	3.27e-06	6.12	0.000	.0000135	.0000265
dwgt	.0000123	2.30e-06	5.36	0.000	7.75e-06	.0000169
displacement	.0000296	.0000187	1.58	0.119	-7.81e-06	.000067
foreign	-.0117756	.0086088	-1.37	0.176	-.0289497	.0053986
_cons	.0053352	.0046748	1.14	0.258	-.0039909	.0146612

◁

Methods and formulas

estimates replay is implemented as an ado-file.

Also see

[R] **estimates** — Save and manipulate estimation results

Title

estimates save — Save and use estimation results

Syntax

estimates save *filename* [, append replace]

estimates use *filename* [, number(*#*)]

estimates esample: [*varlist*] [*if*] [*in*] [*weight*]
    [, replace stringvars(*varlist*) zeroweight]

estimates esample

Menu

estimates save

Statistics > Postestimation > Manage estimation results > Save to disk

estimates use

Statistics > Postestimation > Manage estimation results > Load from disk

Description

estimates save *filename* saves the current (active) estimation results in *filename*.

estimates use *filename* loads the results saved in *filename* into the current (active) estimation results.

In both cases, if *filename* is specified without an extension, .ster is assumed.

estimates esample: (note the colon) resets e(sample). After estimates use *filename*, e(sample) is set to contain 0, meaning that none of the observations currently in memory was used in obtaining the estimates.

estimates esample (without a colon) displays how e(sample) is currently set.

Options

append, used with estimates save, specifies that results be appended to an existing file. If the file does not already exist, a new file is created.

replace, used with estimates save, specifies that *filename* can be replaced if it already exists.

number(*#*), used with estimates use, specifies that the *#*th set of estimation results from *filename* be loaded. This assumes that multiple sets of estimation results have been saved in *filename* by estimates save, append. The default is number(1).

replace, used with `estimates esample:`, specifies that `e(sample)` can be replaced even if it is already set.

`stringvars`(*varlist*), used with `estimates esample:`, specifies string variables. Observations containing variables that contain "" will be omitted from `e(sample)`.

`zeroweight`, used with `estimates esample:`, specifies that observations with zero weights are to be included in `e(sample)`.

Remarks

See [R] **estimates** for an overview of the `estimates` commands.

For a description of `estimates save` and `estimates use`, see *Saving and using estimation results* in [R] **estimates**.

The rest of this entry concerns `e(sample)`.

Remarks are presented under the following headings:

> *Setting e(sample)*
> *Resetting e(sample)*
> *Determining who set e(sample)*

Setting e(sample)

After `estimates use` *filename*, the situation is nearly identical to what it was immediately after you fit the model. The one difference is that `e(sample)` is set to 0.

`e(sample)` is Stata's function to mark which observations among those currently in memory were used in producing the estimates. For instance, you might type

```
. use http://www.stata-press.com/data/r12/auto
(1978 Automobile Data)

. regress mpg weight displ if foreign
(output omitted)
. summarize mpg if e(sample)
(output omitted)
```

and `summarize` would report the summary statistics for the observations `regress` in fact used, which would exclude not only observations for which `foreign = 0` but also any observations for which `mpg`, `weight`, or `displ` was missing.

If you saved the above estimation results and then reloaded them, however, `summarize mpg if e(sample)` would produce

```
. summarize mpg if e(sample)
```

Variable	Obs	Mean	Std. Dev.	Min	Max
mpg	0				

Stata thinks that none of these observations was used in producing the estimates currently loaded.

What else could Stata think? When you `estimates use` *filename*, you do not have to have the original data in memory. Even if you do have data in memory that look like the original data, it might not be. Setting `e(sample)` to 0 is the safe thing to do. There are some postestimation statistics, for instance, that are appropriate only when calculated on the estimation sample. Setting `e(sample)` to 0 ensures that, should you ask for one of them, you will get back a null result.

We recommend that you leave e(sample) set to 0. But what if you really need to calculate that postestimation statistic? Well, you can get it, but you are going to take responsibility for setting e(sample) correctly. Here we just happen to know that all the foreign observations were used, so we can type

        . estimates esample: if foreign

If all the observations had been used, we could simply type

        . estimates esample:

The safe thing to do, however, is to look at the estimation command—estimates describe will show it to you—and then type

        . estimates esample: mpg weight displ if foreign

Resetting e(sample)

estimates esample: will allow you to not only set but also reset e(sample). If e(sample) has already been set (say that you just fit the model) and you try to set it, you will see

        . estimates esample: mpg weight displ if foreign
        no; e(sample) already set
        r(322);

Here you can specify the replace option:

        . estimates esample: mpg weight displ if foreign, replace

We do not recommend resetting e(sample), but the situation can arise where you need to. Imagine that you estimates use *filename*, you set e(sample), and then you realize that you set it wrong. Here you would want to reset it.

Determining who set e(sample)

estimates esample without a colon will report whether and how e(sample) was set. You might see

        . estimates esample
          e(sample) set by estimation command

or

        . estimates esample
          e(sample) set by user

or

        . estimates esample
          e(sample) not set (0 assumed)

Saved results

estimates esample without the colon saves macro r(who), which will contain cmd, user, or zero'd.

Methods and formulas

estimates save, estimates use, estimates esample:, and estimates esample are implemented as ado-files.

Also see

[R] **estimates** — Save and manipulate estimation results

Title

> **estimates stats** — Model statistics

Syntax

> estimates stats $\big[$ *namelist* $\big]$ $\big[$, n(#) $\big]$

where *namelist* is a name, a list of names, _all, or *. A name may be ., meaning the current (active) estimates. _all and * mean the same thing.

Menu

Statistics > Postestimation > Manage estimation results > Table of fit statistics

Description

estimates stats reports model-selection statistics, including the Akaike information criterion (AIC) and the Bayesian information criterion (BIC). These measures are appropriate for maximum likelihood models.

If estimates stats is used for a non–likelihood-based model, such as qreg, missing values are reported.

Option

n(#) specifies the N to be used in calculating BIC; see [R] **BIC note**.

Remarks

If you type estimates stats without arguments, a table for the most recent estimation results will be shown:

```
. use http://www.stata-press.com/data/r12/auto
(1978 Automobile Data)
. logistic foreign mpg weight displ
(output omitted )
. estimates stats
```

Model	Obs	ll(null)	ll(model)	df	AIC	BIC
.	74	-45.03321	-20.59083	4	49.18167	58.39793

Note: N=Obs used in calculating BIC; see **[R] BIC note**

Regarding the note at the bottom of the table, N is an ingredient in the calculation of BIC; see [R] **BIC note**. The note changes if you specify the n() option, which tells estimates stats what N to use. $N = \text{Obs}$ is the default.

Regarding the table itself, ll(null) is the log likelihood for the constant-only model, ll(model) is the log likelihood for the model, df is the number of degrees of freedom, and AIC and BIC are the Akaike and Bayesian information criteria.

Models with smaller values of an information criterion are considered preferable.

estimates stats can compare estimation results:

```
. use http://www.stata-press.com/data/r12/auto
(1978 Automobile Data)
. logistic foreign mpg weight displ
(output omitted )
. estimates store full
. logistic foreign mpg weight
(output omitted )
. estimates store sub
. estimates stats full sub
```

Model	Obs	ll(null)	ll(model)	df	AIC	BIC
full	74	-45.03321	-20.59083	4	49.18167	58.39793
sub	74	-45.03321	-27.17516	3	60.35031	67.26251

Note: N=Obs used in calculating BIC; see **[R] BIC note**

Saved results

estimates stats saves the following in r():

Matrices
    r(S)   matrix with 6 columns (N, ll0, ll, df, AIC, and BIC) and rows corresponding to models in table

Methods and formulas

estimates stats is implemented as an ado-file.

See [R] **BIC note**.

Also see

[R] **estimates** — Save and manipulate estimation results

Title

estimates store — Store and restore estimation results

Syntax

estimates <u>sto</u>re *name* $\left[\, , \text{ nocopy}\right]$

estimates <u>res</u>tore *name*

estimates <u>q</u>uery

estimates dir $\left[\textit{namelist}\right]$

estimates drop *namelist*

estimates clear

where *namelist* is a name, a list of names, _all, or *. _all and * mean the same thing.

Menu

estimates store

Statistics > Postestimation > Manage estimation results > Store in memory

estimates restore

Statistics > Postestimation > Manage estimation results > Restore from memory

estimates dir

Statistics > Postestimation > Manage estimation results > List results stored in memory

estimates drop

Statistics > Postestimation > Manage estimation results > Drop from memory

Description

estimates store *name* saves the current (active) estimation results under the name *name*.

estimates restore *name* loads the results saved under *name* into the current (active) estimation results.

estimates query tells you whether the current (active) estimates have been stored and, if so, the name.

estimates dir displays a list of the stored estimates.

estimates drop *namelist* drops the specified stored estimation results.

estimates clear drops all stored estimation results.

`estimates clear`, `estimates drop _all`, and `estimates drop *` do the same thing. `estimates drop` and `estimates clear` do not eliminate the current (active) estimation results.

Option

`nocopy`, used with `estimates store`, specifies that the current (active) estimation results are to be moved into *name* rather than copied. Typing

    . estimates store hold, nocopy

is the same as typing

    . estimates store hold
    . ereturn clear

except that the former is faster. The `nocopy` option is sometimes used by programmers.

Remarks

`estimates store` stores estimation results in memory so that you can access them later.

    . use http://www.stata-press.com/data/r12/auto
    (1978 Automobile Data)

    . regress mpg weight displ
    (output omitted)

    . estimates store myreg

    . ... you do other things, including fitting other models ...

    . estimates restore myreg

    . regress
    (same output shown again)

After `estimates restore myreg`, things are once again just as they were, estimationwise, just after you typed `regress mpg weight displ`.

`estimates store` stores results in memory. When you exit Stata, those stored results vanish. If you wish to make a permanent copy of your estimation results, see [R] **estimates save**.

The purpose of making copies in memory is 1) so that you can quickly switch between them and 2) so that you can make tables comparing estimation results. Concerning the latter, see [R] **estimates table** and [R] **estimates stats**.

Saved results

`estimates dir` saves the following in `r()`:

Macros
    `r(names)`   names of stored results

Methods and formulas

`estimates store`, `estimates restore`, `estimates query`, `estimates dir`, `estimates drop`, and `estimates clear` are implemented as ado-files.

References

Jann, B. 2005. Making regression tables from stored estimates. *Stata Journal* 5: 288–308.

———. 2007. Making regression tables simplified. *Stata Journal* 7: 227–244.

Also see

[R] **estimates** — Save and manipulate estimation results

Title

> **estimates table** — Compare estimation results

Syntax

$$\underline{\text{est}}\text{imates } \underline{\text{tab}}\text{le } \big[\textit{namelist}\big] \ \big[\text{, } \textit{options}\big]$$

where *namelist* is a name, a list of names, _all, or *. A name may be ., meaning the current (active) estimates. _all and * mean the same thing.

options	Description
Main	
stats(*scalarlist*)	report *scalarlist* in table
star$\big[$ *(#1 #2 #3)* $\big]$	use stars to denote significance levels
Options	
keep(*coeflist*)	report coefficients in order specified
drop(*coeflist*)	omit specified coefficients from table
equations(*matchlist*)	match equations of models as specified
Numerical formats	
b$\big[$ *(%fmt)* $\big]$	how to format coefficients, which are always reported
se$\big[$ *(%fmt)* $\big]$	report standard errors and use optional format
t$\big[$ *(%fmt)* $\big]$	report t or z and use optional format
p$\big[$ *(%fmt)* $\big]$	report p-values and use optional format
stfmt(*%fmt*)	how to format scalar statistics
General format	
varwidth(*#*)	use # characters to display variable names and statistics
modelwidth(*#*)	use # characters to display model names
eform	display coefficients in exponentiated form
label	display variable labels rather than variable names
newpanel	display statistics in separate table from coefficients
style(oneline)	put vertical line after variable names; the default
style(columns)	put vertical line separating every column
style(noline)	suppress all vertical lines
coded	display compact table
Reporting	
display_options	control row spacing and display of omitted variables and base and empty cells
title(*string*)	title for table

title() does not appear in the dialog box.

where

- A *scalarlist* is a list of any or all of the names of scalars stored in e(), plus aic, bic, and rank.

- *#1 #2 #3* are three numbers such as .05 .01 .001.

- A *coeflist* is a list of coefficient names, each name of which may be simple (for example, price), an equation name followed by a colon (for example, mean:), or a full name (for example, mean:price). Names are separated by blanks.

- A *matchlist* specifies how equations from different estimation results are to be matched. If you need to specify a *matchlist*, the solution is usually 1, as in equations(1). The full syntax is

$$matchlist := term \left[, term \ldots \right]$$

$$term := \left[eqname = \right] \#:\#\ldots:\#$$

$$\left[eqname = \right] \#$$

See equations() under *Options* below.

Menu

Statistics > Postestimation > Manage estimation results > Table of estimation results

Description

estimates table displays a table of coefficients and statistics for one or more sets of estimation results.

Options

────┌ Main ┐──

stats(*scalarlist*) specifies one or more scalar statistics to be displayed in the table. *scalarlist* may contain

aic	Akaike's information criterion
bic	Schwarz's Bayesian information criterion
rank	rank of e(V) (# of free parameters in model)

along with the names of any scalars stored in e(). The specified statistics do not have to be available for all estimation results being displayed.

For example, stats(N ll chi2 aic) specifies that e(N), e(ll), e(chi2), and AIC be included. In Stata, e(N) records the number of observations; e(ll), the log likelihood; and e(chi2), the chi-squared test that all coefficients in the first equation of the model are equal to zero.

star and star(*#1 #2 #3*) specify that stars (asterisks) are to be used to mark significance. The second syntax specifies the significance levels for one, two, and three stars. If you specify simply star, that is equivalent to specifying star(.05 .01 .001), which means one star (*) if $p < 0.05$, two stars (**) if $p < 0.01$, and three stars (***) if $p < 0.001$.

The star and star() options may not be combined with se, t, or p option.

⌐ Options ⌐

keep(*coeflist*) and drop(*coeflist*) are alternatives; they specify coefficients to be included or omitted from the table. The default is to display all coefficients.

If keep() is specified, it specifies not only the coefficients to be included but also the order in which they appear.

A *coeflist* is a list of coefficient names, each name of which may be simple (for example, price), an equation name followed by a colon (for example, mean:), or a full name (for example, mean:price). Names are separated from each other by blanks.

When full names are not specified, all coefficients that match the partial specification are included. For instance, drop(_cons) would omit _cons for all equations.

equations(*matchlist*) specifies how the equations of the models in *namelist* are to be matched. The default is to match equations by name. Matching by name usually works well when all results were fit by the same estimation command. When you are comparing results from different estimation commands, however, specifying equations() may be necessary.

The most common usage is equations(1), which indicates that all first equations are to be matched into one equation named #1.

matchlist has the syntax

    term [, *term* ...]

where *term* is

    [*eqname* =] #:#...:#   (syntax 1)

    [*eqname* =] #   (syntax 2)

In syntax 1, each # is a number or a period (.). If a number, it specifies the position of the equation in the corresponding model; 1:3:1 would indicate that equation 1 in the first model matches equation 3 in the second, which matches equation 1 in the third. A period indicates that there is no corresponding equation in the model; 1:.:1 indicates that equation 1 in the first matches equation 1 in the third.

In syntax 2, you specify just one number, say, 1 or 2, and that is shorthand for 1:1...:1 or 2:2...:2, meaning that equation 1 matches across all models specified or that equation 2 matches across all models specified.

Now that you can specify a *term*, you can put that together into a *matchlist* by separating one term from the other by commas. In what follows, we will assume that three names were specified,

    . estimates table alpha beta gamma, ...

equations(1) is equivalent to equations(1:1:1); we would be saying that the first equations match across the board.

equations(1:.:1) would specify that equation 1 matches in models alpha and gamma but that there is nothing corresponding in model beta.

equations(1,2) is equivalent to equations(1:1:1, 2:2:2). We would be saying that the first equations match across the board and so do the second equations.

equations(1, 2:.:2) would specify that the first equations match across the board, that the second equations match for models alpha and gamma, and that there is nothing equivalent to equation 2 in model beta.

If equations() is specified, equations not matched by position are matched by name.

Numerical formats

b(%*fmt*) specifies how the coefficients are to be displayed. You might specify b(%9.2f) to make decimal points line up. There is also a b option, which specifies that coefficients are to be displayed, but that is just included for consistency with the se, t, and p options. Coefficients are always displayed.

se, t, and p specify that standard errors, t or z statistics, and significance levels are to be displayed. The default is not to display them. se(%*fmt*), t(%*fmt*), and p(%*fmt*) specify that each is to be displayed and specifies the display format to be used to format them.

stfmt(%*fmt*) specifies the format for displaying the scalar statistics included by the stats() option.

General format

varwidth(#) specifies the number of character positions used to display the names of the variables and statistics. The default is 12.

modelwidth(#) specifies the number of character positions used to display the names of the models. The default is 12.

eform displays coefficients in exponentiated form. For each coefficient, $\exp(\beta)$ rather than β is displayed, and standard errors are transformed appropriately. Display of the intercept, if any, is suppressed.

label specifies that variable labels be displayed instead of variable names.

newpanel specifies that the statistics be displayed in a table separated by a blank line from the table with coefficients rather than in the style of another equation in the table of coefficients.

style(*stylespec*) specifies the style of the coefficient table.

style(oneline) specifies that a vertical line be displayed after the variables but not between the models. This is the default.

style(columns) specifies that vertical lines be displayed after each column.

style(noline) specifies that no vertical lines be displayed.

coded specifies that a compact table be displayed. This format is especially useful for comparing variables that are included in a large collection of models.

Reporting

display_options: noomitted, vsquish, noemptycells, baselevels, and allbaselevels; see [R] **estimation options**.

The following option is available with estimates table but is not shown in the dialog box:

title(*string*) specifies the title to appear above the table.

Remarks

If you type estimates table without arguments, a table of the most recent estimation results will be shown:

```
. use http://www.stata-press.com/data/r12/auto
(1978 Automobile Data)

. regress mpg weight displ
 (output omitted )
```

```
. estimates table
```

Variable	active
weight	-.00656711
displacement	.00528078
_cons	40.084522

The real use of estimates table, however, is for comparing estimation results, and that requires using it after estimates store:

```
. regress mpg weight displ
(output omitted)
. estimates store base
. regress mpg weight displ foreign
(output omitted)
. estimates store alt
. qreg mpg weight displ foreign
(output omitted)
. estimates store qreg
. estimates table base alt qreg, stats(r2)
```

Variable	base	alt	qreg
weight	-.00656711	-.00677449	-.00595056
displacement	.00528078	.00192865	.00018552
foreign		-1.6006312	-2.1326004
_cons	40.084522	41.847949	39.213348
r2	.6529307	.66287957	

Saved results

estimates table saves the following in r():

Macros
    r(names)   names of results used

Matrices
    r(coef)   matrix M: $n \times 2*m$
                  $M[i, 2j-1] = i$th parameter estimate for model j;
                  $M[i, 2j] = $ variance of $M[i, 2j-1]$; $i=1,...,n$; $j=1,...,m$
    r(stats)   matrix S: $k \times m$ (if option stats() specified)
                  $S[i, j] = i$th statistic for model j; $i=1,...,k$; $j=1,...,m$

Methods and formulas

estimates table is implemented as an ado-file.

Reference

Weiss, M. 2010. Stata tip 90: Displaying partial results. *Stata Journal* 10: 500–502.

Also see

[R] **estimates** — Save and manipulate estimation results

Title

estimates title — Set title for estimation results

Syntax

estimates title: $\left[\,text\,\right]$

estimates title

Menu

Statistics > Postestimation > Manage estimation results > Title/retitle results

Description

estimates title: (note the colon) sets or clears the title for the current estimation results. The title is used by estimates table and estimates stats (see [R] **estimates table** and [R] **estimates stats**).

estimates title without the colon displays the current title.

Remarks

After setting the title, if estimates have been stored, do not forget to store them again:

```
. use http://www.stata-press.com/data/r12/auto
(1978 Automobile Data)
. regress mpg gear turn
  (output omitted )
. estimates store reg
```

Now let's add a title:

```
. estimates title: "My regression"
. estimates store reg
```

Methods and formulas

estimates title: and estimates title are implemented as ado-files.

Also see

[R] **estimates** — Save and manipulate estimation results

Title

estimation options — Estimation options

Description

This entry describes the options common to many estimation commands. Not all the options documented below work with all estimation commands. See the documentation for the particular estimation command; if an option is listed there, it is applicable.

Options

Model

noconstant suppresses the constant term (intercept) in the model.

offset(*varname*) specifies that *varname* be included in the model with the coefficient constrained to be 1.

exposure(*varname*) specifies a variable that reflects the amount of exposure over which the *depvar* events were observed for each observation; ln(*varname*) with coefficient constrained to be 1 is entered into the log-link function.

constraints(*numlist* | *matname*) specifies the linear constraints to be applied during estimation. The default is to perform unconstrained estimation. See [R] **reg3** for the use of constraints in multiple-equation contexts.

> constraints(*numlist*) specifies the constraints by number after they have been defined by using the constraint command; see [R] **constraint**. Some commands (for example, slogit) allow only constraints(*numlist*).

> constraints(*matname*) specifies a matrix containing the constraints; see [P] **makecns**.

> constraints(*clist*) is used by some estimation commands, such as mlogit, where *clist* has the form $\#\left[-\#\right]\left[, \#\left[-\#\right] \dots \right]$.

collinear specifies that the estimation command not omit collinear variables. Usually, there is no reason to leave collinear variables in place, and, in fact, doing so usually causes the estimation to fail because of the matrix singularity caused by the collinearity. However, with certain models, the variables may be collinear, yet the model is fully identified because of constraints or other features of the model. In such cases, using the collinear option allows the estimation to take place, leaving the equations with collinear variables intact. This option is seldom used.

force specifies that estimation be forced even though the time variable is not equally spaced. This is relevant only for correlation structures that require knowledge of the time variable. These correlation structures require that observations be equally spaced so that calculations based on lags correspond to a constant time change. If you specify a time variable indicating that observations are not equally spaced, the (time dependent) model will not be fit. If you also specify force, the model will be fit, and it will be assumed that the lags based on the data ordered by the time variable are appropriate.

Correlation

corr(*correlation*) specifies the within-group correlation structure; the default corresponds to the equal-correlation model, corr(exchangeable).

When you specify a correlation structure that requires a lag, you indicate the lag after the structure's name with or without a blank; for example, `corr(ar 1)` or `corr(ar1)`.

If you specify the fixed correlation structure, you specify the name of the matrix containing the assumed correlations following the word `fixed`, for example, `corr(fixed myr)`.

⌐‾‾‾‾‾‾⌐ Reporting ⌐‾‾

`level(#)` specifies the confidence level, as a percentage, for confidence intervals. The default is `level(95)` or as set by `set level`; see [U] **20.7 Specifying the width of confidence intervals**.

`noskip` specifies that a full maximum-likelihood model with only a constant for the regression equation be fit. This model is not displayed but is used as the base model to compute a likelihood-ratio test for the model test statistic displayed in the estimation header. By default, the overall model test statistic is an asymptotically equivalent Wald test of all the parameters in the regression equation being zero (except the constant). For many models, this option can substantially increase estimation time.

`nocnsreport` specifies that no constraints be reported. The default is to display user-specified constraints above the coefficient table.

`noomitted` specifies that variables that were omitted because of collinearity not be displayed. The default is to include in the table any variables omitted because of collinearity and to label them as "(omitted)".

`vsquish` specifies that the blank space separating factor-variable terms or time-series–operated variables from other variables in the model be suppressed.

`noemptycells` specifies that empty cells for interactions of factor variables not be displayed. The default is to include in the table interaction cells that do not occur in the estimation sample and to label them as "(empty)".

`baselevels` and `allbaselevels` control whether the base levels of factor variables and interactions are displayed. The default is to exclude from the table all base categories.

   `baselevels` specifies that base levels be reported for factor variables and for interactions whose bases cannot be inferred from their component factor variables.

   `allbaselevels` specifies that all base levels of factor variables and interactions be reported.

`cformat(%fmt)` specifies how to format coefficients, standard errors, and confidence limits in the coefficient table.

`pformat(%fmt)` specifies how to format p-values in the coefficient table.

`sformat(%fmt)` specifies how to format test statistics in the coefficient table.

`nolstretch` specifies that the width of the coefficient table not be automatically widened to accommodate longer variable names. The default, `lstretch`, is to automatically widen the coefficient table up to the width of the Results window. To change the default, use `set lstretch off`. `nolstretch` is not shown in the dialog box.

⌐‾‾‾‾‾‾⌐ Integration ⌐‾‾‾

`intmethod(intmethod)` specifies the integration method to be used for the random-effects model. It accepts one of three arguments: `mvaghermite`, the default, performs mean and variance adaptive Gauss–Hermite quadrature first on every and then on alternate iterations; `aghermite` performs mode and curvature adaptive Gauss–Hermite quadrature on the first iteration only; `ghermite` performs nonadaptive Gauss–Hermite quadrature.

intpoints(#) specifies the number of integration points to use for integration by quadrature. The default is intpoints(12); the maximum is intpoints(195). Increasing this value improves the accuracy but also increases computation time. Computation time is roughly proportional to its value.

The following option is not shown in the dialog box:

coeflegend specifies that the legend of the coefficients and how to specify them in an expression be displayed rather than displaying the statistics for the coefficients.

Also see

[U] **20 Estimation and postestimation commands**

Title

> **exit** — Exit Stata

Syntax

exit [, clear]

Description

Typing exit causes Stata to stop processing and return control to the operating system. If the dataset in memory has changed since the last save command, you must specify the clear option before Stata will let you exit.

exit may also be used for exiting do-files or programs; see [P] **exit**.

Stata for Windows users may also exit Stata by clicking on the **Close** button or by pressing *Alt+F4*.

Stata for Mac users may also exit Stata by pressing *Command+Q*.

Stata(GUI) users may also exit Stata by clicking on the **Close** button.

Option

clear permits you to exit, even if the current dataset has not been saved.

Remarks

Type exit to leave Stata and return to the operating system. If the dataset in memory has changed since the last time it was saved, however, Stata will refuse. At that point, you can either save the dataset and then type exit, or type exit, clear:

```
. exit
no; data in memory would be lost
r(4);
. exit, clear
```

Also see

[P] **exit** — Exit from a program or do-file

Title

exlogistic — Exact logistic regression

Syntax

exlogistic *depvar* *indepvars* [*if*] [*in*] [*weight*] [, *options*]

options	Description
Model	
<u>cond</u>vars(*varlist*)	condition on variables in *varlist*
<u>group</u>(*varname*)	groups/strata are stratified by unique values of *varname*
<u>bin</u>omial(*varname* | #)	data are in binomial form and the number of trials is contained in *varname* or in #
<u>est</u>constant	estimate constant term; do not condition on the number of successes
<u>nocons</u>tant	suppress constant term
Terms	
<u>terms</u>(*termsdef*)	terms definition
Options	
<u>mem</u>ory(#[b | k | m | g])	set limit on memory usage; default is memory(10m)
<u>sav</u>ing(*filename*)	save the joint conditional distribution to *filename*
Reporting	
<u>l</u>evel(#)	set confidence level; default is level(95)
coef	report estimated coefficients
<u>t</u>est(*testopt*)	report significance of observed sufficient statistic, conditional scores test, or conditional probabilities test
mue(*varlist*)	compute the median unbiased estimates for *varlist*
midp	use the mid-p-value rule
<u>nolog</u>	do not display the enumeration log

by, statsby, and xi are allowed; see [U] **11.1.10 Prefix commands**.

fweights are allowed; see [U] **11.1.6 weight**.

See [U] **20 Estimation and postestimation commands** for more capabilities of estimation commands.

Menu

Statistics > Exact statistics > Exact logistic regression

Description

exlogistic fits an exact logistic regression model of *depvar* on *indepvars*.

exlogistic is an alternative to logistic, the standard maximum-likelihood–based logistic regression estimator; see [R] **logistic**. exlogistic produces more-accurate inference in small samples because it does not depend on asymptotic results and exlogistic can better deal with one-way causation, such as the case where all females are observed to have a positive outcome.

exlogistic with the group(*varname*) option is an alternative to clogit, the conditional logistic regression estimator; see [R] **clogit**. Like clogit, exlogistic conditions on the number of positive outcomes within stratum.

depvar can be specified in two ways. It can be zero/nonzero, with zero indicating failure and nonzero representing positive outcomes (successes), or if you specify the binomial(*varname* | #) option, *depvar* may contain the number of positive outcomes within each trial.

exlogistic is computationally intensive. Unlike most estimators, rather than calculating coefficients for all independent variables at once, results for each independent variable are calculated separately with the other independent variables temporarily conditioned out. You can save considerable computer time by skipping the parameter calculations for variables that are not of direct interest. Specify such variables in the condvars() option rather than among the *indepvars*; see condvars() below.

Unlike Stata's other estimation commands, you may not use test, lincom, or other postestimation commands after exlogistic. Given the method used to calculate estimates, hypothesis tests must be performed during estimation by using exlogistic's terms() option; see terms() below.

Options

___ Model ___

condvars(*varlist*) specifies variables whose parameter estimates are not of interest to you. You can save substantial computer time and memory moving such variables from *indepvars* to condvars(). Understand that you will get the same results for x1 and x3 whether you type

> . exlogistic y x1 x2 x3 x4

or

> . exlogistic y x1 x3, condvars(x2 x4)

group(*varname*) specifies the variable defining the strata, if any. A constant term is assumed for each stratum identified in *varname*, and the sufficient statistics for *indepvars* are conditioned on the observed number of successes within each group. This makes the model estimated equivalent to that estimated by clogit, Stata's conditional logistic regression command (see [R] **clogit**). group() may not be specified with noconstant or estconstant.

binomial(*varname* | #) indicates that the data are in binomial form and *depvar* contains the number of successes. *varname* contains the number of trials for each observation. If all observations have the same number of trials, you can instead specify the number as an integer. The number of trials must be a positive integer at least as great as the number of successes. If binomial() is not specified, the data are assumed to be Bernoulli, meaning that *depvar* equaling zero or nonzero records one failure or success.

estconstant estimates the constant term. By default, the models are assumed to have an intercept (constant), but the value of the intercept is not calculated. That is, the conditional distribution of the sufficient statistics for the *indepvars* is computed given the number of successes in *depvar*, thus conditioning out the constant term of the model. Use estconstant if you want the estimate of the intercept reported. estconstant may not be specified with group().

noconstant; see [R] **estimation options**. noconstant may not be specified with group().

_____| Terms |_____

`terms`(*termname* = *variable* ... *variable* [, *termname* = *variable* ... *variable* ...]) defines additional
terms of the model on which you want `exlogistic` to perform joint-significance hypothesis tests.
By default, `exlogistic` reports tests individually on each variable in *indepvars*. For instance,
if variables `x1` and `x3` are in *indepvars*, and you want to jointly test their significance, specify
`terms(t1=x1 x3)`. To also test the joint significance of `x2` and `x4`, specify `terms(t1=x1 x3,
t2=x2 x4)`. Each variable can be assigned to only one term.

Joint tests are computed only for the conditional scores tests and the conditional probabilities tests.
See the `test()` option below.

_____| Options |_____

`memory`(*#* [b | k | m | g]) sets a limit on the amount of memory `exlogistic` can use when computing
the conditional distribution of the parameter sufficient statistics. The default is `memory(10m)`,
where m stands for megabyte, or 1,048,576 bytes. The following are also available: b stands for
byte; k stands for kilobyte, which is equal to 1,024 bytes; and g stands for gigabyte, which is
equal to 1,024 megabytes. The minimum setting allowed is `1m` and the maximum is `2048m` or
`2g`, but do not attempt to use more memory than is available on your computer. Also see the first
technical note under example 4 on counting the conditional distribution.

`saving`(*filename* [, `replace`]) saves the joint conditional distribution to *filename*. This distribution
is conditioned on those variables specified in `condvars()`. Use `replace` to replace an existing
file with *filename*. A Stata data file is created containing all the feasible values of the parameter
sufficient statistics. The variable names are the same as those in *indepvars*, in addition to a variable
named `_f_` containing the feasible value frequencies (sometimes referred to as the condition
numbers).

_____| Reporting |_____

`level`(*#*); see [R] **estimation options**. The `level(#)` option will not work on replay because
confidence intervals are based on estimator-specific enumerations. To change the confidence level,
you must refit the model.

`coef` reports the estimated coefficients rather than odds ratios (exponentiated coefficients). `coef` may
be specified when the model is fit or upon replay. `coef` affects only how results are displayed and
not how they are estimated.

`test`(`sufficient` | `score` | `probability`) reports the significance level of the observed sufficient
statistics, the conditional scores tests, or the conditional probabilities tests, respectively. The default
is `test(sufficient)`. If `terms()` is included in the specification, the conditional scores test
and the conditional probabilities test are applied to each term providing conditional inference for
several parameters simultaneously. All the statistics are computed at estimation time regardless of
which is specified. Each statistic may thus also be displayed postestimation without having to refit
the model; see [R] **exlogistic postestimation**.

`mue`(*varlist*) specifies that median unbiased estimates (MUEs) be reported for the variables in *varlist*.
By default, the conditional maximum likelihood estimates (CMLEs) are reported, except for those
parameters for which the CMLEs are infinite. Specify `mue(_all)` if you want MUEs for all the
indepvars.

`midp` instructs `exlogistic` to use the mid-p-value rule when computing the MUEs, significance
levels, and confidence intervals. This adjustment is for the discreteness of the distribution and
halves the value of the discrete probability of the observed statistic before adding it to the p-value.
The mid-p-value rule cannot be applied to MUEs whose corresponding parameter CMLE is infinite.

`nolog` prevents the display of the enumeration log. By default, the enumeration log is displayed, showing the progress of computing the conditional distribution of the sufficient statistics.

Remarks

Exact logistic regression is the estimation of the logistic model parameters by using the conditional distribution of the parameter sufficient statistics. The estimates are referred to as the conditional maximum likelihood estimates (CMLEs). This technique was first introduced by Cox and Snell (1989) as an alternative to using maximum likelihood estimation, which can perform poorly for small sample sizes. For stratified data, exact logistic regression is a small-sample alternative to conditional logistic regression. See [R] **logit**, [R] **logistic**, and [R] **clogit** to obtain maximum likelihood estimates (MLEs) for the logistic model and the conditional logistic model. For a comprehensive overview of exact logistic regression, see Mehta and Patel (1995).

Let Y_i denote a Bernoulli random variable where we observe the outcome $Y_i = y_i$, $i = 1, \ldots, n$. Associated with each independent observation is a $1 \times p$ vector of covariates, $\mathbf{x}_i$. We will denote $\pi_i = \Pr(Y_i \mid \mathbf{x}_i)$ and let the logit function model the relationship between Y_i and $\mathbf{x}_i$,

$$\log\left(\frac{\pi_i}{1 - \pi_i}\right) = \theta + \mathbf{x}_i \boldsymbol{\beta}$$

where the constant term θ and the $p \times 1$ vector of regression parameters $\boldsymbol{\beta}$ are unknown. The probability of observing $Y_i = y_i$, $i = 1, \ldots, n$, is

$$\Pr(\mathbf{Y} = \mathbf{y}) = \prod_{i=1}^{n} \pi_i^{y_i} (1 - \pi_i)^{1 - y_i}$$

where $\mathbf{Y} = (Y_1, \ldots, Y_n)$ and $\mathbf{y} = (y_1, \ldots, y_n)$. The MLEs for θ and $\boldsymbol{\beta}$ maximize the log of this function.

The sufficient statistics for θ and β_j, $j = 1, \ldots, p$, are $M = \sum_{i=1}^{n} Y_i$ and $T_j = \sum_{i=1}^{n} Y_i x_{ij}$, respectively, and we observe $M = m$ and $T_j = t_j$. By default, `exlogistic` tallies the conditional distribution of $\mathbf{T} = (T_1, \ldots, T_p)$ given $M = m$. This distribution will have a size of $\binom{n}{m}$. (It would have a size of 2^n without conditioning on $M = m$.) Denote one of these vectors $\mathbf{T}^{(k)} = (t_1^{(k)}, \ldots, t_p^{(k)})$, $k = 1, \ldots, N$, with combinatorial coefficient (frequency) c_k, $\sum_{k=1}^{N} c_k = \binom{n}{m}$. For each independent variable x_j, $j = 1, \ldots, p$, we reduce the conditional distribution further by conditioning on all other observed sufficient statistics $T_l = t_l$, $l \neq j$. The conditional probability of observing $T_j = t_j$ has the form

$$\Pr(T_j = t_j \mid T_l = t_l, l \neq j, M = m) = \frac{c \, e^{t_j \beta_j}}{\sum_k c_k e^{t_j^{(k)} \beta_j}}$$

where the sum is over the subset of $\mathbf{T}$ vectors such that $(T_1^{(k)} = t_1, \ldots, T_j^{(k)} = t_j^{(k)}, \ldots, T_p^{(k)} = t_p)$ and c is the combinatorial coefficient associated with the observed $\mathbf{t}$. The CMLE for β_j maximizes the log of this function.

Specifying nuisance variables in `condvars()` will reduce the size of the conditional distribution by conditioning on their observed sufficient statistics as well as conditioning on $M = m$. This reduces the amount of memory consumed at the cost of not obtaining regression estimates for those variables specified in `condvars()`.

Inferences from MLEs rely on asymptotics, and if your sample size is small, these inferences may not be valid. On the other hand, inferences from the CMLEs are exact in the sense that they use the conditional distribution of the sufficient statistics outlined above.

For small datasets, it is common for the dependent variable to be completely determined by the data. Here the MLEs and the CMLEs are unbounded. `exlogistic` will instead compute the MUE, the regression estimate that places the observed sufficient statistic at the median of the conditional distribution.

▷ Example 1

One example presented by Mehta and Patel (1995) is data from a prospective study of perinatal infection and human immunodeficiency virus type 1 (HIV-1). We use a variation of this dataset. There was an investigation Hutto et al. (1991) into whether the blood serum levels of glycoproteins CD4 and CD8 measured in infants at 6 months of age might predict their development of HIV infection. The blood serum levels are coded as ordinal values 0, 1, and 2.

```
. use http://www.stata-press.com/data/r12/hiv1
(prospective study of perinatal infection of HIV-1)
. list
```

	hiv	cd4	cd8
1.	1	0	0
2.	0	0	0
3.	1	0	2
4.	1	1	0
5.	0	1	0
	(output omitted)		
46.	0	2	1
47.	0	2	2

We first obtain the MLEs from `logistic` so that we can compare the estimates and associated statistics with the CMLEs from `exlogistic`.

```
. logistic hiv cd4 cd8, coef
```

```
Logistic regression                              Number of obs   =         47
                                                 LR chi2(2)      =      15.75
                                                 Prob > chi2     =     0.0004
Log likelihood = -20.751687                      Pseudo R2       =     0.2751
```

hiv	Coef.	Std. Err.	z	P>|z|	[95% Conf. Interval]	
cd4	-2.541669	.8392231	-3.03	0.002	-4.186517	-.8968223
cd8	1.658586	.821113	2.02	0.043	.0492344	3.267938
_cons	.5132389	.6809007	0.75	0.451	-.8213019	1.84778

```
. exlogistic hiv cd4 cd8, coef
Enumerating sample-space combinations:
observation 1:     enumerations =           2
observation 2:     enumerations =           3
 (output omitted)
observation 46:    enumerations =         601
observation 47:    enumerations =         326
```

```
Exact logistic regression                       Number of obs =           47
                                                Model score   =     13.34655
                                                Pr >= score   =       0.0006
```

hiv	Coef.	Score	Pr(Suff.)	[95% Conf. Interval]
cd4	-2.387632	10	0.0004	-4.699633	-.8221807
cd8	1.592366	12	0.0528	-.0137905	3.907876

exlogistic produced a log showing how many records are generated as it processes each observation. The primary purpose of the log is to provide feedback because generating the distribution can be time consuming, but we also see from the last entry that the joint distribution for the sufficient statistics for cd4 and cd8 conditioned on the total number of successes has 326 unique values (but a size of $\binom{47}{14} = 341{,}643{,}774{,}795$).

The statistics for logistic are based on asymptotics: for a large sample size, each Z statistic will be approximately normally distributed (with a mean of zero and a standard deviation of one) if the associated regression parameter is zero. The question is whether a sample size of 47 is large enough.

On the other hand, the p-values computed by exlogistic are from the conditional distributions of the sufficient statistics for each parameter given the sufficient statistics for all other parameters. In this sense, these p-values are exact. By default, exlogistic reports the sufficient statistics for the regression parameters and the probability of observing a more extreme value. These are single-parameter tests for H_0: $\beta_{cd4} = 0$ and H_0: $\beta_{cd8} = 0$ versus the two-sided alternatives. The conditional scores test, located in the coefficient table header, is testing that both H_0: $\beta_{cd4} = 0$ and H_0: $\beta_{cd8} = 0$. We find these p-values to be in fair agreement with the Wald and likelihood-ratio tests from logistic.

The confidence intervals for exlogistic are computed from the exact conditional distributions. The exact confidence intervals are asymmetrical about the estimate and are wider than the normal-based confidence intervals from logistic.

Both estimation techniques indicate that the incidence of HIV infection decreases with increasing CD4 blood serum levels and increases with increasing CD8 blood serum levels. The constant term is missing from the exact logistic coefficient table because we conditioned out its observed sufficient statistic when tallying the joint distribution of the sufficient statistics for the cd4 and cd8 parameters.

The test() option provides two other test statistics used in exact logistic: the conditional scores test, test(score), and the conditional probabilities test, test(probability). For comparison, we display the individual parameter conditional scores tests.

```
. exlogistic, test(score) coef
```
Exact logistic regression

```
                                        Number of obs =          47
                                        Model score   =    13.34655
                                        Pr >= score   =     0.0006
```

hiv	Coef.	Score	Pr>=Score	[95% Conf. Interval]	
cd4	-2.387632	12.88022	0.0003	-4.699633	-.8221807
cd8	1.592366	4.604816	0.0410	-.0137905	3.907876

For the probabilities test, the probability statistic is computed from (1) in *Methods and formulas* with $\beta = 0$. For this example, the significance of the probabilities tests matches the scores tests so they are not displayed here.

◁

❑ Technical note

Typically, the value of θ, the constant term, is of little interest, as well as perhaps some of the parameters in β, but we need to include all parameters in the model to correctly specify it. By conditioning out the nuisance parameters, we can reduce the size of the joint conditional distribution that is used to estimate the regression parameters of interest. The condvars() option allows you to specify a *varlist* of nuisance variables. By default, exlogistic conditions on the sufficient statistic of θ, which is the number of successes. You can save computation time and computer memory by using the condvars() option because infeasible values of the sufficient statistics associated with the variables in condvars() can be dropped from consideration before all n observations are processed.

Specifying some of your independent variables in condvars() will not change the estimated regression coefficients of the remaining independent variables. For instance, in example 1, if we instead type

```
. exlogistic hiv cd4, condvars(cd8) coef
```

the regression coefficient for cd4 (as well as all associated inference) will be identical.

One reason to have multiple variables in *indepvars* is to make conditional inference of several parameters simultaneously by using the terms() option. If you do not wish to test several parameters simultaneously, it may be more efficient to obtain estimates for individual variables by calling exlogistic multiple times with one variable in *indepvars* and all other variables listed in condvars(). The estimates will be the same as those with all variables in *indepvars*.

❑

❑ Technical note

If you fit a clogit (see [R] **clogit**) model to the HIV data from example 1, you will find that the estimates differ from those with exlogistic. (To fit the clogit model, you will have to create a group variable that includes all observations.) The regression estimates will be different because clogit conditions on the constant term only, whereas the estimates from exlogistic condition on the sufficient statistic of the other regression parameter as well as the constant term.

❑

▷ Example 2

The HIV data presented in table IV of Mehta and Patel (1995) are in a binomial form, where the variable hiv contains the HIV cases that tested positive and the variable n contains the number of individuals with the same CD4 and CD8 levels, the binomial number-of-trials parameter. Here *depvar* is hiv, and we use the binomial(n) option to identify the number-of-trials variable.

```
. use http://www.stata-press.com/data/r12/hiv_n
(prospective study of perinatal infection of HIV-1; binomial form)
. list
```

	cd4	cd8	hiv	n
1.	0	2	1	1
2.	1	2	2	2
3.	0	0	4	7
4.	1	1	4	12
5.	2	2	1	3
6.	1	0	2	7
7.	2	0	0	2
8.	2	1	0	13

Further, the cd4 and cd8 variables of the hiv dataset are actually factor variables, where each has the ordered levels of $(0, 1, 2)$. Another approach to the analysis is to use indicator variables, and following Mehta and Patel (1995), we used a 0–1 coding scheme that will give us the odds ratio of level 0 versus 2 and level 1 versus 2.

```
. gen byte cd4_0 = (cd4==0)
. gen byte cd4_1 = (cd4==1)
. gen byte cd8_0 = (cd8==0)
. gen byte cd8_1 = (cd8==1)
. exlogistic hiv cd4_0 cd4_1 cd8_0 cd8_1, terms(cd4=cd4_0 cd4_1,
> cd8=cd8_0 cd8_1) binomial(n) test(probability) saving(dist) nolog
note: saving distribution to file dist.dta
note: CMLE estimate for cd4_0 is +inf; computing MUE
note: CMLE estimate for cd4_1 is +inf; computing MUE
note: CMLE estimate for cd8_0 is -inf; computing MUE
note: CMLE estimate for cd8_1 is -inf; computing MUE
```

```
Exact logistic regression              Number of obs =          47
Binomial variable: n                   Model prob.   =    3.19e-06
                                       Pr <= prob.   =      0.0011
```

hiv	Odds Ratio	Prob.	Pr<=Prob.	[95% Conf.	Interval]
cd4		.0007183	0.0055		
cd4_0	18.82831*	.007238	0.0072	1.714079	+Inf
cd4_1	11.53732*	.0063701	0.0105	1.575285	+Inf
cd8		.0053212	0.0323		
cd8_0	.1056887*	.0289948	0.0290	0	1.072531
cd8_1	.0983388*	.0241503	0.0242	0	.9837203

```
(*) median unbiased estimates (MUE)
. matrix list e(sufficient)
e(sufficient)[1,4]
     cd4_0  cd4_1  cd8_0  cd8_1
r1       5      8      6      4
. display e(n_possible)
1091475
```

Here we used terms() to specify two terms in the model, cd4 and cd8, that make up the cd4 and cd8 indicator variables. By doing so, we obtained a conditional probabilities test for cd4, simultaneously testing both cd4_0 and cd4_1, and for cd8, simultaneously testing both cd8_0 and cd8_1. The significance levels for the two terms are 0.0055 and 0.0323, respectively.

This example also illustrates instances where the dependent variable is completely determined by the independent variables and CMLEs are infinite. If we try to obtain MLEs, `logistic` will drop each variable and then terminate with a no-data error, error number 2000.

```
. use http://www.stata-press.com/data/r12/hiv_n, clear
(prospective study of perinatal infection of HIV-1; binomial form)
. gen byte cd4_0 = (cd4==0)
. gen byte cd4_1 = (cd4==1)
. gen byte cd8_0 = (cd8==0)
. gen byte cd8_1 = (cd8==1)
. expand n
(39 observations created)
. logistic hiv cd4_0 cd4_1 cd8_0 cd8_1
note: cd4_0 != 0 predicts success perfectly
      cd4_0 dropped and 8 obs not used
note: cd4_1 != 0 predicts success perfectly
      cd4_1 dropped and 21 obs not used
note: cd8_0 != 0 predicts failure perfectly
      cd8_0 dropped and 2 obs not used
outcome = cd8_1 <= 0 predicts data perfectly
r(2000);
```

In the previous example, `exlogistic` generated the joint conditional distribution of T_{cd4_0}, T_{cd4_1}, T_{cd8_0}, and T_{cd8_1} given $M = 14$ (the number of individuals that tested positive), and for reference, we listed the observed sufficient statistics that are stored in the matrix `e(sufficient)`. Below we take that distribution and further condition on $T_{cd4\_1} = 8$, $T_{cd8\_0} = 6$, and $T_{cd8\_1} = 4$, giving the conditional distribution of T_{cd4_0}. Here we see that the observed sufficient statistic $T_{cd4\_0} = 5$ is last in the sorted listing or, equivalently, T_{cd4_0} is at the domain boundary of the conditional probability distribution. When this occurs, the conditional probability distribution is monotonically increasing in β_{cd4_0} and a maximum does not exist.

```
. use dist, clear
. keep if cd4_1==8 & cd8_0==6 & cd8_1==4
(4139 observations deleted)
. list, sep(0)
```

	_f_	cd4_0	cd4_1	cd8_0	cd8_1
1.	1668667	0	8	6	4
2.	18945542	1	8	6	4
3.	55801053	2	8	6	4
4.	55867350	3	8	6	4
5.	17423175	4	8	6	4
6.	1091475	5	8	6	4

When the CMLEs are infinite, the MUEs are computed (Hirji, Tsiatis, and Mehta 1989). For the `cd4_0` estimate, we compute the value $\overline{\beta}_{cd4\_0}$ such that

$$\Pr(T_{cd4\_0} \geq 5 \mid \beta_{cd4\_0} = \overline{\beta}_{cd4\_0}, T_{cd4\_1} = 8, T_{cd8\_0} = 6, T_{cd8\_1} = 4, M = 14) = 1/2$$

using (1) in *Methods and formulas*.

The output is in agreement with example 1: there is an increase in risk of HIV infection for a CD4 blood serum level of 0 relative to a level of 2 and for a level of 1 relative to a level of 2; there is a decrease in risk of HIV infection for a CD8 blood serum level of 0 relative to a level of 2 and for a level of 1 relative to a level of 2.

We also displayed e(n_possible). This is the combinatorial coefficient associated with the observed sufficient statistics. The same value is found in the _f_ variable of the conditional distribution dataset listed above. The size of the distribution is $\binom{47}{14} = 341{,}643{,}774{,}795$. This can be verified by summing the _f_ variable of the generated conditional distribution dataset.

```
. use dist, clear

. summarize _f_, meanonly

. di %15.1f r(sum)
341643774795.0
```

◁

▷ Example 3

One can think of exact logistic regression as a covariate-adjusted exact binomial. To demonstrate this point, we will use exlogistic to compute a binomial confidence interval for m successes of n trials, by fitting the constant-only model, and we will compare it with the confidence interval computed by ci (see [R] ci). We will use the saving() option to retain the dataset containing the feasible values for the constant term sufficient statistic, namely, the number of successes, m, given n trials and their associated combinatorial coefficients $\binom{n}{m}$, $m = 0, 1, \ldots, n$.

```
. input y

           y
1. 1
2. 0
3. 1
4. 0
5. 1
6. 1
7. end

. ci y, binomial
```

| | | | | — Binomial Exact — | |
Variable	Obs	Mean	Std. Err.	[95% Conf. Interval]	
y	6	.6666667	.1924501	.2227781	.9567281

```
. exlogistic y, estconstant nolog coef saving(binom)
note: saving distribution to file binom.dta

Exact logistic regression
                                              Number of obs =          6
```

	Coef.	Suff.	2*Pr(Suff.)	[95% Conf. Interval]	
_cons	.6931472	4	0.6875	-1.24955	3.096017

We use the postestimation program estat predict to transform the estimated constant term and its confidence bounds by using the inverse logit function, invlogit() (see [D] **functions**). The standard error for the estimated probability is computed using the delta method.

. estat predict

	y	Predicted	Std. Err.	[95% Conf. Interval]	
Probability		0.6667	0.1925	0.2228	0.9567

. use binom, replace

. list, sep(0)

	_f_	_cons_
1.	1	0
2.	6	1
3.	15	2
4.	20	3
5.	15	4
6.	6	5
7.	1	6

Examining the listing of the generated data, the values contained in the variable _cons_ are the feasible values of M, and the values contained in the variable _f_ are the binomial coefficients $\binom{6}{m}$ with total $\sum_{m=0}^{6} \binom{6}{m} = 2^6 = 64$. In the coefficient table, the sufficient statistic for the constant term, labeled Suff., is $m = 4$. This value is located at record 5 of the dataset. Therefore, the two-tailed probability of the sufficient statistic is computed as $0.6875 = 2(15 + 6 + 1)/64$.

The constant term is the value of θ that maximizes the probability of observing $M = 4$; see (1) of *Methods and formulas*:

$$\Pr(M = 4|\theta) = \frac{15e^{4\alpha}}{1 + 6e^{\alpha} + 15e^{2\alpha} + 20e^{3\alpha} + 15e^{4\alpha} + 6e^{5\alpha} + e^{6\alpha}}$$

The maximum is at the value $\theta = \log 2$, which is demonstrated in the figure below.

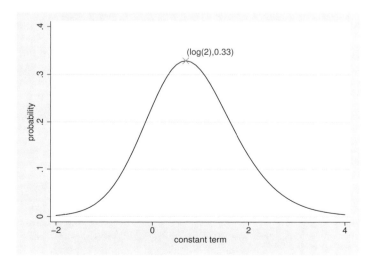

The lower and upper confidence bounds are the values of θ such that $\Pr(M \geq 4|\theta) = 0.025$ and $\Pr(M \leq 4|\theta) = 0.025$, respectively. These probabilities are plotted in the figure below for $\theta \in [-2, 4]$.

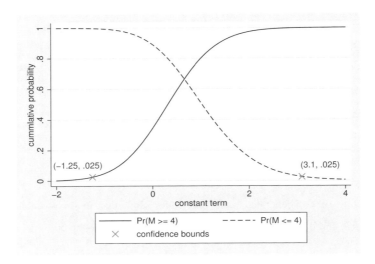

◁

▷ Example 4

This example demonstrates the group() option, which allows the analysis of stratified data. Here the logistic model is

$$\log\left(\frac{\pi_{ik}}{1 - \pi_{ik}}\right) = \theta_k + \mathbf{x}_{ki}\boldsymbol{\beta}$$

where k indexes the s strata, $k = 1, \dots, s$, and θ_k is the strata-specific constant term whose sufficient statistic is $M_k = \sum_{i=1}^{n_k} Y_{ki}$.

Mehta and Patel (1995) use a case–control study to demonstrate this model, which is useful in comparing the estimates from exlogistic and clogit. This study was intended to determine the role of birth complications in people with schizophrenia (Garsd 1988). Siblings from seven families took part in the study, and each individual was classified as normal or schizophrenic. A birth complication index is recorded for each individual that ranges from 0, an uncomplicated birth, to 15, a very complicated birth. Some of the frequencies contained in variable f are greater than 1, and these count different births at different times where the individual has the same birth complications index, found in variable BCindex.

```
. use http://www.stata-press.com/data/r12/schizophrenia, clear
(case-control study on birth complications for people with schizophrenia)
. list, sepby(family)
```

	family	BCindex	schizo	f
1.	1	6	0	1
2.	1	7	0	1
3.	1	3	0	2
4.	1	2	0	3
5.	1	5	0	1
6.	1	0	0	1
7.	1	15	1	1
8.	2	2	1	1
9.	2	0	0	1
10.	3	2	0	1
11.	3	9	1	1
12.	3	1	0	1
13.	4	2	1	1
14.	4	0	0	4
15.	5	3	1	1
16.	5	6	0	1
17.	5	0	1	1
18.	6	3	0	1
19.	6	0	1	1
20.	6	0	0	2
21.	7	2	0	1
22.	7	6	1	1

```
. exlogistic schizo BCindex [fw=f], group(family) test(score) coef
Enumerating sample-space combinations:
observation 1:     enumerations =      2
observation 2:     enumerations =      3
observation 3:     enumerations =      4
observation 4:     enumerations =      5
observation 5:     enumerations =      6
    (output omitted )
observation 21:    enumerations =     72
observation 22:    enumerations =     40

Exact logistic regression              Number of obs     =          29
Group variable: family                 Number of groups  =           7

                                       Obs per group: min =           2
                                                      avg =         4.1
                                                      max =          10

                                       Model score       =     6.32803
                                       Pr >= score       =      0.0167
```

schizo	Coef.	Score	Pr>=Score	[95% Conf. Interval]	
BCindex	.3251178	6.328033	0.0167	.0223423	.7408832

The asymptotic alternative for this model can be estimated using clogit (equivalently, xtlogit, fe) and is listed below for comparison. We must expand the data because clogit will not accept frequency weights if they are not constant within the groups.

```
. expand f
(7 observations created)
. clogit schizo BCindex, group(family) nolog
note: multiple positive outcomes within groups encountered.
Conditional (fixed-effects) logistic regression   Number of obs   =         29
                                                  LR chi2(1)      =       5.20
                                                  Prob > chi2     =     0.0226
Log likelihood = -6.2819819                       Pseudo R2       =     0.2927
```

schizo	Coef.	Std. Err.	z	P>|z|	[95% Conf. Interval]	
BCindex	.3251178	.1678981	1.94	0.053	-.0039565	.654192

Both techniques compute the same regression estimate for the BCindex, which might not be too surprising because both estimation techniques condition on the total number of successes in each group. The difference lies in the p-values and confidence intervals. The p-value testing $H_0 : \beta_{\text{BCindex}} = 0$ is approximately 0.0167 for the exact conditional scores test and 0.053 for the asymptotic Wald test. Moreover, the exact confidence interval is asymmetric about the estimate and does not contain zero.

◁

❑ Technical note

The memory(#) option limits the amount of memory that exlogistic will consume when computing the conditional distribution of the parameter sufficient statistics. memory() is independent of the data maximum memory setting (see set max_memory in [D] **memory**), and it is possible for exlogistic to exceed the memory limit specified in set max_memory without terminating. By default, a log is provided that displays the number of enumerations (the size of the conditional distribution) after processing each observation. Typically, you will see the number of enumerations increase, and then at some point they will decrease as the multivariate shift algorithm (Hirji, Mehta, and Patel 1987) determines that some of the enumerations cannot achieve the observed sufficient statistics of the conditioning variables. When the algorithm is complete, however, it is necessary to store the conditional distribution of the parameter sufficient statistics as a dataset. It is possible, therefore, to get a memory error when the algorithm has completed if there is not enough memory to store the conditional distribution.

❑

❑ Technical note

Computing the conditional distributions and reported statistics requires data sorting and numerical comparisons. If there is at least one single-precision variable specified in the model, exlogistic will make comparisons with a relative precision of 2^{-5}. Otherwise, a relative precision of 2^{-11} is used. Be careful if you use recast to promote a single-precision variable to double precision (see [D] **recast**). You might try listing the data in full precision (maybe %20.15g; see [D] **format**) to make sure that this is really what you want. See [D] **data types** for information on precision of numeric storage types.

❑

Saved results

exlogistic saves the following in e():

Scalars
e(N)	number of observations
e(k_groups)	number of groups
e(n_possible)	number of distinct possible outcomes where sum(sufficient) equals observed e(sufficient)
e(n_trials)	binomial number-of-trials parameter
e(sum_y)	sum of *depvar*
e(k_indvars)	number of independent variables
e(k_terms)	number of model terms
e(k_condvars)	number of conditioning variables
e(condcons)	conditioned on the constant(s) indicator
e(midp)	mid-p-value rule indicator
e(eps)	relative difference tolerance

Macros
e(cmd)	exlogistic
e(cmdline)	command as typed
e(title)	title in estimation output
e(depvar)	name of dependent variable
e(indvars)	independent variables
e(condvars)	conditional variables
e(groupvar)	group variable
e(binomial)	binomial number-of-trials variable
e(level)	confidence level
e(wtype)	weight type
e(wexp)	weight expression
e(datasignature)	the checksum
e(datasignaturevars)	variables used in calculation of checksum
e(properties)	b
e(estat_cmd)	program used to implement estat
e(predict)	program used to implement predict
e(marginsnotok)	predictions disallowed by margins

Matrices
e(b)	coefficient vector
e(mue_indicators)	indicator for elements of e(b) estimated using MUE instead of CMLE
e(se)	e(b) standard errors (CMLEs only)
e(ci)	matrix of e(level) confidence intervals for e(b)
e(sum_y_groups)	sum of e(depvar) for each group
e(N_g)	number of observations in each group
e(sufficient)	sufficient statistics for e(b)
e(p_sufficient)	p-value for e(sufficient)
e(scoretest)	conditional scores tests for *indepvars*
e(p_scoretest)	p-values for e(scoretest)
e(probtest)	conditional probabilities tests for *indepvars*
e(p_probtest)	p-value for e(probtest)
e(scoretest_m)	conditional scores tests for model terms
e(p_scoretest_m)	p-value for e(scoretest_m)
e(probtest_m)	conditional probabilities tests for model terms
e(p_probtest_m)	p-value for e(probtest_m)

Functions
e(sample)	marks estimation sample

Methods and formulas

exlogistic is implemented as an ado-file.

Methods and formulas are presented under the following headings:

Sufficient statistics
Conditional distribution and CMLE
Median unbiased estimates and exact CI
Conditional hypothesis tests
Sufficient-statistic p-value

Sufficient statistics

Let $\{Y_1, Y_2, \ldots, Y_n\}$ be a set of n independent Bernoulli random variables, each of which can realize two outcomes, $\{0, 1\}$. For each $i = 1, \ldots, n$, we observe $Y_i = y_i$, and associated with each observation is the covariate row vector of length p, $\mathbf{x}_i = (x_{i1}, \ldots, x_{ip})$. Denote $\boldsymbol{\beta} = (\beta_1, \ldots, \beta_p)^T$ to be the column vector of regression parameters and θ to be the constant. The sufficient statistic for β_j is $T_j = \sum_{i=1}^n Y_i x_{ij}$, $j = 1, \ldots, p$, and for θ is $M = \sum_{i=1}^n Y_i$. We observe $T_j = t_j$, $t_j = \sum_{i=1}^n y_i x_{ij}$, and $M = m$, $m = \sum_{i=1}^n y_i$. The probability of observing $(Y_1 = y_1, Y_2 = y_2, \ldots, Y_n = y_n)$ is

$$\Pr(Y_1 = y_1, \ldots, Y_n = y_n \mid \boldsymbol{\beta}, \mathbf{X}) = \frac{\exp(m\theta + \mathbf{t}\boldsymbol{\beta})}{\prod_{i=1}^n \{1 + \exp(\theta + \mathbf{x}_i\boldsymbol{\beta})\}}$$

where $\mathbf{t} = (t_1, \ldots, t_p)$ and $\mathbf{X} = (\mathbf{x}_1^T, \ldots, \mathbf{x}_n^T)^T$.

The joint distribution of the sufficient statistics $\mathbf{T}$ is obtained by summing over all possible binary sequences $Y_1, \ldots, Y_n$ such that $\mathbf{T} = \mathbf{t}$ and $M = m$. This probability function is

$$\Pr(T_1 = t_1, \ldots, T_p = t_p, M = m \mid \boldsymbol{\beta}, \mathbf{X}) = \frac{c(\mathbf{t}, m) \exp(m\theta + \mathbf{t}\boldsymbol{\beta})}{\prod_{i=1}^n \{1 + \exp(\theta + \mathbf{x}_i\boldsymbol{\beta})\}}$$

where $c(\mathbf{t}, m)$ is the combinatorial coefficient of $(\mathbf{t}, m)$ or the number of distinct binary sequences $Y_1, \ldots, Y_n$ such that $\mathbf{T} = \mathbf{t}$ and $M = m$ (Cox and Snell 1989).

Conditional distribution and CMLE

Without loss of generality, we will restrict our discussion to computing the CMLE of β_1. If we condition on observing $M = m$ and $T_2 = t_2, \ldots, T_p = t_p$, the probability function of $(T_1 \mid \beta_1, T_2 = t_2, \ldots, T_p = t_p, M = m)$ is

$$\Pr(T_1 = t_1 \mid \beta_1, T_2 = t_2, \ldots, T_p = t_p, M = m) = \frac{c(\mathbf{t}, m) e^{t_1 \beta_1}}{\sum_u c(u, t_2, \ldots, t_p, m) e^{u\beta_1}} \quad (1)$$

where the sum in the denominator is over all possible values of T_1 such that $M = m$ and $T_2 = t_2, \ldots, T_p = t_p$ and $c(u, t_2, \ldots, t_p, m)$ is the combinatorial coefficient of $(u, t_2, \ldots, t_p, m)$ (Cox and Snell 1989). The CMLE for β_1 is the value $\widehat{\beta}_1$ that maximizes the log of (1). This optimization task is carried out by `ml`, using the conditional frequency distribution of $(T_1 \mid T_2 = t_2, \ldots, T_p = t_p, M = m)$ as a dataset. Generating the joint conditional distribution is efficiently computed using the multivariate shift algorithm described by Hirji, Mehta, and Patel (1987).

Difficulties in computing $\widehat{\beta}_1$ arise if the observed $(T_1 = t_1, \ldots, T_p = t_p, M = m)$ lies on the boundaries of the distribution of $(T_1 \mid T_2 = t_2, \ldots, T_p = t_p, M = m)$, where the conditional probability function is monotonically increasing (or decreasing) in β_1. Here the CMLE is plus infinity if it is on the upper boundary, $\Pr(T_1 \leq t_1 \mid T_2 = t_2, \ldots, T_p = t_p, M = m) = 1$, and is minus infinity if it is on the lower boundary of the distribution, $\Pr(T_1 \geq t_1 \mid T_2 = t_2, \ldots, T_p = t_p, M = m) = 1$. This concept is demonstrated in example 2. When infinite CMLEs occur, the MUE is computed.

Median unbiased estimates and exact CI

The MUE is computed using the technique outlined by Hirji, Tsiatis, and Mehta (1989). First, we find the values of $\beta_1^{(u)}$ and $\beta_1^{(l)}$ such that

$$
\begin{aligned}
\Pr(T_1 \leq t_1 \mid \beta_1 = \beta_1^{(u)}, T_2 = t_2, \ldots, T_p = t_p, M = m) = \\
\Pr(T_1 \geq t_1 \mid \beta_1 = \beta_1^{(l)}, T_2 = t_2, \ldots, T_p = t_p, M = m) = 1/2
\end{aligned}
\tag{2}
$$

The MUE is then $\overline{\beta}_1 = \left(\beta_1^{(l)} + \beta_1^{(u)} \right)/2$. However, if T_1 is equal to the minimum of the domain of the conditional distribution, $\beta^{(l)}$ does not exist and $\overline{\beta}_1 = \beta^{(u)}$. If T_1 is equal to the maximum of the domain of the conditional distribution, $\beta^{(u)}$ does not exist and $\overline{\beta}_1 = \beta^{(l)}$.

Confidence bounds for β are computed similarly, except that we substitute $\alpha/2$ for $1/2$ in (2), where $1 - \alpha$ is the confidence level. Here $\beta_1^{(l)}$ would then be the lower confidence bound and $\beta_1^{(u)}$ would be the upper confidence bound (see example 3).

Conditional hypothesis tests

To test $H_0: \beta_1 = 0$ versus $H_1: \beta_1 \neq 0$, we obtain the exact p-value from $\sum_{u \in E} f_1(u) - f_1(t_1)/2$ if the mid-p-value rule is used and $\sum_{u \in E} f_1(u)$ otherwise. Here E is a critical region, and we define $f_1(u) = \Pr(T_1 = u \mid \beta_1 = 0, T_2 = t_2, \ldots, T_p = t_p, M = m)$ for ease of notation. There are two popular ways to define the critical region: the conditional probabilities test and the conditional scores test (Mehta and Patel 1995). The critical region when using the conditional probabilities test is all values of the sufficient statistic for β_1 that have a probability less than or equal to that of the observed t_1, $E_p = \{u : f_1(u) \leq f_1(t_1)\}$. The critical region of the conditional scores test is defined as all values of the sufficient statistic for β_1 such that its score is greater than or equal to that of t_1,

$$
E_s = \left\{ u : (u - \mu_1)^2/\sigma_1^2 \geq (t_1 - \mu_1)^2/\sigma_1^2) \right\}
$$

Here μ_1 and σ_1^2 are the mean and variance of $(T_1 \mid \beta_1 = 0, T_2 = t_2, \ldots, T_p = t_p, M = m)$.

The score statistic is defined as

$$
\left\{ \frac{\partial \ell(\beta)}{\partial \beta} \right\}^2 \left[-E \left\{ \frac{\partial^2 \ell(\beta)}{\partial \beta^2} \right\} \right]^{-1}
$$

evaluated at $H_0: \beta = 0$, where ℓ is the log of (1). The score test simplifies to $(t - E\,[T|\beta])^2/\mathrm{var}(T|\beta)$ (Hirji 2006), where the mean and variance are computed from the conditional distribution of the sufficient statistic with $\beta = 0$ and t is the observed sufficient statistic.

Sufficient-statistic p-value

The p-value for testing $H_0: \beta_1 = 0$ versus the two-sided alternative when $(T_1 = t_1|T_2 = t_2, \ldots, T_p = t_p)$ is computed as $2 \times \min(p_l, p_u)$, where

$$
p_l = \frac{\sum_{u \leq t_1} c(u, t_2, \ldots, t_p, m)}{\sum_u c(u, t_2, \ldots, t_p, m)}
$$

$$
p_u = \frac{\sum_{u \geq t_1} c(u, t_2, \ldots, t_p, m)}{\sum_u c(u, t_2, \ldots, t_p, m)}
$$

It is the probability of observing a more extreme T_1.

References

Cox, D. R., and E. J. Snell. 1989. *Analysis of Binary Data*. 2nd ed. London: Chapman & Hall.

Garsd, A. 1988. Schizophrenia and birth complications. Unpublished manuscript.

Hirji, K. F. 2006. *Exact Analysis of Discrete Data*. Boca Raton: Chapman & Hall/CRC.

Hirji, K. F., C. R. Mehta, and N. R. Patel. 1987. Computing distributions for exact logistic regression. *Journal of the American Statistical Association* 82: 1110–1117.

Hirji, K. F., A. A. Tsiatis, and C. R. Mehta. 1989. Median unbiased estimation for binary data. *American Statistician* 43: 7–11.

Hutto, C., W. P. Parks, S. Lai, M. T. Mastrucci, C. Mitchell, J. Muñoz, E. Trapido, I. M. Master, and G. B. Scott. 1991. A hospital-based prospective study of perinatal infection with human immunodeficiency virus type 1. *Journal of Pediatrics* 118: 347–353.

Mehta, C. R., and N. R. Patel. 1995. Exact logistic regression: Theory and examples. *Statistics in Medicine* 14: 2143–2160.

Also see

[R] **exlogistic postestimation** — Postestimation tools for exlogistic

[R] **binreg** — Generalized linear models: Extensions to the binomial family

[R] **clogit** — Conditional (fixed-effects) logistic regression

[R] **expoisson** — Exact Poisson regression

[R] **logistic** — Logistic regression, reporting odds ratios

[R] **logit** — Logistic regression, reporting coefficients

[U] **20 Estimation and postestimation commands**

Title

> **exlogistic postestimation** — Postestimation tools for exlogistic

Description

The following postestimation commands are of special interest after `exlogistic`:

Command	Description
estat predict	single-observation prediction
estat se	report ORs or coefficients and their asymptotic standard errors

For information about these commands, see below.

The following standard postestimation command is also available:

Command	Description
estat summarize	estimation sample summary

`estat summarize` is not allowed if the `binomial()` option was specified in `exlogistic`.

See [R] **estat** for details.

Special-interest postestimation commands

`estat predict` computes a predicted probability (or linear predictor), its asymptotic standard error, and its exact confidence interval for 1 observation. Predictions are carried out by estimating the constant coefficient after shifting the independent variables and conditioned variables by the values specified in the `at()` option or by their medians. Therefore, predictions must be done with the estimation sample in memory. If a different dataset is used or if the dataset is modified, then an error will result.

`estat se` reports odds ratio or coefficients and their asymptotic standard errors. The estimates are stored in the matrix `r(estimates)`.

Syntax for estat predict

    estat <u>predict</u> [, *options*]

options	Description
<u>pr</u>	probability; the default
xb	linear effect
at(*atspec*)	use the specified values for the *indepvars* and `condvars()`
<u>level</u>(#)	set confidence level for the predicted value; default is `level(95)`
<u>mem</u>ory(#[b | k | m | g])	set limit on memory usage; default is `memory(10m)`
<u>nolog</u>	do not display the enumeration log

These statistics are available only for the estimation sample.

Menu

Statistics > Postestimation > Predictions, residuals, etc.

Options for estat predict

pr, the default, calculates the probability.

xb calculates the linear effect.

at(*varname* = # $\left[\,[\,varname = \#\,]\,[\,\dots\,]\,\right]$) specifies values to use in computing the predicted value. Here *varname* is one of the independent variables, *indepvars*, or the conditioned variables, condvars(). The default is to use the median of each independent and conditioned variable.

level(#) specifies the confidence level, as a percentage, for confidence intervals. The default is level(95) or as set by set level; see [U] **20.7 Specifying the width of confidence intervals**.

memory(#$\left[\,\text{b}\,|\,\text{k}\,|\,\text{m}\,|\,\text{g}\,\right]$) sets a limit on the amount of memory estat predict can use when generating the conditional distribution of the constant parameter sufficient statistic. The default is memory(10m), where m stands for megabyte, or 1,048,576 bytes. The following are also available: b stands for byte; k stands for kilobyte, which is equal to 1,024 bytes; and g stands for gigabyte, which is equal to 1,024 megabytes. The minimum setting allowed is 1m and the maximum is 512m or 0.5g, but do not attempt to use more memory than is available on your computer. Also see *Remarks* in [R] **exlogistic** for details on enumerating the conditional distribution.

nolog prevents the display of the enumeration log. By default, the enumeration log is displayed showing the progress of enumerating the distribution of the observed successes conditioned on the independent variables shifted by the values specified in at() (or by their medians). See *Methods and formulas* in [R] **exlogistic** for details of the computations.

Syntax for estat se

estat se $\left[\,,\ \text{coef}\,\right]$

Menu

Statistics > Postestimation > Reports and statistics

Option for estat se

coef requests that the estimated coefficients and their asymptotic standard errors be reported. The default is to report the odds ratios and their asymptotic standard errors.

Remarks

Predictions must be done using the estimation sample. This is because the prediction is really an estimated constant coefficient (the intercept) after shifting the independent variables and conditioned variables by the values specified in at() or by their medians. The justification for this approach can be seen by rewriting the model as

$$\log\left(\frac{\pi_i}{1-\pi_i}\right) = (\alpha + \mathbf{x}_0\boldsymbol{\beta}) + (\mathbf{x}_i - \mathbf{x}_0)\boldsymbol{\beta}$$

where x_0 are the specified values for the *indepvars* (Mehta and Patel 1995). Because the estimation of the constant term is required, this technique is not appropriate for stratified models that used the group() option.

▷ Example 1

To demonstrate, we return to the example 2 in [R] **exlogistic** using data from a prospective study of perinatal infection and HIV-1. Here there was an investigation into whether the blood serum levels of CD4 and CD8 measured in infants at 6 months of age might predict their development of HIV infection. The blood serum levels are coded as ordinal values 0, 1, and 2. These data are used by Mehta and Patel (1995) as an exposition of exact logistic.

```
. use http://www.stata-press.com/data/r12/hiv_n
(prospective study of perinatal infection of HIV-1; binomial form)
. gen byte cd4_0 = (cd4==0)
. gen byte cd4_1 = (cd4==1)
. gen byte cd8_0 = (cd8==0)
. gen byte cd8_1 = (cd8==1)
. exlogistic hiv cd4_0 cd4_1 cd8_0 cd8_1, terms(cd4=cd4_0 cd4_1,
> cd8=cd8_0 cd8_1) binomial(n) test(probability) saving(dist, replace)
 (output omitted )
. estat predict
Enumerating sample-space combinations:
observation 1:    enumerations =        3
observation 2:    enumerations =       12
observation 3:    enumerations =        5
observation 4:    enumerations =        5
observation 5:    enumerations =        5
observation 6:    enumerations =       35
observation 7:    enumerations =       15
observation 8:    enumerations =       15
observation 9:    enumerations =        9
observation 10:   enumerations =        9
observation 11:   enumerations =        5
observation 12:   enumerations =       18
note: CMLE estimate for _cons is -inf; computing MUE
Predicted value at cd4_0 = 0, cd4_1 = 0, cd8_0 = 0, cd8_1 = 1
```

hiv	Predicted	Std. Err.	[95% Conf. Interval]	
Probability	0.0390*	N/A	0.0000	0.1962

```
(*) identifies median unbiased estimates (MUE); because an MUE
    is computed, there is no SE estimate
```

Because we did not specify values by using the at() option, the median values of the *indepvars* are used for the prediction. By default, medians are used instead of means because we want to use values that are observed in the dataset. If the means of the binary variables cd4_0–cd8_1 were used, we would have created floating point variables in $(0, 1)$ that not only do not properly represent the indicator variables but also would be a source of computational inefficiency in generating the conditional distribution. Because the MUE is computed for the predicted value, there is no standard-error estimate.

From the example discussions in [R] **exlogistic**, the infants at highest risk are those with a CD4 level of 0 and a CD8 level of 2. Below we use the at() option to make a prediction at these blood serum levels.

```
. estat predict, at(cd4_0=1 cd4_1=0 cd8_0=0 cd8_1=0) nolog
note: CMLE estimate for _cons is +inf; computing MUE
Predicted value at cd4_0 = 1, cd4_1 = 0, cd8_0 = 0, cd8_1 = 0
```

hiv	Predicted	Std. Err.	[95% Conf. Interval]
Probability	0.9063*	N/A	0.4637 1.0000

(*) identifies median unbiased estimates (MUE); because an MUE
    is computed, there is no SE estimate

◁

Saved results

estat predict saves the following in r():

Scalars
    r(imue)   1 if r(pred) is an MUE and 0 if a CMLE
    r(pred)   estimated probability or the linear effect
    r(se)   asymptotic standard error of r(pred)

Macros
    r(estimate)   prediction type: pr or xb
    r(level)   confidence level

Matrices
    r(ci)   confidence interval
    r(x)   *indepvars* and condvars() values

Methods and formulas

All postestimation commands listed above are implemented as ado-files using Mata.

Reference

Mehta, C. R., and N. R. Patel. 1995. Exact logistic regression: Theory and examples. *Statistics in Medicine* 14: 2143–2160.

Also see

[R] **exlogistic** — Exact logistic regression

[U] **20 Estimation and postestimation commands**

Title

> **expoisson** — Exact Poisson regression

Syntax

expoisson *depvar indepvars* [*if*] [*in*] [*weight*] [, *options*]

options	Description
Model	
<u>cond</u>vars(*varlist*)	condition on variables in *varlist*
<u>gro</u>up(*varname*)	groups/strata are stratified by unique values of *varname*
<u>e</u>xposure(*varname_e*)	include ln(*varname_e*) in model with coefficient constrained to 1
<u>off</u>set(*varname_o*)	include *varname_o* in model with coefficient constrained to 1
Options	
<u>mem</u>ory(#[b | k | m | g])	set limit on memory usage; default is memory(25m)
<u>sav</u>ing(*filename*)	save the joint conditional distribution to *filename*
Reporting	
<u>l</u>evel(#)	set confidence level; default is level(95)
irr	report incidence-rate ratios
<u>test</u>(*testopt*)	report significance of observed sufficient statistic, conditional scores test, or conditional probabilities test
mue(*varlist*)	compute the median unbiased estimates for *varlist*
midp	use the mid-p-value rule
<u>nolog</u>	do not display the enumeration log

by, statsby, and xi are allowed; see [U] **11.1.10 Prefix commands**.

fweights are allowed; see [U] **11.1.6 weight**.

See [U] **20 Estimation and postestimation commands** for more capabilities of estimation commands.

Menu

Statistics > Exact statistics > Exact Poisson regression

Description

expoisson fits an exact Poisson regression model of *depvar* on *indepvars*. Exact Poisson regression is an alternative to standard maximum-likelihood–based Poisson regression (see [R] **poisson**) that offers more accurate inference in small samples because it does not depend on asymptotic results. For stratified data, expoisson is an alternative to fixed-effects Poisson regression (see xtpoisson, fe in [XT] **xtpoisson**); like fixed-effects Poisson regression, exact Poisson regression conditions on the number of events in each stratum.

Exact Poisson regression is computationally intensive, so if you have regressors whose parameter estimates are not of interest (that is, nuisance parameters), you should specify those variables in the condvars() option instead of in *indepvars*.

Options

condvars(*varlist*) specifies variables whose parameter estimates are not of interest to you. You can save substantial computer time and memory by moving such variables from *indepvars* to condvars(). Understand that you will get the same results for x1 and x3 whether you type

    . expoisson y x1 x2 x3 x4

or

    . expoisson y x1 x3, condvars(x2 x4)

group(*varname*) specifies the variable defining the strata, if any. A constant term is assumed for each stratum identified in *varname*, and the sufficient statistics for *indepvars* are conditioned on the observed number of successes within each group (as well as other variables in the model). The group variable must be integer valued.

exposure(*varname_e*), offset(*varname_o*); see [R] **estimation options**.

memory(#[b|k|m|g]) sets a limit on the amount of memory expoisson can use when computing the conditional distribution of the parameter sufficient statistics. The default is memory(25m), where m stands for megabyte, or 1,048,576 bytes. The following are also available: b stands for byte; k stands for kilobyte, which is equal to 1,024 bytes; and g stands for gigabyte, which is equal to 1,024 megabytes. The minimum setting allowed is 1m and the maximum is 2048m or 2g, but do not attempt to use more memory than is available on your computer. Also see the first technical note under example 3 on counting the conditional distribution.

saving(*filename* [, replace]) saves the joint conditional distribution for each independent variable specified in *indepvars*. There is one file for each variable, and it is named using the prefix *filename* with the variable name appended. For example, saving(mydata) with an independent variable named X would generate a data file named mydata_X.dta. Use replace to replace an existing file. Each file contains the conditional distribution for one of the independent variables specified in *indepvars* conditioned on all other *indepvars* and those variables specified in condvars(). There are two variables in each data file: the feasible sufficient statistics for the variable's parameter and their associated weights. The weights variable is named _w_.

level(#); see [R] **estimation options**. The level(#) option will not work on replay because confidence intervals are based on estimator-specific enumerations. To change the confidence level, you must refit the model.

irr reports estimated coefficients transformed to incidence-rate ratios, that is, $\exp(\beta)$ rather than β. Standard errors and confidence intervals are similarly transformed. This option affects how results are displayed, not how they are estimated or stored. irr may be specified at estimation or when replaying previously estimated results.

test(sufficient|score|probability) reports the significance level of the observed sufficient statistic, the conditional scores test, or the conditional probabilities test. The default is test(sufficient). All the statistics are computed at estimation time, and each statistic may be displayed postestimation; see [R] **expoisson postestimation**.

mue(*varlist*) specifies that median unbiased estimates (MUEs) be reported for the variables in *varlist*. By default, the conditional maximum likelihood estimates (CMLEs) are reported, except for those parameters for which the CMLEs are infinite. Specify mue(_all) if you want MUEs for all the *indepvars*.

`midp` instructs `expoisson` to use the mid-p-value rule when computing the MUEs, significance levels, and confidence intervals. This adjustment is for the discreteness of the distribution by halving the value of the discrete probability of the observed statistic before adding it to the p-value. The mid-p-value rule cannot be MUEs whose corresponding parameter CMLE is infinite.

`nolog` prevents the display of the enumeration log. By default, the enumeration log is displayed, showing the progress of computing the conditional distribution of the sufficient statistics.

Remarks

Exact Poisson regression estimates the model parameters by using the conditional distributions of the parameters' sufficient statistics, and the resulting parameter estimates are known as CMLEs. Exact Poisson regression is a small-sample alternative to the maximum-likelihood ML Poisson model. See [R] **poisson** and [XT] **xtpoisson** to obtain maximum likelihood estimates (MLEs) for the Poisson model and the fixed-effects Poisson model.

Let Y_i denote a Poisson random variable where we observe the outcome $Y_i = y_i$, $i = 1, \ldots, n$. Associated with each independent observation is a $1 \times p$ vector of covariates, $\mathbf{x}_i$. We will denote $\mu_i = E[Y_i \mid \mathbf{x}_i]$ and use the log linear model to model the relationship between Y_i and $\mathbf{x}_i$,

$$\log(\mu_i) = \theta + \mathbf{x}_i \boldsymbol{\beta}$$

where the constant term, θ, and the $p \times 1$ vector of regression parameters, $\boldsymbol{\beta}$, are unknown. The probability of observing $Y_i = y_i$, $i = 1, \ldots, n$, is

$$\Pr(\mathbf{Y} = \mathbf{y}) = \prod_{i=1}^{n} \frac{\mu_i^{y_i} e^{-\mu_i}}{y_i!}$$

where $\mathbf{Y} = (Y_1, \ldots, Y_n)$ and $\mathbf{y} = (y_1, \ldots, y_n)$. The MLEs for θ and $\boldsymbol{\beta}$ maximize the log of this function.

The sufficient statistics for θ and β_j, $j = 1, \ldots, p$, are $M = \sum_{i=1}^{n} Y_i$ and $T_j = \sum_{i=1}^{n} Y_i x_{ij}$, respectively, and we observe $M = m$ and $T_j = t_j$. `expoisson` tallies the conditional distribution for each T_j, given the other sufficient statistics $T_l = t_l$, $l \neq j$ and $M = m$. Denote one of these values to be $t_j^{(k)}$, $k = 1, \ldots, N$, with weight w_k that accounts for all the generated $\mathbf{Y}$ vectors that give rise to $t_j^{(k)}$. The conditional probability of observing $T_j = t_j$ has the form

$$\Pr(T_j = t_j \mid T_l = t_l, l \neq j, M = m) = \frac{w\, e^{t_j \beta_j}}{\sum_k w_k e^{t_j^{(k)} \beta_j}} \tag{1}$$

where the sum is over the subset of $\mathbf{T}$ vectors such that $(T_1^{(k)} = t_1, \ldots, T_j^{(k)} = t_j^{(k)}, \ldots, T_p^{(k)} = t_p)$ and w is the weight associated with the observed $\mathbf{t}$. The CMLE for β_j maximizes the log of this function.

Specifying nuisance variables in `condvars()` prevents `expoisson` from estimating their associated regression coefficients. These variables are still conditional variables when tallying the conditional distribution for the variables in *indepvars*.

Inferences from MLEs rely on asymptotics, and if your sample size is small, these inferences may not be valid. On the other hand, inferences from the CMLEs are exact in that they use the conditional distribution of the sufficient statistics outlined above.

For small datasets, the dependent variable can be completely determined by the data. Here the MLEs and the CMLEs are unbounded. When this occurs, expoisson will compute the MUE, the regression estimate that places the observed sufficient statistic at the median of the conditional distribution.

See [R] **exlogistic** for a more thorough discussion of exact estimation and related statistics.

▷ Example 1

Armitage, Berry, and Matthews (2002, 499–501) fit a log-linear model to data containing the number of cerebrovascular accidents experienced by 41 men during a fixed period, each of whom had recovered from a previous cerebrovascular accident and was hypertensive. Sixteen men received treatment, and in the original data, there are three age groups (40–49, 50–59, $\geq$60), but we pool the first two age groups to simplify the example. Armitage, Berry, and Matthews point out that this was not a controlled trial, but the data are useful to inquire whether there is evidence of fewer accidents for the treatment group and if age may be an important factor. The dependent variable count contains the number of accidents, variable treat is an indicator for the treatment group (1 = treatment, 0 = control), and variable age is an indicator for the age group (0 = 40–59; 1 = $\geq$60).

First, we load the data, list it, and tabulate the cerebrovascular accident counts by treatment and age group.

```
. use http://www.stata-press.com/data/r12/cerebacc
(cerebrovascular accidents in hypotensive-treated and control groups)
. list
```

	treat	count	age
1.	control	0	40/59
2.	control	0	>=60
3.	control	1	40/59
4.	control	1	>=60
5.	control	2	40/59
6.	control	2	>=60
7.	control	3	40/59
	(output omitted)		
35.	treatment	0	40/59
36.	treatment	0	40/59
37.	treatment	0	40/59
38.	treatment	0	40/59
39.	treatment	1	40/59
40.	treatment	1	40/59
41.	treatment	1	40/59

```
. tabulate treat age [fw=count]
```

hypotensive drug treatment	age group 40/59	>=60	Total
control	15	10	25
treatment	4	0	4
Total	19	10	29

Next we estimate the CMLE with expoisson and, for comparison, the MLE with poisson.

```
. expoisson count treat age

Estimating: treat
Enumerating sample-space combinations:
observation 1:    enumerations =           11
observation 2:    enumerations =           11
observation 3:    enumerations =           11
       (output omitted )
observation 39:   enumerations =          410
observation 40:   enumerations =          410
observation 41:   enumerations =           30

Estimating: age
Enumerating sample-space combinations:
observation 1:    enumerations =            5
observation 2:    enumerations =           15
observation 3:    enumerations =           15
       (output omitted )
observation 39:   enumerations =          455
observation 40:   enumerations =          455
observation 41:   enumerations =           30

Exact Poisson regression
                                              Number of obs =           41
```

count	Coef.	Suff.	2*Pr(Suff.)	[95% Conf. Interval]	
treat	-1.594306	4	0.0026	-3.005089	-.4701708
age	-.5112067	10	0.2794	-1.416179	.3429232

```
. poisson count treat age, nolog

Poisson regression                            Number of obs   =           41
                                              LR chi2(2)      =        10.64
                                              Prob > chi2     =       0.0049
Log likelihood =  -38.97981                   Pseudo R2       =       0.1201
```

count	Coef.	Std. Err.	z	P>|z|	[95% Conf. Interval]	
treat	-1.594306	.5573614	-2.86	0.004	-2.686714	-.5018975
age	-.5112067	.4043525	-1.26	0.206	-1.303723	.2813096
_cons	.233344	.2556594	0.91	0.361	-.2677391	.7344271

`expoisson` generates an enumeration log for each independent variable in *indepvars*. The conditional distribution of the parameter sufficient statistic is tallied for each independent variable. The conditional distribution for `treat`, for example, has 30 records containing the weights, w_k, and feasible sufficient statistics, $t_{\text{treat}}^{(k)}$. In essence, the set of points $(w_k, t_{\text{treat}}^{(k)})$, $k = 1, \ldots, 30$, tallied by `expoisson` now become the data to estimate the regression coefficient for `treat`, using (1) as the likelihood. Remember that one of the 30 $(w_k, t_{\text{treat}}^{(k)})$ must contain the observed sufficient statistic, $t_{\text{treat}} = \sum_{i=1}^{41} \text{treat}_i \times \text{count}_i = 4$, and its relative position in the sorted set of points (sorted by $t_{\text{treat}}^{(k)}$) is how the sufficient-statistic significance is computed. This algorithm is repeated for the `age` variable.

The regression coefficients for `treat` and `age` are numerically identical for both Poisson models. Both models indicate that the treatment is significant at reducing the rate of cerebrovascular accidents, $\approx e^{-1.59} \approx 0.204$, or a reduction of about 80%. There is no significant age effect.

The *p*-value for the treatment regression-coefficient sufficient statistic indicates that the treatment effect is a bit more significant than for the corresponding asymptotic Z statistic from `poisson`. However, the exact confidence intervals are wider than their asymptotic counterparts.

◁

▷ Example 2

Agresti (2002) used the data from Laird and Olivier (1981) to demonstrate the Poisson model for modeling rates. The data consist of patient survival after heart valve replacement operations. The sample consists of 109 patients that are classified by type of heart valve (aortic, mitral) and by age (<55, ≥55). Follow-up observations cover lengths from 3 to 97 months, and the time at risk, or exposure, is stored in the variable TAR. The response is whether the subject died. First, we take a look at the data and then estimate the incidence rates (IRs) with expoisson and poisson.

```
. use http://www.stata-press.com/data/r12/heartvalve
(heart valve replacement data)
. list
```

	age	valve	deaths	TAR
1.	< 55	aortic	4	1259
2.	< 55	mitral	1	2082
3.	>= 55	aortic	7	1417
4.	>= 55	mitral	9	1647

The age variable is coded 0 for age <55 and 1 for age ≥55, and the valve variable is coded 0 for the aortic valve and 1 for the mitral valve. The total number of deaths, $M = 21$, is small enough that enumerating the conditional distributions for age and valve type is feasible and asymptotic inferences associated with standard ML Poisson regression may be questionable.

```
. expoisson deaths age valve, exposure(TAR) irr
Estimating: age
Enumerating sample-space combinations:
observation 1:    enumerations =          11
observation 2:    enumerations =          11
observation 3:    enumerations =         132
observation 4:    enumerations =          22

Estimating: valve
Enumerating sample-space combinations:
observation 1:    enumerations =          17
observation 2:    enumerations =          17
observation 3:    enumerations =         102
observation 4:    enumerations =          22

Exact Poisson regression
```

Number of obs = 4

deaths	IRR	Suff.	2*Pr(Suff.)	[95% Conf. Interval]
age	3.390401	16	0.0194	1.182297 11.86935
valve	.7190197	10	0.5889	.2729881 1.870068
ln(TAR)	1	(exposure)		

```
. poisson deaths age valve, exposure(TAR) irr nolog
```

Poisson regression

	Number of obs	=	4
	LR chi2(2)	=	7.62
	Prob > chi2	=	0.0222
Log likelihood = -8.1747285	Pseudo R2	=	0.3178

deaths	IRR	Std. Err.	z	P>|z|	[95% Conf. Interval]	
age	3.390401	1.741967	2.38	0.017	1.238537	9.280965
valve	.7190197	.3150492	-0.75	0.452	.3046311	1.6971
_cons	.0018142	.0009191	-12.46	0.000	.0006722	.0048968
ln(TAR)	1	(exposure)				

The CMLE and the MLE are numerically identical. The death rate for the older age group is about 3.4 times higher than the younger age group, and this difference is significant at the 5% level. This means that for every death in the younger group each month, we would expect about three deaths in the older group. The IR estimate for valve type is approximately 0.72, but it is not significantly different from one. The exact Poisson confidence intervals are a bit wider than the asymptotic CIs.

You can use ir (see [ST] **epitab**) to estimate IRs and exact CIs for one covariate, and we compare these CIs with those from expoisson, where we estimate the incidence rate by using age only.

```
. ir deaths age TAR
```

	age of patient Exposed	Unexposed	Total
number of deaths	16	5	21
time at risk	3064	3341	6405
Incidence rate	.0052219	.0014966	.0032787
	Point estimate	[95% Conf. Interval]	
Inc. rate diff.	.0037254	.00085	.0066007
Inc. rate ratio	3.489295	1.221441	12.17875 (exact)
Attr. frac. ex.	.7134092	.1812948	.9178898 (exact)
Attr. frac. pop	.5435498		

(midp) Pr(k>=16) =	0.0049 (exact)	
(midp) 2*Pr(k>=16) =	0.0099 (exact)	

```
. expoisson deaths age, exposure(TAR) irr midp nolog
```

Exact Poisson regression

Number of obs = 4

deaths	IRR	Suff.	2*Pr(Suff.)	[95% Conf. Interval]	
age	3.489295	16	0.0099	1.324926	10.64922
ln(TAR)	1	(exposure)			

mid-p-value computed for the probabilities and CIs

Both ir and expoisson give identical IRs and p-values. Both report the two-sided exact significance by using the mid-p-value rule that accounts for the discreteness in the distribution by subtracting $p_{1/2} = \Pr(T = t)/2$ from $p_l = \Pr(T \le t)$ and $p_g = \Pr(T \ge t)$, computing $2 \times \min(p_l - p_{1/2}, p_g - p_{1/2})$. By default, expoisson will not use the mid-p-value rule (when you exclude the midp option), and here the two-sided exact significance would be $2 \times \min(p_l, p_g) = 0.0158$. The confidence intervals differ because expoisson uses the mid-p-value rule when computing the confidence intervals, yet ir does not. You can verify this by executing `expoisson` without the `midp` option for this example; you will get the same CIs as `ir`.

You can replay `expoisson` to view the conditional scores test or the conditional probabilities test by using the `test()` option.

```
. expoisson, test(score) irr
Exact Poisson regression
```
                                                      Number of obs =   4

deaths	IRR	Score	Pr>=Score	[95% Conf. Interval]
age	3.489295	6.76528	0.0113	1.324926	10.64922
ln(TAR)	1	(exposure)			

```
mid-p-value computed for the probabilities and CIs
```

All the statistics for `expoisson` are defined in *Methods and formulas* of [R] **exlogistic**. Apart from enumerating the conditional distributions for the logistic and Poisson sufficient statistics, computationally, the primary difference between `exlogistic` and `expoisson` is the weighting values in the likelihood for the parameter sufficient statistics.

◁

▷ Example 3

In this example, we fabricate data that will demonstrate the difference between the CMLE and the MUE when the CMLE is not infinite. A difference in these estimates will be more pronounced when the probability of the coefficient sufficient statistic is skewed when plotted as a function of the regression coefficient.

```
. clear
. input y x

            y          x
  1. 0 2
  2. 1 1
  3. 1 0
  4. 0 0
  5. 0 .5
  6. 1 .5
  7. 2 .01
  8. 3 .001
  9. 4 .0001
 10. end
. expoisson y x, test(score)
Enumerating sample-space combinations:
observation 1:    enumerations =         13
observation 2:    enumerations =         91
observation 3:    enumerations =        169
observation 4:    enumerations =        169
observation 5:    enumerations =        313
observation 6:    enumerations =        313
observation 7:    enumerations =       1469
observation 8:    enumerations =       5525
observation 9:    enumerations =       5479
Exact Poisson regression
```
                                                      Number of obs =   9

y	Coef.	Score	Pr>=Score	[95% Conf. Interval]
x	-1.534468	2.955316	0.0810	-3.761718	.0485548

```
. expoisson y x, test(score) mue(x) nolog
Exact Poisson regression
                                           Number of obs =          9
```

y	Coef.	Score	Pr>=Score	[95% Conf. Interval]
x	-1.309268*	2.955316	0.0810	-3.761718 .0485548

(*) median unbiased estimates (MUE)

We observe (x_i, y_i), $i = 1, \ldots, 9$. If we condition on $m = \sum_{i=1}^{9} y_i = 12$, the conditional distribution of $T_x = \sum_i Y_i x_i$ has a size of 5,479 elements. For each entry in this enumeration, a realization of $Y_i = y_i^{(k)}$, $k = 1, \ldots, 5,479$, is generated such that $\sum_i y_i^{(k)} = 12$. One of these realizations produces the observed $t_x = \sum_i y_i x_i \approx 1.5234$.

Below is a graphical display comparing the CMLE with the MUE. We plot $\Pr(T_x = t_x \mid M = 12, \beta_x)$ versus β_x, $-6 \leq \beta_x \leq 1$, in the upper panel and the cumulative probabilities, $\Pr(T_x \leq t_x \mid M = 12, \beta_x)$ and $\Pr(T_x \geq t_x \mid M = 12, \beta_x)$, in the lower panel.

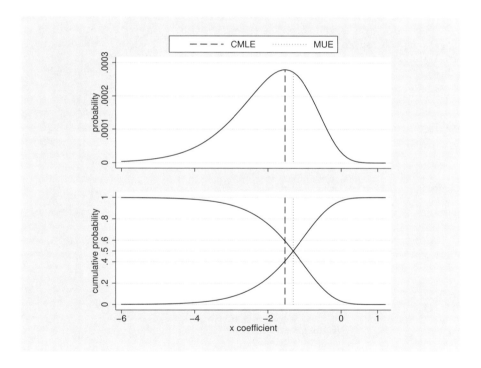

The location of the CMLE, indicated by the dashed line, is at the mode of the probability profile, and the MUE, indicated by the dotted line, is to the right of the mode. If we solve for the $\beta_x^{(u)}$ and $\beta_x^{(l)}$ such that $\Pr(T_x \leq t_x \mid M = 12, \beta_x^{(u)}) = 1/2$ and $\Pr(T_x \geq t_x \mid M = 12, \beta_x^{(l)}) = 1/2$, the MUE is $(\beta_x^{(u)} + \beta_x^{(l)})/2$. As you can see in the lower panel, the MUE cuts through the intersection of these cumulative probability profiles.

◁

❏ Technical note

The memory(#) option limits the amount of memory that expoisson will consume when computing the conditional distribution of the parameter sufficient statistics. memory() is independent of the data maximum memory setting (see set max_memory in [D] **memory**), and it is possible for expoisson to exceed the memory limit specified in set max_memory without terminating. By default, a log is provided that displays the number of enumerations (the size of the conditional distribution) after processing each observation. Typically, you will see the number of enumerations increase, and then at some point they will decrease as the multivariate shift algorithm (Hirji, Mehta, and Patel 1987) determines that some of the enumerations cannot achieve the observed sufficient statistics of the conditioning variables. When the algorithm is complete, however, it is necessary to store the conditional distribution of the parameter sufficient statistics as a dataset. It is possible, therefore, to get a memory error when the algorithm has completed if there is not enough memory to store the conditional distribution.   ❏

❏ Technical note

Computing the conditional distributions and reported statistics requires data sorting and numerical comparisons. If there is at least one single-precision variable specified in the model, expoisson will make comparisons with a relative precision of 2^{-5}. Otherwise, a relative precision of 2^{-11} is used. Be careful if you use recast to promote a single-precision variable to double precision (see [D] **recast**). You might try listing the data in full precision (maybe %20.15g; see [D] **format**) to make sure that this is really what you want. See [D] **data types** for information on precision of numeric storage types.   ❏

Saved results

expoisson saves the following in e():

Scalars

e(N)	number of observations
e(k_groups)	number of groups
e(relative_weight)	relative weight for the observed e(sufficient) and e(condvars)
e(sum_y)	sum of *depvar*
e(k_indvars)	number of independent variables
e(k_condvars)	number of conditioning variables
e(midp)	mid-p-value rule indicator
e(eps)	relative difference tolerance

Macros

e(cmd)	expoisson
e(cmdline)	command as typed
e(title)	title in estimation output
e(depvar)	name of dependent variable
e(indvars)	independent variables
e(condvars)	conditional variables
e(groupvar)	group variable
e(exposure)	exposure variable
e(offset)	linear offset variable
e(level)	confidence level
e(wtype)	weight type
e(wexp)	weight expression
e(datasignature)	the checksum
e(datasignaturevars)	variables used in calculation of checksum
e(properties)	b V
e(estat_cmd)	program used to implement estat
e(predict)	program used to implement predict
e(marginsnotok)	predictions disallowed by margins

Matrices
     e(b)   coefficient vector
     e(mue_indicators)   indicator for elements of e(b) estimated using MUE instead of CMLE
     e(se)   e(b) standard errors (CMLEs only)
     e(ci)   matrix of e(level) confidence intervals for e(b)
     e(sum_y_groups)   sum of e(depvar) for each group
     e(N_g)   number of observations in each group
     e(sufficient)   sufficient statistics for e(b)
     e(p_sufficient) p-value for e(sufficient)
     e(scoretest)   conditional scores tests for *indepvars*
     e(p_scoretest) p-values for e(scoretest)
     e(probtest)   conditional probability tests for *indepvars*
     e(p_probtest) p-value for e(probtest)

Functions
     e(sample)   marks estimation sample

Methods and formulas

expoisson is implemented as an ado-file.

Let $\{Y_1, Y_2, \ldots, Y_n\}$ be a set of n independent Poisson random variables. For each $i = 1, \ldots, n$, we observe $Y_i = y_i \geq 0$, and associated with each observation is the covariate row vector of length p, $\mathbf{x}_i = (x_{i1}, \ldots, x_{ip})$. Denote $\boldsymbol{\beta} = (\beta_1, \ldots, \beta_p)^T$ to be the column vector of regression parameters and θ to be the constant. The sufficient statistic for β_j is $T_j = \sum_{i=1}^n Y_i x_{ij}$, $j = 1, \ldots, p$, and for θ is $M = \sum_{i=1}^n Y_i$. We observe $T_j = t_j$, $t_j = \sum_{i=1}^n y_i x_{ij}$, and $M = m$, $m = \sum_{i=1}^n y_i$. Let κ_i be the exposure for the ith observation. Then the probability of observing $(Y_1 = y_1, Y_2 = y_2, \ldots, Y_n = y_n)$ is

$$ \Pr(Y_1 = y_1, \ldots, Y_n = y_n \mid \boldsymbol{\beta}, \mathbf{X}, \boldsymbol{\kappa}) = \frac{\exp(m\theta + \mathbf{t}\boldsymbol{\beta})}{\exp\{\sum_{i=1}^n \kappa_i \exp(\theta + \mathbf{x}_i\boldsymbol{\beta})\}} \prod_{i=1}^n \frac{\kappa_i^{y_i}}{y_i!} $$

where $\mathbf{t} = (t_1, \ldots, t_p)$, $\mathbf{X} = (\mathbf{x}_1^T, \ldots, \mathbf{x}_n^T)^T$, and $\boldsymbol{\kappa} = (\kappa_1, \ldots, \kappa_n)^T$.

The joint distribution of the sufficient statistics $(\mathbf{T}, M)$ is obtained by summing over all possible sequences $Y_1 \geq 0, \ldots, Y_n \geq 0$ such that $\mathbf{T} = \mathbf{t}$ and $M = m$. This probability function is

$$ \Pr(T_1 = t_1, \ldots, T_p = t_p, M = m \mid \boldsymbol{\beta}, \mathbf{X}, \boldsymbol{\kappa}) = \frac{\exp(m\theta + \mathbf{t}\boldsymbol{\beta})}{\exp\left\{\sum_{i=1}^n \kappa_i \exp(\theta + \mathbf{x}_i\boldsymbol{\beta})\right\}} \left(\sum_{\mathbf{u}} \prod_{i=1}^n \frac{\kappa_i^{u_i}}{u_i!} \right) $$

where the sum $\sum_{\mathbf{u}}$ is over all nonnegative vectors $\mathbf{u}$ of length n such that $\sum_{i=1}^n u_i = m$ and $\sum_{i=1}^n u_i \mathbf{x}_i = \mathbf{t}$.

Conditional distribution

Without loss of generality, we will restrict our discussion to the conditional distribution of the sufficient statistic for β_1, T_1. If we condition on observing $M = m$ and $T_2 = t_2, \ldots, T_p = t_p$, the probability function of $(T_1 \mid \beta_1, T_2 = t_2, \ldots, T_p = t_p, M = m)$ is

$$ \Pr(T_1 = t_1 \mid \beta_1, T_2 = t_2, \ldots, T_p = t_p, M = m) = \frac{\left(\sum_{\mathbf{u}} \prod_{i=1}^n \frac{\kappa_i^{u_i}}{u_i!} \right) e^{t_1 \beta_1}}{\sum_{\mathbf{v}} \left(\prod_{i=1}^n \frac{\kappa_i^{v_i}}{v_i!} \right) e^{\beta_1 \sum_i v_i x_{i1}}} \tag{2} $$

where the sum $\sum_{\mathbf{u}}$ is over all nonnegative vectors $\mathbf{u}$ of length n such that $\sum_{i=1}^{n} u_i = m$ and $\sum_{i=1}^{n} u_i \mathbf{x}_i = \mathbf{t}$, and the sum $\sum_{\mathbf{v}}$ is over all nonnegative vectors $\mathbf{v}$ of length n such that $\sum_{i=1}^{n} v_i = m$, $\sum_{i=1}^{n} v_i x_{i2} = t_2, \ldots, \sum_{i=1}^{n} v_i x_{ip} = t_p$. The CMLE for β_1 is the value that maximizes the log of (1). This optimization task is carried out by `ml` (see [R] **ml**), using the conditional distribution of $(T_1 \mid T_2 = t_2, \ldots, T_p = t_p, M = m)$ as a dataset. This dataset consists of the feasible values and weights for T_1,

$$\left\{ \left(s_1, \prod_{i=1}^{n} \frac{\kappa_i^{v_i}}{v_i!} \right) : \sum_{i=1}^{n} v_i = m, \sum_{i=1}^{n} v_i x_{i1} = s_1, \sum_{i=1}^{n} v_i x_{i2} = t_2, \ldots, \sum_{i=1}^{n} v_i x_{ip} = t_p \right\}$$

Computing the CMLE, MUE, confidence intervals, conditional hypothesis tests, and sufficient statistic p-values is discussed in *Methods and formulas* of [R] **exlogistic**. The only difference between the two techniques is the use of the weights; that is, the weights for exact logistic are the combinatorial coefficients, $c(\mathbf{t}, m)$, in (1) of *Methods and formulas* in [R] **exlogistic**. `expoisson` and `exlogistic` use the same `ml` likelihood evaluator to compute the CMLEs as well as the same ado-programs and Mata functions to compute the MUEs and estimate statistics.

References

Agresti, A. 2002. *Categorical Data Analysis.* 2nd ed. Hoboken, NJ: Wiley.

Armitage, P., G. Berry, and J. N. S. Matthews. 2002. *Statistical Methods in Medical Research.* 4th ed. Oxford: Blackwell.

Cox, D. R., and E. J. Snell. 1989. *Analysis of Binary Data.* 2nd ed. London: Chapman & Hall.

Hirji, K. F., C. R. Mehta, and N. R. Patel. 1987. Computing distributions for exact logistic regression. *Journal of the American Statistical Association* 82: 1110–1117.

Laird, N. M., and D. Olivier. 1981. Covariance analysis of censored survival data using log-linear analysis techniques. *Journal of the American Statistical Association* 76: 231–240.

Also see

Title

> **expoisson postestimation** — Postestimation tools for expoisson

Description

The following postestimation command is of special interest after `expoisson`:

Command	Description
estat se	report coefficients or IRRs and their asymptotic standard errors

For information about this command, see below.

The following standard postestimation command is also available:

Command	Description
estat summarize	estimation sample summary

See [R] **estat** for details.

Special-interest postestimation command

`estat se` reports regression coefficients or incidence-rate asymptotic standard errors. The estimates are stored in the matrix `r(estimates)`.

Syntax for estat se

> estat se [, irr]

Menu

Statistics > Postestimation > Reports and statistics

Option for estat se

`irr` requests that the incidence-rate ratios and their asymptotic standard errors be reported. The default is to report the coefficients and their asymptotic standard errors.

Remarks

▷ Example 1

To demonstrate `estat se` after `expoisson`, we use the British physicians smoking data.

```
. use http://www.stata-press.com/data/r12/smokes
(cigarette smoking and lung cancer among British physicians (45-49 years))
. expoisson cases smokes, exposure(peryrs) irr nolog
Exact Poisson regression
                                              Number of obs =          7
```

cases	IRR	Suff.	2*Pr(Suff.)	[95% Conf. Interval]
smokes	1.077718	797.4	0.0000	1.04552 1.111866
ln(peryrs)	1	(exposure)		

```
. estat se, irr
```

cases	IRR	Std. Err.
smokes	1.077718	.0168547

◁

Methods and formulas

All postestimation commands listed above are implemented as ado-files.

Also see

[R] **expoisson** — Exact Poisson regression

[U] **20 Estimation and postestimation commands**

Title

fracpoly — Fractional polynomial regression

Syntax

Fractional polynomial regression

fracpoly $\left[\, , \text{ fracpoly\_options} \right]$: regression_cmd $\left[\text{yvar}_1 \left[\text{yvar}_2 \right] \right]$

$\text{xvar}_1 \left[\# \left[\#... \right] \right] \left[\text{xvar}_2 \left[\# \left[\#... \right] \right] \right] \left[... \right] \left[\text{xvarlist} \right] \left[\text{if} \right] \left[\text{in} \right] \left[\text{weight} \right]$

$\left[\, , \text{ regression\_cmd\_options} \right]$

Display table showing the best fractional polynomial model for each degree

fracpoly, <u>com</u>pare

Create variables containing fractional polynomial powers

fracgen *varname* # $\left[\# \ ... \right] \left[\text{if} \right] \left[\text{in} \right] \left[\, , \text{ fracgen\_options} \right]$

fracpoly_options	Description
Model	
<u>degree</u>(#)	degree of fractional polynomial to fit; default is degree(2)
Model 2	
<u>nosca</u>ling	suppress scaling of first independent variable
<u>nocon</u>stant	suppress constant term
<u>pow</u>ers(*numlist*)	list of fractional polynomial powers from which models are chosen
<u>cent</u>er(*cent_list*)	specification of centering for the independent variables
all	include out-of-sample observations in generated variables
Reporting	
log	display iteration log
<u>com</u>pare	compare models by degree
display_options	control column formats and line width

regression_cmd_options	Description
Model 2	
regression_cmd_options	options appropriate to the regression command in use

All weight types supported by *regression_cmd* are allowed; see [U] **11.1.6 weight**.
See [U] **20 Estimation and postestimation commands** for more capabilities of estimation commands.

543

where

> *cent_list* is a comma-separated list with elements *varlist*: {mean | # | no}, except that the first element may optionally be of the form {mean | # | no} to specify the default for all variables.

> *regression_cmd* may be clogit, glm, intreg, logistic, logit, mlogit, nbreg, ologit, oprobit, poisson, probit, qreg, regress, rreg, stcox, stcrreg, streg, or xtgee.

fracgen_options	Description
Main	
center(no | mean | #)	center *varname* as specified; default is center(no)
noscaling	suppress scaling of *varname*
restrict([*varname*] [*if*])	compute centering and scaling using specified subsample
replace	replace variables if they exist

Menu

fracpoly

Statistics > Linear models and related > Fractional polynomials > Fractional polynomial regression

fracgen

Statistics > Linear models and related > Fractional polynomials > Create fractional polynomial powers

Description

fracpoly fits fractional polynomials (FPs) in $xvar_1$ as part of the specified regression model. After execution, fracpoly leaves variables in the dataset named I$xvar$__1, I$xvar$__2, ..., where *xvar* represents the first four letters of the name of $xvar_1$. The new variables contain the best-fitting FP powers of $xvar_1$.

Covariates other than $xvar_1$, which are optional, are specified in $xvar_2$, ..., and *xvarlist*. They may be modeled linearly and with specified FP transformations. Fractional polynomial powers are specified by typing numbers after the variable's name. A variable name typed without numbers is entered linearly.

fracgen creates new variables named *varname*_1, *varname*_2, ..., containing FP powers of *varname* by using the powers (# [#...]) specified.

See [R] **fracpoly postestimation** for information on fracplot and fracpred.

See [R] **mfp** for multivariable FP model fitting.

Options for fracpoly

_____| Model |_____

degree(#) determines the degree of FP to be fit. The default is degree(2), that is, a model with two power terms.

_____| Model 2 |_____

noscaling suppresses scaling of $xvar_1$ and its powers.

noconstant suppresses the regression constant if this is permitted by *regression_cmd*.

powers(*numlist*) is the set of FP powers from which models are to be chosen. The default is powers(-2,-1,-.5,0,.5,1,2,3) (0 means log).

center(*cent_list*) defines the centering for the covariates *xvar*$_1$, *xvar*$_2$, ..., *xvarlist*. The default is center(mean). A typical item in *cent_list* is *varlist*:{mean | # | no}. Items are separated by commas. The first item is special because *varlist*: is optional, and if omitted, the default is (re)set to the specified value (mean or # or no). For example, center(no, age:mean) sets the default to no and sets the centering for age to mean.

regression_cmd_options are options appropriate to the regression command in use. For example, for stcox, *regression_cmd_options* may include efron or some alternate method for handling tied failures.

all includes out-of-sample observations when generating the best-fitting FP powers of *xvar*$_1$, *xvar*$_2$, etc. By default, the generated FP variables contain missing values outside the estimation sample.

___

| Reporting |

log displays deviances and (for regress) residual standard deviations for each FP model fit.

compare reports a closed-test comparison between FP models.

display_options: cformat(%*fmt*), pformat(%*fmt*), sformat(%*fmt*), and nolstretch; see [R] **estimation options**.

Options for fracgen

| Main |

center(no | mean | #) specifies whether *varname* is to be centered; the default is center(no).

noscaling suppresses scaling of *varname*.

restrict([*varname*] [*if*]) specifies that centering and scaling be computed using the subsample identified by *varname* and if.

The subsample is defined by the observations for which *varname* $\neq$ 0 that also meet the if conditions. Typically, *varname* = 1 defines the subsample and *varname* = 0 indicates observations not belonging to the subsample. For observations whose subsample status is uncertain, *varname* should be set to a missing value; such observations are dropped from the subsample.

By default, fracgen computes the centering and scaling by using the sample of observations identified in the [*if*] [*in*] options. The restrict() option identifies a subset of this sample.

replace specifies that any existing variables named *varname*_1, *varname*_2, ... may be replaced.

Remarks

Remarks are presented under the following headings:

> *Introduction*
> *fracpoly*
> *Centering*
> *Output with the compare option*
> *fracgen*
> *Models with several continuous covariates*
> *Examples*

Introduction

Regression models based on FP functions of a continuous covariate are described by Royston and Altman (1994b). Detailed examples using an earlier and rather more complex version of this set of commands are presented by Royston and Altman (1994a).

FPs increase the flexibility afforded by the family of conventional polynomial models. Although polynomials are popular in data analysis, linear and quadratic functions are severely limited in their range of curve shapes, whereas cubic and higher-order curves often produce undesirable artifacts, such as edge effects and waves.

A polynomial of degree m may be written as

$$\beta_0 + \beta_1 x + \beta_2 x^2 + \cdots + \beta_m x^m$$

whereas FP of degree m has m integer and/or fractional powers $p_1 < \cdots < p_m$,

$$\beta_0 + \beta_1 x^{(p_1)} + \beta_2 x^{(p_2)} + \cdots + \beta_m x^{(p_m)}$$

where for a power, p,

$$x^{(p)} = \begin{cases} x^p & \text{if } p \neq 0 \\ \log x & \text{if } p = 0 \end{cases}$$

x must be positive. An FP of first degree ($m = 1$) involves one power or log transformation of x.

This family of FP functions may be extended in a mathematically natural way to include repeated powers. An FP of degree m with exactly m repeated powers of p is defined as

$$\beta_0 + \beta_1 x^{(p)} + \beta_2 x^{(p)} \log x + \cdots + \beta_m x^{(p)} (\log x)^{m-1}$$

For example, an FP of second degree ($m = 2$) with repeated powers of 0.5 is

$$\beta_0 + \beta_1 x^{0.5} + \beta_2 x^{0.5} \log x$$

A general FP may include some unique and some repeated powers. For example, one with powers $(-1, 1, 3, 3)$ is

$$\beta_0 + \beta_1 x^{-1} + \beta_2 x + \beta_3 x^3 + \beta_4 x^3 \log x$$

The permitted powers are restricted to the set $\{-2, -1, -0.5, 0, 0.5, 1, 2, 3\}$ because our experience using FPs in data analysis indicates that including extra powers in the set is not often worthwhile.

Now we consider using FPs in regression modeling. If the values of the powers $p_1, \ldots, p_m$ were known, the FP would resemble a conventional multiple linear regression model with coefficients $\beta_0, \beta_1, \ldots, \beta_m$. However, the powers are not (usually) known and must be estimated, together with the coefficients, from the data. Estimation involves a systematic search for the best power or combination of powers from the permitted set. For each possible combination, a linear regression model as just described is fit, and the corresponding deviance (defined as minus twice the log likelihood) is noted. The model with the lowest deviance is deemed to have the best fit, and the corresponding powers and regression coefficients constitute the final FP model. In practice, $m = 2$ is often sufficient (Royston and Sauerbrei 2008, 76).

fracpoly

fracpoly finds and reports a multiple regression model comprising the best-fitting powers of *xvar*$_1$ together with other covariates specified by *xvar*$_2$, . . . , *xvarlist*. The model that is fit depends on the type of *regression_cmd* used.

The regression output for the best-fitting model may be reproduced by typing *regression_cmd* without variables or options. predict, test, etc., may be used after fracpoly; the results will depend on *regression_cmd*.

The standard errors of the fitted values (as estimated after use of fracpoly by using predict or fracpred with the stdp option) are somewhat too low because no allowance has been made for the estimation of the powers.

If *xvar*$_1$ has any negative or zero values, fracpoly subtracts the minimum of *xvar* from *xvar* and then adds the rounding (or counting) interval. The interval is defined as the smallest positive difference between the ordered values of *xvar*. After this change of origin, the minimum value of *xvar*$_1$ is positive, so FPs (which require *xvar*$_1 > 0$) can be used. Unless the noscaling option is used, fracpoly scales the resulting variable by a power of 10 calculated from the data. The scaling is designed to improve numerical stability when fitting FP models.

After execution, fracpoly leaves in the dataset variables named I*xvar*__1, I*xvar*__2, . . . , which are the best-fitting FP powers of *xvar*$_1$ (calculated, if necessary, after a change in origin and scale as just described, and if centering is specified, with a constant added to or subtracted from the values after FP transformation). Other variables, whose names follow the same convention, are left in the dataset if *xvar*$_2$ has been specified.

Centering

As discussed by Garrett (1995, 1998), covariate centering is a sensible, indeed often essential, step when reporting and interpreting the results of multiple regression models. For this and other reasons, centering has been introduced as the default option in fracpoly. As written, the familiar straight-line regression function $E[y|x] = \beta_0 + \beta_1 x$ is "centered" to 0 in that $\beta_0 = E[y|0]$. This is fine if $x = 0$ is a sensible base point. However, the sample values of x may not even encompass 0 (this is usually the case when FP models are contemplated). Then β_0 is a meaningless intercept, and the standard error of its estimate $\widehat{\beta}_0$ will be large. For FP model $E[y|x] = \beta_0 + \beta_1 x^{(p)}$, the point $x^{(p)} = 0$ may even correspond to $x = \infty$ (consider $p < 0$). The scheme adopted by fracpoly is to center on the mean of x. For example, for the FP $E[y|x] = \beta_0 + \beta_1 x^p + \beta_1 x^q$, fracpoly actually fits the model

$$E[y|x] = \beta_0 + \beta_1 \left(x^p - \overline{x}^p\right) + \beta_2 \left(x^q - \overline{x}^q\right)$$

where $\overline{x}$ is the sample mean of the x values and $E[y|\overline{x}] = \beta_0$, giving β_0 a respectable interpretation as the predicted value of y at the mean of x. This approach has the advantage that plots of the fitted values and 95% confidence intervals for $E[y|x]$ as a function of x, even within a multiple regression model, are always sensible (provided that the other predictors are suitably centered—otherwise, the confidence limits can be alarmingly wide).

Sometimes centering on the mean is not appropriate, an example being a binary covariate where often you will want to center on the lower value, usually 0 (that is, not center). You should then use the center() option to override the default. An example is center(x1:mean,x2-x5:no,x6:1).

Output with the compare option

If the compare option is used, fracpoly displays a table showing the best FP model for each degree $k < m$ (including the model without x and the model linear in x). Deviance differences between each FP model and the degree m model are also reported along with the corresponding p-values (Royston and Altman 1994b; Royston and Sauerbrei 2008).

The compare option implements a closed-test approach to selecting an FP model. It has the advantage of preserving the type I error probability at a nominal value. For example, suppose a nominal 5% significance level was chosen, and the test of FP2 versus the null model (that is, omitting x) was not significant. No further tests among FP models would then be done, and x would be considered nonsignificant, regardless of the results of any further model comparisons.

fracgen

The basic syntax of fracgen is

> fracgen *varname* # $\big[$ # ... $\big]$

Each power (represented by # in the syntax diagram) should be separated by a space. fracgen creates new variables called *varname*_1, *varname*_2, etc. Each variable is labeled according to its power, preliminary linear transformation, and centering, if applied.

Positive or negative powers of *varname* are defined in the usual way. A power of zero is interpreted as log.

Models with several continuous covariates

fracpoly estimates powers for FP models in just one continuous covariate (*xvar*$_1$), though other covariates of any kind (*xvar*$_2$, ..., *xvarlist*) may be included as linear or predefined FP terms. An algorithm was suggested by Royston and Altman (1994b) for the joint estimation of FP models in several continuous covariates. It was later refined by Sauerbrei and Royston (1999) and is implemented in the Stata command mfp. See [R] **mfp** as well as Royston and Ambler (1998) and Royston and Sauerbrei (2008).

Examples

▷ Example 1

Consider the serum immunoglobulin G (IgG) dataset from Isaacs et al. (1983), which consists of 298 independent observations in young children. The dependent variable sqrtigg is the square root of the IgG concentration, and the independent variable age is the age of each child. (Preliminary Box–Cox analysis shows that a square root transformation removes the skewness in IgG.) The aim is to find a model that accurately predicts the mean of sqrtigg given age. We use fracpoly to find the best FP model of degree 2 (the default option) and graph the resulting fit and 95% confidence interval:

```
. use http://www.stata-press.com/data/r12/igg
(Immunoglobulin in children)

. fracpoly: regress sqrtigg age
........
-> gen double Iage__1 = age^-2-.1299486216 if e(sample)
-> gen double Iage__2 = age^2-7.695349038 if e(sample)
```

Source	SS	df	MS			
Model	22.2846976	2	11.1423488			
Residual	50.9676492	295	.172771692			
Total	73.2523469	297	.246640898			

```
                                        Number of obs =      298
                                        F(  2,   295) =    64.49
                                        Prob > F      =   0.0000
                                        R-squared     =   0.3042
                                        Adj R-squared =   0.2995
                                        Root MSE      =   .41566
```

sqrtigg	Coef.	Std. Err.	t	P>|t|	[95% Conf. Interval]	
Iage__1	-.1562156	.027416	-5.70	0.000	-.2101713	-.10226
Iage__2	.0148405	.0027767	5.34	0.000	.0093757	.0203052
_cons	2.283145	.0305739	74.68	0.000	2.222974	2.343315

```
Deviance:   319.45. Best powers of age among 44 models fit: -2 2.

. fracplot age, msize(small)
```

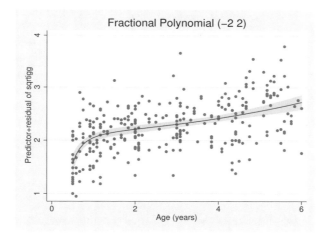

The new variables created by fracpoly contain the best-fitting FP powers of age, as centered by fracpoly. For example, Iage__1 is centered by subtracting the mean of age raised to the power -2. In general, the variables created by fracpoly are centered and possibly scaled, which is reflected in the estimated regression coefficients and intercept. Centering does have its advantages (see the *Centering* section earlier in this entry); however, sometimes you may want estimation for uncentered variables. To obtain regression results for uncentered and unscaled FP variables, specify options center(no) and noscaling to fracpoly. For a more detailed discussion, see Royston and Sauerbrei (2008, sec. 4.11).

The fitted curve has an asymmetric S shape. This model has powers $(-2, 2)$ and deviance 319.45. As many as 44 models have been quietly fit in the search for the best powers. Now let's look at models of degree ≤ 4:

```
. fracpoly, degree(4) compare: regress sqrtigg age
..............................................................................
> ..............................................................................
> ........
-> gen double Iage__1 = ln(age)-1.020308063 if e(sample)
-> gen double Iage__2 = age^3-21.34727694 if e(sample)
-> gen double Iage__3 = age^3*ln(age)-21.78079878 if e(sample)
-> gen double Iage__4 = age^3*ln(age)^2-22.22312461 if e(sample)
```

Source	SS	df	MS		Number of obs =	298
					F(4, 293) =	32.63
Model	22.5754541	4	5.64386353		Prob > F =	0.0000
Residual	50.6768927	293	.172958678		R-squared =	0.3082
					Adj R-squared =	0.2987
Total	73.2523469	297	.246640898		Root MSE =	.41588

sqrtigg	Coef.	Std. Err.	t	P>|t|	[95% Conf. Interval]	
Iage__1	.8761824	.1898721	4.61	0.000	.5024962	1.249869
Iage__2	-.1922029	.0684934	-2.81	0.005	-.3270044	-.0574015
Iage__3	.2043794	.074947	2.73	0.007	.0568767	.3518821
Iage__4	-.0560067	.0212969	-2.63	0.009	-.097921	-.0140924
_cons	2.238735	.0482705	46.38	0.000	2.143734	2.333736

```
Deviance:   317.74. Best powers of age among 494 models fit: 0 3 3 3.
Fractional polynomial model comparisons:
```

age	df	Deviance	Res. SD	Dev. dif.	P (*)	Powers
Not in model	0	427.539	.49663	109.795	0.000	
Linear	1	337.561	.42776	19.818	0.006	1
m = 1	2	327.436	.420554	9.692	0.140	0
m = 2	4	319.448	.415658	1.705	0.794	-2 2
m = 3	6	319.275	.416243	1.532	0.473	-2 1 1
m = 4	8	317.744	.415883	—	—	0 3 3 3

```
(*) P-value from deviance difference comparing reported model with m = 4 model
```

It appears that the degree 4 FP model is not significantly different from the other FP models (at the 5% level).

Let's compare the curve shape from the $m = 2$ model with that from a conventional quartic polynomial, whose fit turns out to be significantly better than a cubic (not shown). We use the ability of `fracpoly` both to generate the required powers of `age`, namely, $(1, 2, 3, 4)$ for the quartic and $(-2, 2)$ for the second-degree FP, and to fit the model. We fit both models and graph the resulting curves:

```
. fracpoly: regress sqrtigg age 1 2 3 4
-> gen double Iage__1 = age-2.774049213 if e(sample)
-> gen double Iage__2 = age^2-7.695349038 if e(sample)
-> gen double Iage__3 = age^3-21.34727694 if e(sample)
-> gen double Iage__4 = age^4-59.21839681 if e(sample)
```

Source	SS	df	MS		
Model	22.5835458	4	5.64588646		
Residual	50.668801	293	.172931061		
Total	73.2523469	297	.246640898		

```
Number of obs =     298
F(  4,  293) =   32.65
Prob > F      =  0.0000
R-squared     =  0.3083
Adj R-squared =  0.2989
Root MSE      =  .41585
```

sqrtigg	Coef.	Std. Err.	t	P>|t|	[95% Conf. Interval]	
Iage__1	2.047831	.4595962	4.46	0.000	1.143302	2.952359
Iage__2	-1.058902	.2822803	-3.75	0.000	-1.614456	-.5033479
Iage__3	.2284917	.0667591	3.42	0.001	.0971037	.3598798
Iage__4	-.0168534	.0053321	-3.16	0.002	-.0273475	-.0063594
_cons	2.240012	.0480157	46.65	0.000	2.145512	2.334511

```
Deviance:  317.70.

. predict fit1
(option xb assumed; fitted values)

. fracpoly: regress sqrtigg age -2 2
-> gen double Iage__1 = age^-2-.1299486216 if e(sample)
-> gen double Iage__2 = age^2-7.695349038 if e(sample)
```

Source	SS	df	MS		
Model	22.2846976	2	11.1423488		
Residual	50.9676492	295	.172771692		
Total	73.2523469	297	.246640898		

```
Number of obs =     298
F(  2,  295) =   64.49
Prob > F      =  0.0000
R-squared     =  0.3042
Adj R-squared =  0.2995
Root MSE      =  .41566
```

sqrtigg	Coef.	Std. Err.	t	P>|t|	[95% Conf. Interval]	
Iage__1	-.1562156	.027416	-5.70	0.000	-.2101713	-.10226
Iage__2	.0148405	.0027767	5.34	0.000	.0093757	.0203052
_cons	2.283145	.0305739	74.68	0.000	2.222974	2.343315

```
Deviance:  319.45.

. predict fit2
(option xb assumed; fitted values)
```

```
. scatter sqrtigg fit1 fit2 age, c(. l l) m(o i i) msize(small)
> clpattern(. -_.) ytitle("Square root of IgG") xtitle("Age, years")
```

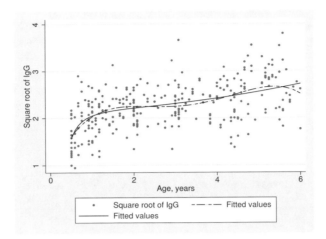

The quartic curve has an unsatisfactory wavy appearance that is implausible for the known behavior of IgG, the serum level of which increases throughout early life. The FP curve increases monotonically and is therefore biologically the more plausible curve. The two models have approximately the same deviance.

◁

▷ Example 2

Data from Smith et al. (1992) contain times to complete healing of leg ulcers in a randomized controlled clinical trial of two treatments in 192 elderly patients. Several covariates were available, of which an important one is mthson, the number of months since the recorded onset of the ulcer. Because the response variable is time to an event of interest and some (in fact, about one-half) of the times are censored, using Cox regression to analyze the data is appropriate. We consider FPs in mthson, adjusting for four other covariates: age; ulcarea, the area of tissue initially affected by the ulcer; deepppg, a binary variable indicating the presence or absence of deep vein involvement; and treat, a binary variable indicating treatment type. We fit FPs of degree 1 and 2:

```
. use http://www.stata-press.com/data/r12/legulcer, clear
(Leg ulcer clinical trial)
. stset ttevent, fail(cens)
  (output omitted )
```

```
. fracpoly, compare: stcox mthson age ulcarea deepppg treat, nohr
-> gen double Iage__1 = age-73.453125 if e(sample)
-> gen double Iulca__1 = ulcarea-1326.203125 if e(sample)
-> gen double Itrea__1 = treat-1 if e(sample)
........
-> gen double Imths__1 = X^.5-.4930242557 if e(sample)
-> gen double Imths__2 = X^.5*ln(X)+.6973304564 if e(sample)
   (where: X = (mthson+1)/100)
          failure _d:  censored
    analysis time _t:  ttevent

Iteration 0:    log likelihood = -422.65089
Iteration 1:    log likelihood = -390.49313
Iteration 2:    log likelihood = -383.44258
Iteration 3:    log likelihood = -374.28707
Iteration 4:    log likelihood = -369.31417
Iteration 5:    log likelihood = -368.38104
Iteration 6:    log likelihood = -368.35448
Iteration 7:    log likelihood = -368.35446
Refining estimates:
Iteration 0:    log likelihood = -368.35446

Cox regression -- Breslow method for ties

No. of subjects =          192             Number of obs    =         192
No. of failures =           92
Time at risk    =        13825
                                           LR chi2(6)       =      108.59
Log likelihood  =    -368.35446            Prob > chi2      =      0.0000
```

_t	Coef.	Std. Err.	z	P>|z|	[95% Conf. Interval]	
Imths__1	-2.81425	.6996385	-4.02	0.000	-4.185516	-1.442984
Imths__2	1.541451	.4703143	3.28	0.001	.6196521	2.46325
Iage__1	-.0261111	.0087983	-2.97	0.003	-.0433556	-.0088667
Iulca__1	-.0017491	.000359	-4.87	0.000	-.0024527	-.0010455
deepppg	-.5850499	.2163173	-2.70	0.007	-1.009024	-.1610758
Itrea__1	-.1624663	.2171048	-0.75	0.454	-.5879838	.2630513

```
Deviance:   736.71. Best powers of mthson among 44 models fit: .5 .5.
Fractional polynomial model comparisons:
```

mthson	df	Deviance	Dev. dif.	P (*)	Powers
Not in model	0	754.345	17.636	0.001	
Linear	1	751.680	14.971	0.002	1
m = 1	2	738.969	2.260	0.323	-.5
m = 2	4	736.709	—	—	.5 .5

```
(*) P-value from deviance difference comparing reported model with m = 2 model
```

The best-fit FP of degree 2 has powers $(0.5, 0.5)$ and deviance 736.71. However, this model does not fit significantly better than the FP of degree 1 (at the 5% level), which has power -0.5 and deviance 738.97. We prefer the model with $m = 1$, for which the partial linear predictor is shown on the next page.

```
. quietly fracpoly, degree(1): stcox mthson age ulcarea deepppg treat, nohr
. fracplot, ytitle(Partial linear predictor) m(i) ciopts(bcolor(white))
```

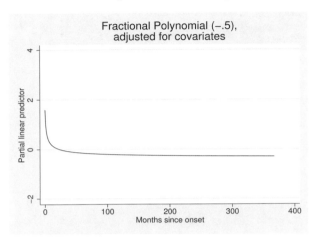

The hazard for healing is much higher for patients whose ulcer is of recent onset than for those who have had an ulcer for many months.

`fracpoly` has automatically centered the predictors on their mean values, but because in Cox regression there is no constant term, we cannot see the effects of centering in the table of regression estimates. The effects would be present if we were to graph the baseline hazard or survival function because these functions are defined with all predictors set equal to 0.

A more appropriate analysis of this dataset, if one wished to model all the predictors, possibly with FP functions, would be to use `mfp`; see [R] **mfp**.

◁

Saved results

In addition to what *regression_cmd* saves, `fracpoly` saves the following in e():

Scalars
e(fp_N)	number of nonmissing observations
e(fp_dev)	deviance for FP model of degree m
e(fp_df)	FP model degrees of freedom
e(fp_d0)	deviance for model without $xvar_1$
e(fp_s0)	residual SD for model without $xvar_1$
e(fp_dlin)	deviance for model linear in $xvar_1$
e(fp_slin)	residual SD for model linear in $xvar_1$
e(fp_d1), e(fp_d2), ...	deviances for FP models of degree $1,2,...,m$
e(fp_s1), e(fp_s2), ...	residual SDs for FP models of degree $1,2,...,m$

Macros
e(fp_cmd)	fracpoly
e(cmdline)	command as typed
e(fp_depv)	$yvar_1$ ($yvar_2$)
e(fp_rhs)	$xvar_1$
e(fp_base)	variables in $xvar_2$, ..., $xvarlist$ after centering and FP transformation
e(fp_xp)	I$xvar\_\_1$, I$xvar\_\_2$, etc.
e(fp_fvl)	variables in model finally estimated
e(fp_wgt)	weight type or ""
e(fp_wexp)	weight expression if 'e(fp_wgt)' != ""
e(fp_pwrs)	powers for FP model of degree m
e(fp_x1), e(fp_x2), ...	$xvar_1$ and variables in model
e(fp_k1), e(fp_k2), ...	powers for FP models of degree $1,2,...,m$

Residual SDs are stored only when *regression_cmd* is `regress`.

Methods and formulas

`fracpoly` and `fracgen` are implemented as ado-files.

The general definition of an FP, accommodating possible repeated powers, may be written for functions $H_1(x), \ldots, H_m(x)$ as

$$\beta_0 + \sum_{j=1}^{m} \beta_j H_j(x)$$

where $H_1(x) = x^{(p_1)}$ and for $j = 2, \ldots, m$,

$$H_j(x) = \begin{cases} x^{(p_j)} & \text{if } p_j \neq p_{j-1} \\ H_{j-1}(x) \log x & \text{if } p_j = p_{j-1} \end{cases}$$

For example, an FP of degree 3 with powers $(1, 3, 3)$ has $H_1(x) = x$, $H_2(x) = x^3$, and $H_3(x) = x^3 \log x$ and equals $\beta_0 + \beta_1 x + \beta_2 x^3 + \beta_3 x^3 \log x$.

An FP model of degree m is taken to have $2m + 1$ degrees of freedom (df): one for β_0 and one for each β_j and its associated power. Because the powers in an FP are chosen from a finite set rather than from the entire real line, the df defined in this way are approximate.

The deviance D of a model is defined as -2 times its maximized log likelihood. For normal-errors models, we use the formula

$$D = n \left(1 - \bar{l} + \log \frac{2\pi \text{RSS}}{n} \right)$$

where n is the sample size, $\bar{l}$ is the mean of the lognormalized weights ($\bar{l} = 0$ if the weights are all equal), and RSS is the residual sum of squares as fit by `regress`.

The `compare` option causes `fracpoly` to report a table comparing FP models of degree $k < m$ to the degree m FP model. Suppose that we are comparing FP regression models with degrees k and m. The p-values reported by `compare` are calculated differently for normal and nonnormal regressions. Let D_k and D_m be the deviances of the models with degrees k and m, respectively. For normal-errors models such as `regress`, a variance ratio F is calculated as

$$F = \frac{n_2}{n_1}\left\{\exp\left(\frac{D_k - D_m}{n}\right) - 1\right\}$$

where n_1 is the numerator df (here, $2m - 2k$) and n_2 is the denominator df (equal to rdf $- 2m$, where rdf is the residual df for the regression model involving only the covariates in $xvar_2$, if any, but not x). The p-value is obtained by referring F to an F distribution on $(2, \text{rdf})$ df.

For nonnormal models (`clogit`, `glm`, `logistic`, ... `stcox`, or `streg`; not `regress`), the p-value is obtained by referring $D_k - D_m$ to a χ^2 distribution on $2m - 2k$ df. These p-values for comparing models are approximate and are typically somewhat conservative (Royston and Altman 1994b).

The component-plus-residual values graphed by `fracplot` are calculated as follows: let the data consist of triplets $(y_i, x_i, \mathbf{z}_i)$, $i = 1, \ldots, n$, where $\mathbf{z}_i$ is the vector of covariates for the ith observation, after applying possible FP transformation and centering as described earlier. Let $\widehat{\eta}_i = \widehat{\beta}_0 + \{\mathbf{H}(x_i) - \mathbf{H}(x_0)\}'\widehat{\boldsymbol{\beta}} + \mathbf{z}_i'\widehat{\boldsymbol{\gamma}}$ be the linear predictor from the FP model, as given by the `fracpred` command, or equivalently, by the `predict` command with the `xb` option, after the use of `fracpoly`. Here $\mathbf{H}(x_i) = \{H_1(x_i), \ldots, H_m(x_i)\}'$ is the vector of FP functions described above, $\mathbf{H}(x_0) = \{H_1(x_0), \ldots, H_m(x_0)\}'$ is the vector of centering to x_0 (x_0 is often chosen to be the mean of the x_i), $\widehat{\boldsymbol{\beta}}$ is the estimated parameter vector, and $\widehat{\boldsymbol{\gamma}}$ is the estimated parameter vector for the covariates. The values $\widehat{\eta}_i^* = \widehat{\beta}_0 + \{\mathbf{H}(x_i) - \mathbf{H}(x_0)\}'\widehat{\boldsymbol{\beta}}$ represent the behavior of the FP model for x at fixed values $\mathbf{z} = \mathbf{0}$ of the (centered) covariates. The ith component-plus-residual is defined as $\widehat{\eta}_i^* + d_i$, where d_i is the deviance residual for the ith observation. For normal-errors models, $d_i = \sqrt{w_i}(y_i - \widehat{\eta}_i)$, where w_i is the case weight (or 1, if *weight* is not specified). For logistic, Cox, and generalized linear regression models, see [R] **logistic**, [R] **probit**, [ST] **stcox**, and [R] **glm**, respectively, for the formula for d_i. The formula for `poisson` models is the same as that for `glm` with `family(poisson)`. For `stcox`, d_i is the partial martingale residual (see [ST] **stcox postestimation**).

Acknowledgment

`fracpoly` and `fracgen` were written by Patrick Royston, MRC Clinical Trials Unit, London.

References

Becketti, S. 1995. sg26.2: Calculating and graphing fractional polynomials. *Stata Technical Bulletin* 24: 14–16. Reprinted in *Stata Technical Bulletin Reprints*, vol. 4, pp. 129–132. College Station, TX: Stata Press.

Garrett, J. M. 1995. sg33: Calculation of adjusted means and adjusted proportions. *Stata Technical Bulletin* 24: 22–25. Reprinted in *Stata Technical Bulletin Reprints*, vol. 4, pp. 161–165. College Station, TX: Stata Press.

——. 1998. sg33.1: Enhancements for calculation of adjusted means and adjusted proportions. *Stata Technical Bulletin* 43: 16–24. Reprinted in *Stata Technical Bulletin Reprints*, vol. 8, pp. 111–123. College Station, TX: Stata Press.

Isaacs, D., D. G. Altman, C. E. Tidmarsh, H. B. Valman, and A. D. Webster. 1983. Serum immunoglobulin concentrations in preschool children measured by laser nephelometry: Reference ranges for IgG, IgA, IgM. *Journal of Clinical Pathology* 36: 1193–1196.

Royston, P. 1995. sg26.3: Fractional polynomial utilities. *Stata Technical Bulletin* 25: 9–13. Reprinted in *Stata Technical Bulletin Reprints*, vol. 5, pp. 82–87. College Station, TX: Stata Press.

Royston, P., and D. G. Altman. 1994a. sg26: Using fractional polynomials to model curved regression relationships. *Stata Technical Bulletin* 21: 11–23. Reprinted in *Stata Technical Bulletin Reprints*, vol. 4, pp. 110–128. College Station, TX: Stata Press.

———. 1994b. Regression using fractional polynomials of continuous covariates: Parsimonious parametric modelling. *Applied Statistics* 43: 429–467.

Royston, P., and G. Ambler. 1998. sg81: Multivariable fractional polynomials. *Stata Technical Bulletin* 43: 24–32. Reprinted in *Stata Technical Bulletin Reprints*, vol. 8, pp. 123–132. College Station, TX: Stata Press.

———. 1999a. sg112: Nonlinear regression models involving power or exponential functions of covariates. *Stata Technical Bulletin* 49: 25–30. Reprinted in *Stata Technical Bulletin Reprints*, vol. 9, pp. 173–179. College Station, TX: Stata Press.

———. 1999b. sg81.1: Multivariable fractional polynomials: Update. *Stata Technical Bulletin* 49: 17–23. Reprinted in *Stata Technical Bulletin Reprints*, vol. 9, pp. 161–168. College Station, TX: Stata Press.

———. 1999c. sg112.1: Nonlinear regression models involving power or exponential functions of covariates: Update. *Stata Technical Bulletin* 50: 26. Reprinted in *Stata Technical Bulletin Reprints*, vol. 9, p. 180. College Station, TX: Stata Press.

———. 1999d. sg81.2: Multivariable fractional polynomials: Update. *Stata Technical Bulletin* 50: 25. Reprinted in *Stata Technical Bulletin Reprints*, vol. 9, p. 168. College Station, TX: Stata Press.

Royston, P., and W. Sauerbrei. 2008. *Multivariable Model-building: A Pragmatic Approach to Regression Analysis Based on Fractional Polynomials for Modelling Continuous Variables*. Chichester, UK: Wiley.

Sauerbrei, W., and P. Royston. 1999. Building multivariable prognostic and diagnostic models: Transformation of the predictors by using fractional polynomials. *Journal of the Royal Statistical Society, Series A* 162: 71–94.

Smith, J. M., C. J. Dore, A. Charlett, and J. D. Lewis. 1992. A randomized trial of Biofilm dressing for venous leg ulcers. *Phlebology* 7: 108–113.

Also see

Title

---
fracpoly postestimation — Postestimation tools for fracpoly

Description

The following postestimation commands are of special interest after `fracpoly`:

Command	Description
fracplot	plot data and fit from most recently fit fractional polynomial model
fracpred	create variable containing prediction, deviance residuals, or SEs of fitted values

For information about these commands, see below.

The following standard postestimation commands are also available if available after *regression_cmd*:

Command	Description
estat	AIC, BIC, VCE, and estimation sample summary
estimates	cataloging estimation results
lincom	point estimates, standard errors, testing, and inference for linear combinations of coefficients
linktest	link test for model specification
lrtest	likelihood-ratio test
nlcom	point estimates, standard errors, testing, and inference for nonlinear combinations of coefficients
predict	predictions, residuals, influence statistics, and other diagnostic measures
predictnl	point estimates, standard errors, testing, and inference for generalized predictions
test	Wald tests of simple and composite linear hypotheses
testnl	Wald tests of nonlinear hypotheses

See the corresponding entries in the *Base Reference Manual* for details.

Special-interest postestimation commands

`fracplot` plots the data and fit, with 95% confidence limits, from the most recently fit fractional polynomial (FP) model. The data and fit are plotted against *varname*, which may be $xvar_1$ or another of the covariates ($xvar_2, \ldots$, or a variable from *xvarlist*). If *varname* is not specified, $xvar_1$ is assumed.

`fracpred` creates *newvar* containing the fitted index or deviance residuals for the whole model, or the fitted index or its standard error for *varname*, which may be $xvar_1$ or another covariate.

Syntax for predict

The behavior of `predict` following `fracpoly` is determined by *regression_cmd*. See the corresponding *regression_cmd* **postestimation** entry for available `predict` options.

Also see information on `fracpred` below.

Syntax for fracplot and fracpred

Plot data and fit from most recently fit fractional polynomial model

  `fracplot` [*varname*] [*if*] [*in*] [, *fracplot_options*]

Create variable containing the prediction, deviance residuals, or SEs of fitted values

  `fracpred` *newvar* [, *fracpred_options*]

fracplot_options	Description
Plot	
marker_options	change look of markers (color, size, etc.)
marker_label_options	add marker labels; change look or position
Fitted line	
`lineopts`(*cline_options*)	affect rendition of the fitted line
CI plot	
`ciopts`(*area_options*)	affect rendition of the confidence bands
Add plots	
`addplot`(*plot*)	add other plots to the generated graph
Y axis, X axis, Titles, Legend, Overall	
twoway_options	any options other than `by()` documented in [G-3] ***twoway_options***

fracpred_options	Description
<u>f</u>or(*varname*)	compute prediction for *varname*
<u>d</u>resid	compute deviance residuals
<u>s</u>tdp	compute standard errors of the fitted values *varname*

`fracplot` is not allowed after `fracpoly` or `mfp` with `clogit`, `mlogit`, or `stcrreg`. `fracpred`, `dresid` is not allowed after `fracpoly` or `mfp` with `clogit`, `mlogit`, or `stcrreg`.

Menu

fracplot

Statistics > Linear models and related > Fractional polynomials > Fractional polynomial regression plot

fracpred

Statistics > Linear models and related > Fractional polynomials > Fractional polynomial prediction

Options for fracplot

⌐ Plot ⌐

marker_options affect the rendition of markers drawn at the plotted points, including their shape, size, color, and outline; see [G-3] **marker_options**.

marker_label_options specify if and how the markers are to be labeled; see [G-3] **marker_label_options**.

⌐ Fitted line ⌐

lineopts(*cline_options*) affect the rendition of the fitted line; see [G-3] **cline_options**.

⌐ CI plot ⌐

ciopts(*area_options*) affect the rendition of the confidence bands; see [G-3] **area_options**.

⌐ Add plots ⌐

addplot(*plot*) provides a way to add other plots to the generated graph. See [G-3] **addplot_option**.

⌐ Y axis, X axis, Titles, Legend, Overall ⌐

twoway_options are any of the options documented in [G-3] **twoway_options**, excluding by(). These include options for titling the graph (see [G-3] **title_options**) and for saving the graph to disk (see [G-3] **saving_option**).

Options for fracpred

for(*varname*) specifies (partial) prediction for variable *varname*. The fitted values are adjusted to the value specified by the center() option in fracpoly.

dresid specifies that deviance residuals be calculated.

stdp specifies calculation of the standard errors of the fitted values *varname*, adjusted for all the other predictors at the values specified by center().

Remarks

fracplot actually produces a component-plus-residual plot. For normal-error models with constant weights and one covariate, this amounts to a plot of the observations with the fitted line inscribed. For other normal-error models, weighted residuals are calculated and added to the fitted values.

For models with additional covariates, the line is the partial linear predictor for the variable in question (*xvar*$_1$ or a covariate) and includes the intercept β_0.

For generalized linear and Cox models, the fitted values are plotted on the scale of the "index" (linear predictor). Deviance residuals are added to the (partial) linear predictor to give component-plus-residual values. These values are plotted as small circles.

See [R] **fracpoly** for examples using `fracplot`.

Methods and formulas

All postestimation commands listed above and `fracplot` and `fracpred` are implemented as ado-files.

See *Methods and formulas* in [R] **fracpoly** for notation.

The component-plus-residual values graphed by `fracplot` are calculated as follows: Let the data consist of triplets $(y_i, x_i, \mathbf{z}_i)$, $i = 1, \ldots, n$, where $\mathbf{z}_i$ is the vector of covariates for the ith observation, after applying possible fractional polynomial transformation and adjustment as described earlier. Let $\widehat{\eta}_i = \widehat{\beta}_0 + \{\mathbf{H}(x_i) - \mathbf{H}(x_0)\}' \widehat{\boldsymbol{\beta}} + \mathbf{z}_i' \widehat{\boldsymbol{\gamma}}$ be the linear predictor from the FP model, as given by the `fracpred` command or, equivalently, by the `predict` command with the `xb` option, following `fracpoly`. Here $\mathbf{H}(x_i) = \{H_1(x_i), \ldots, H_m(x_i)\}'$ is the vector of FP functions described above, $\mathbf{H}(x_0) = \{H_1(x_0), \ldots, H_m(x_0)\}'$ is the vector of adjustments to x_0 (often, x_0 is chosen to be the mean of the x_i), $\widehat{\boldsymbol{\beta}}$ is the estimated parameter vector, and $\widehat{\boldsymbol{\gamma}}$ is the estimated parameter vector for the covariates. The values $\widehat{\eta}_i^* = \widehat{\beta}_0 + \{\mathbf{H}(x_i) - \mathbf{H}(x_0)\}' \widehat{\boldsymbol{\beta}}$ represent the behavior of the FP model for x at fixed values $\mathbf{z} = \mathbf{0}$ of the (adjusted) covariates. The ith component-plus-residual is defined as $\widehat{\eta}_i^* + d_i$, where d_i is the deviance residual for the ith observation. For normal-errors models, $d_i = \sqrt{w_i}(y_i - \widehat{\eta}_i)$, where w_i is the case weight (or 1, if *weight* is not specified). For logistic, Cox, and generalized linear regression models, see [R] **logistic**, [R] **probit**, [ST] **stcox**, and [R] **glm** for the formula for d_i. The formula for `poisson` models is the same as that for `glm` with `family(poisson)`. For `stcox`, d_i is the partial martingale residual (see [ST] **stcox postestimation**).

`fracplot` plots the values of d_i and the curve represented by $\widehat{\eta}_i^*$ against x_i. The confidence interval for $\widehat{\eta}_i^*$ is obtained from the variance–covariance matrix of the entire model and takes into account the uncertainty in estimating β_0, $\boldsymbol{\beta}$, and $\boldsymbol{\gamma}$ (but not in estimating the FP powers for x).

`fracpred` with the `for(`*varname*`)` option calculates the predicted index at $x_i = x_0$ and $\mathbf{z}_i = \mathbf{0}$; that is, $\widehat{\eta}_i = \widehat{\beta}_0 + \{\mathbf{H}(x_i) - \mathbf{H}(x_0)\}' \widehat{\boldsymbol{\beta}}$. The standard error is calculated from the variance–covariance matrix of $(\widehat{\beta}_0, \widehat{\boldsymbol{\beta}})$, again ignoring estimation of the powers.

Acknowledgment

`fracplot` and `fracpred` were written by Patrick Royston of the MRC Clinical Trials Unit, London.

Also see

[R] **fracpoly** — Fractional polynomial regression

[U] **20 Estimation and postestimation commands**

Title

> **frontier** — Stochastic frontier models

Syntax

frontier *depvar* $\big[$*indepvars*$\big]$ $\big[$*if*$\big]$ $\big[$*in*$\big]$ $\big[$*weight*$\big]$ $\big[$, *options*$\big]$

options	Description
Model	
<u>nocons</u>tant	suppress constant term
<u>d</u>istribution(<u>h</u>normal)	half-normal distribution for the inefficiency term
<u>d</u>istribution(<u>e</u>xponential)	exponential distribution for the inefficiency term
<u>d</u>istribution(<u>t</u>normal)	truncated-normal distribution for the inefficiency term
<u>uf</u>rom(*matrix*)	specify untransformed log likelihood; only with d(tnormal)
cm(*varlist* $\big[$, <u>nocons</u>tant $\big]$)	fit conditional mean model; only with d(tnormal); use noconstant to suppress constant term
Model 2	
<u>constraints</u>(*constraints*)	apply specified linear constraints
<u>col</u>linear	keep collinear variables
<u>uhet</u>(*varlist* $\big[$, <u>nocons</u>tant $\big]$)	explanatory variables for technical inefficiency variance function; use noconstant to suppress constant term
<u>vhet</u>(*varlist* $\big[$, <u>nocons</u>tant $\big]$)	explanatory variables for idiosyncratic error variance function; use noconstant to suppress constant term
cost	fit cost frontier model; default is production frontier model
SE	
vce(*vcetype*)	*vcetype* may be oim, opg, <u>boot</u>strap, or <u>jack</u>knife
Reporting	
<u>level</u>(#)	set confidence level; default is level(95)
<u>nocnsre</u>port	do not display constraints
display_options	control column formats, row spacing, line width, and display of omitted variables and base and empty cells
Maximization	
maximize_options	control the maximization process; seldom used
<u>coefl</u>egend	display legend instead of statistics

indepvars and *varlist* may contain factor variables; see [U] **11.4.3 Factor variables**.
bootstrap, by, jackknife, rolling, and statsby are allowed; see [U] **11.1.10 Prefix commands**.
Weights are not allowed with the bootstrap prefix; see [R] **bootstrap**.
fweights, iweights, and pweights are allowed; see [U] **11.1.6 weight**.
coeflegend does not appear in the dialog box.
See [U] **20 Estimation and postestimation commands** for more capabilities of estimation commands.

Menu

Statistics > Linear models and related > Frontier models

Description

frontier fits stochastic production or cost frontier models; the default is a production frontier model. It provides estimators for the parameters of a linear model with a disturbance that is assumed to be a mixture of two components, which have a strictly nonnegative and symmetric distribution, respectively. frontier can fit models in which the nonnegative distribution component (a measurement of inefficiency) is assumed to be from a half-normal, exponential, or truncated-normal distribution. See Kumbhakar and Lovell (2000) for a detailed introduction to frontier analysis.

Options

> _Model_

noconstant; see [R] **estimation options**.

distribution(*distname*) specifies the distribution for the inefficiency term as half-normal (hnormal), exponential, or truncated-normal (tnormal). The default is hnormal.

ufrom(*matrix*) specifies a $1 \times K$ matrix of untransformed starting values when the distribution is truncated-normal (tnormal). frontier can estimate the parameters of the model by maximizing either the log likelihood or a transformed log likelihood (see *Methods and formulas*). frontier automatically transforms the starting values before passing them on to the transformed log likelihood. The matrix must have the same number of columns as there are parameters to estimate.

cm(*varlist* [, noconstant]) may be used only with distribution(tnormal). Here frontier will fit a conditional mean model in which the mean of the truncated-normal distribution is modeled as a linear function of the set of covariates specified in *varlist*. Specifying noconstant suppresses the constant in the mean function.

> _Model 2_

constraints(*constraints*), collinear; see [R] **estimation options**.

By default, when fitting the truncated-normal model or the conditional mean model, frontier maximizes a transformed log likelihood. When constraints are applied, frontier will maximize the untransformed log likelihood with constraints defined in the untransformed metric.

uhet(*varlist* [, noconstant]) specifies that the technical inefficiency component is heteroskedastic, with the variance function depending on a linear combination of *varlist*$_u$. Specifying noconstant suppresses the constant term from the variance function. This option may not be specified with distribution(tnormal).

vhet(*varlist* [, noconstant]) specifies that the idiosyncratic error component is heteroskedastic, with the variance function depending on a linear combination of *varlist*$_v$. Specifying noconstant suppresses the constant term from the variance function. This option may not be specified with distribution(tnormal).

cost specifies that frontier fit a cost frontier model.

SE

vce(*vcetype*) specifies the type of standard error reported, which includes types that are derived from asymptotic theory and that use bootstrap or jackknife methods; see [R] **vce_option**.

Reporting

level(*#*); see [R] **estimation options**.

nocnsreport; see [R] **estimation options**.

display_options: <u>noomitted</u>, vsquish, <u>noempty</u>cells, <u>base</u>levels, <u>allbase</u>levels, cformat(*% fmt*), pformat(*% fmt*), sformat(*% fmt*), and nolstretch; see [R] **estimation options**.

Maximization

maximize_options: <u>diff</u>icult, <u>tech</u>nique(*algorithm_spec*), <u>iter</u>ate(*#*), [<u>no</u>]log, <u>tr</u>ace, gradient, showstep, <u>hess</u>ian, <u>showtol</u>erance, <u>tol</u>erance(*#*), <u>ltol</u>erance(*#*), <u>nrtol</u>erance(*#*), <u>nonrtol</u>erance, and from(*init_specs*); see [R] **maximize**. These options are seldom used.

Setting the optimization type to technique(bhhh) resets the default *vcetype* to vce(opg).

The following option is available with frontier but is not shown in the dialog box:

coeflegend; see [R] **estimation options**.

Remarks

Stochastic production frontier models were introduced by Aigner, Lovell, and Schmidt (1977) and Meeusen and van den Broeck (1977). Since then, stochastic frontier models have become a popular subfield in econometrics. Kumbhakar and Lovell (2000) provide a good introduction.

frontier fits three stochastic frontier models with distinct parameterizations of the inefficiency term and can fit stochastic production or cost frontier models.

Let's review the nature of the stochastic frontier problem. Suppose that a producer has a production function $f(\mathbf{z}_i, \beta)$. In a world without error or inefficiency, the ith firm would produce

$$q_i = f(\mathbf{z}_i, \beta)$$

Stochastic frontier analysis assumes that each firm potentially produces less than it might due to a degree of inefficiency. Specifically,

$$q_i = f(\mathbf{z}_i, \beta)\xi_i$$

where ξ_i is the level of efficiency for firm i; ξ_i must be in the interval $(0, 1]$. If $\xi_i = 1$, the firm is achieving the optimal output with the technology embodied in the production function $f(\mathbf{z}_i, \beta)$. When $\xi_i < 1$, the firm is not making the most of the inputs $\mathbf{z}_i$ given the technology embodied in the production function $f(\mathbf{z}_i, \beta)$. Because the output is assumed to be strictly positive (that is, $q_i > 0$), the degree of technical efficiency is assumed to be strictly positive (that is, $\xi_i > 0$).

Output is also assumed to be subject to random shocks, implying that

$$q_i = f(\mathbf{z}_i, \beta)\xi_i\exp(v_i)$$

Taking the natural log of both sides yields

$$\ln(q_i) = \ln\{f(\mathbf{z}_i, \beta)\} + \ln(\xi_i) + v_i$$

Assuming that there are k inputs and that the production function is linear in logs, defining $u_i = -\ln(\xi_i)$ yields

$$\ln(q_i) = \beta_0 + \sum_{j=1}^{k} \beta_j \ln(z_{ji}) + v_i - u_i \tag{1}$$

Because u_i is subtracted from $\ln(q_i)$, restricting $u_i \geq 0$ implies that $0 < \xi_i \leq 1$, as specified above.

Kumbhakar and Lovell (2000) provide a detailed version of the above derivation, and they show that performing an analogous derivation in the dual cost function problem allows us to specify the problem as

$$\ln(c_i) = \beta_0 + \beta_q \ln(q_i) + \sum_{j=1}^{k} \beta_j \ln(p_{ji}) + v_i + u_i \tag{2}$$

where q_i is output, z_{ji} are input quantities, c_i is cost, and the p_{ji} are input prices.

Intuitively, the inefficiency effect is required to lower output or raise expenditure, depending on the specification.

❑ Technical note

The model that `frontier` actually fits is of the form

$$y_i = \beta_0 + \sum_{j=1}^{k} \beta_j x_{ji} + v_i - su_i$$

where

$$s = \begin{cases} 1, & \text{for production functions} \\ -1, & \text{for cost functions} \end{cases}$$

so, in the context of the discussion above, $y_i = \ln(q_i)$, and $x_{ji} = \ln(z_{ji})$ for a production function; and for a cost function, $y_i = \ln(c_i)$, and the x_{ji} are the $\ln(p_{ji})$ and $\ln(q_i)$. You must take the natural logarithm of the data before fitting a stochastic frontier production or cost model. `frontier` performs no transformations on the data.

❑

Different specifications of the u_i and the v_i terms give rise to distinct models. `frontier` provides estimators for the parameters of three basic models in which the idiosyncratic component, v_i, is assumed to be independently $N(0, \sigma_v)$ distributed over the observations. The basic models differ in their specification of the inefficiency term, u_i, as follows:

  exponential: the u_i are independently exponentially distributed with variance σ_u^2

  hnormal: the u_i are independently half-normally $N^+(0, \sigma_u^2)$ distributed

  tnormal: the u_i are independently $N^+(\mu, \sigma_u^2)$ distributed with truncation point at 0

For half-normal or exponential distributions, `frontier` can fit models with heteroskedastic error components, conditional on a set of covariates. For a truncated-normal distribution, `frontier` can also fit a conditional mean model in which the mean is modeled as a linear function of a set of covariates.

▷ Example 1

For our first example, we demonstrate the half-normal and exponential models by reproducing a study found in Greene (2003, 505), which uses data originally published in Zellner and Revankar (1969). In this study of the transportation-equipment manufacturing industry, observations on value added, capital, and labor are used to estimate a Cobb–Douglas production function. The variable lnv is the log-transformed value added, lnk is the log-transformed capital, and lnl is the log-transformed labor. OLS estimates are compared with those from stochastic frontier models using both the half-normal and exponential distribution for the inefficiency term.

```
. use http://www.stata-press.com/data/r12/greene9
. regress lnv lnk lnl
```

Source	SS	df	MS
Model	44.1727741	2	22.086387
Residual	1.22225984	22	.055557265
Total	45.3950339	24	1.89145975

Number of obs = 25
F(2, 22) = 397.54
Prob > F = 0.0000
R-squared = 0.9731
Adj R-squared = 0.9706
Root MSE = .23571

lnv	Coef.	Std. Err.	t	P>|t|	[95% Conf. Interval]	
lnk	.2454281	.1068574	2.30	0.032	.0238193	.4670368
lnl	.805183	.1263336	6.37	0.000	.5431831	1.067183
_cons	1.844416	.2335928	7.90	0.000	1.359974	2.328858

```
. frontier lnv lnk lnl
Iteration 0:   log likelihood =  2.3357572
Iteration 1:   log likelihood =  2.4673009
Iteration 2:   log likelihood =  2.4695125
Iteration 3:   log likelihood =  2.4695222
Iteration 4:   log likelihood =  2.4695222
```

Stoc. frontier normal/half-normal model

Number of obs = 25
Wald chi2(2) = 743.71
Prob > chi2 = 0.0000

Log likelihood = 2.4695222

lnv	Coef.	Std. Err.	z	P>|z|	[95% Conf. Interval]	
lnk	.2585478	.098764	2.62	0.009	.0649738	.4521218
lnl	.7802451	.1199399	6.51	0.000	.5451672	1.015323
_cons	2.081135	.281641	7.39	0.000	1.529128	2.633141
/lnsig2v	-3.48401	.6195353	-5.62	0.000	-4.698277	-2.269743
/lnsig2u	-3.014599	1.11694	-2.70	0.007	-5.203761	-.8254368
sigma_v	.1751688	.0542616			.0954514	.3214633
sigma_u	.2215073	.1237052			.074134	.6618486
sigma2	.0797496	.0426989			-.0039388	.163438
lambda	1.264536	.1678684			.9355204	1.593552

Likelihood-ratio test of sigma_u=0: chibar2(01) = 0.43   Prob>=chibar2 = 0.256

```
. predict double u_h, u
```

```
. frontier lnv lnk lnl, distribution(exponential)
Iteration 0:   log likelihood = 2.7270659
Iteration 1:   log likelihood = 2.8551532
Iteration 2:   log likelihood = 2.8604815
Iteration 3:   log likelihood = 2.8604897
Iteration 4:   log likelihood = 2.8604897
```

```
Stoc. frontier normal/exponential model          Number of obs   =        25
                                                 Wald chi2(2)    =    845.68
Log likelihood =  2.8604897                       Prob > chi2     =    0.0000
```

lnv	Coef.	Std. Err.	z	P>|z|	[95% Conf. Interval]	
lnk	.2624859	.0919988	2.85	0.004	.0821717	.4428002
lnl	.7703795	.1109569	6.94	0.000	.5529079	.9878511
_cons	2.069242	.2356159	8.78	0.000	1.607444	2.531041
/lnsig2v	-3.527598	.4486176	-7.86	0.000	-4.406873	-2.648324
/lnsig2u	-4.002457	.9274575	-4.32	0.000	-5.820241	-2.184674
sigma_v	.1713925	.0384448			.1104231	.2660258
sigma_u	.1351691	.0626818			.0544692	.3354317
sigma2	.0476461	.0157921			.016694	.0785981
lambda	.7886525	.087684			.616795	.9605101

```
Likelihood-ratio test of sigma_u=0: chibar2(01) = 1.21  Prob>=chibar2 = 0.135
. predict double u_e, u

. list state u_h u_e
```

	state	u_h	u_e
1.	Alabama	.2011338	.14592865
2.	California	.14480966	.0972165
3.	Connecticut	.1903485	.13478797
4.	Florida	.51753139	.5903303
5.	Georgia	.10397912	.07140994
6.	Illinois	.12126696	.0830415
7.	Indiana	.21128212	.15450664
8.	Iowa	.24933153	.20073081
9.	Kansas	.10099517	.06857629
10.	Kentucky	.05626919	.04152443
11.	Louisiana	.20332731	.15066405
12.	Maine	.22263164	.17245793
13.	Maryland	.13534062	.09245501
14.	Massachusetts	.15636999	.10932923
15.	Michigan	.15809566	.10756915
16.	Missouri	.10288047	.0704146
17.	NewJersey	.09584337	.06587986
18.	NewYork	.27787793	.22249416
19.	Ohio	.22914231	.16981857
20.	Pennsylvania	.1500667	.10302905
21.	Texas	.20297875	.14552218
22.	Virginia	.14000132	.09676078
23.	Washington	.11047581	.07533251
24.	WestVirginia	.15561392	.11236153
25.	Wisconsin	.14067066	.0970861

The parameter estimates and the estimates of the inefficiency terms closely match those published in Greene (2003, 505), but the standard errors of the parameter estimates are estimated differently (see the technical note below).

The output from `frontier` includes estimates of the standard deviations of the two error components, σ_v and σ_u, which are labeled `sigma_v` and `sigma_u`, respectively. In the log likelihood, they are parameterized as $\ln\sigma_v^2$ and $\ln\sigma_u^2$, and these estimates are labeled `/lnsig2v` and `/lnsig2u` in the output. `frontier` also reports two other useful parameterizations. The estimate of the total error variance, $\sigma_S^2 = \sigma_v^2 + \sigma_u^2$, is labeled `sigma2`, and the estimate of the ratio of the standard deviation of the inefficiency component to the standard deviation of the idiosyncratic component, $\lambda = \sigma_u/\sigma_v$, is labeled `lambda`.

At the bottom of the output, `frontier` reports the results of a test that there is no technical inefficiency component in the model. This is a test of the null hypothesis $H_0 : \sigma_u^2 = 0$ against the alternative hypotheses $H_1 : \sigma_u^2 > 0$. If the null hypothesis is true, the stochastic frontier model reduces to an OLS model with normal errors. However, because the test lies on the boundary of the parameter space of σ_u^2, the standard likelihood-ratio test is not valid, and a one-sided generalized likelihood-ratio test must be constructed; see Gutierrez, Carter, and Drukker (2001). For this example, the output shows LR = 0.43 with a p-value of 0.256 for the half-normal model and LR = 1.21 with a p-value of 0.135 for the exponential model. There are several possible reasons for the failure to reject the null hypothesis, but the fact that the test is based on an asymptotic distribution and the sample size was 25 is certainly a leading candidate among those possibilities.

◁

❑ Technical note

`frontier` maximizes the log-likelihood function of a stochastic frontier model by using the Newton–Raphson method, and the estimated variance–covariance matrix is calculated as the inverse of the negative Hessian (matrix of second partial derivatives); see [R] **ml**. When comparing the results with those published using other software, be aware of the difference in the optimization methods, which may result in different, yet asymptotically equivalent, variance estimates.

❑

▷ Example 2

Often the error terms may not have constant variance. `frontier` allows you to model heteroskedasticity in either error term as a linear function of a set of covariates. The variance of either the technical inefficiency or the idiosyncratic component may be modeled as

$$\sigma_i^2 = \exp(\mathbf{w}_i\boldsymbol{\delta})$$

The default constant included in $\mathbf{w}_i$ may be suppressed by appending a `noconstant` option to the list of covariates. Also, you can simultaneously specify covariates for both σ_{u_i} and σ_{v_i}.

In the example below, we use a sample of 756 observations of fictional firms producing a manufactured good by using capital and labor. The firms are hypothesized to use a constant returns-to-scale technology, but the sizes of the firms differ. Believing that this size variation will introduce heteroskedasticity into the idiosyncratic error term, we estimate the parameters of a Cobb–Douglas production function. To do this, we use a conditional heteroskedastic half-normal model, with the size of the firm as an explanatory variable in the variance function for the idiosyncratic error. We also perform a test of the hypothesis that the firms use a constant returns-to-scale technology.

```
. use http://www.stata-press.com/data/r12/frontier1, clear

. frontier lnoutput lnlabor lncapital, vhet(size)
Iteration 0:    log likelihood = -1508.3692
Iteration 1:    log likelihood =  -1501.583
Iteration 2:    log likelihood = -1500.3942
Iteration 3:    log likelihood = -1500.3794
Iteration 4:    log likelihood = -1500.3794
```

Stoc. frontier normal/half-normal model		Number of obs	=	756
		Wald chi2(2)	=	9.68
Log likelihood = -1500.3794		Prob > chi2	=	0.0079

lnoutput	Coef.	Std. Err.	z	P>|z|	[95% Conf. Interval]	
lnoutput						
lnlabor	.7090933	.2349374	3.02	0.003	.2486244	1.169562
lncapital	.3931345	.5422173	0.73	0.468	-.6695919	1.455861
_cons	1.252199	3.14656	0.40	0.691	-4.914946	7.419344
lnsig2v						
size	-.0016951	.0004748	-3.57	0.000	-.0026256	-.0007645
_cons	3.156091	.9265826	3.41	0.001	1.340023	4.97216
lnsig2u						
_cons	1.947487	.1017653	19.14	0.000	1.748031	2.146943
sigma_u	2.647838	.134729			2.396514	2.925518

```
. test _b[lnlabor] + _b[lncapital] = 1

 ( 1)  [lnoutput]lnlabor + [lnoutput]lncapital = 1

           chi2(  1) =    0.03
         Prob > chi2 =    0.8622
```

The output above indicates that the variance of the idiosyncratic error term is a function of firm size. Also, we failed to reject the hypothesis that the firms use a constant returns-to-scale technology.

◁

❑ Technical note

In small samples, the conditional heteroskedastic estimators will lack precision for the variance parameters and may fail to converge altogether.

❑

▷ Example 3

Let's turn our attention to the truncated-normal model. Once again, we will use fictional data. For this example, we have 1,231 observations on the quantity of output, the total cost of production for each firm, the prices that each firm paid for labor and capital services, and a categorical variable measuring the quality of each firm's management. After taking the natural logarithm of the costs (lncost), prices (lnp_k and lnp_l), and output (lnout), we fit a stochastic cost frontier model and specify the distribution for the inefficiency term to be truncated normal.

```
. use http://www.stata-press.com/data/r12/frontier2

. frontier lncost lnp_k lnp_l lnout, distribution(tnormal) cost

Iteration 0:    log likelihood = -2386.9523
Iteration 1:    log likelihood = -2386.5146
Iteration 2:    log likelihood = -2386.2704
Iteration 3:    log likelihood = -2386.2504
Iteration 4:    log likelihood = -2386.2493
Iteration 5:    log likelihood = -2386.2493

Stoc. frontier normal/truncated-normal model      Number of obs   =      1231
                                                  Wald chi2(3)    =      8.82
Log likelihood = -2386.2493                       Prob > chi2     =    0.0318
```

lncost	Coef.	Std. Err.	z	P>|z|	[95% Conf.	Interval]
lnp_k	.3410717	.2363861	1.44	0.149	-.1222366	.80438
lnp_l	.6608628	.4951499	1.33	0.182	-.3096131	1.631339
lnout	.7528653	.3468968	2.17	0.030	.0729601	1.432771
_cons	2.602609	1.083004	2.40	0.016	.4799595	4.725259
/mu	1.095705	.881517	1.24	0.214	-.632037	2.823446
/lnsigma2	1.5534	.1873464	8.29	0.000	1.186208	1.920592
/ilgtgamma	1.257862	.2589522	4.86	0.000	.7503255	1.765399
sigma2	4.727518	.8856833			3.274641	6.825001
gamma	.7786579	.0446303			.6792496	.8538846
sigma_u2	3.681119	.7503408			2.210478	5.15176
sigma_v2	1.046399	.2660035			.5250413	1.567756

```
H0: No inefficiency component:            z =   5.595          Prob>=z = 0.000
```

In addition to the coefficients, the output reports estimates for several parameters. `sigma_v2` is the estimate of σ_v^2. `sigma_u2` is the estimate of σ_u^2. `gamma` is the estimate of $\gamma = \sigma_u^2/\sigma_S^2$. `sigma2` is the estimate of $\sigma_S^2 = \sigma_v^2 + \sigma_u^2$. Because γ must be between 0 and 1, the optimization is parameterized in terms of the inverse logit of γ, and this estimate is reported as `ilgtgamma`. Because σ_S^2 must be positive, the optimization is parameterized in terms of $\ln(\sigma_S^2)$, whose estimate is reported as `lnsigma2`. Finally, `mu` is the estimate of μ, the mean of the truncated-normal distribution.

In the output above, the generalized log-likelihood test for the presence of the inefficiency term has been replaced with a test based on the third moment of the OLS residuals. When $\mu = 0$ and $\sigma_u = 0$, the truncated-normal model reduces to a linear regression model with normally distributed errors. However, the distribution of the test statistic under the null hypothesis is not well established, because it becomes impossible to evaluate the log likelihood as σ_u approaches zero, prohibiting the use of the likelihood-ratio test.

However, Coelli (1995) noted that the presence of an inefficiency term would negatively skew the residuals from an OLS regression. By identifying negative skewness in the residuals with the presence of an inefficiency term, Coelli derived a one-sided test for the presence of the inefficiency term. The results of this test are given at the bottom of the output. For this example, the null hypothesis of no inefficiency component is rejected.

In the example below, we fit a truncated model and detect a statistically significant inefficiency term in the model. We might question whether the inefficiency term is identically distributed over all firms or whether there might be heterogeneity across firms. `frontier` provides an extension to the truncated normal model by allowing the mean of the inefficiency term to be modeled as a linear function of a set of covariates. In our dataset, we have a categorical variable that measures the quality of a firm's management. We refit the model, including the `cm()` option, specifying a set of binary indicator variables representing the different categories of the quality-measurement variable as covariates.

```
. frontier lncost lnp_k lnp_l lnout, distribution(tnormal) cm(i.quality) cost
Iteration 0:   log likelihood = -2386.9523
Iteration 1:   log likelihood =  -2384.936
Iteration 2:   log likelihood = -2382.3942
Iteration 3:   log likelihood =  -2382.324
Iteration 4:   log likelihood = -2382.3233
Iteration 5:   log likelihood = -2382.3233
```

Stoc. frontier normal/truncated-normal model		Number of obs	=	1231
		Wald chi2(3)	=	9.31
Log likelihood = -2382.3233		Prob > chi2	=	0.0254

lncost	Coef.	Std. Err.	z	P>|z|	[95% Conf. Interval]	
lncost						
lnp_k	.3611204	.2359749	1.53	0.126	-.1013819	.8236227
lnp_l	.680446	.4934935	1.38	0.168	-.2867835	1.647675
lnout	.7605533	.3466102	2.19	0.028	.0812098	1.439897
_cons	2.550769	1.078911	2.36	0.018	.4361417	4.665396
mu						
quality						
2	.5056067	.3382907	1.49	0.135	-.1574309	1.168644
3	.783223	.376807	2.08	0.038	.0446947	1.521751
4	.5577511	.3355061	1.66	0.096	-.0998288	1.215331
5	.6792882	.3428073	1.98	0.048	.0073981	1.351178
_cons	.6014025	.990167	0.61	0.544	-1.339289	2.542094
/lnsigma2	1.541784	.1790926	8.61	0.000	1.190769	1.892799
/ilgtgamma	1.242302	.2588968	4.80	0.000	.734874	1.749731
sigma2	4.67292	.8368852			3.289611	6.637923
gamma	.7759645	.0450075			.6758739	.8519189
sigma_u2	3.62602	.7139576			2.226689	5.025351
sigma_v2	1.0469	.2583469			.5405491	1.553251

The conditional mean model was developed in the context of panel-data estimators, and we can apply `frontier`'s conditional mean model to panel data.

◁

Saved results

`frontier` saves the following in `e()`:

Scalars

e(N)	number of observations
e(df_m)	model degrees of freedom
e(k)	number of parameters
e(k_eq)	number of equations in e(b)
e(k_eq_model)	number of equations in overall model test
e(k_dv)	number of dependent variables
e(chi2)	χ^2
e(ll)	log likelihood
e(ll_c)	log likelihood for H_0: $\sigma_u = 0$
e(z)	test for negative skewness of OLS residuals
e(sigma_u)	standard deviation of technical inefficiency
e(sigma_v)	standard deviation of v_i
e(p)	significance
e(chi2_c)	LR test statistic
e(p_z)	p-value for z
e(rank)	rank of e(V)
e(ic)	number of iterations
e(rc)	return code
e(converged)	1 if converged, 0 otherwise

Macros

e(cmd)	frontier
e(cmdline)	command as typed
e(depvar)	name of dependent variable
e(function)	production or cost
e(wtype)	weight type
e(wexp)	weight expression
e(title)	title in estimation output
e(chi2type)	Wald; type of model χ^2 test
e(dist)	distribution assumption for u_i
e(het)	heteroskedastic components
e(u_hetvar)	*varlist* in uhet()
e(v_hetvar)	*varlist* in vhet()
e(vce)	*vcetype* specified in vce()
e(vcetype)	title used to label Std. Err.
e(opt)	type of optimization
e(which)	max or min; whether optimizer is to perform maximization or minimization
e(ml_method)	type of ml method
e(user)	name of likelihood-evaluator program
e(technique)	maximization technique
e(properties)	b V
e(predict)	program used to implement predict
e(asbalanced)	factor variables fvset as asbalanced
e(asobserved)	factor variables fvset as asobserved

Matrices

    e(b)   coefficient vector
    e(Cns)   constraints matrix
    e(ilog)   iteration log (up to 20 iterations)
    e(gradient)   gradient vector
    e(V)   variance–covariance matrix of the estimators
    e(V_modelbased)   model-based variance

Functions

    e(sample)   marks estimation sample

Methods and formulas

frontier is implemented as an ado-file.

Consider an equation of the form

$$y_i = \mathbf{x}_i \boldsymbol{\beta} + v_i - su_i$$

where y_i is the dependent variable, $\mathbf{x}_i$ is a $1 \times k$ vector of observations on the independent variables included as indent covariates, $\boldsymbol{\beta}$ is a $k \times 1$ vector of coefficients, and

$$s = \begin{cases} 1, & \text{for production functions} \\ -1, & \text{for cost functions} \end{cases}$$

The log-likelihood functions are as follows.

Normal/half-normal model:

$$\ln L = \sum_{i=1}^{N} \left\{ \frac{1}{2} \ln\left(\frac{2}{\pi}\right) - \ln \sigma_S + \ln \Phi\left(-\frac{s\epsilon_i \lambda}{\sigma_S}\right) - \frac{\epsilon_i^2}{2\sigma_S^2} \right\}$$

Normal/exponential model:

$$\ln L = \sum_{i=1}^{N} \left\{ -\ln \sigma_u + \frac{\sigma_v^2}{2\sigma_u^2} + \ln \Phi\left(\frac{-s\epsilon_i - \frac{\sigma_v^2}{\sigma_u}}{\sigma_v}\right) + \frac{s\epsilon_i}{\sigma_u} \right\}$$

Normal/truncated-normal model:

$$\ln L = \sum_{i=1}^{N} \left\{ -\frac{1}{2} \ln(2\pi) - \ln \sigma_S - \ln \Phi\left(\frac{\mu}{\sigma_S \sqrt{\gamma}}\right) \right.$$
$$\left. + \ln \Phi\left[\frac{(1-\gamma)\mu - s\gamma\epsilon_i}{\{\sigma_S^2 \gamma (1-\gamma)\}^{1/2}}\right] - \frac{1}{2}\left(\frac{\epsilon_i + s\mu}{\sigma_S}\right)^2 \right\}$$

where $\sigma_S = (\sigma_u^2 + \sigma_v^2)^{1/2}$, $\lambda = \sigma_u/\sigma_v$, $\gamma = \sigma_u^2/\sigma_S^2$, $\epsilon_i = y_i - \mathbf{x}_i\beta$, and $\Phi()$ is the cumulative distribution function of the standard normal distribution.

To obtain estimation for u_i, you can use either the mean or the mode of the conditional distribution $f(u|\epsilon)$.

$$E\left(u_i \mid \epsilon_i\right) = \mu_{*i} + \sigma_* \left\{ \frac{\phi(-\mu_{*i}/\sigma_*)}{\Phi(\mu_{*i}/\sigma_*)} \right\}$$

$$M\left(u_i \mid \epsilon_i\right) = \begin{cases} \mu_{*i}, & \text{if } \mu_{*i} \geq 0 \\ 0, & \text{otherwise} \end{cases}$$

Then the technical efficiency ($s = 1$) or cost efficiency ($s = -1$) will be estimated by

$$\begin{aligned} E_i &= E\left\{\exp(-su_i) \mid \epsilon_i\right\} \\ &= \left\{ \frac{1 - \Phi\left(s\sigma_* - \mu_{*i}/\sigma_*\right)}{1 - \Phi\left(-\mu_{*i}/\sigma_*\right)} \right\} \exp\left(-s\mu_{*i} + \frac{1}{2}\sigma_*^2\right) \end{aligned}$$

where μ_{*i} and σ_* are defined for the normal/half-normal model as

$$\mu_{*i} = -s\epsilon_i\sigma_u^2/\sigma_S^2$$
$$\sigma_* = \sigma_u\sigma_v/\sigma_S$$

for the normal/exponential model as

$$\mu_{*i} = -s\epsilon_i - \sigma_v^2/\sigma_u$$
$$\sigma_* = \sigma_v$$

and for the normal/truncated-normal model as

$$\mu_{*i} = \frac{-s\epsilon_i\sigma_u^2 + \mu\sigma_v^2}{\sigma_S^2}$$
$$\sigma_* = \sigma_u\sigma_v/\sigma_S$$

In the half-normal and exponential models, when heteroskedasticity is assumed, the standard deviations, σ_u or σ_v, will be replaced in the above equations by

$$\sigma_i^2 = \exp(\mathbf{w}_i\delta)$$

where $\mathbf{w}$ is the vector of explanatory variables in the variance function.

In the conditional mean model, the mean parameter of the truncated normal distribution, μ, is modeled as a linear combination of the set of covariates, $\mathbf{w}$.

$$\mu = \mathbf{w}_i\delta$$

Therefore, the log-likelihood function can be rewritten as

$$
\ln L = \sum_{i=1}^{N} \left[-\frac{1}{2} \ln (2\pi) - \ln \sigma_S - \ln \Phi \left(\frac{\mathbf{w}_i \boldsymbol{\delta}}{\sqrt{\sigma_S^2 \gamma}} \right) \right.
$$
$$
\left. + \ln \Phi \left\{ \frac{(1-\gamma) \mathbf{w}_i \boldsymbol{\delta} - s\gamma \epsilon_i}{\sqrt{\sigma_S^2 \gamma (1-\gamma)}} \right\} - \frac{1}{2} \left(\frac{\epsilon_i + s \mathbf{w}_i \boldsymbol{\delta}}{\sigma_S} \right)^2 \right]
$$

The z test reported in the output of the truncated-normal model is a third-moment test developed by Coelli (1995) as an extension of a test previously developed by Pagan and Hall (1983). Coelli shows that under the null of normally distributed errors, the statistic

$$
z = \frac{m_3}{\left(\frac{6m_2^3}{N} \right)^{1/2}}
$$

has a standard normal distribution, where m_3 is the third moment from the OLS regression. Because the residuals are either negatively skewed (production function) or positively skewed (cost function), a one-sided p-value is used.

References

Aigner, D., C. A. K. Lovell, and P. Schmidt. 1977. Formulation and estimation of stochastic frontier production function models. *Journal of Econometrics* 6: 21–37.

Caudill, S. B., J. M. Ford, and D. M. Gropper. 1995. Frontier estimation and firm-specific inefficiency measures in the presence of heteroscedasticity. *Journal of Business and Economic Statistics* 13: 105–111.

Coelli, T. J. 1995. Estimators and hypothesis tests for a stochastic frontier function: A Monte Carlo analysis. *Journal of Productivity Analysis* 6: 247–268.

Gould, W. W., J. S. Pitblado, and B. P. Poi. 2010. *Maximum Likelihood Estimation with Stata*. 4th ed. College Station, TX: Stata Press.

Greene, W. H. 2003. *Econometric Analysis*. 5th ed. Upper Saddle River, NJ: Prentice Hall.

Gutierrez, R. G., S. Carter, and D. M. Drukker. 2001. sg160: On boundary-value likelihood-ratio tests. *Stata Technical Bulletin* 60: 15–18. Reprinted in *Stata Technical Bulletin Reprints*, vol. 10, pp. 269–273. College Station, TX: Stata Press.

Kumbhakar, S. C., and C. A. K. Lovell. 2000. *Stochastic Frontier Analysis*. Cambridge: Cambridge University Press.

Meeusen, W., and J. van den Broeck. 1977. Efficiency estimation from Cobb–Douglas production functions with composed error. *International Economic Review* 18: 435–444.

Pagan, A. R., and A. D. Hall. 1983. Diagnostic tests as residual analysis. *Econometric Reviews* 2: 159–218.

Petrin, A., B. P. Poi, and J. Levinsohn. 2004. Production function estimation in Stata using inputs to control for unobservables. *Stata Journal* 4: 113–123.

Stevenson, R. E. 1980. Likelihood functions for generalized stochastic frontier estimation. *Journal of Econometrics* 13: 57–66.

Zellner, A., and N. S. Revankar. 1969. Generalized production functions. *Review of Economic Studies* 36: 241–250.

Also see

Title

> **frontier postestimation** — Postestimation tools for frontier

Description

The following postestimation commands are available after `frontier`:

Command	Description
contrast	contrasts and ANOVA-style joint tests of estimates
estat	AIC, BIC, VCE, and estimation sample summary
estimates	cataloging estimation results
lincom	point estimates, standard errors, testing, and inference for linear combinations of coefficients
linktest	link test for model specification
lrtest	likelihood-ratio test
margins	marginal means, predictive margins, marginal effects, and average marginal effects
marginsplot	graph the results from margins (profile plots, interaction plots, etc.)
nlcom	point estimates, standard errors, testing, and inference for nonlinear combinations of coefficients
predict	predictions, residuals, influence statistics, and other diagnostic measures
predictnl	point estimates, standard errors, testing, and inference for generalized predictions
pwcompare	pairwise comparisons of estimates
test	Wald tests of simple and composite linear hypotheses
testnl	Wald tests of nonlinear hypotheses

See the corresponding entries in the *Base Reference Manual* for details.

Syntax for predict

predict $\left[\, type\, \right]$ *newvar* $\left[\, if\, \right]$ $\left[\, in\, \right]$ $\left[\,$, *statistic* $\right]$

predict $\left[\, type\, \right]$ $\{$ *stub** | *newvar*$_{xb}$ *newvar*$_v$ *newvar*$_u$ $\}$ $\left[\, if\, \right]$ $\left[\, in\, \right]$, <u>sc</u>ores

statistic	Description
Main	
xb	linear prediction; the default
stdp	standard error of the prediction
u	estimates of minus the natural log of the technical efficiency via $E\left(u_i \mid \epsilon_i\right)$
m	estimates of minus the natural log of the technical efficiency via $M\left(u_i \mid \epsilon_i\right)$
te	estimates of the technical efficiency via $E\left\{\exp(-su_i) \mid \epsilon_i\right\}$
	$s = \begin{cases} 1, & \text{for production functions} \\ -1, & \text{for cost functions} \end{cases}$

These statistics are available both in and out of sample; type `predict ... if e(sample) ...` if wanted only for the estimation sample.

Menu

Statistics > Postestimation > Predictions, residuals, etc.

Options for predict

     ⌐ Main ⌐

xb, the default, calculates the linear prediction.

stdp calculates the standard error of the linear prediction.

u produces estimates of minus the natural log of the technical efficiency via $E\left(u_i \mid \epsilon_i\right)$.

m produces estimates of minus the natural log of the technical efficiency via $M\left(u_i \mid \epsilon_i\right)$.

te produces estimates of the technical efficiency via $E\left\{\exp(-su_i) \mid \epsilon_i\right\}$.

scores calculates equation-level score variables.

     The first new variable will contain $\partial \ln L / \partial(\mathbf{x}_i\boldsymbol{\beta})$.

     The second new variable will contain $\partial \ln L / \partial(\texttt{lnsig2v})$.

     The third new variable will contain $\partial \ln L / \partial(\texttt{lnsig2u})$.

Methods and formulas

All postestimation commands listed above are implemented as ado-files.

Also see

[R] **frontier** — Stochastic frontier models

[U] **20 Estimation and postestimation commands**

Title

> **fvrevar** — Factor-variables operator programming command

Syntax

> fvrevar [*varlist*] [*if*] [*in*] [, <u>sub</u>stitute <u>ts</u>only <u>l</u>ist stub(*stub*)]

You must tsset your data before using fvrevar if *varlist* contains time-series operators; see [TS] **tsset**.

Description

> fvrevar creates an equivalent, temporary variable list for a *varlist* that might contain factor variables, interactions, or time-series–operated variables so that the resulting variable list can be used by commands that do not otherwise support factor variables or time-series–operated variables. The resulting list also could be used in a program to speed execution at the cost of using more memory.

Options

> substitute specifies that equivalent, temporary variables be substituted for any factor variables, interactions, or time-series–operated variables in *varlist*. substitute is the default action taken by fvrevar; you do not need to specify the option.

> tsonly specifies that equivalent, temporary variables be substituted for only the time-series–operated variables in *varlist*.

> list specifies that all factor-variable operators and time-series operators be removed from *varlist* and the resulting list of base variables be returned in r(varlist). No new variables are created with this option.

> stub(*stub*) specifies that fvrevar generate named variables instead of temporary variables. The new variables will be named *stub#*.

Remarks

> fvrevar might create no new variables, one new variable, or many new variables, depending on the number of factor variables, interactions, and time-series operators appearing in *varlist*. Any new variables created are temporary. The new, equivalent varlist is returned in r(varlist). The new varlist corresponds one to one with the original *varlist*.

▷ Example 1

> Typing

> . use http://www.stata-press.com/data/r12/auto
> . fvrevar i.rep78 mpg turn

creates five temporary variables corresponding to the levels of rep78. No new variables are created for variables mpg and turn because they do not contain factor-variable or time-series operators.

The resulting variable list is

```
. display "'r(varlist)'"
__000000 __000001 __000002 __000003 __000004  mpg turn
```

(Your temporary variable names may be different, but that is of no consequence.)

Temporary variables automatically vanish when the program concludes.

◁

▷ Example 2

Suppose we want to create temporary variables for specific levels of a factor variable. To do this, we can use the parenthesis notation of factor-variable syntax.

```
. fvrevar i(2,3)bn.rep78 mpg
```

creates two temporary variables corresponding to levels 2 and 3 of rep78. Notice that we specified that neither level 2 nor 3 be set as the base level by using the bn notation. If we did not specify bn, level 2 would have been treated as the base level.

The resulting variable list is

```
. display "'r(varlist)'"
__00000E __00000F mpg
```

We can see the results by listing the new variables alongside the original value of rep78.

```
. list rep78 'r(varlist)'  in 1/5
```

	rep78	__00000E	__00000F	mpg
1.	3	1	0	22
2.	3	1	0	17
3.	.	.	.	22
4.	3	1	0	20
5.	4	0	1	15

If we had needed only the base-variable names, we could have specified

```
. fvrevar i(2,3)bn.rep78 mpg, list
. display "'r(varlist)'"
mpg rep78
```

The order of the list will probably differ from that of the original list; base variables are listed only once.

◁

▷ Example 3

Now let's assume we have a *varlist* containing both an interaction and time-series–operated variables. If we want to create temporary variables for the entire equivalent *varlist*, we can specify fvrevar with no options.

```
. generate t = _n
. tsset t
. fvrevar c.turn#i(2,3).rep78 L.mpg
```

The resulting variable list is

```
. display "'r(varlist)'"
__00000I __00000K __00000M
```

If we want to create temporary variables only for the time-series–operated variables, we can specify the tsonly option.

```
. fvrevar c.turn#i(2,3).rep78 L.mpg, tsonly
```

The resulting variable list is

```
. display "'r(varlist)'"
2b.rep78#c.turn 3.rep78#c.turn __00000M
```

Notice that fvrevar returned the expanded factor-variable list with the tsonly option.

◁

❏ Technical note

fvrevar, substitute avoids creating duplicate variables. Consider

```
. fvrevar i.rep78 turn mpg i.rep78
```

i.rep78 appears twice in the varlist. fvrevar will create only one set of new variables for the five levels of rep78 and will use these new variables once in the resulting r(varlist). Moreover, fvrevar will do this even across multiple calls:

```
. fvrevar i.rep78 turn mpg
. fvrevar i.rep78
```

i.rep78 appears in two separate calls. At the first call, fvrevar creates five temporary variables corresponding to the five levels of rep78. At the second call, fvrevar remembers what it has done and uses the same temporary variables for i.rep78.

❏

Saved results

fvrevar saves the following in r():

Macros
    r(varlist)   the modified variable list or list of base-variable names

Also see

[P] **syntax** — Parse Stata syntax

[TS] **tsrevar** — Time-series operator programming command

[P] **unab** — Unabbreviate variable list

[U] **11 Language syntax**

[U] **11.4.4 Time-series varlists**

[U] **18 Programming Stata**

Title

> **fvset** — Declare factor-variable settings

Syntax

Declare base settings

> fvset <u>b</u>ase *base_spec varlist*

Declare design settings

> fvset <u>design</u> *design_spec varlist*

Clear the current settings

> fvset clear *varlist*

Report the current settings

> fvset report $\left[\,varlist\,\right]$ $\left[\,,\,\underline{b}ase(base\_spec)\,\underline{design}(design\_spec)\,\right]$

base_spec	Description
default	default base
<u>first</u>	lowest level value; the default
<u>last</u>	highest level value
<u>frequent</u>	most frequent level value
<u>none</u>	no base
#	nonnegative integer value

design_spec	Description
default	default base
<u>asbal</u>anced	accumulate using $1/k$, k = number of levels
<u>asobs</u>erved	accumulate using observed relative frequencies; the default

Description

fvset declares factor-variable settings. Factor-variable settings identify the base level and how to accumulate statistics over levels.

fvset base specifies the base level for each variable in *varlist*. The default for factor variables without a declared base level is first.

fvset design specifies how to accumulate over the levels of a factor variable. The margins command is the only command aware of this setting; see [R] **margins**. By default, margins assumes that factor variables are asobserved, meaning that they are accumulated by weighting by the number of observations or the sum of the weights if weights have been specified.

fvset clear removes factor-variable settings for each variable in *varlist*. fvset clear _all removes all factor-variable settings from all variables.

fvset report reports the current factor-variable settings for each variable in *varlist*. fvset without arguments is a synonym for fvset report.

Options

base(*base_spec*) restricts fvset report to report only the factor-variable settings for variables with the specified *base_spec*.

design(*design_spec*) restricts fvset report to report only the factor-variable settings for variables with the specified *design_spec*.

Remarks

▷ Example 1

Using our auto dataset, we include factor variable i.rep78 in a regression:

```
. use http://www.stata-press.com/data/r12/auto
(1978 Automobile Data)

. regress mpg i.rep78, baselevels
```

Source	SS	df	MS		Number of obs =		69
					F(4, 64) =		4.91
Model	549.415777	4	137.353944		Prob > F =		0.0016
Residual	1790.78712	64	27.9810488		R-squared =		0.2348
					Adj R-squared =		0.1869
Total	2340.2029	68	34.4147485		Root MSE =		5.2897

mpg	Coef.	Std. Err.	t	P>|t|	[95% Conf. Interval]	
rep78						
1	0	(base)				
2	-1.875	4.181884	-0.45	0.655	-10.22927	6.479274
3	-1.566667	3.863059	-0.41	0.686	-9.284014	6.150681
4	.6666667	3.942718	0.17	0.866	-7.209818	8.543152
5	6.363636	4.066234	1.56	0.123	-1.759599	14.48687
_cons	21	3.740391	5.61	0.000	13.52771	28.47229

We specified the baselevels option so that the base level would be included in the output. By default, the first level is the base level. We can change the base level to 2:

```
. fvset base 2 rep78
. regress mpg i.rep78, baselevels
```

Source	SS	df	MS				
Model	549.415777	4	137.353944				
Residual	1790.78712	64	27.9810488				
Total	2340.2029	68	34.4147485				

```
                                  Number of obs =      69
                                  F(  4,    64) =    4.91
                                  Prob > F      =  0.0016
                                  R-squared     =  0.2348
                                  Adj R-squared =  0.1869
                                  Root MSE      =  5.2897
```

mpg	Coef.	Std. Err.	t	P>|t|	[95% Conf.	Interval]
rep78						
1	1.875	4.181884	0.45	0.655	-6.479274	10.22927
2	0	(base)				
3	.3083333	2.104836	0.15	0.884	-3.896559	4.513226
4	2.541667	2.247695	1.13	0.262	-1.948621	7.031954
5	8.238636	2.457918	3.35	0.001	3.32838	13.14889
_cons	19.125	1.870195	10.23	0.000	15.38886	22.86114

Let's set rep78 to have no base level and fit a cell-means regression:

```
. fvset base none rep78
. regress mpg i.rep78, noconstant
```

Source	SS	df	MS				
Model	31824.2129	5	6364.84258				
Residual	1790.78712	64	27.9810488				
Total	33615	69	487.173913				

```
                                  Number of obs =      69
                                  F(  5,    64) =  227.47
                                  Prob > F      =  0.0000
                                  R-squared     =  0.9467
                                  Adj R-squared =  0.9426
                                  Root MSE      =  5.2897
```

mpg	Coef.	Std. Err.	t	P>|t|	[95% Conf.	Interval]
rep78						
1	21	3.740391	5.61	0.000	13.52771	28.47229
2	19.125	1.870195	10.23	0.000	15.38886	22.86114
3	19.43333	.9657648	20.12	0.000	17.504	21.36267
4	21.66667	1.246797	17.38	0.000	19.1759	24.15743
5	27.36364	1.594908	17.16	0.000	24.17744	30.54983

◁

▷ Example 2

By default, margins accumulates a margin by using the observed relative frequencies of the factor levels.

```
. regress mpg i.foreign
```

Source	SS	df	MS
Model	378.153515	1	378.153515
Residual	2065.30594	72	28.6848048
Total	2443.45946	73	33.4720474

	Number of obs =	74
	F(1, 72) =	13.18
	Prob > F =	0.0005
	R-squared =	0.1548
	Adj R-squared =	0.1430
	Root MSE =	5.3558

mpg	Coef.	Std. Err.	t	P>|t|	[95% Conf. Interval]	
1.foreign	4.945804	1.362162	3.63	0.001	2.230384	7.661225
_cons	19.82692	.7427186	26.70	0.000	18.34634	21.30751

```
. margins
```

Predictive margins   Number of obs   =   74
Model VCE   : OLS

Expression   : Linear prediction, predict()

	Margin	Delta-method Std. Err.	z	P>|z|	[95% Conf. Interval]	
_cons	21.2973	.6226014	34.21	0.000	20.07702	22.51757

Let's set `foreign` to always accumulate using equal relative frequencies:

```
. fvset design asbalanced foreign
. regress mpg i.foreign
```

Source	SS	df	MS
Model	378.153515	1	378.153515
Residual	2065.30594	72	28.6848048
Total	2443.45946	73	33.4720474

	Number of obs =	74
	F(1, 72) =	13.18
	Prob > F =	0.0005
	R-squared =	0.1548
	Adj R-squared =	0.1430
	Root MSE =	5.3558

mpg	Coef.	Std. Err.	t	P>|t|	[95% Conf. Interval]	
1.foreign	4.945804	1.362162	3.63	0.001	2.230384	7.661225
_cons	19.82692	.7427186	26.70	0.000	18.34634	21.30751

```
. margins
```

Adjusted predictions   Number of obs   =   74
Model VCE   : OLS

Expression   : Linear prediction, predict()
at   : foreign   (asbalanced)

	Margin	Delta-method Std. Err.	z	P>|z|	[95% Conf. Interval]	
_cons	22.29983	.6810811	32.74	0.000	20.96493	23.63472

Suppose that we issued the `fvset design` command earlier in our session and that we cannot remember which variables we set as `asbalanced`. We can retrieve this information by using the `fvset report` command:

```
. fvset report, design(asbalanced)
Variable     Base    Design
foreign              asbalanced
```

◁

❑ Technical note

margins is aware of a factor variable's design setting only through the estimation results it is working with. The design setting is stored by the estimation command; thus changing the design setting between the estimation command and margins will have no effect. For example, the output from the following two calls to margins yields the same results:

```
. fvset clear foreign
. regress mpg i.foreign

      Source |       SS       df       MS              Number of obs =      74
-------------+------------------------------           F(  1,    72) =   13.18
       Model |  378.153515        1  378.153515        Prob > F      =  0.0005
    Residual |  2065.30594       72  28.6848048        R-squared     =  0.1548
-------------+------------------------------           Adj R-squared =  0.1430
       Total |  2443.45946       73  33.4720474        Root MSE      =  5.3558

-------------------------------------------------------------------------------
         mpg |      Coef.   Std. Err.      t    P>|t|     [95% Conf. Interval]
-------------+-----------------------------------------------------------------
   1.foreign |   4.945804   1.362162     3.63   0.001     2.230384    7.661225
       _cons |   19.82692   .7427186    26.70   0.000     18.34634    21.30751
-------------------------------------------------------------------------------

. margins
Predictive margins                                Number of obs   =        74
Model VCE    : OLS

Expression   : Linear prediction, predict()

-------------------------------------------------------------------------------
             |            Delta-method
             |     Margin   Std. Err.      z    P>|z|     [95% Conf. Interval]
-------------+-----------------------------------------------------------------
       _cons |    21.2973   .6226014    34.21   0.000     20.07702    22.51757
-------------------------------------------------------------------------------

. fvset design asbalanced foreign
. margins
Predictive margins                                Number of obs   =        74
Model VCE    : OLS

Expression   : Linear prediction, predict()

-------------------------------------------------------------------------------
             |            Delta-method
             |     Margin   Std. Err.      z    P>|z|     [95% Conf. Interval]
-------------+-----------------------------------------------------------------
       _cons |    21.2973   .6226014    34.21   0.000     20.07702    22.51757
-------------------------------------------------------------------------------
```

❑

Saved results

fvset saves the following in r():

Macros
    r(varlist)   *varlist*
    r(baselist)   base setting for each variable in *varlist*
    r(designlist)   design setting for each variable in *varlist*

Methods and formulas

fvset is implemented as an ado-file.

GOMBRICH
THE
RENAISSANCE

Volume 2: Symbolic Images